COASTAL DISASTER RISK MANAGEMENT IN BANGLADESH

This book addresses disaster and disaster risk reduction (DRR) practices, constraints and capacity in the context of coastal Bangladesh.

Located in the lower riparian of the Bay of Bengal, Bangladesh has to face frequent disasters such as floods, cyclones, river erosion, salinity intrusion as well as drought. Drawing together a range of multidisciplinary perspectives, *Coastal Disaster Risk Management in Bangladesh* explores the connection between climate change and DRR issues in this region. The editors reorganize disaster studies around social and physical changes that can reduce these risks and put at risk populations on a stronger footing by making risk reduction the focus. These include measures to improve disaster preparedness, to boost recovery by creating better disaster planning and programs, and physical and social initiatives to improve disaster resilience. Also, analyzing the gender perspective, the volume also utilizes the local knowledge framework to consider whether these populations have resilient knowledge that needs to be incorporated into initiatives based on advanced technology and perspectives.

This book will be of interest to academics, researchers, students, policymakers and practitioners in the field of disaster, DRR and governance, climate change, climate change adaptation (CCA) and the environment.

Mahbuba Nasreen is Professor and former Director at the Institute of Disaster Management and Vulnerability Studies, University of Dhaka, Bangladesh.

Khondoker Mokaddem Hossain is Professor and founding Director at the Institute of Disaster Management and Vulnerability Studies, University of Dhaka, Bangladesh.

Mohammed Moniruzzaman Khan is Associate Professor at the Institute of Disaster Management and Vulnerability Studies, University of Dhaka, Dhaka, Bangladesh.

ROUTLEDGE EXPLORATIONS IN ENVIRONMENTAL STUDIES

Nature and Bureaucracy
The Wildness of Managed Landscapes
David Jenkins

Art, Farming and Food for the Future
Transforming Agriculture
Barbara L. Benish and Nathalie Blanc

Sustainable Places
Addressing Social Inequality and Environmental Crisis
David Adamson, Lorena Axinte, Mark Lang and Terry Marsden

Urban Aerial Pesticide Spraying Campaigns
Government Disinformation, Industry Profits, and Public Harm
Manuel Vallée

Addicted to Growth
Societal Therapy for a Sustainable Wellbeing Future
Robert Costanza

Coastal Disaster Risk Management in Bangladesh
Vulnerability and Resilience
Edited by Mahbuba Nasreen, Khondoker Mokaddem Hossain and Mohammed Moniruzzaman Khan

For more information about this series, please visit: www.routledge.com/Routledge-Explorations-in-Environmental-Studies/book-series/REES

COASTAL DISASTER RISK MANAGEMENT IN BANGLADESH

Vulnerability and Resilience

Edited by Mahbuba Nasreen, Khondoker Mokaddem Hossain and Mohammed Moniruzzaman Khan

Routledge
Taylor & Francis Group
LONDON AND NEW YORK

earthscan
from Routledge

First published 2023
by Routledge
4 Park Square, Milton Park, Abingdon, Oxon OX14 4RN

and by Routledge
605 Third Avenue, New York, NY 10158

Routledge is an imprint of the Taylor & Francis Group, an informa business

British Library Cataloguing-in-Publication Data
A catalogue record for this book is available from the British Library

ISBN: 978-1-032-18221-6 (hbk)
ISBN: 978-1-032-18223-0 (pbk)
ISBN: 978-1-003-25349-5 (ebk)

DOI: 10.4324/9781003253495

Typeset in Bembo
by Newgen Publishing UK

This book is dedicated to the loving memory of our parents who have sacrificed everything to make and keep us smiling.

CONTENTS

FIGURES

TABLES

CONTRIBUTORS

Md. Nasif Ahsan is Professor, Economics Discipline, Khulna University, Bangladesh.

Shahana Akter is Environmental Science Discipline, Khulna University, Khulna-9208, Bangladesh.

Mosammat Rowshan Ara is Professor, Sociology Discipline, Khulna University, Khulna, Bangladesh.

Prabal Barua is Programme Manager, Young Power in Social Action (YPSA), Chattogram, Bangladesh.

Sanjoy Kumar Chanda is Associate Professor, Sociology Discipline, Khulna University, Bangladesh and PhD Research Fellow, School of HealtCrae, University of Leeds, UK.

Syed Ashik-E-Elahi is Assistant Professor, Department of Sociology, University of Barishal, Bangladesh and PhD Fellow, Bangladesh University of Professionals (BUP), Bangladesh.

Zawad Ibn Farid is Lecturer at the Institute of Disaster Management and Vulnerability Studies, University of Dhaka, Dhaka, Bangladesh.

C. Emdad Haque is affiliated with Natural Resources Institute, University of Manitoba, Winnipeg, Manitoba, Canada.

Md. Farhad Hasan is School of Geography and Sustainable Communities, University of Wollongong, Australia.

Md. Kamrul Hasan is Staff Member, Western Sydney University, Australia.

M.M. Abdullah Al Mamun Sony is Doctoral Student, Gaza Marton Doctoral School of Legal Studies, University of Debrecen, Debrecen, Hungary.

Fahreen Hossain is Research Associate, Gender Team, REACH, Oxford-IWFM-IDMVS, Dhaka.

Khondoker Mokaddem Hossain is Professor and founding Director, Institute of Disaster Management and Vulnerability Studies, University of Dhaka, Bangladesh.

Saleemul Huq is affiliated with International Center for Climate Change Adaptation and Development (ICCCAD), Dhaka, Bangladesh.

G.M. Tarekul Islam is Professor, Institute of Water and Flood Management (IWFM), Bangladesh University of Engineering and Technology (BUET), Bangladesh.

Saadi Islam is Public Relations, American International University-Bangladesh (AIUB), Dhaka, Bangladesh.

Syed Labib Ul Islam is Railway Track Engineer, JPZ Consulting (Bangladesh) Ltd.

Md. Masud-All-Kamal is Associate Professor in Department of Sociology, University of Chittagong, Chattogram, Bangladesh.

Mohammed Moniruzzaman Khan is Associate Professor at the Institute of Disaster Management and Vulnerability Studies, University of Dhaka, Bangladesh.

Joris van Loenhout is Post-Doctoral Researcher, Centre for Research on the Epidemiology of Disasters (CRED), Institute of Health and Society (IRSS), Université Catholique de Louvain (UCL), Brussels, Belgium.

Md. Abdul Malak is Associate Professor, Department of Geography and Environment, Jagannath University, Dhaka, Bangladesh.

Bishawjit Mallick is Technische Universität Dresden (TUD), Faculty of Environmental Sciences, Chair of Environmental Development and Risk Management, Dresden, Germany and Institute of Behavioral Science (IBS), University of Colorado Boulder, USA.

Dwijen Mallick is Research Fellow, Centre for Advanced Studies (BCAS), Dhaka.

Nurunnahar Mazumder is Associate Professor, Department of Law, Jagannath University, Dhaka.

Abu-Hena Mostofa Kamal is Assistant Professor, Humanities Department, Khulna University of Engineering and Technology, Bangladesh.

Mahbuba Nasreen is Professor and former Director, Institute of Disaster Management and Vulnerability Studies, University of Dhaka, Dhaka, Bangladesh.

Sharmind Neelormi is Associate Professor, Economics, Jahangirnagar University.

Rupkatha Priodarshini is affiliated with International Center for Climate Change Adaptation and Development (ICCCAD), Dhaka, Bangladesh.

Mohammad Abdul Quader is Professor, Department of Geography and Environment, Jagannath University, Dhaka, Bangladesh.

M. Abul Kalam Azad Assistant Professor, Institute of Disaster Management and Vulnerability Studies, University of Dhaka, Dhaka 1000.

Mamunur Rahman is Professor, Economics Discipline, Khulna University, Bangladesh.

Md. Habibur Rahman is from Graduate School of Agriculture, Kyoto University, Kyoto 606-8502, Japan.

Md. Mujibor Rahman is affiliated with Environmental Science Discipline, Khulna University, Khulna-9208, Bangladesh.

Md. Sadequr Rahman is Associate Professor, Department of Sociology, University of Barishal, Bangladesh.

Rezaur Rahman is Professor, Institute of Water and Flood Management (IWFM), Bangladesh University of Engineering and Technology (BUET), Bangladesh.

Syed Hafizur Rahman is Professor, Department of Environmental Sciences, Jahangirnagar University.

Nahid Rezwana is Associate Professor, Department of Geography and Environment, Dhaka University, Bangladesh.

Tuhin Roy is Professor, Sociology Discipline, Khulna University, Khulna, Bangladesh.

Choyon Kumar Saha is Assistant Professor, Department of Sociology, Jagannath University, Dhaka, Bangladesh and Global Governance Program, Balsillie School of International Affairs, University of Waterloo, Canada.

Debarati Guha-Sapir is Director, Centre for Research on the Epidemiology of Disasters (CRED), Institute of Health and Society (IRSS), Université Catholique de Louvain (UCL), Brussels, Belgium.

Monishankar Sarkar is Assistant Professor, Institute of Disaster Management and Vulnerability Studies, University of Dhaka, Dhaka 1000.

M. Salim Uddin is affiliated with Department of Geography and Environmental Management, University of Waterloo, Ontario, Canada, Centre for Asian Climate and Environmental Policy Studies, Ontario, Canada and Bangladesh Institute of Social Research Trust (BISR), Dhaka, Bangladesh.

Pascaline Wallemacq is Researcher, Centre for Research on the Epidemiology of Disasters (CRED), Institute of Health and Society (IRSS), Université Catholique de Louvain (UCL), Brussels, Belgium.

Dilara Zahid is Associate Professor, Institute of Disaster Management and Vulnerability Studies, University of Dhaka, Dhaka 1000.

Sabrina Zaman is former Research Fellow, Oxford REACH, IDMVS, Dhaka.

DECLARATION

This book is our own work and acknowledges all results and quotations from the published or unpublished study of others. There were no changes to intellectual content or substance of this book because of advice from third parties (paid or unpaid) who contributed by collecting data or provided copyediting or proof-reading services. We completely acknowledged all such contributions. To our knowledge, this original work does not break any law or infringe any third party's copyright or intellectual property right.

PREFACE

Located in the lower riparian of the Bay of Bengal, Bangladesh has to face frequent disasters such as floods, cyclones, river erosion, salinity intrusion drought and others. The country is also at risk of earthquake. The most common disasters that affect the country are floods and cyclones, which increase vulnerability of the people living in hazardous locations. Academics, researchers, practitioners, policymakers and disaster managers around the world relate frequent disasters with climate change. Large numbers of the stakeholders pay attention to the adaptation practices of the people in coastal Bangladesh. Lessons have also been learnt from the climate change adaptation (CCA) practices of the coastal people. There are a number of documents regarding the realities in coastal region; however, there is no single book that addresses disaster and disaster risk reduction (DRR) practices, constraints and capacity inclusively thoroughly the context of disaster and DRR in coastal Bangladesh. Though it is challenging, this book perhaps is the first attempt to draw a nexus between climate change and DRR issues in coastal region of Bangladesh from comprehensive and multidisciplinary perspectives. Divided into several chapters' guidance and emphasis on issues trajectory and paradigm shift of the concept of disaster, capacity and agency, DRR interventions (GO/NGOs), in connection with climate change, have been focused. This book will also extensively give sense of the range and diversity based on practical experiences on key issues such as community participation and indigenous knowledge, policies and regulations, innovation and technology, eco-system-based DRR and social inclusion in disaster and DRR in disaster-prone coastal areas in Bangladesh. The primary targets of the book are academics, researchers, students (both undergraduate and postgraduate), policymakers and program planners, managers and practitioners in the field of disaster, DRR and governance, climate change, CCA and environment to provide better and comprehensive understanding of the disaster from society (coastal Bangladesh) where disaster is an event of everyday life.

ACKNOWLEDGMENTS

Our foremost appreciation, deepest sense of gratitude and indebtedness to all the authors for their scholastic contribution throughout the period that has made this book possible. We are particularly indebted to them for their professionalism, persistent effort, coordination and cooperation whenever required. The innumerable discussions between the authors where we shared our ideas and experiences truly helped shape our arguments for this work. We thank the Institute of Disaster Management and Vulnerability Studies (IDMVS), University of Dhaka for imbuing in us a deep sense of pro-people social study as well as their constant support and approach.

We thank all the mentors who guided us for building our academic careers from child to maturity, who taught us to see social problems through the lens of the others. We are indebted to sociologists, social scientists and disaster management specialists we came across. Our sincere gratitude to the people living in coastal region for sharing their realities with us. We owe them for making us what we are producing today. We would like to extend our heartfelt thanks to our family and feel privileged to have supportive and loving family members without whose sacrifices this book would not have been possible to publish.

ABBREVIATIONS

ADB	Asian Development Bank
ADW	Alternative drying and wetting
ANC	*Antenatal care*
ANOVA	Analysis of variance
APEASE	Affordability, practicability, effectiveness, acceptability, side effects and safety, and equity
AR5	Fifth Assessment Report
AVAS	Association of Voluntary Action for Society
BADC	Bangladesh Agricultural Development Corporation
BAL	Bangladesh Awami League
BARD	Bangladesh Academy for Rural Development
BARI	Bangladesh Agriculture Research Institute
BBC	British Broadcasting Corporation
BBS	Bangladesh Bureau of Statistics
BCT	Behaviour Change Technique
BCTT	Behaviour Change Technique Taxonomy
BCW	Behavioural Change Wheel
BDP	Bangladesh Delta Plan
BDT	Bangladeshi currency
BFD	Bangladesh Forest Department
BFFEA	Bangladesh Frozen Foods Exporters Association
BFQ	Better Fisheries Quality
BFRI	Bangladesh Fisheries Research Institute
BID	Best Interest of Child
BIDS	Bangladesh Institute of Development Studies
BIWTA	Bangladesh Inland Water Transport Authority
BMD	Bangladesh Meteorological Department

BMI	Body mass index
BNP	Bangladesh Nationalist Party
BOD	Biological oxygen demand
BP	Blood pressure
BSF	Biosand filter
BUET	Bangladesh University of Engineering and Technology
BWDB	Bangladesh Water Development Board
CAA	Country's climate change adaptation
Cap-net	International Network for Capacity Building in Integrated Water Resource Management
CBDM	Community-based disaster management
CBDRR	Community-based disaster risk reduction
CC	Community clinic
CDMP	Comprehensive Disaster Management Programme
CEP	Coastal Embankment Project
CEGIS	Center for Environmental and Geographic Information Services
Cfu	Colony-forming units
CFS	Child-friendly space
CHCP	Community health care provider
CKD	Chronic kidney diseases
COD	Chemical oxygen demand
CODEC	Community Development Centre
COM-B	Capability, Opportunity and Motivation-Behaviour
COR	Conservation of resources
CPDRRI	Community prioritization in DRR intervention
CPP	Cyclone preparedness program
CPWG	Child Protection Working Group
CRA	Community risk assessment
CRC	Convention on the Rights of the Child
CSO	Civil Society Organization
CTC	Close-to-community
CVC	Cultivable *Vibrio*counts
CVD	Cardiovascular disease
DAE	Department of Agricultural Extension
DBP	Diastolic blood pressure
DDM	Department of Disaster Management
DDMC	District Disaster Management Committee
DFID	Department for International Development
DM	Disaster management
DMA	Disaster Management Act
DMC	Disaster Management Committee
DoF	Department of Fisheries
DoE	Department of Environment

DPB	Disaster preparedness behavior
DPHE	Department of Public Health Engineering
DPSIR	Driver-Pressure-State-Impact-Response
DRM	Disaster risk management
DRR	Disaster risk reduction
DRRO	District Relief and Rehabilitation Office
dS m	Deci-Siemens per meter
DTW	Deep tube well
DUHC	Dacope Upazila Health Complex
DVS	Disaster vulnerability score
DWEF	Davos World Economic Forum
EC	Electrical conductivity
ECA	Ecologically critical area
ECC	Environmental clearance certificate
EIA	Environmental impact assessment
EiE	Education in emergency
EPI	Extended Programme on Immunization
EPBB	Export Promotion Bureau of Bangladesh
EU	The European Union
EWS	Early warning system
FAO	Food and Agriculture Organization
FC	Fecal coliforms
FEMA	Federal Emergency Management Agency
FFW	Food for Work
FGD	Focus group discussion
FHH	Female-headed households
FIP	Forest Investment Plan
FRSS	Fisheries Resources Survey System
FYP	Five Year Plan
FWA	Family Welfare Assistant
GBM	Ganges-Brahmaputra-Meghna
GDL	Gender division of labour
GED	General Economics Division
GIDs	Gender-inclusive dimensions
GIS	Geographic Information System
GO	Government organization
GoB	Government of Bangladesh
HA	Health Assistant
HFA	Hyogo Framework for Action
HH	Household
HIV	Human immunodeficiency virus
HS	Hemorrhagic stroke
HT	High tide
HTN	Hypertension

ICDDR'B	International Centre for Diarrheal Disease Research in Bangladesh
ICT	Information and Communications Technology
ICZM	Integrated Coastal Zone Management
IDI	In-depth interview
IDNDR	International Decade for Natural Disaster Reduction
IGA	Income-generating activities
ILO	International Labour Organization
IMD	India Meteorological Department
IMDMCC	Inter-ministerial Disaster Management Coordination Committee
IMF	International Monetary Fund
INGO	International nongovernmental organization
IPCC	Intergovernmental Panel on Climate Change
IPM	Integrated Pest Management
IS	Ischemic stroke
ISDR	International Strategy for Disaster Reduction
IUCN	International Union for Conservation of Nature
IWA	International Water Association
IWRM	Integrated Water Resource Management
JRC	Joint Rivers Commission
KII	Key informant interview
KJDRP	Khulna-Jashore Drainage Rehabilitation Project
km/h	Kilometer per hour
LDCs	Least developed countries
LGED	Local Government Engineering Department
LT	Low tide
MAR	Managed aquifer recharge
MBBS	Bachelor of Medicine and a Bachelor of Surgery
MDGs	Millennium Development Goals
MEDC	More economically developed countries
MHH	Male-headed households
mmol/L	Millimoles per litre
MoAF	Ministry of Agriculture and Forest
MoDMR	Ministry of Disaster Management and Relief
MoFDM	Ministry of Food and Disaster Management
MoFL	Ministry of Fisheries and Livestock
MoLGRD&C	Ministry of Local Government, Rural Development and Cooperatives
MoR&R	Ministry of Relief and Rehabilitation
MoWCA	Ministry of Women and Children Affairs
MoWR	Ministry of Water Resources
MRC	Medical Research Council
mS/cm	Milli Siemens per centimeter

MUAC	Mid-upper arm circumference
NAPA	National Adaptation Programme of Action
NbS	Nature-based solutions
NCS	National Conservation Strategy
NDMC	National Disaster Management Council
NDMP	National Disaster Management Policy
NDNP	NijhumDwip National Park
NDPD	National Disaster Preparedness Day
NGO	Nongovernment organization
NPDM	National Plan for Disaster Management
NWDP	National Women Development Policy
NWMP	National Water Resources Management Plan
NWRC	National Water Resources Council
NWSI	National Water Security Index
OCHA	Office of the Coordination of Humanitarian Affairs
ODA	Overseas Development Administration
ORT	Oral rehydration therapy
PIC	Project Implementation Committee
PNC	Post-natal care
PPS	Probability proportionate sampling
PPT	Parts per thousand
PRA	Participatory rural appraisal
PRSP	Poverty reduction strategic plan
PSF	Pond sand filter
PWS	Pipeline water supply
RO	Reverse osmosis
RRI	River Research Institute
RRAP	Risk reduction action plan
SAPs	Structural Adjustment Programmes
SBP	Systolic blood pressure
μS/cm	Micro Siemens per centimeter
SDG	Sustainable Development Goal
SFDRR	Sendai Framework for Disaster Risk Reduction
SLR	Sea-level rise
SOD	Standing Orders on Disaster
SMRC	Salinity Management and Research Center
SPM	Suspended particulate matter
SPSS	Statistical Package for the Social Sciences
SRDI	Soil Resource Development Institute
SSI	Semi-structured interview
SST	Sea surface temperature
TAB	Total aerobic bacteria
TDF	Theoretical domains framework
TDS	Total dissolved solids

TEC	Technology Executive Committee
TNA	Technology Needs Assessment
TRM	Tidal River Management
TS	Total solids
TSDF	Treatment,storage and disposal facility
UDMC	Union Disaster Management Committee
UGC	University Grant Commission
UHC	Upazila Health Complex
UHI	Union Health Inspector
UIC	Upazila Irrigation Committee
UK	United Kingdom
UN	United Nations
UNDP	United Nations Development Program
UNEP	United Nations Environment Programme
UNESCAP	United Nations Economic and Social Commission for Asia and the Pacific
UNICEF	United Nations Children's Fund
UNIDO	United Nations Industrial Development Organization
UNISDR	United Nations International Strategy for Disaster Reduction
UNO	Upazila Nirbahi Officer
UN-Water	United Nations Water
UP	Union Parishad
UPC	Union Parishad Chairman
USA	United States of America
USAID	United States Agency for International Development
USD	United States Dollar
UzDMC	Upazilla Disaster Management Committee
VGF	Vulnerable Group Feeding
WARPO	Water Resources Planning Organization
WASA	Water and Sewerage Authority
WATSAN	Water and Sanitation
WHO	World Health Organization

GLOSSARY OF TERMS

Beel	Beel is the local term for natural depression in the landscape that holds some water at the deepest pocket round the year.
Chula	A kind of stove made of clay.
District	Bangladesh has eight administrative divisions. Districts are administrative unit under division and there are 64 districts in Bangladesh.
Easha azan	Nighttime prayer for Muslims.
Fajar azan	Morning prayer for Muslims.
Gher	It is a local term, which is understood as the shrimp ponds that are bounded the little bit embankments but are always close to tidal channels so that the saline water intrudes in this pond and are present only in the dry season.
Hectare	A metric unit of square measure, equal to 100 acres. 1 Hectare = 10,000 square meters.
Kabiraj	Persons practicing Ayurveda or customs involving medicines.
Keora	(*Sonneratia apetala*) is a type of mangrove species.
Killa	A killa is an earthen platform that offers shelter to about 400 livestock animals, primarily cattle and poultry. To protect cattle from cyclonic wind and storm surges, 196 killas have been constructed in cyclone-prone regions, out of a total of 872 killas constructed across Bangladesh till now.
Lead time	The time it takes for individuals to arrive at a safe location (i.e., cyclone shelter) after receiving a warning or becoming aware of a particular disaster is known as lead time.
Livelihood	Jobs or other sources of income.
Machan	Machan is an indigenous structure made of bamboo or wood. It is a platform prepared for sleeping, which is tied

with bamboo or wooden pillars. People stay and keep all the belongings on it during disasters like storm surges in the coastal areas of Bangladesh.

Mouza A local term used synonymously with village.

Mujib Killa The 'Mujib Killa' was the name given to these earthen mounds, which were built in various parts of the country following a cyclone, which made a landfall in the southern region of the country in 1970. The construction of such earthen mounds was in higher ground level than the food level. The initiative was taken under the leadership of Bangabandhu Sheikh Mujibur Rahman, the founding father of Bangladesh, to safeguard the resources, notably livestock, from cyclones, storm surges and floods, and per the name of the Sheikh Mujibor Rahman, it became known as 'Mujib Killas'. However, killaswhich already have been constructed or are in the process of being constructed are referred to as 'Mujib Killas'.

Nossimon A form of local transportation in the villages.

Ona A mobile survey platform that provides a hosted server. It helps to upload, edit, view and submit the forms easily.

Parda Parda or purda (meaning 'veil' or 'curtain') is a prominent religious and social practice of female seclusion in several Muslim communities. It takes place in two ways: physical division of the sexes and the demand that women wrap their bodies to hide their complexion and shape.

Pataton In the coastal areas of Bangladesh, people prepare houses in such a way that they can use the upper part of the house as a shelter to save their lives and belongings. This type of structure is called Pataton.

Polder A tract of low land reclaimed from a body of water (Merriam-Webster dictionary). In the 1960s, 123 polders were constructed in order to protect the low-lying coastal areas from tidal flooding and salinity intrusion in southern Bangladesh (low-surface areas with earthen embankments), including 49 sea-facing polders.

Political influence A significant influence by the political leaders in decision-making process.

Pond sand filter A specialized unit made of concrete, sand and other materials to desalinize surface water.

Purdah Purdah is a custom in some societies, where Muslim women in particular either stay in a particular part of the house or cover their faces and bodies to avoid being seen by men.

Rainwater harvesting A simple process by which rain water can be preserved in a safe storage for household uses.

Tubewell A kind of metallic water well used for pumping underground water.

Union Parishad Union Parishad (Union Council) is, in essence, Bangladesh's smallest rural legislative and local government entity. It is the oldest and smallest local government system of the country. It has been in operation for more than a century, assisting in the development of rural areas around the nation.

Upazila An 'Upazila' is a 'sub-district' in Bangladesh. It was originally known as 'Thana'. It is an administrative area that functions as a sub-unit of a particular district of the country. After divisions and district tiers of local government, this is Bangladesh's third biggest administrative division.

1

INTRODUCING COASTAL REGION OF BANGLADESH

Mahbuba Nasreen, Mohammed Moniruzzaman Khan and Khondoker Mokaddem Hossain

Bangladesh: The Natural Setting

The People's Republic of Bangladesh (hereinafter Bangladesh) is one of the largest deltas in the world and is geographically highly vulnerable to disasters. The land area of this country is 147,750 square km consisting almost 80 percent of floodplains. Only 13 percent of its territory, in the north and east, is hilly and high above sea level. The adverse effects of climate change is such as cyclones, salinity intrusion, extreme temperatures, heavy monsoon downpours, storm surges and flooding aggravated to a great extent to put the country at the top of the list of those most vulnerable to disasters (Denissen, 2012).

Most of its west, north and east sides are surrounded by India, a part of its east side abuts Myanmar and its south side fronts onto the Bay of Bengal. Its density of population is 834/square km, which is very high. Its territory is located on the Ganges Delta, which is formed by the Ganges River running from the west and the Brahmaputra (Jamuna) River running from the north-east and most of it is flat lowlands up to 10m above sea level. Bangladesh is a liberal progressive country with ethnic, linguistic and religious diversity. Administratively it is divided into eight divisions, 64 districts, 574 *upazilas*, 5,400 union *parishad*, and 87,310 villages, and the percentage of people living in rural and urban areas is 72 percent and 28 percent, respectively. The total number of populations is 143 million and its population growth rate is 1.34 percent per year (BBS, 2011).

After becoming independent from the British colony of sub-continent India in 1947, Bangladesh was established as a province of Pakistan and was known as East Pakistan. West Pakistan was the powerbase of Pakistan including the capital city. There was discriminatory division of power between the two provinces. Immediately after being free from the British colony, West Pakistan imposed Urdu as the sole state language even upon East Pakistan, although the mother language

DOI: 10.4324/9781003253495-1

of East Pakistan was Bangla. Anger at the enforcement of language along with other discriminatory socio-economic and political issues led to a movement in 1952, which started as a language movement but developed into an independence movement that finally shaped the Liberation War of Bangladesh from March 1971 to 16 December 1971. Bangladesh achieved independence and was officially born as a new state guided by a constitution that asserted the principles of nationalism, socialism, democracy and secularism (Schendal, 2009).

The history of Bangladesh is little known to the wider world; in the West media emphasizes its poverty, underdevelopment, disasters, political instability, popular unrest and economic uncertainty. Bangladesh has a history of striking policies but poor implementation since late seventies, she losses in public sector enterprises etc. not only worries national planners but also international donor agencies (Kochanek, 1996; Sobhan, 2012; Ferdousi, 2015).

Disasters: Global Context

Over the last several decades the world has experienced very frequent/unusual natural and technological disasters that have impacted massively on the lives and livelihoods of people all over the world. The IFRC (2010) records that in 2013, 330 natural disasters resulted in an overall death toll of 22,452, and affected over 96.5 million people, with an estimated economic damage of US$118.6 billion worldwide. Research indicates (Dasgupta et al., 2010, 2012) that the number of disasters increased on average 6 percent every year between 1962 and 1992 and affected 200 million people annually during the 1990s. Also, there was four-fold increase in disasters from the late 1960s.

Although there are disasters all over the world, occurring with varying frequency and level of impact, the Asian nations held the top position as the most disaster-afflicted region in the world during 1975–2006 (Saha, 2015). Asian Development Bank (ADB) (2010) reported that during this time (1975–2006) 89 percent of the people in Asia were affected and 44 percent of the total damage was economic. In 2008 due to cyclone *Nargis* in Myanmar over 80,000 people died and about seven million people were affected. The economic loss was over US$10 billion and almost one million people were compelled to live without sufficient safe shelter, food, medical care, drinking water etc. (UN, 2009 as cited Saha, 2015).

The Indian Ocean earthquake and tsunami in 2004 caused over 220,000 deaths (Kron, 2012), and displaced 500,000 persons in Sri Lanka alone. In Sri Lanka following the tsunami incidents of sexual harassment such as rape, gang rape, physical abuse of women and girls, and a sense of insecurity and fear in the camps were reported (Chadwick, 2002). Fisher (2010) argues that lack of security and chaotic conditions in shelter centers created further atmosphere in which girls and women were vulnerable to sexual violence.

In Florida after Hurricane Andrew in 1991 the rate of violence increased and there was a 50 percent increase in spousal abuse calls including family violence, spousal abuse, child abuse and domestic violence (Shaw et al., 1995; Morrow, 1997;

Wilson et al., 1998). In New Orleans and Mississippi in 2005 Hurricane Katrina caused 1,836 fatalities, made 1,000,000 homeless, resulted in 7,500 cases of health-related illness and injuries, and caused economic damage of over US$125 billion. At the same time, it exposed 80 percent of people to severe flood and inundation (Tierney, 2006; Brakenridge et al., 2012; Kron, 2012).

An increase in disasters, including cyclones, floods, earthquakes, rising sea levels etc. all over the world, is causing drying out of wetlands, irrigation crises, threats to coastal people, and imbalance in coastal ecosystems, destruction of fisheries and infrastructure, increased death tolls and other economic losses. Flood disasters alone resulted in 175,000 death tolls worldwide and affected 2.2 billion people from 1975 to 2002 (Jonkman, 2005 as cited in Saha, 2015) and cost approximately US$470 billion (HSBC, 2011). As reported by the secretariat of ISDR (2004) in the last ten years 478,100 people were killed, more than 2.5 billion people were affected and about US$690 billion of economic losses were caused by flooding. Mental depression, distress, disparity and discrimination, unemployment, disability, trauma, illness, dislocation of social networks, infectious diseases, shortage of drinking water, sanitation, malnutrition and disruption of healthcare are increasing in society due to disaster. Events were trigged by hydro-meteorological hazards in such a severe way that 97 percent of total people were affected and 60 percent have had economic losses (Shaw, 2006).

The magnitudes of effects from a disaster always vary depending on class, gender, age, race and ethnicity (Saha, 2015). The people of the poorest regions of the world and the most vulnerable groups experience double challenges due to social exclusion, type of residence and scarce access to resources even in wealthy industrialized countries (Peacock et al., 1997; Fothergil & Peek, 2004). Disaster experts find that women, children, disabled and elderly people are the most vulnerable groups among those who face the adverse effects of catastrophic events (Cutter et al., 2003; Peek & Stough, 2010 as cited in Saha, 2015). The response and recovery of women varies from that of men due to different experiences at disastrous times (Fothergill, 2004). In particular female-headed households and single mothers face critical vulnerability because of the structural positions in society and their responsibilities and routine activities (Tobin-Gurley et al., 2010) and pregnant women are less likely to have access to enough family and medical care (Zahran et al., 2013 as cited in Saha, 2015). They also disrupt social arrangements and inflict challenges to recovery, environmental remediation and mitigation of future disasters. Apart from those problems, disasters fuel up social disorder, crime and domestic violence (Tierney et al., 2006; Zahran et al., 2009; Saha, 2015), break up family relations (Peek et al., 2011 as cited in Saha, 2015), cause uncertainty, maladjustment to new places and environment, sadness, loneliness, marginalization of different groups, income and wealth inequality (Nel & Righarts, 2008 as cited in Saha, 2015).

Since ages Bangladesh is facing the challenges of disasters due to natural or anthropogenic causes. Prior to independence this part of land was recognized only as a poor and disaster-prone one. Without independence, the country would have never been identified with successes that are achieved within the 50years of independence: Bangladesh has awarded with dignified identity as a 'developing country'. It is also a well-accepted phenomenon that initiatives for managing disasters are intrinsically linked with the independent Bangladesh (Nasreen, 2021).

Disasters in Bangladesh

Bangladesh is identified as the country of the highest risk, and vulnerability to climate change has placed the country as one of the most susceptible countries of the world. It leads the top ten countries in the Asia-Pacific region for floods; fifth for storms; and eighth for earthquakes due to its physical exposure (UN ESCAP, 2010). Bangladesh ranks as one of the most vulnerable countries to natural hazards, particularly to rises in sea level, because of its geographical location, land characteristics, multiplicity of rivers, monsoon climate and climatic variability (Huq & Karim, 1999; Zaman, 1999). World Bank (2006) reports that Bangladesh will be a victim of environmental catastrophe due to rising sea levels. About 33 million people will be evicted because of permanent flooding in many areas of coastal Bangladesh. By the year 2030, an extra 14 percent of the country will become vulnerable to floods, and current vulnerable areas will experience more severe flooding.

Climate change has been proved as the major cause for increasing intensity and frequency of disasters. Bangladesh has been playing a crucial role to raise voice in climate change negotiations and to taking initiatives to mitigate and adapt with changing climate (Nasreen, 2021). The scale of the climatic disasters that Bangladesh has been facing for a long time will continue due to the sea-level geography of the coastal area, the high density of population and a lack of coastal protection, along with its location at the head of the triangular-shaped Bay of Bengal. About 40 percent of the total global storm surges and

TABLE 1.1 Major cyclones that hit the Bangladesh coast

Year (Storm name, where given)	*Maximum wind speed (km/h)*	*Maximum storm surge height (meters)*	*Death toll*
11 May 1965	161	7.6	19,300
15 December 1965	217	3.6	870
1 October 1966	139	6.7	850
12 November 1970 (Bhola)	224	10.0	300,000
25 May 1985	154	4.6	11,000
29 April 1991 (Gorky)	225	7.6	139,000
19 May 1997	232	4.6	150
15 November 2007 (*Sidr*)	223	4.0	3,360
15 May 2009 (*Aila*)	92	3.50	190
16 May 2013 (Mahasen)	120	3.7	17
9 November 2019 (Bulbul)	120	3.0	25
19 May 2020 (Amphan)	160	3.0	26

Source: National Plan for Disaster Management 2010–2020.

deadliest cyclones, in terms of deaths and casualties, in the past 50 years have been recorded in Bangladesh. It is estimated that about ten million Bangladeshi citizens are impacted by natural hazards annually (Khan & Azad, 2012). CRED (2015) estimates that from 1979 to 2010 over 234 disasters occurred that caused the death of 191,836 people and directly affected 229 million people badly and caused economic damage of about US$5.6 billion. Research indicates (Huq et al., 2004; Nahar et al., 2014) that Bangladesh as a developing country has multiple challenges, including rapid population growth, illiteracy, poverty etc. Disasters, particularly cyclone, floods, coastal bank erosion, rising sea levels and tidal surges, pose serious threats to its economic development. Annually 25 percent of the country is flooded, and every four to five years about 60 percent of the total area is flooded (GoB, 2010a). The deadliest disasters of Bangladesh over the last three decades are provided in Table 1.1.

Disaster Risk Reduction in Bangladesh: A Shift of Paradigm

In the attempts to reduce disaster risks, disaster risk reduction (DRR) concept in Bangladesh has been emerged as a shift of paradigm from 'reactive relief-based measures to proactive integrated management approach'. Preparedness, mitigation, response and recovery have become an integrated part of DRR mechanisms in most of the disaster-prone countries. Different policies, legislations and frameworks were developed for achieving DRR goals across the world. The international commitments, such as the Yokohama Strategy and Plan of Action (1990–1999), which is considered as the first global strategy to address the disaster and disaster risk management, was followed by Bangladesh. Disaster Management Bureau was established within this time period. In this connection, Bangladesh is the first country within South Asia to develop Standing Order on Disasters (SOD) in 1997. Next remarkable development is the Hyogo Framework of Action (HFA, 2005–2015), aimed to promote DRR as a policy priority, stepping towards disaster risk governance. Countries started to take initiatives to building resilient nations and communities through identifying some gaps of HFA. Based on the feedback from member countries of United Nations, Sendai Framework for Disaster Risk Reduction (SFDRR) (2015–2030) was developed focusing 'disaster resilience' ensuring adequate means of implementation. The scope of SFDRR in disaster-prone countries, including Bangladesh, has been extended through addressing environmental, technological and biological hazards and risks in addition to 'natural' and anthropogenic or human-induced hazards and disasters. Comprehensive Disaster Management Programme (CDMP) was established with the support of development partners and was implemented in two phases. CDMP initiated professionalization of disaster management. Disaster management has started to become institutionalized as part of academic disciplines in several universities. Over the last decades good number of institutional structures to achieve technical monitoring, capacity building, preparedness and response in reducing disaster risks has been initiated. The Ministry of Disaster Management and Relief (MoDMR) with the line agency, Department of Disaster Management (DDM) is coordinating national disaster management efforts across relevant ministries and agencies. The national DRR drivers are directed by Vision 2021 and 2041; Five Year Plan (FYP) (the Seventh FYP first had a background document on DRR and climate change adaptation); Delta Plan, 2100; commitments towards achieving goals of two aligned international drivers: 'Transforming Our World: the 2030 Agenda for Sustainable Development' (2016–2030) and SFDRR (2016–2030).

The major drivers on DRR in Bangladesh are National Plan for Disaster Management (2010; revised in 2016–2020); SOD (developed in 1997; revised in 2010 and 2019); Disaster Management Act (2012); Disaster Management Policy (2015); National Earthquake Contingency Plan and some other relevant documents published by the DDM and MoDMR. The revised SOD (2019) has, for the first time, included experts of several sectors to be nominated by the government such as water resources, climate change, earthquake, disaster management, gender and social inclusion among others in the National Disaster Management Advisory Committee (NDMAC). Bangladesh is also recognized as pioneer in promoting gender and DRR agenda within South Asia and has received global recognition for gender mainstreaming efforts.

Major Disasters in the Coastal Area in Bangladesh

Out of 64 districts of Bangladesh the coastal zone covers 16 districts, an area of 47,211 square km, 32 percent of the country's geographical area, where 35 million people, that is, 28 percent of the country's total population, live in 6.85 million households (Huq et al., 2004; BBS, 2011), which are extremely vulnerable to climatic hazards. In the coastal area of Bangladesh there have been 70 major cyclones in the last 200 years. And in the last 35 years, almost 900,000 people have died as a result of these disasters (Mallick & Vogt, 2009). The coastal area of Bangladesh is different from the rest of the country for socio-cultural and political reasons, which often limit the access of the people to endowed resources. The socio-cultural and political features of the coastal area perpetuate risks (Shamsuddoha & Chowdhury, 2007). The increasing number and intensification of disasters every year, including cyclones, floods, storm surges and tidal surges, combined with rising sea levels makes the livelihoods of coastal people vulnerable. On top of that, changes in upstream river discharge, lack of strong embankments, water logging, flash floods upstream, shortages of proper canals and sluice gates for controlling the water, the presence of salinity and unplanned shrimp cultivation are making the life of the coastal community harder and more exposed to vulnerability. During pre-monsoon and post-monsoon seasons, violent storms, storm surges and tropical cyclones are very common, which is why the coastal zone of Bangladesh is termed as a geographical 'death trap' (Shamsuddoha & Chowdhury, 2007).

The coastal zone in Bangladesh is composed of 19 sea-facing districts and nearly 600 km of coastlines and people in the region have to encounter frequent cyclones usually in the months of May and November. In addition, flood, tidal surge, riverbank erosion, long-term water logging, overflow of river water, thunderstorm and tidal flooding (Choudhury et al., 2019; Haque & Jahan, 2015; Juran & Trivedi, 2015) have also become a part of life of the people in the region.

Cyclones cause deaths of human and others species, devastate and destroy resources. However, based on the lessons learned from the earlier experiences, disaster management efforts have been able to gradually reduce the number of death toll from three digits to two. As a consequence of cyclone and other anthropogenic activities people in the coastal region face several challenges, mostly related to water, such as salinity intrusion in soil and water, water logging and river erosion. Soil and water salinity are the major concerns of this region over the last three decades as their livelihoods are dependent on the two resources. Study (Sarker, 2020) indicates that livelihood of coastal communities is closely connected to water-related parameters.

In the last decade cyclone *Sidr* hit the coastal area of Bangladesh on 15 November 2007. As cyclone *Sidr* approached, 40,000 Red Crescent volunteers were deployed to direct residents from the 15 affected districts. There were around 3,406 confirmed deaths and over 55,000 people sustained permanent physical injuries (Paul, 2009; Nahar et al., 2014) along with economic loss, and damage to houses, crops and vegetable lands, wood and fruit trees, domestic animals, community infrastructures and coastal defences (Haq et al., 2012; Saha, 2015). Ten million people were affected, 1.5 million houses were damaged (Paul, 2009) and 186,883 hectares of crop areas were totally damaged (GoB, 2010b). Cyclone *Aila* hit the south-western coastline of Bangladesh and neighbouring districts of India severely on 25 May 2009 and the hard-core poor were the worst victims (Unnayan Onneshan, 2009). Compared to cyclone *Sidr* that struck the coastal area of Bangladesh in 2007, *Aila* did not result in a high death toll, but the aftermath of the cyclone was horrendous as half a million people migrated due to permanent waterlogged conditions in the affected areas (Unnayan Onneshan, 2009) and the storm displaced around two million people, most of whom returned to their places of origin and started spontaneous and assisted recovery. The catastrophic cyclone *Aila* killed 190 people along with almost 8,000 probably dead (missing/never found) persons (Mallick & Vogt, 2012), affecting more than 3.9 million people, including disruption to livelihoods and destruction of infrastructure (IFRC, 2010). Approximately 352 unions and 62 *upazilas* in 14 districts were affected (Mallick & Vogt, 2012). Cyclone *Aila* completely damaged 68,385 acres of crop land, and partially damaged 239,212 acres, killed over one million livestock (including cows, goats and chickens, which were people's subsistence as well as their means to cash), destroyed/washed away shrimp farms and homestead gardens depriving people not just of their prime income source, but also non-economic revenues and assets (Mallick & Vogt, 2012). Also, in the 14 districts almost 258.5 km of roads were too damaged to be used and 509 km of embankments were destroyed and 2,155 km were damaged (Kumar et al., 2010 as cited in Mallick & Vogt, 2012). The Cyclone 'Amphan' of 13 May 2020 left a trail of devastation in the region with wind speed of up to 180km/h and over 10 feet of tidal surges (Figure 1.1).

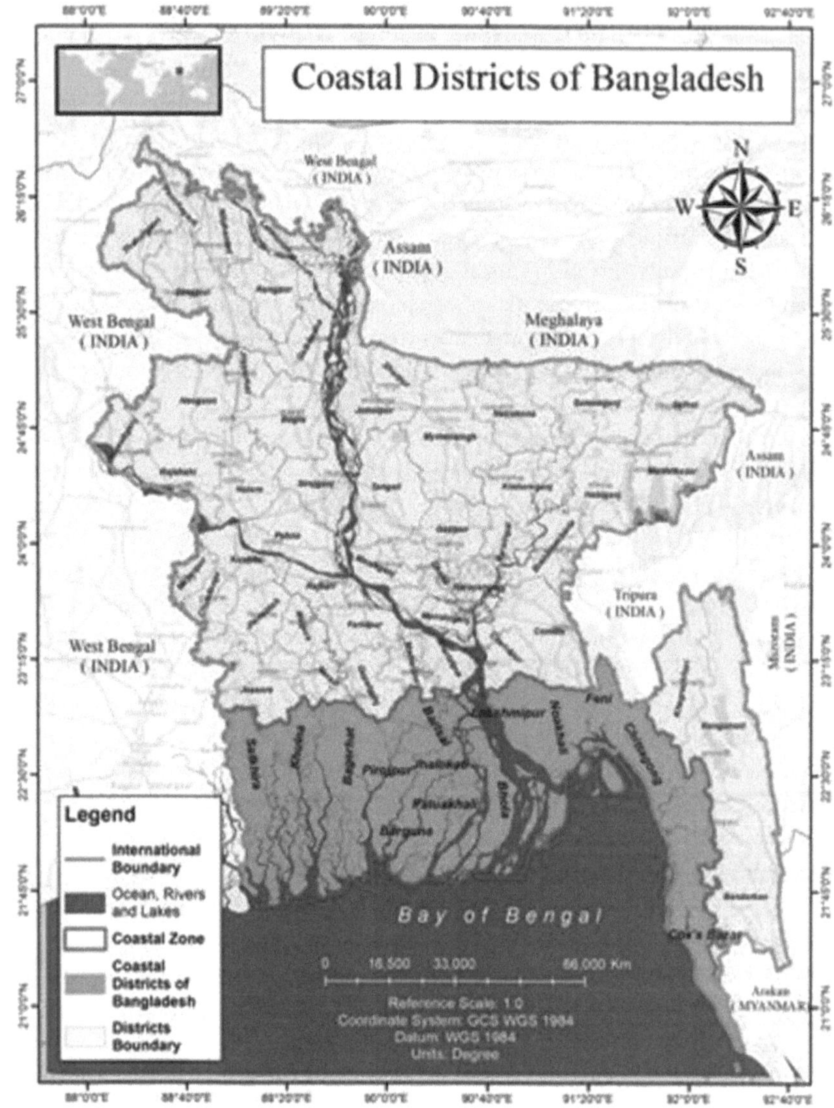

FIGURE 1.1 Coastal districts of Bangladesh

Vulnerabilities and Adaptation to Climate Change by Coastal People

Climate change has already posed a threat to the agricultural activities, one of the major livelihoods in Bangladesh. Studies (Azad et al., 2013; SAKTEE, 2020) have identified different adaptation measures taken by the people in coastal Bangladesh. The unique geo-physical nature of coastal region has made her more prone to disasters than the other parts of the country. In addition to the most common disaster, that is, cyclone, the region is highly susceptible to tidal surge, salinity intrusion, erratic rainfall, heat wave, drought, thunderstorm, soil erosion and flooding. Starting from the devastating Bhola cyclone of 1970, to the 'Amphan' of 13 May 2020 tell the stories of vulnerabilities of coastal region in Bangladesh. Severe cyclones cause damage to lives and livelihoods, destroying standing crops, infrastructures and environment in Satkhira region of the country.

Climate change is expected to aggravate the situation of the most common disasters in Bangladesh, that is, flood, cyclone, and salinity intrusion in coastal belts (IPCC 5th Assessment Report). It is also projected that the sea-level rise may increase up to 98 cm by 2100. All of these changing climates will have a profound effect on the lives and livelihoods of agro-based economy. The agricultural practices are already at risk and people in the disaster-prone areas have been trying to adopt with the changing environment.

Salinity intrusion in coastal belt marked as the severe threat for the people in the coastal regions. In the highly stratified society, marginalized people such as poor, women and special occupational group depending on agriculture and/or fisheries are facing the challenges more than others. The coastal communities have already started to practice alternative agricultural activities as adaptation approach. It is a recognized fact that disasters are inevitable and people have to leave with these through increasing their capacities. The gender-based resilience, with special focus on agricultural adaptation to climate change, has already been explored and identified by researchers (Azad et al., 2013; Nasreen, 2105; Nasreen, 2017; Nasreen, 2018; SAKTEE, 2021). The adaptation practices and resilience mechanisms include identification of suitable cropping pattern, choice of seed, homestead gardening including waterlogged areas, fish cage cultivation, irrigation and water management, soil management, crop intensification, suitable transplanting and others. The book has been designed to identify the vulnerabilities of coastal region as well as finding out the potentials of people living in the hazardous location to adapt with the adversities.

References

Alston, M., (2015). *Women and Climate Change in Bangladesh*. New York: Routledge.

Asian Development Bank [ADB], (2004). *Country Environmental Analysis*. Dhaka.

Azad, A. K., Hossain, K. M., & Nasreen, M. (2013). Flood-Induced Vulnerabilities and Problems Encountered by Women in Northern Bangladesh. *International Journal of Disaster Risk Science*, 4(4), 190–199. https://doi.org/10.1007/s13753-013-0020-z

Bangladesh Bureau of Statistics [BBS], (2011). *Population and Housing Census*. Dhaka: Statistics Division and Information Division.

Brakenridge, G. R., Syvitski, J. P. M., Overeem, I., Higgins, S. A., Kettner, A. J., Stewart-Moore, J. A., & Westerhoff, R., (2012). Global Mapping of Storm Surges and the Assessment of Coastal Vulnerability. *Natural Hazards*, 66(3), 1295–1312.

Centre for Research on the Epidemiology of Disasters [CRED], (2015). *The Human Cost of Natural Disasters 2015: A Global Perspective*. Brussels: CRED.

Chadwick, B., (2002). *Beyond Conflict to Consensus:An Introductory Learning Manual.* Retrieved from www.managingwholes.com/chadwick.htm on April 7, 2015.

Choudhury, M. U. I., Uddin, M. S., &Haque, C. E. (2019). "Nature Brings Us Extreme Events, Some People Cause Us Prolonged Sufferings": The Role of Good Governance in Building Community Resilience to Natural Disasters in Bangladesh. *Journal of Environmental Planning and Management*, 62(10), 1761–1781.

Cutter, S. L., Boruff, B. J., &Shirley, W. L., (2003). Social Vulnerability to Environmental Hazards. *Social Science Quarterly*, 84, 242–261.

Dasgupta, S., Siriner, I., & Sarathi, P. De (Eds.), (2010). *Women's Encounter with Disaster.* London: Front Page Publications.

Dasgupta, S., Yeates, N., Driskell, R., & Braun, Y. A. (Eds.), (2012). *Women's Encounter with Globalization.* London: FrontPage Publications.

Denissen, A. K., (2012). *Climate Change and Its Impacts on Bangladesh.* Retrieved from www.ncdo.nl. on 17 Janauary 2019.

Ferdousi, F., (2015). Factors Affecting Implementation of New Public Management in Bangladesh. *International Journal of Management Sciences*, 5(5), 315–328.

Fisher, S., (2010). Violence against Women and Natural Disasters: Findings from Post-Tsunami Sri Lanka. *Violence against Women*, 16, 902–918.

Fothergill, A., &Peek, L., (2004). Poverty and Disasters in the United States: A Review of Recent Sociological Findings. *Natural Hazards*, 32, 89–110.

Government of the People's Republic of Bangladesh [GoB], (2010a). *Standing Order on Disaster.* Dhaka: Disaster Management Bureau.

Government of the People's Republic of Bangladesh [GoB], (2010b). *National Plan for Disaster Management 2010–2015.* Dhaka: Disaster Management Bureau.

Haq, Z., Robbani, M., Ali, M., Hasan, M., Uddin, J., & Karim R., (2012) Damage and Management of Cyclone Sidr-Affected Homestead Tree Plantations: A Case Study from Patuakhali, Bangladesh. *Natural Hazards*, 64, 1305–1322.

Haque, A., & Jahan, S. (2015). Impact of Flood Disasters in Bangladesh: A Multi-Sector Regional Analysis. *International Journal of Disaster Risk Reduction*, 13, 266–275.

HSBC, (2011, October). *Climate Investment Update.* HSBC Global Research.

Huq, S., &Karim, Z., (1999). *Vulnerability and Adaptation to Climate Change for Bangladesh.* The Netherlands: Kluwer Academic.

Huq, S., Reid, H., Konate, M., Rahman, A., Sokona, Y., & Crick, F., (2004). Mainstreaming Adaptation to Climate Change in Least Developed Countries (LDCs). *Climate Policy*, 4(1), 25–43.

International Federation of Red Cross and Red Crescent Societies [IFRC], (2010). Dhaka: Cyclone *Aila.*

Jonkman, S.N. (2005). Global Perspectives on Loss of Human Life Caused by Floods. *Natural Hazards*, 34, 151–175. https://doi.org/10.1007/s11069-004-8891-3

Juran, L., & Trivedi, J. (2015). Women, Gender Norms, and Natural Disasters in Bangladesh. *Geographical Review*, 105(4), 601–611.

Khan, M. M., & Azad, A. K., (2012). The Nexus between Local Politics and Salinity Intrusion in Coastal Area of Bangladesh: Unveiling the Development Myth. *The Journal of Social Development*, 24(1), 65–76.

Kochanek, S. A. (1996). Bangladesh in 1996: The 25th Year of Independence. *Asian Survey*, 37(2), 136–142.

Kron, W., (2012). Coasts: The High-Risk Areas of the World. *Natural Hazards*, 66(3), 1363–1382.

Kumar-Range, S., (2001, November). *Environmental Management and Disaster Risk Reduction: A Gender Perspective.* Prepared for DAW/ISDR expert group meeting on 'Environmental

Management and the Mitigation of Natural Disasters: A Gender Perspective' in Ankara, Turkey.

Kumar, U., Baten, M.A., Al-Masud, A., Osman, K.S., & Rahman, M.M. (2010). *Cyclone Aila: One Year on Natural Disaster to Human Sufferings*. Unnayan Onneshan: Dhaka, Bangladesh, Socio-Economic Impacts of Cyclone Aila on Migrant and Non-Migrant Households in the Southwestern Coastal Areas of Bangladesh. Available from: https://www.researchgate.net/publication/337304823_Socio-Economic_Impacts_of_Cyclone_Aila_on_Migrant_and_Non-Migrant_Households_in_the_Southwestern_Coastal_Areas_of_Bangladesh

Mallick. B., & Vogt, J., (2009). Analysis of Disaster Vulnerability for Sustainable Coastal Zone Management: A Case of Cyclone *Sidr* 2007 in Bangladesh. In Climate Change: Global Risks, Challenges and Decisions, Conference Series, *Earth and Environmental Science* 6.

Mallick, B., & Vogt, J., (2012). Cyclone, Coastal Society and Migration: Empirical Evidence from Bangladesh. *International Development Plan Review*, 34(3), 217–240.

Mallick, B., & Vogt, J., (2014). Population Displacement after Cyclone and Its Consequences: Empirical Evidence from Coastal Bangladesh. *Natural Hazards*, 73(2), 191–212.

Morrow, B. H., (1997). Stretching the Bond: The Families of Andrew. In W.G. Peacock, B. H. Morrow, & H. Gladwin (Eds.), *Hurricane Andrew: Ethnicity, Gender and the Sociology of Disasters*(pp.141–170). London: Routledge.

Nahar, N., Blomstedt, Y., Wu, B., Kandarina, I., Trisnantoro, L., & Kinsman, J., (2014). Increasing the Provision of Mental Health Care for Vulnerable, Disaster-Affected People in Bangladesh. *BMC Public Health*, 14, 708–721.

Nasreen, M., (2012). *Women and Girls Vulnerable or Resilient?* Dhaka: Institute of Disaster Management and Vulnerability Studies, University of Dhaka.

Nasreen, M., (2015). Effectiveness of Resilient Livelihood Framework, Christian Aid.

Nasreen, M., (2017). Bangladesh National Conservation Strategy: Gender Issues. MoEFCC & IUCN.

Nasreen, M. (2021). Fifty Years of Independence: Fifty Years for Disaster Risk Governance in Bangladesh, *The Daily Star*, 26 March. Retrieved from www.thedailystar.net/supplements/celebrating-50-years-bangladesh/news/fifty-years-disaster-risk-governance-bangladesh-2067109

Nasreen, M. et al., (2014). Rapid Assessment on the Situation of Sexual and Reproductive Health during Emergencies, Sponsored by Family Planning Association of Bangladesh & SPRINT South Asia.

Nasreen, M. et al., (2018). Integration of Science, Technology and Indigenous Knowledge into Community-Based Disaster Risk Management in Bangladesh, Research Report, International Center for Collaborative Research on Disaster Risk Reduction (ICCR-DRR), Beijing Normal University, China, February.

Nasreen, M., Hossain, K. M., & Azad, A. K. (2016). Assessing Vulnerabilities and Resilience of Persons with Disabilities during Disasters: Case Study through Gender Lens, Department of Disaster Management, MoDMR, GoB.

Nel, P., & Righarts, M. (2008). Natural Disasters and the Risk of Violent Civil Conflict. *International Studies Quarterly*, 52(1), 159–185. https://doi.org/10.1111/j.1468-2478.2007.00495.x

Paul, B. K., (2009). Why Relatively Fewer People Died? The Case of Bangladesh's Cyclone *Sidr*. *Natural Hazards*, 50(2), 289–304.

Peacock, W. G., Morrow, B. H., &Gladwin, H., (1997). *Hurricane Andrew: Ethnicity, Gender, and the Sociology of Disasters*. New York: Routledge.

Peek, L., Morrissey, B., & Marlatt, H. (2011). Disaster Hits Home: A Model of Displaced Family Adjustment After Hurricane Katrina. *Journal of Family Issues*, 32(10), 1371–1396. doi:10.1177/0192513X11412496

Peek, L. & Stough, L.M. (2010). Children with Disabilities in the Context of Disaster: A Social Vulnerability Perspective. *Child Development*, 81(4), 1260–1270.

Saha, C. K., (2015). Dynamics of Disaster-Induced Risk in Southwestern Coastal Bangladesh: An Analysis on Tropical Cyclone *Aila* 2009. *Natural Hazards*, 75(1), 727–754.

Scaling Climate Change Adaptation Knowledge and Technologies for Empowering Women, and to Enhance (SAKTEE) Social Equity and Disaster Resilience in Bangladesh, (2020). Government of Bangladesh (GoB), University of Manitoba, Bangladesh Centre for Advanced Study (BCAS), International Centre for Climate Change and Development (ICCCAD).

Schendel, W.V., (2009). *A History of Bangladesh*. Cambridge: Cambridge University Press.

Shamsuddoha, M., &Chowdhury, R. K., (2007). *Climate Change Impact and Disaster Vulnerabilities in the Areas of Bangladesh*. Dhaka: COAST Trust.

Shaw, R., (2006). *Indian Ocean Tsunami and Aftermath: Need for Environment Disaster Synergy in the Reconstruction Process*. Retrieved from www.emaraldinsight.com/0965-3562.htm on 2 June 2015

Shaw, J., Applegate, B., Tanner, S., Perez, D., Rothe, E., Bowen A.C., & Lahey, B., (1995). Psychological Effects of Hurricane Andrew on an Elementary School Population. *Journal of the American Academy of Child and Adolescent Psychiatry*, 34(9), 1185–1192.

Sobhan, R., (2012). *Challenging the Injustice of Poverty*. Delhi: Sage.

Thieme, S., & Siegmann, K. (2010). Coping on Women's Backs: Social Capital-Vulnerability Links through a Gender Lens. *Current Sociology*, 58(5), 719–737.

Tierney, K. (2006, March). *Hurricane Katrina: Catastrophic Impacts and Alarming Lessons*. In Conference Paper Series Prepared for Berkeley Symposium on Real Estate, Catastrophic Risk, and Public Policy.

Tobin-Gurley, J., Peek, L., &Loomis, J., (2010). Displaced Single Mothers in the Aftermath of Hurricane Katrina: Resource Needs and Resource Acquisition. *International Journal of Mass Emergencies Disasters*, 28(2), 170–206.

United Nations Economic and Social Commission for Asia and the Pacific (UN ESCAP). (2010). *Asia Pacific Disaster Report 2010*. Bangkok: UN ESCAP.

Unnayan Onneshan, (2009). *Initial Assessment Report: With Focus on Khulna District*. Dhaka: Unnayan Onneshan.

Unnayan Onneshan, (2010). *Cyclone Aila: One Year on Natural Disaster to Human Sufferings*. Dhaka: Unnayan Onneshan.

Wilson, J., Phillips, B. D., & Neal, D. M., (1998). Domestic Violence after Disaster. In E. Enarson & B. H. Morrow (Eds.), *The Gendered Terrain of Disaster: Through Women's Eye* (pp. 81–94). Westport, CT: Praeger Publishers.

World Bank (2006). Bangladesh Country Environmental Analysis. *Bangladesh Development Series, 12*. Dhaka: World Bank Office.

Zahran, S., Peek, L., Snodgrass, J.G., Weiler, S., & Hempel, L. (2013). Abnormal Labor Outcomes as a Function of Maternal Exposure to a Catastrophic Hurricane Event During Pregnancy. *Natural Hazards*, 66(1), 61–76. doi:10.1007/s11069-011-0065-5

Zahran, S., Shelley, T. C., Peek, L., &Brody, S. D., (2009). Natural Disasters and Social Order: Modeling Crime Outcomes in Florida. *International Journal of Mass Emergencies Disasters*, 27(1), 26–52.

Zaman, M. Q., (1999). Vulnerability, Disaster, and Survival in Bangladesh: Three Case Studies. In A. Oliver-Smith & S. M. Hoffman (Eds.), *The Angry Earth: Disaster in Anthropological Perspective* (pp. 192–212). New York: Routledge.

PART I

Understanding Vulnerability and Risk of Coastal Bangladesh

2

EFFECTS OF SALINITY ON HEALTH DUE TO ENVIRONMENTAL EXPOSURE

Experiences from Bangladesh

Md. Sadequr Rahman, Khondoker Mokaddem Hossain, Joris van Loenhout, Pascaline Wallemacq, and Debarati Guha-Sapir

Introduction

Bangladesh is struggling with multiple risks, like natural and anthropogenic hazards, socioeconomic and environmental challenges, and health threats due to salinity intrusion. The geographical setting and meteorological characteristics have made the country vulnerable to different geo-hazards and hydro-meteorological hazards. These hazards often lead to various disasters. The major disasters concerned in the country are floods, cyclones, droughts, tidal surges, tornadoes, nor-wester, earthquakes, river erosion, arsenic contamination of groundwater, water and soil salinity, heat wave, cold wave, water logging, fire, building collapse, epidemic, and various forms of pollution. The 'World Risk Report 2018 has identified that Bangladesh is the sixth topmost natural disaster-prone country among 171 countries in the world (UNISDR 2018). According to the German Watch on Global Climate Risk Index 2019, Bangladesh is ranked as the seventh most affected and vulnerable country due to climatic disasters (National Plan for Disaster Management 2021–2025). The Global Assessment Report (GAR) 2011 of the UNISDR claimed that regarding the number of people exposed to flood, cyclones, and sea surges, Bangladesh is ranked first out of 162 countries (UNISDR 2011). It is estimated that about 2.8 million hectares of land of Bangladesh are affected by salinity—that is, almost one-third of the 9 million hectares of total national cultivated area, and about one-fifth of the total area of Bangladesh (Salim et al. 2007; Khan et al. 2008). Saltwater or salinity intrudes in the coastal belt from the flow of seawater through natural disasters like cyclones, tidal surge, flood, river flow, estuarine circulation, and so on. Besides, climate change, sea-level rise, precipitation, and water and land management practices exacerbate the salinity (Rahman et al. 2000; Nahian et al. 2018). It has been well-documented that salinity has adverse effects on crop productivity and food grain production (Khan et al. 2008). Salinity brings a threat to the

DOI: 10.4324/9781003253495-3

well-being of coastal communities of this country. It is also very threatening to agriculture, coastal biodiversity, and human health. Salinity intrusion is extending into the non-coastal areas, and it has reached more than 100 km inland in Bangladesh (Talukder et al. 2016a).

Salinity (surface water and groundwater) poses negative health risks for the people living in southwest coastal areas of Bangladesh and reduced agricultural production of its 35 million inhabitants (almost one-third of the total population) directly or indirectly (Clarke et al. 2015; Ayers et al. 2017; Khan et al. 2011). Drinking saline water can cause hypertension in the long run. In southwest Bangladesh, high drinking water salinity has also been linked to relatively higher rates of pre-eclampsia and gestational hypertension in the dry season than the wet season (Khan et al. 2011b).

Sources or Causes of Salinity

Bangladesh is one of the most vulnerable countries regarding climate change and sea-level rise that lead to the challenging risk of water and soil salinity for coastal Bangladesh (Dasgupta et al. 2015). Salinity mainly affects land and water in the coastal areas. The main causes of salinity intrusion, propagation, and expansion in the coastal belt of Bangladesh are critical geographical setting, the upstream diversion, and reduction of freshwater flow (over-extraction of freshwater) in Ganges and Gorai river, faulty management of coastal polders, saltwater intrusion due to sea-level rise, cyclone and storm surge, backwater effect, the expansion of shrimp culture and the increased practice of agricultural irrigation, along with natural and environmental processes like river salinity, loss of wetlands and interaction with groundwater, tidal surge, and storm surge inundation and water logging, and so on (Rahman et al. 2000; Nahian et al. 2018).

Due to increasing salinity in the water and soil, the people of the coastal region are suffering from the scarcity of safe drinking water, irrigation, agriculture, and other uses. Soil salinity is considered a major constraint to food grain production in coastal areas of the country. The increase of the level of water salinity affects livelihood and insecure water availability and pushes poor peoples' lives to a more vulnerable position than earlier. Water salinity also increases in soil salinity, which further deteriorates soil fertility, loses agricultural productivity, and brings significant pressure on food security. Saline water is needed to cultivate shrimp cultures. So, saline water is intruded into the shrimp pond or '*gher*'. As a result, the increase of salinity in surface and groundwater is also expanding because of continuous shrimp cultivation in the fresh agricultural land (Mahmuduzzaman et al. 2014).

High salinity in water and soil also decreases the crop yields. For example, when irrigation water exceeds 5 parts per thousand (ppt) salinity, crop yields decrease as much as 50% (Clarke et al. 2015; Ayers et al. 2017; Khan et al. 2011a). Ayers et al. (2016) explored the causes of salinization and arsenic contamination of groundwater resources, which are widespread problems in tidal delta plain of southwest Bangladesh, and found that salts in the shallow aquifer groundwater were derived from connate water. Soil salinity is affected by water salinity, drainage, evaporation, saline intrusion, and other effects of climate change. In Bangladesh, soil salinity comes from cyclone and tidal inundation, adoption of brackish water for shrimp cultivation, and the capillary rise of salts from shallow groundwater in the dry season (Clarke et al. 2015).

Soil is mainly classified into two categories: saline and sodic. Saline soils are widely affected across the continents like in Central and South Asia, West Asia and North Africa, Latin America, Africa, and Australia, and so on. Sodic soils are also found on all the main continents but are particularly concentrated in Australia, Europe, and Latin America. Salinization occurs due to natural processes or because of human activity. It may result from the mobilization of salts dissolved in rainwater and their concentration in water and soil, often far from their origins (Stenhouse and Kijne 2006).

FAO (1996) has estimated about 397 million hectares of saline soils and 434 million hectares of sodic soils globally. According to Oldeman et al. (1991), more than 76 million hectares of agricultural land are salt affected; the majority of it is found in Asia and Africa. In Bangladesh, about 30% of the cultivable land is in coastal areas where salinity is affected by tidal flooding during the wet season, direct inundation by storm surges, and movement of saline ground and surface water during the dry season (Haque 2006; Dasgupta et al. 2015).

Scientific studies show that salinity levels have already increased in domestic ponds, groundwater supplies, and agricultural land in Bangladesh through the

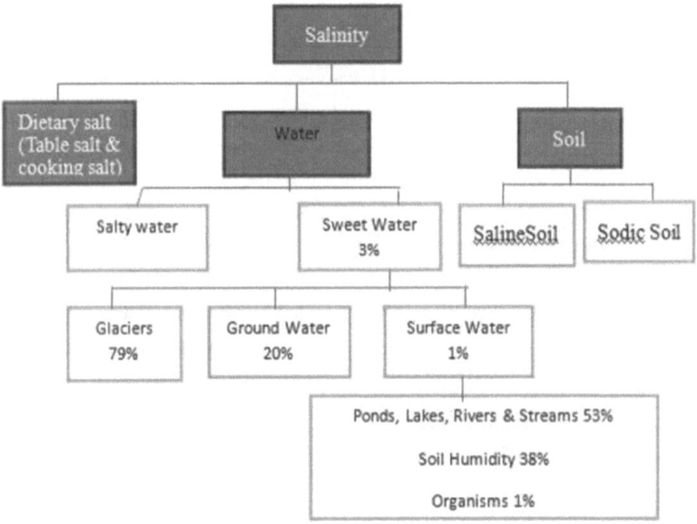

FIGURE 2.1 Sources of salinity

Source: Khanom and Salehin (2012); Ayers et al. (2016, 2017); Khan et al. (2011); Stenhouse and Kijne (2006); Rasheed et al. (2016); Naser et al. (2017b); Lecture sheet of Joos van den Noortgate [WASH & MSF, CRED].

various estuaries and water inlets intertwined with major rivers (Colligan 2011). According to the Ministry of Environment and Forest (MOEF) 2006, about 20 million of the 37 million of coastal Bangladeshi people are susceptible to the challenges of excessive salinity (Khan et al. 2011b). Dasgupta et al. (2015) carried out a study regarding soil salinity of 140 sub-districts (upazilas) in four regions of southern Bangladesh: Barisal (38 upazilas), Chittagong (30), Dhaka (13), and Khulna (59) for the period 2001–2009. They showed that the soil salinity in Barisal is relatively higher than their riverine counterparts (Dasgupta et al. 2015). Salinityis found mostly in water, soil, and diet as table salt or cooking salt, which are shown in Figure 2.1.

Salinity Standard

Groundwater Salinity in Bangladesh

Groundwater salinity of coastal region comprises soluble salts such as the alkali element Na, alkaline earths Mg and Sr, and halogen Cl, which are all highly correlated (Ayers et al. 2016). The standard way of measuring salinity is milligram of total dissolved solids (TDS) per litre. It is expressed using the Electrical Conductivity (in reference to 25°C) often expressed either in milliSiemens per centimetre (mS/cm) or microSiemens per centimetre (μS/cm) (Duncan et al. 2000). In southwest Bangladesh groundwater specific conductivity (SpC) ranges from 1 to 10 (mS/cm), equivalent to salinities of 0.5–5.2 parts per thousand (ppt, seawater is 35 ppt). Tidal channels are an important potential source of high salinity water. Tidal channel water is a mixture of seawater from the Bay of Bengal and freshwater, with salinities up to ~15 ppt during the dry season (Ayers et al. 2016; George 2013). The average salinity of groundwater in the southwest area is ~10 ppt, roughly one-third of that of seawater and like the salinity of tidal channel water in the dry season (Ayers et al. 2016).

Salinity is the measure of concentration of dissolved salts in water such as chloride anions, usually expressed in parts per thousand (1 ppt = 1,000 mg/l = 1,000 ppm). According to the Environment Conservation Rules, 1997 in Bangladesh, the recommended range of chloride in drinking water is 150–600 ppm (mg/l). A higher value of salinity up to 1,000 ppm for problematic areas including the coastal belt is considered for drinking (GoB 1997; Duncan et al. 2000). FAO has classified the water based on the quantity of salt and regarding this, a modification of Bangladesh Standard for Drinking water was shown in Table 2.1.

Soil Salinity in Bangladesh

Climate change poses major soil salinization risk for coastal Bangladesh (Khanom and Salehin 2012). Soil is linked with the salinity through nearby rivers, land elevation, temperature, and rainfall (Rhoades et al. 1999). Soil salinity is particularly high in the Khulna coastal region. The salinity level at the Bay of Bengal during the dry

TABLE 2.1 Classification of water as a function of salinity (in milliSiemens/cm and mg TDS/l)

Type	*mS/cm*	*mg TDS/l*	*Drinking or irrigation water*
Non-saline or fresh water	<0.8	<600	Drinking and irrigation water
Slightly saline	0.8–2	600–1,500	Irrigation water
Moderately saline	2–10	1,500–7,000	Primary drainage water and groundwater
Highly saline	10–25	7,000–15,000	Secondary drainage water and groundwater
Very highly saline	25–45	15,000–35,000	Seawater is 35,000 TDS mg/l
Brine	>45	>45,000	

Dietary sodium is consumed mainly as salt: NaCl = Na (g) × 2.541 mmol sodium corresponds to 23.0 mg, 1 mmol chloride to 35.5 mg; 1 g NaCl consists of ~17 mmol sodium and chloride each.

Source: Duncan et al. (2000).

TABLE 2.2 Characteristics of saline and sodic soils

Type of soil	*Electrical conductivity (dS/m)*	*Exchangeable sodium percentage*	*Sodium absorption ration*	*pH*
Saline	>4	<15	<13	<8.5
Sodic	<4	>15	>13	>8.5

season is also comparatively high; and saline water intrudes via tidal effects through the major rivers: the Baleswar, Jamuna, lower Meghna, Malancha, Pussur, Sibsa, and Tetulia. The topography of the region is very flat; strong tidal effects sometimes travel 200 km upstream from the coast (Khanom and Salehin 2012).

Salinity is measured by electrical conductivity, sodium content, and pH, which is shown in Table 2.2. Saline soils are comprising mainly excess soluble salts like chlorides, sulphates and bicarbonates of sodium, calcium, and magnesium that have direct deleterious effects on plant growth. On the other hand, alkaline or sodic soils have excess sodium of the soil, which adversely affects the soil structure and properties and creates the conditions that are unfavourable for plant growth, development, and nutrient uptake (Stenhouse and Kijne 2006).

Materials and Methods of the Study

Data Sources: Systematic literature reviews were used for this study by searching the databases such as PubMed, Science Direct, and grey literature like Government of Bangladesh, the International Centre for Diarrheal Disease Research (ICDDR'B), and NGO/INGO regarding salinity on health or hypertension-related diseases in Bangladesh. Moreover, some secondary sources like books, journals, articles, reports, workshop, newspapers, internet websites etc. and other documents were also used as support.

Search Methodology for Identification of Studies: We used the following major search keywords: 'Salinity', 'saline', 'salt', 'health', 'diseases', 'hypertension', and 'Bangladesh'. We connected keywords related to the same concept using the Boolean operator OR, and different concepts using the logical operator and. For the databases with peer-reviewed journals, in PubMed the advanced search was as follows: (Saline or Salinity or Salt) and (Health or Diseases) and Bangladesh. And on 31.07.2018 it was found 190 in total, and 146 articles from 2000 AD to till date. After screening the titles, the abstract and full text were selected for 68 articles. After reading the abstract and full text for relevance, articles were selected for systematic literature review. Besides, we searched for these keywords as advanced search in Science Direct; we connected them in the expert search. In ScienceDirect, we searched this process: (Saline or Salinity or Salt) and (Health or Diseases) and Bangladesh, and on 01.08.2018 it was found 421 in total, and 331 articles from 2000 AD to till date. After screening the titles, the abstract and full text were found for 42 articles. Further, we also filtered the search strategies in grey literature databases as follows: To collect salinity-related information from Government of Bangladesh and ICDDR'B, various reports, surveys or assessment or document categories, health for sectors; health- or disease-related information regarding salinity and climate change like drinking saltwater and hypertension, blood pressure, kidney disease, anaemia, diarrhoea, dehydration, maternal and child health, eclampsia or pre-eclampsia, pregnancy-related complicacy, and so on. Finally, after deleting the duplicate and irrelevant articles, for systematic literature review, we reviewed a total of 27 articles (case-control, cross-sectional, cohort study, and randomized clinical study) of salinity on health or particularly hypertension/cardiovascular diseases in Bangladesh.

Inclusion and Exclusion Criteria: Peer-reviewed and grey full-text manuscripts in English published after 2000 AD to 01.08.2018 with no limits for study design were included as inclusion criteria. We included studies examining the effect of saltwater intake or salinity on health or hypertension or cardiovascular diseases in Bangladesh due to natural disasters. We excluded editorial notes, letters, systematic reviews and viewpoint articles, previous version of the same grey literature (e.g. if the same report is published in 2013 and 2014, the 2013 report was excluded) as well as one selected article from any one by excluding various publications (e.g. if the same report is available in Pubmed and Science Direct, only one is counted and other report was excluded). Figure 2.2 shows the sources of literature that we reviewed systematically/ narratively by searching the literature for peer-reviewed articles using the databases PubMed and ScienceDirect, and for gray literature using salinity-related information of Government of Bangladesh, the International Centre for Diarrheal Disease Research (ICDDR'B), and NGO/INGO.

Data Analysis: We reviewed systematically by summarizing, comparing, and contrasting the extracted data of the studies. The following section presents the results of the narrative literature review of the (some exact) eligible studies.

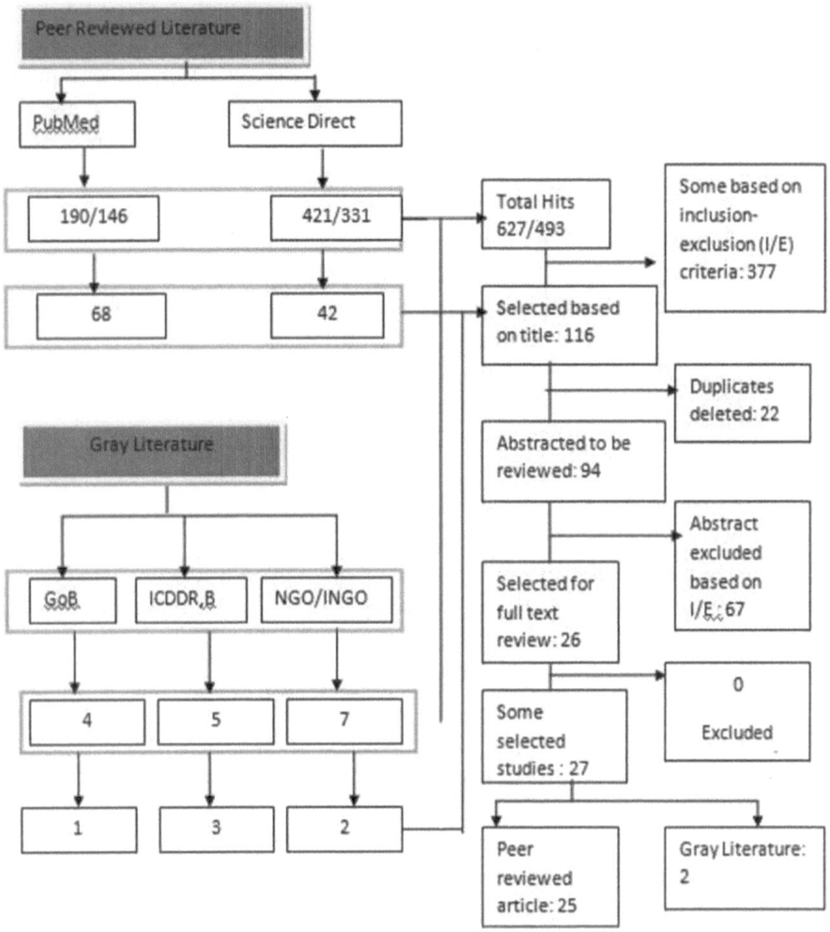

FIGURE 2.2 Research methodology decision tree for exclusion/inclusion

Source: Gul et al. (2015).

Effects of Salinity

The effects of higher salinity on health were seen in low-income countries (Rahaman et al. 2019). The government of Bangladesh and Caritas Development Institute (CDI) reported a range of health risks among coastal people with potential links to increased salinity exposure through drinking, cooking, and bathing, including hypertension and miscarriage among pregnant women, skin diseases, acute respiratory infection, and diarrhoeal diseases (Vineis et al. 2011).

Salinity has directly or indirectly affected human health or animals' health. The main health problems of coastal belt people in Bangladesh were reported in various literature as follows: hypertension or high blood pressure, stroke, heart diseases, osteoporosis, malaria, diarrhoea, dehydration, skin diseases (itch, boil/blister), malnutrition, acute illness, respiratory illness such as asthma, stomach cancer, obesity, kidney diseases, chronic illness, trauma and other injuries, infection or communicable diseases, non-communicable diseases (NCDs), pregnancy-related complicacy or gynaecological problems and injuries, anaemia, pre-eclampsia, and so on (Colligan 2011).

It is well-documented that a high salt intake in children also influences blood pressure and may predispose an individual to the development of a number of diseases including high blood pressure, osteoporosis, and respiratory illnesses such as asthma, stomach cancer, and obesity. Due to natural disasters-induced salinity intrusion, it is seen that the premature birth and miscarriage rate are more severe in salinity-prone coastal regions than other areas. It is known that children, elderly, women, pregnant women, and disable people are comparatively more at risk. Women are the most vulnerable and worst sufferers regarding their socioeconomic or physical status (Downie 2007).

Salinity causes direct and indirect effects on the environment and plant growth, where agricultural production is almost impossible due to severe salinity (Stenhouse and Kijne 2006). It can also cause loss of biodiversity, impacts on wildlife, and disruption of ecosystems, leading to loss of ecosystem function (Barrett-Lennard et al. 2005). Homestead production like kitchen vegetables and domestic livestock is reduced as a result of the increased salinity of domestic water. In addition to the losses of agricultural production, salinity causes other economic loss through its direct effects on infrastructure and drinking water. Salt is highly active chemically and can cause destruction of infrastructures like roads and buildings (Abdel-Dayem 2004).

Fisheries will be greatly hampered if the saltwater is present in groundwater level and open water bodies (including ponds). Diseases of fish may increase. Production of sweet water fish (like carps) will shrink and extinct if the sea level rises. Sweet water fish or carp fishes will die or become contagious for intolerable excessive salinity, and human beings will get these diseases by eating these contaminated fishes (Abdullah et al. 2021).

Dietary Salt Intake and Hypertension

It was first recognized a century ago and now well established in ecological, epidemiological, and experimental human studies that excessive dietary salt (sodium or sodium chloride) intake is strongly associated with blood pressure or hypertension (Alderman 2000; Schmidt 2017). World Health Organization (WHO, 2013) identified hypertension or high blood pressure as a silent killer, global public health crisis, and the number one risk factor for premature death worldwide, accounting for 9 million deaths per year. Chrysant et al. (1999) carried out a research to measure the association of salt consumption and blood pressure in an entire population of Bangladesh and showed that reduction of dietary sodium creates the lower blood pressure in at most 30% to 60% of hypertensive subjects and 25% to 50% of normotensive subjects. Rasheed et al. (2014) assessed the salt consumption among adults (400 participants) in Chakaria and found that females, housewives, and coastal people consumed high salt for low level of awareness compared to male and labour-intensive people that leads to hypertension. They also found that salt consumption and excess drinking water salinity is a risk factor for hypertension.

Rasheed et al. (2016) carried out further a cross-sectional survey among 400 adults in Chakaria and explored that salt consumption was likely at high risk of hypertension among the coastal people in Bangladesh. It is found that salt is very harmful for the health of pregnant and lactating mothers. Sometimes the legs and feet swelled up for pregnant woman having a habit of taking table salt. Excessive salt intake among pregnant women increases blood pressure that leads to pre-eclampsia or eclampsia. Such excessive salt intake can turn many people at risk of hypertension, diabetes or kidney diseases, and even death (Rasheed et al. 2016).

Talukder et al. (2016a) conducted a cross-sectional survey and assessed spot urine samples and high sodium consumption among young adults of 282 participants (19–25 years) during May–June 2014 in Koyra of Khulna district in southwestern coastal Bangladesh. They found that married people consume salt 2.5 times higher than the non-married group, significantly 80% higher in female married, and higher among the middle wealth quintile group. They also found that almost one-fifth of the young population had blood pressure above the normal range (SBP >120 mmHg or DBP >80 mmHg), which implies the potential of being hypertensive in later life and risk of developing other chronic diseases (cardiovascular and kidney diseases).Sakamoto (2017) carried out a survey of 100 sample tube wells and 569 family members of Chakaria Upazila in Cox's Bazar District and showed that older people usually consume more salt, particularly cooking salt.

Zaman et al. (2017) conducted a cross-sectional study on a total of 200 (100 rural people from Ekhlaspur village of Chandpur district and 100 urban people at National Heart Foundation Hospital and Research Institute, Dhaka) people aged 20 years or older in 2009 to reveal the association between salt consumption and hypertension in Bangladesh. People prefer salty taste and a habit of adding salt during meals (during cooking and while eating a meal), and preservation of fish with salt in Bangladeshi people is very common. Many people eat sour seasonal fruit with salts. This dietary practice often leads to hypertension.

Alderman (2000) studied about 14,000 adults in NHANES data of US population to further explore the positive association of sodium intake with increasing blood pressure to strokes and heart attacks and cardiovascular morbidity and mortality. Those in the lowest quartile of sodium intake were 20% more likely to die of a cardiovascular cause than those in the highest quartile of sodium consumption (Alderman et al. 1998; Alderman 2000). They found that by 100-mmol increase in dietary sodium intake, stroke incidence, stroke mortality, and cardiovascular disease mortality were evident: 0.99 (0.81–1.21), 0.82 (0.55–1.22), and 1.00 (0.84–1.19) among non-overweight and 1.39 (1.09–1.77), 1.98 (1.25–3.14), and 1.45 (1.20–1.75) among overweight participants significantly (Alderman 2000).

The prevalence of hypertension varies considerably by country: 20% in the United States, and 25%–50% in different regions in Europe. The exact prevalence of hypertension in Bangladesh is not known. In Bangladesh, approximately 20% of adults and 40%–65% of elderly people suffer from hypertension (Islam and Majumder 2012). Particularly, 18% of the adults aged 25 years and above in Bangladesh have been found to be hypertensive (Talukder et al. 2016a; WHO 2013). Zaman and Rouf (1999) studied meta-analysis and Zaman et al. (2017) studied two times differently and found that the prevalence of hypertension was 11.3%, 18.6%, whereas Alamgir et al. (2009) conducted a survey and found that the prevalence of hypertension was 20.1%. According to the Bangladesh Society of Medicine in collaboration with the Directorate General of Health Services and WHO (2010) survey aged ≥25 years, the prevalence of hypertension is 17.9% in general, 18.5% in men, and 17.3% in women. Hypertension is more common in elder people, 65% in general, 75% in urban area, and 53% in rural area (WHO 2010; Moni et al. 2010). Alam et al. (2013) suggested by another study that about 17.1% of the total population aged more than 20 years are hypertensive. On the other hand, Bangladesh Demographic and Health Survey (BDHS) found hypertension prevalence higher among women (31.9%) than men (19.4%) participants of 35 years and older age group (NIPORT 2013).

Drinking Water Salinity and Hypertension

Drinking water sodium or salinity is also a risk factor for hypertension or blood pressure–related diseases in salinity-prone coastal areas that leads to cardiovascular morbidity and mortality. Women are associated with high blood pressure or hypertension due to high drinking saltwater, salt, or sodium intake (Khan et al. 2008, 2011, 2014; Nahian et al. 2018, Clarke et al. 2015; Rasheed et al. 2016; Naser et al. 2017a; Vineis et al. 2011; Scheelbeek et al. 2016b; Adhikary 2016). Some studies like Khan et al. (2008, 2011), Khan (2014), and Rasheed et al. (2016) assessed the relations between drinking water salinity and the rates of pre-eclampsia, eclampsia, and gestational hypertension in pregnant women. They also mentioned that pregnant women were identified as having 'hypertension in pregnancy' (SBP > 140 mmHg or DBP > 90 mmHg after the 20th week of pregnancy and not before pregnancy), pre-eclampsia (high BP in pregnancy), and eclampsia due to drinking water salinity. Hypertension in pregnancy is related with increased rates of adverse maternal and foetal outcomes including impaired liver function, low platelet count, intrauterine growth retardation, preterm birth, and maternal and prenatal deaths in both acute and long term. Vineis et al. (2011) evaluated the potential damage of economic and infrastructural, livelihoods and increases health risks through saline contamination of drinking water regarding the parameters of independent variable (IV)—climate change, salinity intrusion, and dependent variable (DV)—shrimp culture, soil salinity, low productivity, hypertension or high blood pressure, miscarriage, skin diseases, respiratory diseases, mosquito-borne diseases, diarrhoea, and stroke. They found that water salinity led a large proportion of the population in developing pre-hypertension (SBP=120–139 mmHg or DBP=80–89 mmHg) and hypertension (SBP > 140 mmHg or DBP > 90 mmHg). Hypertension and miscarriage among pregnant women, skin diseases, acute respiratory infection, and diarrhoeal diseases were prevalent because of increased salinity in drinking, cooking, and bathing water. Scheelbeek et al. (2017) explored the relationship between drinking water salinity (sodium concentration) and blood pressure in a coastal population in Bangladesh regarding the confounders drinking water salinity, dietary salt intake, age, sex (model 1), physical activity, body mass index, smoking (model 2), temperature, disease; marital status; religion; HH members; education; use of *paan, hukka, and gul*; insecticides and chemical; and lifestyles. It is the first cohort study on drinking water sodium and blood pressure in (non-pregnant) adults in a salinity-affected coastal area. Khan and her team found that salinity in drinking water is associated with increased risk of pre-eclampsia and gestational hypertension in this population and the number of hypertensions was almost eight times higher than in non-coastal regions of Bangladesh. The higher prevalence of hypertension or raised blood pressure was observed in pregnant women during the dry season than in the rainy season (Clarke et al 2015; Khan et al. 2011; Naser et al. 2017a). Nath et al. (2012) conducted case-control study from 290 respondents of salinity-prone area (Bagerhat district) and non-salinity area (Tangail) to assess the blood pressure status regarding salinity and non-salinity area in Bangladesh. About 20% of the non-saline respondents' SBP was 110–120 mmHg and 26% of saline area had SBP >135–140 mmHg. About 19% of the non-saline respondents' DBP was ≤ 85 mmHg. About one-third had ≥ 90 mmHg, among them 31% were from saline area. They observed that mean of SBP and DBP among the people of salinity area (Bagerhat) was more than that of non-salinity (Tangail) area. Khan et al. (2008) conducted a survey of 561 pregnant women in 2008 of Gynaecology department in Chalna Upazilla Health Complex in the southwestern coastal areas and examined the relations between drinking water salinity and the rates of pre-eclampsia, eclampsia, and gestational hypertension in pregnant women. They noted that 21% of the respondents and between the ages of 16 and 40 years were diagnosed regarding hypertensive disorder due to drinking water salinity. Khan et al. (2011) further showed that women who drank shallow tube well water were more likely to have urine sodium > 100 mmol/d than women who drank rainwater (odds ratio [OR] 2.05; 95% confidence interval [CI], 1.11–3.80).

Talukder et al. (2016a)conducted a cross-sectional study among 253 participants aged 19–25 years from four villages during May–June 2014 in Koyra of Khulna in Bangladesh to examine the association between high salinity potable water and blood pressure in young adults in coastal Bangladesh. They reported that 65% of the study population of Koyra were exposed to highly saline drinking water (Bangladesh standard 600 mg/l and above) and their mean salinity in drinking water was 817 mg/l. This study is the first population-based study to describe the blood pressure of young adults in low-lying coastal communities regarding climate change-induced sea-level rise and salinity-affected exposure. Rasheed et al. (2014) found that more than three-fourths of the study population consumed more than biologically needed, and half consumed greater than the 5 g of salt daily; excess drinking water salinity and salt consumption are the risk factors for hypertension. Talukder et al. (2016a) and Sheelbeek et al. (2016a) showed that salinity (sodium) was higher in tube well (ground) water users (two times more) than pond (surface) water users.

On the contrary, Talukder et al. (2017) further extracted data on blood pressure or prevalence of hypertension and calculated pooled estimates to evaluate the relationship between sodium in drinking water and raises in blood pressure systematically. The pooled standardized mean difference (SMD) in seven studies (12 datasets) comparing the low- and high-water sodium exposure groups for SBP was 0.08 (95% CI: −0.17, 0.34) and for DBP was 0.23 (95% CI: 0.09, 0.36). Of the three studies, they assessed the association between high-water sodium in pregnant women and odds of hypertension; two studies showed the higher risk of hypertension. They noted that there was an association between water sodium and blood pressure, more consistently for DBP, and other related chronic diseases (e.g. cardiovascular and kidney diseases) in low-lying coastal areas.

Nahian et al. (2018) explored the association among socio-ecological systems (SESs), drinking water salinity, and health crises, more specifically hypertension or blood pressure, through cross-sectional household survey from more than 1,500 household in coastal Bangladesh. They found that hypertension is more observed in female, at higher age group (35−59 years) and among those who drink higher-concentration saline water (≥3,000 mg/l). Drinking water salinity was significantly associated with high blood pressure or (pre-hypertension and hypertension) which leads to cardiovascular diseases. Due to slightly saline (1,000−2,000 mg/l) and moderately saline (≥2,000 mg/l) concentration drinking water, people exposed to higher risk of hypertension: 17% ($p < 0.1$) and 42% ($p < 0.05$), respectively, than those who consumed fresh water (<1,000 mg/l). Women had 31% more risk of being hypertensive than men. Besides, 35−59 years of age group respondents were about 2.4 times more likely to be hypertensive compared to below 35 years age group. For the 35 years and above age group, both pre-hypertension and hypertension were found higher than national rural statistics (50.1%) for saline water categories (53.8% for slightly and 62.5% for moderate saline).

Scheelbeek et al. (2016a) reported that drinking tube well water, pond water, and combined sources were significantly associated with higher systolic pressure (+3.62 mmHg [CI: 1.20, 6.04] and +4.85 mmHg [2.55, 7.25] for pond and tube well users, respectively) compared with drinking rainwater after adjustment for known and context-specific confounders. It was found that drinking water with a high sodium concentration from saline contamination is associated with higher blood pressures in normotensive pregnant women in coastal areas of Bangladesh. Scheelbeek et al. (2016b) and Nahian et al. (2018) explored that high salt or sodium intake is strongly associated with increased risk of hypertension of aged group. Naser et al. (2017a) observed that brackish or salty groundwater in southern coastal area of Bangladesh is associated with high blood pressure among the adult population, and pre-eclampsia and gestational hypertension among pregnant women.

Naser et al. (2017a) and Talukder et al. (2016a,2016b) found that more than two-thirds of young adult of 19–25 years age drink high salt water linked with hypertension and almost one-fifth of the young population had blood pressure above normal range (SBP >120 mmHg or DBP>80 mmHg), which implies potential of being hypertensive in later life and risk of developing other chronic diseases (such as cardiovascular and kidney diseases).

Adhikary (2016) conducted an exploratory study from 55 women in Paikgasa of Khulna district in Bangladesh to explore the knowledge, perception, and practices of women of childbearing age, regarding the health impacts of water salinity. He showed that respondents shared their knowledge and perception. Due to intake of saltwater, few diseases such as cardiovascular diseases like hypertension, heart failure, respiratory infections, abdominal gas, and burning pain in the stomach, indigestion, diarrhoea, dysentery, skin disease, maternal and reproductive complicacy, fever, appendicitis, and even cancer were observed. Regarding maternal and reproductive health-related issues, respondents reported that under-nutrition, vomiting during pregnancy, itching, ulcers and pain around the genitalia, vaginal discharges, and increased number of caesarean section deliveries have arisen.

Khan et al. (2008) found that the prevalence of pregnancy-induced SBP and DBP among these areas women are higher of 6.8% and 5.4%, respectively, than a non-coastal area, which is studied by Sayeed et al. (2005), and the rate of hypertension of these areas population is strikingly higher than the 2.65% seen in Matuail, which is observed by Sameena Chowdhury (Gynaecologist-Obstetrician, personal communication) in another non-coastal rural community of Bangladesh. Khan et al. (2014) conducted further a case-control study between October 2009 and April 2011 at Dacope of Khulna in Bangladesh among 202 pregnant women aged 13–45 years with pre-eclampsia or gestational hypertension, enrolled from the community served by the Upazilla Health Complex, Dacope and 1,006 matched controls from the same area, among a population of approximately 143,000 people. Urinary sodium and sodium levels in drinking water were high and found high blood pressure (>140/90 mmHg) and/or oedema. There were 162 cases of pre-eclampsia, 33 cases of gestational hypertension, seven cases of eclampsia. They found that mean water sodium levels of drinking water sources were significantly higher in the cases (728 mg/l) than the controls (440 mg/l) ($p<0.001$). Both SBP and DBP levels were associated with the type of drinking water source. Salinity in drinking water of women who drank filtered pond water, pond water, or tube well water was associated with an increased risk of pre-eclampsia and gestational hypertension.

Khan et al. (2008, 2014) and Talukder et al. (2016b) conducted studies in coastal Bangladesh with pregnant women and young adults that suggested an association between drinking water salinity and hypertension. Besides, there is a positive association between saline drinking water and increased blood pressure (Talukder et al. 2017). In Bangladesh about 20 million people are at high risk of hypertension due to climate change-induced saline intrusion in water and high salt consumption (Rasheed et al. 2016). In Bangladesh, hypertension deaths reached over 16,600 or 2.28% of total deaths. The age-adjusted death rate is 17.92 per 100,000 of population (WHO 2011) ranking Bangladesh 77thin the world (Nahian et al. 2018; WLE 2014).

Naser et al. (2017a) showed that salty groundwater in southern coastal area of Bangladesh is associated with high blood pressure among the adult population and pre-eclampsia and gestational hypertension among pregnant women. Managed aquifer recharge (MAR) is an approach to promote freshwater recharge to increase storage by infiltration of surface water (e.g. rainwater and pond water) to aquifers, which is a probable solution for the salinity problem in southwest coastal Bangladesh.

According to Lim et al. (2012), hypertension is the leading risk factor for coronary heart disease and stroke and for the global burden of disease (7% of global disability-adjusted life years [DALYs]). Hypertension increases the risks of overall mortality, cardiovascular diseases, myocardial infarction, stomach cancer, chronic respiratory disease, obesity, diabetes, renal disease, kidney disease including nephrosclerosis, and retinopathy. It is influenced by a genetic, medical disorder, lifestyle, and environmental factors, primarily by the latter two, which differ between developing and low-income developing countries (Alderman 2000; Islam and Majumder 2012; Rust and Ekmekcioglu 2017; Ibrahim and Damasceno 2012; Dennison-Himmelfarb et al. 2014). It is noted that high blood pressure is associated with genetic predispositions, age, overweight and obesity, low physical activity, and chronic stress. Hypertension occurs increasingly both in aged people and pregnant women for some risk factors, such as high sodium intake, lack of physical activity, obesity, stress, and excessive alcohol intake (Mathers et al. 2009). It is estimated to reach high blood pressure in about 1.56 billion adults by around 2025 with the highest percentage in low-income developing countries (Nahian et al. 2018; Ibrahim and Damasceno 2012; WHF 2017).

Saltwater Intake and Cholera

Cholera is an ancient disease, serious epidemic, and most often centred in tropical areas and specific seasons. The association between cholera and water was first established epidemiologically in the 1850s in London, United Kingdom. Vibrios, including *V. cholerae*, can be found in virtually any coastal water body, especially in the tropics and subtropics. Some scholars showed the associations between contaminated drinking water and cholera. There are five studies that analysed the association between cholera and salinity of water in Bangladesh. Grant et al. (2015) measured the salinity levels of water sources and water palatability by linking cholera outbreaks in coastal Bangladesh with ingestion of saline-rich cholera-infected river water. A simple random sample (50 people in the monsoon season and 50 people during the dry season—25 males and 25 females) of age between 18 and 60 years of the coastal Upazila of Shyamnagar, Satkhira, Bangladesh was carried out in September 2011 and March 2012. Some studies linked the spread of cholera with the role of hygiene, sanitation, population density and food, but particularly in Bangladesh, cholera broke out through drinking water (Grant et al. 2015). Cholera is endemic in Bangladesh where it causes year-round cases. Cholera cases occurring in the southern coastal areas peak at the end of the dry season (usually between late March and May), while cholera in the northern and middle-belt regions of the country peak post-monsoon (September–November). According to Thompson et al. (2004), *Vibrio cholerae*-like toxigenic *Vc*O1 and O139 are causative agents of cholera disease whereas other Vibrios (e.g. *V. parahaemolyticus*, *V. vulnificus*, *V. mimicus*, etc.) are responsible for diarrhoea, gastroenteritis, necrotizing fasciitis, and various further skin diseases and septicaemia affecting the human population worldwide. Some Vibrios (*V. campbellii*, *V. harveyi*, *V. alginolyticus*, etc.) also can cause diseases in fish, shrimp, corals, and other aquatic organisms (Lara et al. 2009). Vibrios are aquatic bacteria of marine and estuarine origin. Akanda et al. (2011), Grant et al. (2015), Lara et al. (2009), Lipp et al. (2002), Huq et al. (1984), and Shishir et al. (2018) all argued that cholera epidemics are related to plankton blooms (*V. cholerae* reservoirs), and several physico-chemical factors in water have been linked to the survival of the pathogen, including moderate salinity, high nutrient content with warm temperature salinity, and slight salinity. Climate, weather, and environment influence the pathogenic *Vibrio* spp. They argued that *V. cholerae* grows differently at different places like at 25,000 mg/l salinity, at higher temperatures than 5°C and 15°C, in warm waters of moderate salinity, in pond surface water by estuarine river circulation via tidal water, in coastal areas, and seawater by the cyclone, flood in the tropics and subtropics.

They argue that during the dry season low river flow leads to an intrusion of saline water from the Bay of Bengal to deeper inland. This creates *V. cholerae* growth and cholera peak in the coastal region. The post-monsoon peak in cholera cases is driven by floodwaters that carry the *V. cholerae* from the polluted rivers to the flooded inland areas, where it then proliferates in waterlogged areas, contaminating water sources including those used for drinking water (Akanda et al. 2011). In the coastal areas of Bangladesh, tube well water, rainwater, the natural reservoirs, ponds, lakes, etc. are the sources of drinking water despite salt. But the pond water in Bangladesh is heavily contaminated with faecal coliforms (FCs), and other pathogenic bacteria and several previous cholera outbreaks also present there.

Shishir et al. (2018) analyse the prevalence of *V. cholerae* (*Vc*—pathogenic and non-pathogenic) and spore formers (Sf) including *Bt* from the alternative drinking waters and the co-association among *Bacillus*-like Sf and *Vc*. A total of 100 water samples (500 ml each sample) [pond surface water (SW)-48, household (HH) water-29, and pond sand filter (PSF)-23] were selected randomly from Mathbaria sub-district in Pirojpur where 26% *Vc*-positive samples were predominant. The ponds were mostly connected to the canals and used for washing utensils and bathing purposes. So, the tidal waters from the estuarine that intruded into the ponds through the canals might have contributed to the transportation of *Vc* reservoir, that is, the planktons. They indicate that the estimation of *Vc* were observed as 33%, 26%, and 13.8% in pond SW, PSF, and HH reservoirs, respectively, and the pond SW where the salinity was predominant was riskier with higher prevalence (more than 20% *Vc*O1) of *Vc* than that obtained from PSF and HH reservoirs. Cholera outbreaks in the spring (from March to May) predominantly in the dry season in such coastal areas, and the inhabitants are most susceptible to the scarcity of drinking water due to the high salinity in the shallow and deep tube well water.

Lara et al. (2009) investigate the effect of changes in the environmental setting and human waste input on the abundance and distribution of *Vibrio* spp. via salinity. Microbiological and hydrological data with salinity (immediately before and after cyclone, flood, and landslide) were collected randomly from 150 vibrio strains of the Karnaphuli estuary in Chittagong in May–June 2007. They showed that after the cyclone, flood, and landslide events, cultivable *Vibrio* counts, turbidity, and FCs increased in the high salinity zone, and high vibrio counts occurred in patches of turbid water having higher salinity.

Lipp et al. (2002) explored the associations among *V. cholerae*, climate, and health due to salinity intrusion and found that climate, weather, and environment influence the pathogenic *Vibrio* spp., in estuarine ecosystems. *V. cholerae* grows in warm waters of moderate salinity. Vibrios, including *V. cholerae*, can be found in virtually any coastal water body, especially in the tropics and subtropics. *V. cholera* is highly concentrated on zooplankton shells and in their gut, the risk of ingesting an infectious dose increases when untreated surface waters are used for consumption.

Huq et al. (1984) collected both adult and immature copepods and water samples. Water temperature and salinity or brackish water and estuarine aquatic environment influence significantly of *V. cholerae* copepods and multiplication on copepods. They pointed out that *V. cholerae* cells grow and survive longer in the presence of live copepods when incubated at higher temperatures than 15°C and survive longest at 25°C and 30°C temperatures. Salinity alone did not influence growth of *V. cholerae* at temperatures as low as 10°C, but when the temperature was increased to 15°C, the influence of salinity was significant.

In Bangladesh, cholera is endemic; seasonal, explosive outbreaks of the disease begin around November, after the monsoon season. The water temperature often exceeds 25°C for prolonged periods during the summer, and zooplankton blooms occur after the monsoons (Huq et al. 1984). A combination of high-water temperature, ingestion of raw, unpurified water, and an abundance of live copepods is to be optimum for the survival and multiplication of *V. cholerae* in the aquatic environment. Bangladesh may have higher numbers of *V. cholerae* cells in water in the environment as endemic. *V. cholerae* becomes hyper-infective after passing through the human digestive tract, indicating a propensity for person-to-person spread. About 47% fewer cholera cases were observed in children whose mothers received a cholera vaccine than in children of unvaccinated mothers, and no significant rise in breast-milk antibodies was seen.

High Salt or Saltwater Intake and Kidney diseases

Epidemiological studies reported that there is an association between dietary salt intake and urinary albumin excretion, independent of blood pressure (du Cailar et al. 2002). We found three studies are directly linked with drinking water salinity or dietary salt intake and kidney disease in Bangladeshi people (Naser et al.

2017b) or British Bangladeshi people (de Brito-Ashurst et al. 2011, 2013). Though drinking saline water is associated with high blood pressure among communities in southwest coastal Bangladesh, but the effect of water salinity on kidney health is almost unknown. Naser et al. (2017b) tried to show the association between drinking water sodium or salinity and urinary protein excretion and kidney disease in southwest coastal Bangladesh. They collected 24-hour urine from 1,185 participants from three districts (Khulna, Satkhira, and Bagerhat) of southwest coastal Bangladesh between November 17, and December 21, 2016, and measured urinary sodium, total protein, and salinity or electrical conductivity of drinking water. They found that 37% participants have mild proteinuria (total protein: 300–500 mg/24 h) and 20% have moderate or substantial proteinuria (total protein: >500 mg/24 h). Drinking water sodium and urinary sodium are associated with increased protein excretion, which is a powerful biomarker for progressive kidney disease in southwest coastal Bangladesh. Talukder et al. (2017), Rasheed et al. (2016), Talukder et al. (2016a), and Naser et al. (2017b) showed the direct association between high dietary salt and drinking saltwater intake and hypertension or blood pressure directly, but they relate this salt with kidney diseases.

deBrito-Ashurst et al. (2011) conducted a study with triangulation methodology and purposive sampling technique was applied in 28 chronic kidney disease (CKD) immigrant adult Bangladeshi patients in East London. In total, 20 female patients were participated in four focus group discussion (FGD), whose mean age and standard deviation were 60 and 8 years, respectively. British Bangladeshi population has an unusually high salt intake, twice that of the general population, which might contribute to CKD and hypertension. Remarkably, 90% of the respondents agreed that "excess salt intake led to an imbalance and ill health, which is bad for heart and raised blood pressure or hypertension. An intake of less than 6 g salt/d (~100 mmol Na/d) is recommended in the UK and Europe based on the association of salt intake with blood pressure and cardiovascular disease rather than the much lower intake required preventing sodium deficiency. Bangladeshi high dietary salt, traditional eating patterns, and dietary consumption had been modified, but the main barriers to changing dietary habits were 'social norms', 'knowledge and skills', and 'behavioural factors'". Participants had been advised to reduce salt intake by dieticians.

deBrito-Ashurst et al. (2013) carried out a further study among 56 participants, randomized 48 adult participants of Bangladeshi origin with CKD and BP >130/80 mmHg or on antihypertensive medication who were randomly allocated between June 2008 and July 2009 renal unit based in acute care hospital in East London regarding high salt intake and kidney diseases. They considered dietary salt intake, age, sex, height, weight, body composition (total body water), physical activity, medical care, diabetes, haemoglobin (HbA1c) samples, blood pressure, and kidney disease as confounders. Bangladeshi and South Asian people suffer a high risk of CKD due to high dietary salt intake. Other ethnic communities such as black African and Afro-Caribbean populations have a high prevalence of CKD and hypertension, and they were advised to reduce salt intake. deBrito-Ashurst et al.

(2013) noted that more than 50% of Bangladeshi patients suffer kidney damage, heart, brain, raised frequency of stroke and myocardial infarction, and higher cardiovascular mortality.

According to Trehan et al. (2003), the rates of kidney disease of British Bangladeshi people are three to five times higher than that of the European UK population due to the habit of dietary salt (de Brito-Ashurst et al. 2011). Some papers linked higher salt and kidney diseases in globally. Cerasola et al. (2010) noted that urinary albumin levels are a risk factor for the development and progression of kidney disease and are also strong predictors for cardiovascular risk. He and MacGregor (2009) showed that a reduction in salt intake of about 3.2 g/d over six weeks leads to significant reductions in blood pressure and urinary albumin excretion.

High Salt Intake and Stroke

The relation between high salt intake and stroke events was found in some of the studies, though most of the experimental, epidemiological, migration, and intervention studies showed a direct relationship between dietary salt intake levels and blood pressure or hypertension, and this hypertension leads to stroke events. Few studies focus on the effect of salt intake on total stroke events or haemorrhagic stroke attacks or ischaemic stroke events, but they didn't pay attention to the relationship between drinking water salinity and stroke. There are two studies (Alam et al. 2012, Riaz et al. 2015) that found the direct relationship between dietary salt intake and stroke events in Bangladesh and Li et al. (2012) studied systematic review and meta-analysis in global studies and found the direct relation between salt intake and stroke, but Vineies et al. (2011), Chrysant et al. (1999), Alderman (2000), and Rasheed et al. (2016) showed the relation between salt intake and blood pressure or hypertension, and this hypertension leads to stroke event.

According to the WHO (2012), 62% of all strokes are attributable to high blood pressure, and it is estimated that about 5.7 million deaths were occurred in 2005, and 8.7% of these deaths were in low-income and middle-income countries. Stroke events occur when the patients show the clinical signs and symptoms of focal (or global) disturbance of cerebral function lasting >24 hours and suggestive of subarachnoid haemorrhage, intracerebral haemorrhage, or cerebral infarction. Typically, the first two types are haemorrhagic stroke (HS), and the last is known as ischaemic stroke (IS). Stroke attack refers to all strokes (both first and recurrent) within 28 days (Li et al 2012).

Alam et al. (2012) studied from 110 cases (55 ischaemic and 55 haemorrhagic strokes) of Mymensingh Medical College Hospital, Bangladesh in January 2009 to December 2010 and found that in haemorrhagic and ischaemic stroke, 52.72% and 41.80% had abnormal sodium level, and 38.18% and 27.22% had abnormal potassium level, respectively, and 60% had abnormal chloride level in both groups. Riaz et al. (2015) studied 70 cases of HS, 105 cases of IS, and a total of 171 controls from Shaheed Suhrawardy Medical College Hospital and Dhaka Medical College

Hospital between January and June 2011 and showed that table salt intake appeared as a significant predictor for both ischaemic and haemorrhagic stroke (OR 8.15; 95% CI, 3.95–16.8, OR 9.9; 3.32–29.6), respectively, with some covariates. Li et al. (2012) studied systematic review and meta-analysis from 12 studies with 225,693 participants from six different countries (Japan-4; Finland-2, United States-2, China-2, Netherlands-1, and Norseland-1), of 8,135 stroke events during 1966–2012 and showed that high salt intake was associated with risk of stroke event (OR 1.34; 95% CI, 1.19–1.51), stroke death (1.40; 1.21–1.63), and stroke onset (1.11; 1.00–1.24), ischaemic stroke death (2.15; 1.57–2.95), not associated with risk of ischaemic stroke onset (1.07, 0.95–1.2), with no significant evidence of publication bias. High salt intake is associated with significantly increased risk of stroke event. Ikehara et al. (2012) and Umesawa et al. (2008) found that there were significant association between high salt intake and the death as a result of stroke (OR 1.22; 95% CI, 1.00–1.49 for women and 1.55; 1.21–2.00 for men and women). Kono et al. (2011) showed that high salt intake is related to stroke recurrence (OR 1.98; 1.02–4.22).

But Liang et al. (2011) and Ellekjaer et al. (1992) showed in their studies that there is no significant association between high salt intake and ischaemic stroke onset. Selhub et al. (1995) studied among Japanese men and found no relation between salt intake and the incidence of stroke (Riaz et al. 2015). All the studies linked with high salt intake and stroke disease. Besides, few studies were found in Bangladesh perspectives. Some articles reviewed the impact of high salt intake on other diseases in global perspectives, but lacked the relationship in Bangladesh perspectives, which are discussed below.

High Salt Intake and Other Diseases

High dietary salt intake is associated with an increased risk for gastric cancer and cardiovascular diseases. Besides, in the last years it is evident that there is a link between high dietary salt intake and risk for overweight and obesity.

Stomach Cancer

D'Elia et al. (2012) showed that dietary salt intake is directly associated with risk of gastric cancer, with increasing risk across consumption levels. Furthermore, Parkin (2011) presented that 24% of stomach cancer cases in the UK in 2010 were suggested to be attributed to high salt consumption. Ge et al. (2012) hypothesized that higher gastric cancer risk is occurred by high dietary salt intake. According to Norat et al. (2015) a significant positive association was observed for stomach cancer and salted food intake.

Obesity

Some studies, like Alexy et al. (2012) and Grimes et al. (2013), found a positive association between urinary sodium excretion/dietary salt intake and beverage consumption and obesity. Ma et al. (2015) studied in a collection of 458 British children and 785 adults in which it was found that salt intake was higher in overweight and obese individuals. A 1-g/d increase in salt intake was associated with an increase in the risk of obesity by 28% (CI 95%: 1.12–1.45) in children and 26% (CI 95 %: 1.16–1.37) in adults after adjusting for various factors.

Besides, Johnson et al. (2012) and Akther et al. (2016) showed that diarrhoea outbreaks caused by waterborne pathogens including *V. parahaemolyticus* may be influenced by various climatic factors, for example, sea surface temperature, salinity, and turbidity. *V. parahaemolyticus* incubated at 37°C for 3 to 4 hours and stored at 80°C with 15% glycerol strains isolated from the surface water of estuarine environment of Bangladesh (studied area Patuakhali), where thousands of people suffer from recurrent diarrhoea and are at high risk of waterborne diseases due to saline water intrusion resulted from the changing climate.

Finally, the negative effects of high sodium/salt intake are not only restricted to the cardiovascular and renal systems but may also adversely affect our metabolism and energy intake. But most of the articles were found in global perspectives, only few were in Bangladesh perspectives. Nineteen of the 27 selected articles focused on the association between high salt intake by drinking water or dietary salt and raised blood pressure or hypertension. This increasing blood pressure leads to other chronic diseases like cardiovascular diseases, stroke or heart attack, pre-eclampsia and eclampsia of pregnant women, impaired liver function or kidney disease, low platelet count, intrauterine growth retardation, preterm birth, and maternal and perinatal deaths. Besides cholera break out in warm temperatures and plankton in high salinity areas, surface of ponds or estuary of rivers, canals etc. due to salinity intrusion.

Recommendation

Drinking water salinity and dietary salt intake are potential risk factors of blood pressure or hypertension, stroke, heart disease, kidney diseases, and cholera. So, we should follow the prevention guidelines (Dennison-Himmelfarb 2014). The WHO guideline value for drinking water salinity is 250 mg of chloride per litre (mg/l) for chloride (Cl) and 200 mg/l for sodium (Na), and is based on taste threshold, not on health consideration. Ahmed and Rahman (2000) suggested that the acceptable level for chloride adopted by Bangladesh is 1,000 mg/l chloride for coastal areas.

The WHO recommends an intake of sodium chloride <5 g/d in general and for children aged 18 months to three years sodium <2 g/d. Dietary Guidelines for Americans (2010) recommend sodium <2.3 g/d or ~6 g NaCl/d in general. American Heart Association (2016) recommends <1.5 g/d of sodium for special group (US Department of Health and Human Services 2011). In addition to dietary assessment the gold standard for measuring salt intake is the analysis of sodium

TABLE 2.3 Sodium intake recommendation

Life stage	*AI (g/d) (DRI 2004)*	*UL (g/d) (DRI 2004)*	*Estimate value for min. intake (g/d) (DGE et al. 2015)*
Infants:			Infants:
0–6 month	0.12	Not determinable	0–<4 month: 0.1
7–12 month	0.37		4–<12 month: 0.18
Children:			Children:
1–3 year	1.0	1.5	1–<4 year: 0.3
			4–<7 year: 0.41
			7–<10 year: 0.46
			10–<13 year: 0.51
			13–<15 year: 0.55
Adults:			Adolescents/Adults:
9–>70 year	1.5	2.3	0.55
Pregnancy	1.5	2.3	
Lactation	1.5	2.3	

Source: DRI (2004), Institute of Medicine (2005), and DGE et al. (2015).

excretion in the 24-hour urine. Spot urine samples are appropriate alternatives for monitoring sodium intake (Rust and Ekmekcioglu 2017). However, salt intake in Bangladeshi population is much higher like 15 g/d and 21 g/d on industrial output and hospital patient data reported, respectively, than recommended by them (Zaman et al. 2017; Penney 2016). Table 2.3 shows that how much salt (sodium AI and potassium UL) should intake a man per day for sound health.

Human body contains about 0.15% by weight sodium and chloride, respectively. For balance of the body the amount of salt consumption or drinking water salinity must be reduced. In addition, regular physical activity (endurance, dynamic resistance, and isometric resistance training) is very important (Rust and Ekmekcioglu 2017). Coastal embankment should be extended to all coastal areas so that the saltwater cannot be entered through cyclone, floods, storm surges, tide and ebb, and estuarine circulation. Existing embankments should be repaired and maintained. Shrimp cultivation should be stopped or cultivated altered way. Initiatives to collect and store drinking water should also be considered. Harvesting rainwater and MAR can be an option to increase storage by infiltration of surface water (e.g. rainwater and pond water) to aquifers in southwest coastal Bangladesh. MAR water can potentially reduce exposure to saline water and the health effects (Naser et al. 2017b). There is a lack of awareness about the consumption and use of saline water and dietary salt intake in the coastal belt of Bangladesh, which increase the risk of health. Intervention should be considered to modify knowledge, awareness levels, and also practices. Government of Bangladesh, donor agencies, politicians, and planners in Bangladesh should take this into account in future planning of coastal zones to desalinization process.

Conclusion

The research tried to depict the actual flows of present scenario of salinity and vulnerability or health risk of southern coastal belt people in Bangladesh. Saltwater or salinity intrudes in the coastal areas of Bangladesh from the flow of seawater through the environmental exposures or natural disasters like cyclone, tidal surge, flood, river flow, estuarine circulation etc. Besides, climate change, sea-level rises, precipitation, and water and land management practices exacerbate the salinity. Surface water, groundwater, and soil contaminated by salinity in these coastal areas (Khan et al. 2008). About 53% of the coastal areas are affected by salinity in Bangladesh. Salinity causes unfavourable environment and hydrological situation that restrict the normal crop production throughout the year. Climate change can be impacted negatively in the future as the land becomes increasingly saline, health risk becomes greater, loss of livelihood and hardship may be arisen, the productivity of the land decreases, future food insecurity becomes a possibility, and the people become generally more vulnerable. In Bangladesh salinity intrusion is a time-varying event, maximum in dry season, and minimum during the monsoon (June–October) by pushback of the rivers' discharge at the salinity front in estuarine and floodplains (Mahmuduzzaman et al. 2014).

Excessive sodium or high dietary salt intake is an important risk factor of hypertension and other NCDs. Hypertension increases the risks of overall mortality, cardiovascular diseases, myocardial infarction, stomach cancer, chronic respiratory disease, obesity, diabetes, renal disease, kidney disease including nephrosclerosis, retinopathy, and so on. (Alderman 2000; Islam and Majumder 2012; Rust and Ekmekcioglu 2017; Ibrahim and Damasceno 2012; Dennison-Himmelfarb et al. 2014). In Bangladesh 18% of the adults aged 25 years and above have been found to be hypertensive (Talukder et al. 2016a; WHO 2013). About 20 million people in Bangladesh are at high risk of hypertension due to climate change-induced saline intrusion in water and high salt consumption (Rasheed et al. 2016). Women, especially pregnant women, are associated with high blood pressure or hypertension with increased gestational hypertension and pre-eclampsia due to high drinking brackish water, salt or excess sodium intake in southwest coastal Bangladesh (Khan et al. 2011, 2014). This raised blood pressure leads to other chronic diseases like cardiovascular diseases, stroke or heart attack, pre-eclampsia and eclampsia of pregnant women, impaired liver function or kidney disease, low platelet count, intrauterine growth retardation, preterm birth, and maternal and perinatal deaths. Besides, cholera break out in warm temperatures and plankton in high salinity areas, surface of ponds or estuary of rivers, canals etc. due to salinity intrusion. According to the WHO (2012), 62% of all strokes are attributable to high blood pressure, and most of the cases occur due to dietary salt intake (Rust and Ekmekcioglu 2017; WHO 2013). In fine, we can say that in general higher level of salinity (sodium) was prevalent in young adult, pregnant women, tube wells (ground), during the dry season than adult or aged, male, pond (surface) and rainwater, and the rainy season in this study. However, for the sound health of southern coastal people in Bangladesh it is a prerequisite that high sodium intake, dietary salt intake, or drinking brackish water should be reduced.

References

Abdel-Dayem AM, & Shalaby H. 2004. Numerical simulation of salt gradient solar ponds. *International Journal of Computational Engineering, Imperial College Press, 5*(3), 673–679.

Abdullah HM, Ahmed SM, Khan BM, Mohana NT, Ahamed T, & Islam I. (2021). Agriculture and fisheries production in a regional blending and dynamic fresh and saline water systems in the coastal area of Bangladesh, *Environmental Challenges*, 4. Available from: https://doi.org/10.1016/j.envc.2021.100089.https://www.sciencedirect.com/science/article/pii/S2667010021000688)

Adhikary G. (2016). Impacts of saline water on reproductive health in coastal Bangladesh: Women's perspective. Institute of Disaster Management and Vulnerability Studies, University of Dhaka. Available from: http://gobeshona.net/wp-content/uploads/2016/01/ P4_Health.pdf

Akanda AS, Jutla AS, Alam M, et al. (2011). Hydroclimatic influences on seasonal and spatial cholera transmission cycles: Implications for public health intervention in the Bengal Delta. *Water Resources and Research, 47*(5), 1–11. https://doi.org/10.1029/2010WR009914

Akther F, Neogi SB, Chowdhury WB, Sadique A, Islam A, Akhter MZ, et al. (2016). Major tdh+ Vibrio parahaemolyticus serotype changes temporally in the Bay of Bengal estuary of Bangladesh. *Infection, Genetics and Evolution, 41*, 153–159.

Alam MN, Uddin MJ, Rahman KM, Ahmed S, Akhter M, Nahar N et al. (2012). Electrolyte changes in stroke. *Mymensingh Medical Journal, 21*(4), 594–599. PMID: 23134903.

Alam DS, Chowdhury MA, Siddiquee AT, Pervin S, Ahmed S, and Niessen LW. (2013). Raising awareness leads to high provider visits and blood pressure reduction among hypertensive patients in Bangladesh: A population-based cohort study. *American Heart Association, 127*(12), 371–385.

Alamgir AKM, Ali SMK, & Haque KMHSS. (2009). Extra salt intake as determinant of high blood pressure. *Chest Heart Journal, 33*, 127–133.

Alderman MH. (2000). Salt, blood pressure, and human health. *Dallas: American Heart Association, 36*, 890–893.

Alderman MH, Cohen H, & Madhavan S. (1998). Dietary sodium intake and mortality: The National Health and Nutrition Examination Survey (NHANES I). *The Lancet, 351*, 781–785.

Alexy U, Cheng G, Libuda, L, Hilbig, A, & Kersting M. (2012). 24h-Sodium excretion and hydration status in children and adolescents-Results of the DONALD Study. *Clinical Nutrition, 31*(1), 78–84.

Ayers JC, Goodbred S, George G, Fry D, Benneyworth L, Hornberger G, et al. (2016). Sources of salinity and arsenic in groundwater in southwest Bangladesh. *Geochemical Transactions, 17*(4), 1–22. DOI: 10.1186/s12932-016-0036-6.

Ayers JC, George G, Fry D, Benneyworth L, Wilson C, Auerbach L, et al. (2017). Salinization and arsenic contamination of surface water in southwest Bangladesh. *Geochemical Transactions, 18*(4), 1–23.Avaiable from: https://doi.org/10.1186/s12932-017-0042-3

Barrett-Lennard EG, George RJ, Hamilton G, Norman HC, & Masters DG (2005). Multidisciplinary approaches suggest profitable and sustainable farming systems for valley floors at risk of salinity. *Australian Journal of Experimental Agriculture, 45*(11), 1415–1424.

Cerasola G, Nardi E, Palermo A, Mulè G, & Cottone S. (2010). Epidemiology and pathophysiology of left ventricular abnormalities in chronic kidney disease: A review. *Journal of Nephrology*, *24*(1), 1–10. doi: 10.5301/jn.2010.2030. PMID: 20437402.
Choudhury AM. (2009). *Protecting Bangladesh from Natural Disasters.* Dhaka 1209, Bangladesh: Academic Press and Publishers Library.
Chrysant GS, Bakir S, & Oparil S. (1999). Dietary salt reduction in hypertension: What is the evidence and why is it still controversial? *Progress in Cardiovascular Diseases*, *42*(1), 23–38.
Clarke D, Williams S, Jahiruddin M, Parksa K, &Salehin M. (2015). Projections of on-farm salinity in coastal Bangladesh. *The Royal Society of Chemistry*, *17*, 1127–1136.
Colligan E. (2011). Salinity issues in Bangladesh. *Georgetown Journal of International Affairs.* Available from: www.georgetownjournalofinternationalaffairs.org/online-edition/guide-to-salinity-issues-in-bangladesh
Dasgupta S, Hossain MM, Huq M, & Wheeler D. (2015). Climate change and soil salinity: The case of coastal Bangladesh. *Royal Swedish Academy of Sciences*, *44*, 815–26.
de Brito-Ashurst I, Perry L, Sanders TAB, Thomas J, Yaqoob MM, & Dobbie H. (2011). Barriers and facilitators of dietary sodium restriction amongst Bangladeshi chronic kidney disease patients. *Journal of Human Nutrition and Dietetics*, *24*(1), 86–95.
de Brito-Ashurst I, Perry L, Sanders TAB, Thomas JE, Dobbie H, Varagunam M, et al. (2013). The role of salt intake and salt sensitivity in the management of hypertension in South Asian people with chronic kidney disease: A randomised controlled trial. *Heart*, *99*, 1256–1260.
D'Elia L, Rossi G, Ippolito R, Cappuccio FP, & Strazzullo P. (2012). Habitual salt intake and risk of gastric cancer: A meta-analysis of prospective studies. *Clinical Nutrition*, *31*(4), 489–498.
Dennison-Himmelfarb C, Handler J, Lackland D, Lefevre M, Mackenzie T, Ogedegbe O, et al. (2014). Evidence-based guideline for the management of high blood pressure in adults report from the panel members appointed to the Eighth Joint National Committee (JNC 8). 2013–2014.
Deutsche Gesellschaftfür Ernährung (DGE) (ed): Fettzufuhr und Präventionausgewählterern ährungsmitbedingterKrankheiten – Evidenzbasierte Leitlinie, ed 2. Bonn, 2015. https://www.dge.de/wissenschaft/leitlinien/ (accessed March 25, 2015).
Downie J. (2007). The Problems of water salinity. New Age [Internet]. 2007 Mar 22. Available from: http://www.newagebd.com/2007/mar/22/oped.html
DRI. (2004). *Dietary Reference Intakes for Sodium and Potassium.* Washington, DC: National Academies Press (US).
Du Cailar G, Ribstein J, & Mimran, A. (2002). Dietary sodium and target organ damage in essential hypertension. *American Journal of Hypertension*, *15*(3), 222–229.
Duncan RR, Carrow RN, & Huck M. (2000). Understanding water quality and guidelines to management: An overview of challenges for water usage on golf courses for the 21st century. *USA: Green Section Record*, *38*, 14–24.
Ellekjaer EF, Wyller TB, Sverre JM, & Holmen J. (1992). Lifestyle factors and risk of cerebral infarction. *Stroke*, *23*, 829–834.
EM-DAT, Guha-Sapir D, Hoyois P, Wallemacq P, & Below R. (2017). *EMDAT—The Emergency Events Database, The International Disaster Database.* Brussels, Belgium: Centre for Research on the Epidemiology of Disasters (CRED), Institute of Health and Society (IRSS), UniversitéCatholique de Louvain, Available from: www.emdat.be/database
FAO Irrigation and Drainage Paper. (1996). *The Use of Saline Waters for Crop Production.* Rome, Italy: Publication Division, Food and Agriculture Organization of the United Nations, 48. Available from: www.fao.org/docrep/T0667E/t0667e00.htm

Ge S, Feng X, Shen L, et al. (2012). Association between habitual dietary salt intake and risk of gastric cancer: A systematic review of observational studies. *Gastroenterology Research and Practice*, 2012, 808120. doi:10.1155/2012/808120.

George G. (2013). Characterization of salinity sources in southwestern Bangladesh evaluated through surface water and groundwater geochemical analyses. Vanderbilt Univ [Internet]. 2013; Available from: www.etd.library.vanderbilt.edu/ available/etd-11272013-113249/

Government of the People's Republic of Bangladesh. (1997). The Environment Conservation Rules, 1997 [Internet]. 1997. Available from: https://www.elaw.org/system/files/Bangladesh+-- +Environmental+Conservation+Rules,+1997.pdf

Government of the People's Republic of Bangladesh. Ministry of Disaster Management and Relief. (2020). National Plan for Disaster Management (2021–2025): Action for Disaster Risk Management towards Resilient Nation.

GrantSL, Tamason CC, Hoque BA, & Jensen PKM. (2015). Drinking cholera: Salinity levels and palatability of drinking water in coastal Bangladesh. *Tropical Medicine and International Health*, *20*(4), 455–461.

Grimes CA, Riddell LJ, Campbell KJ, & Nowson CA. (2013). Dietary salt intake, sugar-sweetened beverage consumption, and obesity risk. *Pediatrics*, *131*(1), 14–21. doi: 10.1542/peds.2012-1628.

Gul S, Nisa NT, Shah TA, Gupta S, Jan A, & Ahmad S. (2015). Middle East: Research productivity and performance across nations. *Scientometrics*, *105*(2), 1157–1166.

He FJ, &MacGregor GA. (2007). Salt, blood pressure and cardiovascular disease. *Current Opinion in Cardiology*, *22*(4), 298–305.

He FJ & MacGregor GA. (2009). A comprehensive review on salt and health and current experience of worldwide salt reduction programmes. *Journal of Human Hypertension*, *23*, 363–384.

Haque SA. (2006). Salinity problem and crop production in coastal regions of Bangladesh. *Pakistan Journal of Botany*, *38*, 1359–1365.

Huq A, West PA, Small EB, Huq MI, &Colwell RR. (1984). Influence of water temperature, salinity, and ph on survival and growth of toxigenic Vibrio cholera Serovar O1 associated with live copepods in laboratory microcosms. *Applied Environmental Microbiology*, *48*(2), 420–424.

Ibrahim M, & Damasceno A. (2012). Hypertension in developing countries. *The Lancet*, *380*, 611–619.

Institute of Medicine. (2005). Does the Built Environment Influence Physical Activity?: Examining the Evidence--Special Report 282. Transportation Research Board.

IPCC 2007. (2009). IPCC Fourth Assessment Report. Asia: Climate Change 2007: Impacts, Adaptation andVulnerability. Contribution ofWorking Group II to the Fourth Assessment Report of Intergovernmental Panel on Climate Change. Cambridge: Cambridge University Press.

Ikehara S, Iso H, Date C, et al. (2012). Salt preference and mortality from stroke and coronary heart disease for Japanese men and women: The JACC study. *Preventive Medicine*, *54*, 32–37.

Islam AKMM & Majumder AAS. (2012). Hypertension in Bangladesh: A review. *Indian Heart Journal*, *64*(3), 319–323. DOI:10.106/S0019–4832(12)60096-0

James PR, & Nelson-Piercy C. (2004). Management of hypertension before, during, and after pregnancy. *Heart, BMJ Publishing Group*, *90*(12), 1499–1504.

Khan AE, Ireson A, Kovats S, Mojumder SK, Khusru A, & Rahman A, et al. (2011). Drinking water salinity and maternal health in coastal Bangladesh: Implications of climate change. *Environmental Health Perspectives*, *119*(9), 1328–1332.

Khan AE, Mojumder SK, Kovats S, & Vineis P. (2008). Saline contamination of drinking water in Bangladesh. *The Lancet, 371*(9610), 385.

Khan AE, Scheelbeek PFD, Shilpi AB, Chan Q, Mojumder SK, Rahman A, et al. (2014). Salinity in drinking water and the risk of (pre)eclampsia and gestational hypertension in coastal Bangladesh: A case-control study. *PLoS One, 9*(9), 1–9.

Khanom S, & Salehin M. (2012). Salinity constraints to different water uses in coastal area of Bangladesh: A case study. *Bangladesh Journal of Scientific Research, 25*, 33–41.

Kono Y, Yamada S, Kamisaka K, et al. (2011). Recurrence risk after noncardioembolic mild ischemic stroke in a Japanese population. *Cerebrovascular Disease, 31*, 365–372.

Lara RJ, Neogi SB, Islam MS, Mahmud ZH, Yamasaki S, & Nair GB. (2009). Influence of catastrophic climatic events and human waste on Vibrio distribution in the Karnaphuli Estuary, Bangladesh. *EcoHealth, 6*(2), 279–286.

Liang W, Lee AH, & Binns CW. (2011). Dietary intake of minerals and the risk of ischemic stroke in Guangdong Province, China, 2007–2008. *Preventing Chronic Diseases, 8*, A38.

Lim SS, Vos T, Flaxman AD, Danaei G, Shibuya K, Adair-Rohani H, et al. (2012). A comparative risk assessment of burden of disease and injury attributable to 67 risk factors and risk factor clusters in 21 regions, 1990-2010: a systematic analysis for the Global Burden of Disease Study 2010. *The Lancet, 380* (9859), 2224–2260. doi: 10.1016/S0140-6736(12)61766-8

Lipp EK, Huq A, & Colwell RR. (2002). Effects of global climate on infectious disease: The cholera model. *Clinical Microbiology Reviews, 15*(4).

Li XY, Cai XL, Bian PD, & Hu LR. (2012). High salt intake and stroke: meta-analysis of the epidemiologic evidence. *CNS Neuroscience & Therapeutics, 18*(8), 691–701. doi: 10.1111/j.1755-5949.2012.00355.x. Epub 2012 Jun 28. PMID: 22742770; PMCID: PMC6493555.

Ma Y, He FJ, & MacGregor GA. (2015). High salt intake: independent risk factor for obesity? *Hypertension, 66*(4), 843–849. doi: 10.1161/HYPERTENSIONAHA.115.05948.

Mahmuduzzaman M, Ahmed ZU, Nuruzzaman AKM, & Ahmed FRS. (2014). Causes of salinity intrusion in coastal belt of Bangladesh. *International Journal of Plant Research, 4*(4A), 8–13.

Mathers C, Stevens G, & Mascarenhas M. (2009). *Global Health Risks: Mortality and Burden of Disease Attributable to Selected Major Risks.* Geneva: World Health Organization.

Moni MA, Rahman MA, Haque MA, et al. (2010). Blood pressure in relation to selected anthropometric measurements in senior citizens. *Mymensingh Medical Journal, 19*, 254–258.

Nahian MA, Ahmed A, Lázár AN, Hutton CW, Salehin M, & Streatfield PK. (2018). Drinking water salinity associated health crisis in coastal Bangladesh. *ELEMENTA: Science of the Anthropocene, 6*(2).

Naser AM, Rahman M, Unicomb L, Doza S, Ahmed KM, Uddin MN, et al. (2017a). Drinking water salinity and kidney health in southwest coastal Bangladesh: Baseline findings of a community-based stepped-wedge randomised trial. *Lancet*, 389(Suppl. 2), April 17, S15. https://doi.org/10.1016/S0140-6736(17)31127-3

Naser AM, Unicomb L, Doza S, Ahmed KM, Rahman M, Uddin MN, et al. (2017b). Stepped-wedge cluster-randomised controlled trial to assess the cardiovascular health effects of a managed aquifer recharge initiative to reduce drinking water salinity in southwest coastal Bangladesh: Study design and rationale. *Bangladesh Medical Journal*, 7(9), 1–11. Available from: http://dx.doi.org/10.1136/bmjopen-2016-015205

Nath P, Siddique ZS, Basher A, Bhuiyan MR, Rahman MH, Rubel MM, et al. (2012). Status of blood pressure among individuals consuming saline water. *Mymensingh Medical Journal, 21*(4), 627–632.

Niport, Mitra and Associates, & ICF International. (2013). *Bangladesh Demographic and Health Survey 2011*. Dhaka, Bangladesh and Maryland, USA: National Institute of Population Research and Training (NIPORT), Mitra and Associates & ICF International.

Norat T, Scoccianti C, Boutron-Ruault MC, Anderson A, Berrino F, Cecchini M, et al. (2015). European Code against cancer. 4th Edition, Diet and cancer. *Cancer Epidemiology*, 39, S56–S66. https://doi.org/10.1016/j.canep.2014.12.016

Oldeman LR, Hakkeling RTA, & Sombroek WG. (1991). *World Map of the status of human-induced soil degradation: An explanatory note*. UNEP and ISRIC: Wageningen.

Parkin DM (2011). Cancers attributable to consumption of alcohol in the UK in 2010. *British Journal of Cancer*, 105, S14–S18. http://dx.doi.org/10.1038/bjc.2011.476

Penney S. (2016). Dropping the salt: Practical steps countries are taking to prevent chronic non-communicable diseases through population-wide dietary salt reduction. *The Public Health Agency of Canada (PHAC)*. Available from: www1.paho.org/english/ad/dpc/nc/salt-mtg-phac-paper.pdf

Rahaman MA, Rahman MM, & Nazimuzzaman M. (2019) Impact of salinity on infectious disease outbreaks: Experiences from the global coastal region. In: Leal Filho W, Wall T, Azul AM, Brandli L, & Özuyar PG (eds) *Good Health and Well-Being*. Encyclopedia of the UN Sustainable Development Goals. Cham: Springer. Available from: https://doi.org/10.1007/978-3-319-95681-7_106

Rahman M, Hassan M, Islam M, & Shamsad S. (2000). Environmental impact assessment on water quality deterioration caused by the decreased Ganges outflow and saline water intrusion in south-western Bangladesh. *Environmental Geology*, *40*, 31–40.

Rasheed S, Jahan S, Sharmin T, Hoque S, Khanam MA, Land MA, et al. (2014). How much salt do adults consume in climate vulnerable coastal Bangladesh? *BMC Public Health*, *14*(584). Available from: www.biomedcentral.com/1471-2458/14/584

Rasheed S, Siddique AK, Sharmin T, Hasan AMR, Hanifi SMA, Iqbal M, et al. (2016). Salt intake and health risk in climate change vulnerable coastal Bangladesh: What role do beliefs and practices play? *PLOS One*, *11*(4).

Rhoades JD, Corwin DL, & Lesch, SM. (1999). *Geospatial Measurements of Soil Electrical Conductivity to Assess Soil Salinity and Diffuse Salt Loading from Irrigation*. American Geophysical Union: Washington D.C., 108(5). Doi:10.1029/GM108p0197.

Riaz BK, Chowdhury SH, Karim MN, Feroz S, Selim S, & Rahman MR. (2015). Risk factors of hemorrhagic and ischemic stroke among hospitalized patients in Bangladesh: A case control study. *Bangladesh Medical Research Council*, *41*(1), 29–34.

Rust P, & Ekmekcioglu C. (2017). Impact of salt intake on the pathogenesis and treatment of hypertension. *Advances in Experimental Medicine and Biology Advances in Internal Medicine*, *956*, 61–84. Available from: DOI 10.1007/5584_2016_147

Sakamoto M. (2017). Saline drinking water and salt in diet: An approximate picture of the situation in a coastal area of southeastern Bangladesh. *International Journal of Disaster Risk Science*, *8*, 109–120.

Salim M, Maruf B, Chowdhury A, Shamsudoha, & Babul A. (2007). Increasing salinity threatens productivity of Bangladesh. *Coastal Association for Social Transmission Trust (COAST)*. Dhaka. Available from: www.coastbd.org/Conference%20260507/Pdf%20English/3.pdf

Sayeed MA, Mahtab H, Khanam PA, et al. (2005). Diabetes and hypertension in pregnancy in a rural community of Bangladesh: A population-based study. *Diabetic Medicine*, *22*, 1267–1271.

Schmidt CW. (2017). Another side of a low-salt diet: Reductions in the salinity of drinking water may lower blood pressure. *Environmental Health Perspectives*, *125*(6).

Scheelbeek PFD, Chowdhury MAH, Haines A, Alam DS, Hoque MA, Butler AP, et al. (2016a). High concentrations of sodium in drinking water and raised blood pressure in coastal deltas affected by episodic seawater inundations. *The Lancet: Global Health, 4*(18, Special issue). Available from: www.thelancet.com/lancetgh

Scheelbeek PFD, Khan AE, Mojumder SK, Elliott P, & Vineis P. (2016b). Drinking water sodium and elevated blood pressure of healthy pregnant women in salinity-affected coastal areas. *American Heart Associations, 68*, 464–470.

Scheelbeek PFD, Chowdhury MAH, Haines A, Alam DS, Hoque MA, Butler AP, et al. (2017). Drinking water salinity and raised blood pressure: Evidence from a cohort study in coastal Bangladesh. *Environmental Health Perspectives, 125*.

Selhub J, Jacques PF, Bostom AG, D'Agostino RB, Wilson PW, Belanger AJ, et al. (1995). Association between plasma homocysteine concentrations and extracranial carotid-artery stenosis. *New England Journal of Medicine*, 332, 286–291.

Shishir MA, Mamun MA, Mian MM, Ferdous UT, Akter NJ, Suravi RS, et al. (2018). Prevalence of Vibrio cholerae in coastal alternative supplies of drinking water and association with Bacillus-like spore formers. *Frontiers in Public Health, 6*(50). Available from: www.frontiersin.org

Stenhouse J, & Kijne JW. (2006). Prospects for productive use of saline water in West Asia and North Africa [Internet]. *Colombo, Sri Lanka: Comprehensive Assessment Research Report 11*. Available from: www.iwmi.org/assessment

Talukder MRR, Rutherford S, Phung D, Malek A, Khan S, & Chu C. (2016a). Drinking water contributes to high salt consumption in young adults in coastal Bangladesh. *The International Water Association (IWA) Publishing, 14*(2), 293–305.

Talukder MRR, Rutherford S, Phung D, Islam MZ, & Chu C. (2016b). The effect of drinking water salinity on blood pressure in young adults of coastal Bangladesh. *Environmental Pollution, 214*, 248–254.

Talukder MRR, Rutherford S, Huang C, Phung D, Islam M, & Chu C. (2017). Drinking water salinity and risk of hypertension: A systematic review and meta-analysis. *Archives of Environmental & Occupational Health, 72*, 126–138.

Thompson FL, Lida T, & Swings J. (2004). Biodiversity of vibrios. *Microbiology Molecular Biology Reviews, 68*(3), 403–431. doi: 10.1128/MMBR.68.3.403-431.2004

Trehan A, Winterbottom J, Lane B, Foley RN, Venning MC, Coward RG, Macleod A, & Gokal, R. (2003). End-stage renal disease in Indo-Asians in the North-West of England. *QJM : Monthly Journal of the Association of Physicians*, 96(7), 499–504.

Umesawa M, Iso H, Date C, et al. (2008). Relations between dietary sodium and potassium intakes and mortality from cardiovascular disease: The Japan Collaborative Cohort Study for Evaluation of Cancer Risks. *American Journal of Clinometric Nutrients, 88*, 195–202.

UNISDR. (2011). The 2011 Global Assessment Report on Disaster Risk Reduction (GAR11). Available from: www.preventionweb.net/english/hyogo/gar/2011/en/home/index. html

UNISDR. (2018). *World Risk Report Analysis and Prospects 2018* [Internet]. Berlin, Germany: Bündnis Entwicklung Hilft, 1–56. Available from: http://weltrisikobericht.de/wp-content/ uploads/2018/11/WRR_2017_E2.pdf

US Department of Health and Human Services, & US Department of Agriculture. (2011). *Dietary guidelines for Americans, 2010*. 7th Ed. Washington DC US Department of Health and Human Services US Department of Agriculture [Internet]. 2011. Available from: http://health. gov/dietaryguidelines/2010.asp

Vineis P, Chan Q, & Khan AE. (2011). Climate change impacts on water salinity and health. *Elsevier Ltd. Saudi Arab, 1*, 5–10.

WHF. (2017). *Hypertension: Hypertension and cardiovascular disease* [Internet]. World Heart Federation; [cited 2017 Feb 26]. Available from: www.world-heart-federation.org/cardiovascular- health/cardiovascular-disease-risk-factors/ hypertension

WHO. (2010). Bangladesh NCD risk factor survey 2010, WHO. Available from: http://www.whoban.org/LinkFiles/Publication_NCD_ FACT_SHEET.pdf.pdf

WHO. (2011). *Non-Communicable Disease Risk Factor Survey Bangladesh 2010*. Dhaka, Bangladesh: World Health Organization.

WHO. (2012). *Guideline: Sodium Intake for Adults and Children*. Geneva, Switzerland: World Health Organization.

WHO. (2013). *A Global Brief on Hypertension: Silent Killer, Global Public Health Crisis. Geneva, Switzerland: World Health Organization.*

WLE. (2014). *Bangladesh: Hypertension* [Internet]. 2014 [cited 2017 Feb 26]. Available from: www. worldlifeexpectancy.com/bangladesh-hypertension

Wolfram G, Bechthold A, Boeing H, Ellinger S, Hauner H, Kroke A, Leschik-Bonnet E, … Dinter J. (2015). *German Nutrition Society. Evidence-based guideline of the German Nutrition Society: Fat intake and prevention of selected nutrition-related diseases. Annals of Nutrition and Metabolism, 67*(3), 141–204. doi: 10.1159/000437243. Epub 2015 Sep 29. PMID: 26414007.

Yanez-Pagans M. (2017). *Bangladesh—Poverty and Groundwater Salinity Survey 2016* [Internet]. Report No.: 2887. World Bank Report. Available from: http://microdata.worldbank.org

Zaman MM, & Rouf MA. (1999). Prevalence of hypertension in Bangladeshi adult population. *Journal of Human Hypertension, 13*, 547–549.

Zaman MM, Choudhury SR, Ahmed J, Khandaker RK, Rouf MA, & Malik A. (2017). Salt intake in an adult population of Bangladesh. *Global Heart, 12*(3), 265–266.

3

DOES SALINITY INTRUSION INCREASE THE RATE OF CHRONIC ILLNESSES FOR THE COASTAL COMMUNITIES?

A Case Study from Dacope Upazila of Coastal Region of Bangladesh

Mohammad Abdul Quader, Md. Farhad Hasan, Md. Habibur Rahman, and Md. Abdul Malak

Introduction

One of the most dynamic natural systems on the planet is the coastal zone. It supports a diverse of ecosystems that offer critical habitat for fisheries and other food supplies, as well as other ecosystem services (Ericson et al. 2006; Vineis et al. 2011). However, coastal zones are a crucial battle front in the onslaught of climate change with its impacts including those associated with increasing sea-level rise (SLR), itself caused by global warming. Flooding, storm surge, saltwater intrusion, changes in surface water quality, and groundwater characteristics are all examples of these effects (Nicholls & Cazenave 2010),which directly impact the community in different ways such as the displacement of people, loss of livelihoods (in agriculture and aquaculture), damage or destruction of coastal infrastructure, and the submergence of low-lying areas as well as the imposition of socio-economic costs (IPCC 2013). Additionally, the Fifth Assessment Report (AR5) of the Intergovernmental Panel on Climate Change (IPCC) reported that the observed global mean SLR of 0.0032 m per year could result in an SLR of nearly 1m by 2100 if greenhouse gas emissions continue to rise at the current rate (IPCC 2013). Such a rise has been largely attributed to thermal expansion of the oceans and increased glacial melt (IPCC 2019). By the end of the century, ice melting from the Greenland and Antarctic ice sheets might contribute an extra 0.3 m of SLR (NRC 2012).

The AR5 also suggested that the implications of SLR will most certainly be worse for less developed countries (LDCs) and small island states; nevertheless, most SLR-related impact assessments have focused solely on industrialized nations, with scant data on poor and coastal countries (Brecht et al. 2012). Moreover, the World Bank (2013)study reported that, if the developed and industrialized countries could

DOI: 10.4324/9781003253495-4

limit global warming to 2°C, it may limit SLR to 0.7m by the end of this century. Nevertheless, while some earlier reports (e.g., IPCC 1990) anticipated an SLR of ≤1 mby 2100 even in the case of a 'business as usual' scenario (with a 'best estimate' of 44 cm by 2070, after which the SLR rate is expected to accelerate), more recently this estimate has been revised with the most likely upper limit by 2100 to be as high as 1.1 m (or higher, depending on whichRepresentative Concentration Pathway [RCP] emission scenariopathway is taken globally), with SLR continuing for not hundreds but potentially thousands of years (IPCC 2019). Another recent report warns that given failure to adhere to the Paris Agreement of 2016, global temperature rise may be at least 2.7° by 2100 (UNEP 2021), exacerbating climate change impacts, including SLR.

According to the World Health Organization (WHO 2021), Climate change is the single most serious hazard to humanity's health. It affects the social and environmental determinants of health such as access to clean and safe drinking water. Flooding and storm surges associated with SLR have already increased the burden on global coastal communities trying to secure or maintain access to clean, safe drinking water (IPCC 2014). SLR-related studies suggest that human health should be a consideration in the managed retreat process that must be undertaken to reduce economic and other impacts; however, the effect of increasing SLR on public health—specifically the possible impacts on disease transmission and mental health—have received less attention in the scientific literature than might have been expected by a number of researchers (Craig 2010; Ebi et al. 2013; Swaminathan et al. 2017;Veenema et al. 2017;Allen et al. 2019). A study by Dvorak et al. (2018)identified three categories of diseases that are associated with water and whose transmission (frequency, location, and duration of contagion) can be affected by SLR: mosquito-borne diseases (such as malaria and dengue), naturalized organisms (*Vibrio* spp. and toxic algae), and faecal–oral diseases (such as typhoid, cholera, and diarrhoea).Furthermore, increasing sea levels force saltwater further into freshwater streams, rivers, and estuaries, possibly polluting freshwater supplies and destroying riparian ecosystems that defend them (Parker 2014).

Saltwater intrusion decreases the accessibility of safe drinking water and increases the risk of vector borne, waterborne, and respiratory diseases, as well as causing mental health consequences in both developed and developing countries (Ramasamy & Surendran 2011; IPCC 2014). WHO reported that consumption of salt (through drinking, cooking, and bathing) higher than the recommended level in coastal zones is associated with sever health effects such as hypertension and stroke, skin diseases, diarrhoeal diseases, acute respiratory infection, and miscarriage among pregnant women (WHO 2017). As a result, WHO considers lowering population salt intake a public health priority. Moreover, recent research by Chakraborty et al. (2019) observed higher rates of diarrhoea and abdominal pain (as well as cardiovascular conditions) among patients who lived in areas with higher salinity drinking water sources than those who lived in areas with lower salinity drinking water sources. The authors also noted that salt ingestion via drinking water salinity can be a notable contributor (even exceeding dietary consumption via other foods) to adverse health consequences. Therefore, the aim of this study conducted among members of a coastal community in southwestern Bangladesh is to describe the possible negative impact of salinity intrusion (due to SLR) on the chronic illness resulting from drinking the water available.

Health Impacts of Sea-Level Rise on Coastal Communities in Bangladesh

The 710-km-long coastline of Bangladesh is part of the world's largest delta called the Bengal Delta or Ganges-Brahmaputra-Meghna (GBM) Delta and the Bay of Bengal (Figure 3.1). The coastal zone covers 32% of the country's total land mass and encompasses 19 districts of its 64 districts. Around 39 million or 29% of the total population (density 1,000 per sq km) live in this zone (BBS 2011).Portions of Satkhira district are located in the non-active Ganges delta, whereas parts of Khulna district are located in both active and inactive Ganges deltas, and pieces of Bagerhat district are located in the active Ganges delta (Islam 2004).Coastal people are exposed to both anthropogenic hazards, and climate change-induced events like flooding, cyclone and storm surges, SLR, and salinity intrusion (Parvin et al. 2008).The coastal zone facing the Bay of Bengal lies about 1.5–11.8 m above the mean sea level and is particularly vulnerable to SLR. One of the most significant repercussions of SLR is coastal salt intrusion, which has already damaged 53% of coastal areas (Haque et al. 2019).

The low socio-economic situation of coastal inhabitants, in addition to the geophysical features of the shoreline, contributes to their sensitivity to SLR consequences (Alam & Rahman 2014).The Climate Change Cell of Bangladesh (2016) reported that projected SLR would be about 0.88 m (above the mean sea level) by the year 2100.Furthermore, according to a World Bank (2013) study, about 20 million people have been affected by salinity intrusion in their drinking water, and that salinity intrusion may cause an increase in the incidence of water-borne diseases by forcing people to use even more contaminated sources, such as pond water, and by higher disease incidence associated with higher drinking water salinity levels (Chakraborty et al. 2019; Jabed et al. 2020; Rahman et al. 2017). Therefore, one of the major consequences of SLR for coastal people is the scarcity of saline-free drinking water, which poses threat to the people's health and even their survival.

The coastal population in Bangladesh heavily relies on river water, tube wells or groundwater, and pond water for washing and bathing, as well as for drinking. Alam and Rahman (2017, 2019) reported that natural disasters like cyclones entailed community suffering by creating a scarcity of safe water as tube wells were broken, damaged, or submerged, and ponds and other water bodies were polluted by the onrush of saltwater, compelling people to drink contaminated water with a higher than desirable level of salinity. While earlier views associated salt with a matter of palatability rather than health, more recent research (Chakraborty et al. 2019;

Jabed et al. 2020; Rahman et al. 2017) indicates there may be mild to severe health consequences across a broad spectrum of the population, including infants.

Nevertheless, the salinity standards for drinking water in Bangladesh remains a total dissolved solid (TDS) of 600 mg/l (Chakraborty et al. 2019). Those affected by a cyclone-induced scarcity of tube wells and lack of access to other water bodies uncontaminated by disease and salinity consequently suffer from various waterborne diseases. Drinking water contaminated by disease-causing microbes or pathogens can result in numerous illnesses and diseases including diarrhoea, dysentery, cholera, and skin diseases as well as allergic conditions (Alam & Rahman 2019). Those drinking water with an increased salinity level due to saltwater intrusion or flooding can suffer from various conditions associated with increased salinity. These include hypertension (Chakraborty et al. 2019).

Based on findings from the existing literature (Abedin et al. 2014), for the coastal people of Bangladesh, there are two major adverse effects of SLR and salinity intrusion: (a) a decline in people's food security with the contraction of various livelihood sectors, for example, agriculture, forestry, and fisheries (Haider & Hossain 2013; Mustari & Karim 2014; Rahman et al. 2017), and (b) scarcity of safe drinking water, which affects people's health, especially that of women (Abedin et al. 2014; Alam & Rahman 2017, 2019).

Increasing salinity levels in drinking water (the WHO drinking water standard for sodiumchloride (NaCl) concentration is 200–250 mg/l) contribute to higher salt consumption leading to the increasing risk of hypertension and infant morbidity and mortality (Dasgupta et al. 2016; Talukder et al. 2016a). Coastal residents who consume saline water suffer from a variety of health issues, including high blood pressure, cardiovascular illness, poor newborn development, and a higher infant death rate (Dasgupta et al. 2016; Talukder et al. 2016a; Rahman et al. 2017). Furthermore, if it is contaminated by pathogens or results in the population accessing other pathogen-contaminated sources of drinking water, people contract diarrhoea, cholera, and skin diseases. The saline water also unsuitable for irrigation. Moreover, groundwater is also not safe, since arsenic poisoning is a major concern across the country, particularly in the coastal areas, due to its presence in the underlying alluvial deltaic strata of the Meghalaya-Ganges Basin that have been accessed by tube wells constructed since the 1970s in attempts to source a reliable water supply (Ahmad et al. 2018). In rural regions, around 97% of the population relies on tube wells for drinking water (Flanagan et al. 2012).

There has been some notable scientific research works in the coastal areas of Bangladesh to identify the impact of increases in salinity from SLR on public health issues. Many conformed their hypothesized relationships. For example, Khan et al. (2011) looked at the possibility of salinity playing a role in explaining the peculiar seasonal pattern of hypertension in pregnancy in Dacope Upazila. In Dacope Upazila, Khan et al. (2014) investigated the link between salt in drinking water and the risk of (pre) eclampsia and gestational hypertension. Grant et al. (2015) investigated the possibility of linking cholera outbreaks with ingestion of saline-rich cholera-infected river water by measuring the salinity levels of common water sources in Shyamnagar Upazila and exploring perceptions of water palatability among the local population but concluded that it was unlikely for several reasons. Other research pointed to more direct relationships.

Dasgupta et al. (2016) investigated the long-term effects of prenatal salinity exposure in coastal Bangladesh, following detailed medical research that connects the salinity of drinking water consumed all through pregnancy to maternal hypertension, pre-eclampsia, and post-partum infant morbidity and mortality. Talukder et al. (2016a)investigated the impact of high salt in drinking water on blood pressure in a youthful population of 19–25 years old in coastal Koyra Upazila. Talukder et al. (2016b) investigated the level of salt consumption among young adults in Koyra Upazila, as well as the factors that contribute to it, to inform policymakers about the need for proper intervention strategies to prevent the potential medium- to long-term health risks associated with increased sodium intake. The effects of sodium in drinking water on blood pressure in coastal communities of Dacope, Batiaghata, and Paikghaccha Upazilas were studied by Scheelbeek et al. (2017). In nine coastal regions, Nahian et al. (2018) investigated the relationship between drinking water salinity and related health issues, notably hypertension.

Chakraborty et al. (2019) investigated the association between salinity exposure with hospital visits for a variety of health conditions in three coastal sub-districts namely Mathbaria and Zianagar Upazilas (Perojpur District) and Mongla Upazila (Bagerhat District). Naser et al. (2019) assessed the association between drinking water salinity, urinary macro-mineral (sodium, calcium, potassium, and magnesium) excretions, and blood pressure of coastal population in Bagerhat, Khulna, and Satkhira Districts. Jabed et al. (2020) conducted research in Chattogram District to investigate people's perceptions of the effects of water salinity on human health. Shuvo et al. (2020) investigated the effects of soil and water salinity on food habits and health risks in Dacope and Paikgacha Upazilas (Khulna District) and Kalapara Upazila (Khulna District) (Patuakhali District). Khan et al. (2020) investigated the link between drinking water salinity and hypertension in three coastal Upazilas: Taltoli (Barguna District), Morrelganj (Bagerhat District), and Shyamnagar (Satkhira District). Therefore, the present study was conducted in coastal DacopeUpazila (sub-district) under the Khulna District of southwestern Bangladesh to fulfil the following two research objectives:

1. To determine the negative impact of salinity intrusion on community health; and
2. To determine the environmental parameters associated with various diseases.

Materials and Methods

The Study Area

Dacope Upazila(sub-district) is situated in the Khulna District of southwestern Bangladesh (Figure 3.1). With an area of 991.58 sq km, the sub-district is located

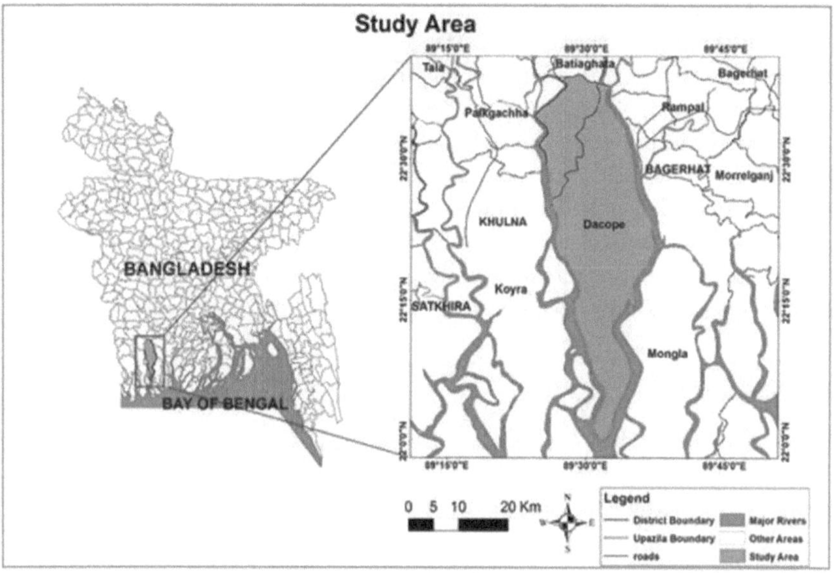

FIGURE 3.1 Location map of Dacope Upazila in the southwestern Bangladesh showing the study area inside the map

between 22°24'and 22°40' N and between 89°24'and 89°35' E. It is bounded by Batiaghata Upazila on the north, Sundarbans mangrove forest (covering an area of11,790 sq km) on the south, Rampal and Batiaghata Upazilas on the east, and Paikgachha Upazila on the west. The sub-district has a population of 158,309 from 29,600 households (http://dakop.khulna.gov.bd).Sibsa, Bhadra, Chunkuri, Pasur, Dhaki, Manga, Dhangmari, and Jhapjapiya rivers flow through the sub-district. Currently, the sub-district consists of 106 villages, a municipality, and nine Union Parishads (lowest tier of administrative entity). Paddy cultivation and shrimp farming are the major income sources for the people. The sub-district consists of 27,228 hectares of agricultural land for cultivating mainly paddy and Bagdah/giant tiger shrimp (*Penaeus monodon*) and Galdah/giant freshwater shrimp (*Macrobrachiumrosenbergii*) (http://dakop.khulna.gov.bd). In the past, the area was part of the Sundarbans mangrove forest, but the forests in the location were partly destroyed to create settlements and agricultural land.

Over the last two decades, salinity intrusion has become a serious concern for Dacope Upazila. Salinity intrusion in both surface and groundwater is a major danger to people's safe water. Repeated disasters, the devastating cyclones *Sidr* (2007) and *Aila* (2009), caused severe damage in Khulna District (including Dacope and its surrounding sub-districts) particularly to safe drinking water sources (Alam & Rahman 2019). Salinity intrusion has already impacted human health and agriculture (including farming, fisheries, forest resources, as well as livestock rearing).As a result of saltwater in rivers being introduced into wetland shrimp agriculture, the sub-district is facing a drop in freshwater owing to the depletion of local freshwater ponds and compartments of safe groundwater aquifers (Khan et al. 2011). Through seepage, the salinity of nearby freshwater ponds and shallow aquifers rises, posing a threat to drinking water safety (Abedin et al. 2014; Khan et al. 2014). Moreover, the presence of natural arsenic in groundwater has exacerbated the scarcity of safe drinking water in the region (Abedin et al. 2014; Khan et al. 2014). Due to the decline in access to safe water and greater exposure to higher levels of salinity, people are affected by different kinds of waterborne diseases, common among them are diarrhoea, dysentery, arsenicosis, skin diseases, and hypertension-related cardiovascular diseases, while tuberculosis is also common (Islam et al. 2019). So, for this study, DacopeUpazila has been selected purposively as the study area.

Data Collection

This research has relied on secondary data sources. Data related to the number of salinities affected people between 2013 and 2019 were collected from the DacopeUpazila Health Complex (DUHC), the only hospital in the sub-district. Some key informant interviews were also conducted with the nurses and medical officers and other staff of the health complex. Everyday people all over the sub-district attend the DUHC for healthcare services, and it has a list of people affected by waterborne and other diseases. For the study, we visited the DUHC and collected this authentic list from the office.

Data on water quality and salinity-related parameters were also collected for use in this study. These comprised arsenic level, salinity level, TDS concentration, pH, total solid (TS) concentration, electrical conductivity (EC), biological oxygen demand (BOD), chemical oxygen demand (COD), and iron and groundwater table data of Dacope and other sub-districts of Khulna District. Table 3.1 shows the standards for environmental parameters for Bangladesh. Moreover, journal articles, reports published by different NGOs and the Water Resources Planning Organisation, British Geological Survey, and Bangladesh Bureau of Statistics were used for literature review for this study.

TABLE 3.1 Description of the environmental parameters used for this study

Environmental parameter	*Standard level for Bangladesh*	*WHO guideline*
Arsenic (As)	0.05 mg/l	0.01 mg/l
Salinity	1,000 mg/l chloride	250 mg/l chloride
Total dissolved solids (TDS)	250 mg/l	1,000 mg/l
pH	6.5–8.5	6.5–8.5
Total solid (TS)	-	-
Electrical conductivity (EC)	1,200µs/cm	1,000 µs/cm
Biological oxygen demand (BOD)	0.2 mg/l	-
Chemical oxygen demand (COD)	4 mg/l	-
Iron (Fe)	0.3–1.0 mg/l	-
Groundwater table	-	-

Data Analysis

Data analyses were conducted using IBM SPSS Statistics (version 23.0) and Microsoft Excel 2016. Multiple regression analysis and one-way analysis of variance (ANOVA) tests were conducted.

Results of the Study

Major Waterborne Diseases in DacopeUpazila

Arsenicosis

Contaminated groundwater is the most common cause of arsenicosis. Drinking arsenic-laced water for an extended length of time can cause human poisoning. From 2013 to 2019, the total number of arsenicosis patients in the DacopeUpazila was found to be843. Female patients numbered higher (434 patients) than the male patients (409 patients). Red or inflamed skin; skin changes, such as new warts or lesions; stomach discomfort, nausea and vomiting, diarrhoea; altered heart rhythm; muscular cramps; and tingling of fingers and toes were all recorded as typical arsenicosis symptoms.

Diarrhoea

Between 2013 and 2019, 31,164 patients affected by diarrhoea reported to and/or were admitted to the DUHC. The highest number of patients (13,494) were oral rehydration therapy (ORT) patients followed by infield patients (8,037) (Table 3.2). The best ways to prevent diarrhoea are to drink safe water (clean or purified water), to use proper food handling techniques, and to maintain personal hygiene, for example, proper hand-washing habits. Antibiotics are rarely recommended by doctors in the research region to prevent diarrhoea since they might create extra difficulties.

Pneumonia and Tuberculosis

Pneumonia is an acute respiratory illness that mostly affects the lungs and is more native to tropical and subtropical regions. When a healthy individual breathes, the lungs are made up of little sacs called alveoli that fill with air (Ferdous et al. 2018). This is impeded by pneumonia. It is the most infectious cause of mortality in children in the world, accounting for 15% of all fatalities in children under the age of five in 2017 (Ferdous et al. 2018). It accounts for 15% of deaths of children under

TABLE 3.2 Number of diarrhoea patients in Dacope Upazila 2013–2019

Infield patients	*ORT total patients*	*Emergency patients*	*Outdoor patients*	*Indoor patients*
8,037	13,494	2,665	3,703	3,265

TABLE 3.3 Number of pneumonia patients in Dacope Upazila 2013–2019

Positive pneumonia	*Mild pneumonia*	*Moderate pneumonia*	*Seriously treated pneumonia*
12,764	4,457	1,407	1,583

TABLE 3.4 Relationship between the waterborne diseases and tuberculosis and environmental parameters

Name of disease	*Adjusted R2*	*Degree of freedom*	*Level of significance*
Arsenicosis	0.402	10	0.670
Diarrhoea	0.744	06	0.316
Tuberculosis	0.953	06	0.102

five in Bangladesh (Ferdous et al. 2018). Pneumonia can be a complication of near drowning. Saltwater inhalation may cause serious injury in the lung and alveoli, which makes breathing painful and limits oxygen intake and can result in pneumonia (Aston 2017).

Pneumonia is a complication experienced because of many infectious diseases and can have a bacterial, viral, or even fungal origin. As in much of the developing world, tuberculosis (from *Mycobacterium tuberculosis*) is a common cause of pneumonia in the coastal area of Bangladesh and specifically in the DacopeUpazila. Between 2013 and 2019, there were 20,211 patients admitted in the DUHC due to pneumonia. Among them, 12,764 patients were positive for pneumonia, and 1,583 patients were treated for serious pneumonia (Table 3.3).

Results from Comparing Environmental Parameters and the Number of Patients with Waterborne Diseases or Tuberculosis

Multiple regression analysis was conducted to assess the relationship between the number of patients with arsenicosis, diarrhoea, or tuberculosis and the environmental parameters. In the case of arsenicosis, the results suggest that there was no significant positive relationship between the patients of arsenicosis and the environmental parameters (t = 0.528, significance level [P/p] = 0.670). The adjusted R^2 value (0.402) indicates that there is a relationship of about 40% between the number of patients seen by the DUHC who had raised (higher than recommended) arsenic level in their blood and the presence of arsenic (As) in the water source/s (Table 3.1). In the case of diarrhoea, results from the multiple regression analysis showed that there was no significant positive relationship between the number of patients with diarrhoea and the environmental parameters (t = 2.525, significance level [P/p] = 0.316). The adjusted R^2 value (0.744) indicates that there is a relationship of about 74% between the number of patients with the increased of diarrhoea and other environmental parameter data (Table 3.4).

The multiple regression analysis also showed that there was no significant positive relationship between the patients of tuberculosis and the environmental parameters ($t = 6.691$, significance level $= 0.102$). The adjusted R^2 value (0.953), however, indicates that there is a relationship of about 95% between the number of patients with tuberculosis and the environmental parameter data collected (Table 3.4).

Adaptation to Saline Water: Coping Strategies Adopted

To reduce the incidence and impact of chronic illness occurring both directly (e.g., hypertension) and less directly (e.g., diarrhoea, by residents sourcing contaminated pond water and untreated water storage)as a result of salinity intrusion, people in the DacopeUpazila have taken different kinds of adaptation and coping strategies that are followed by individuals as well as communities to combat safe drinking water scarcity caused by salinity intrusion in wells and increased presence in surface water.Among the strategies adopted have been the installation of low-cost rainwater harvesting plants (using plastic containers, concrete containers, and/or earthenware containers), low-cost water treatment plants (desalination plants), pond sand filters (PSFs), tube wells with raised platform to prevent surface saline contamination via flooding, pond dewatering and cleaning, conservation of pond water solely for drinking water (as washing hands and legs and ablution are strictly forbidden), digging of new ponds, installing deep tube wells etc. People, mostly by women and children, travel 0.5–3 km to fetch water from safe water sources. NGOs have also distributed chemical additives (Clotech to be used diluted as 'bleach', chlorating, and alum) for purificating before its use for drinking and cooking. Local government institutions, international NGOs, and national and local NGOs are involved in supplying safe drinking water in coastal DacopeUpazila. They have provided technical and financial assistance to the community people to ensure saline-free and bacteria level-reduced safe drinking water.

Discussion

One in every three people lacks access to clean drinking water throughout the world, while Goal 6 of the Sustainable Development Goals aims to provide universal and equitable access to safe and cheap drinking water for everyone by 2030 (https://sdgs.un.org/goals). One of the biggest repercussions of SLR in Bangladesh is salinity intrusion in the coastal zone, which obstructs access to drinkable water. Salinity has already harmed 53% of the coastal regions (Haque et al. 2019). Existing studies have shown that the rate of salinity intrusion is faster than that predicted just a decade ago. The infiltration of salty water is likely to continue, with the area under 1 ppt salinity forecast to rise 18.22% by 2050, and the area under 5 ppt salinity expected to rise 24% (CEGIS and DoE 2011). Freshwater has a salinity level of≤0.5ppt; estuary water salinity ranges from 0.5 to 30ppt, the latter close to that or pure sea water. Khan et al. (2011, 2014) argued that salinity intrusion is anticipated to grow in the future due to further lower river flows, greater upstream extraction, and longer-term climate change-induced declines in dry season rainfall. As a result, agricultural production would decline, food insecurity will rise, and salt in drinking water sources will rise as a result of the damaged groundwater (Hoque et al. 2013).

Salinity intrusion has been highlighted as a growing public health hazard impacting thousands of homes in Asian coastal locations such as Bangladesh, China, and Vietnam every year (Chakraborty et al. 2019; Li & Chen 2019). Changes in Upper River flow and local land-use actions intensify salinity issues produced by periodic storms and consequent seawater inundations, which are presumably (at least in part) connected to climate change (Scheelbeek et al. 2016). The present study evaluated the health problems related to the impacts of consumption of saline water and consumption of pathogen-affected water sources by that and displaced consumption. It evaluated the possibility of a relationship between environmental parameters and three chronic diseases (arsenicosis, diarrhoea, and tuberculosis) in DacopeUpazila. Such health-related problems have also been reported from south-western coastal region of Bangladesh (Khan et al. 2011; Abedin et al. 2014; Khan et al. 2014; Scheelbeek et al. 2016; Talukder et al. 2016a, 2016b; Chakraborty et al. 2019).

The present study identified that local people perceived that there existed a very limited range of government and NGO services to meet their health care demands arising from problems directly and indirectly from salinity intrusion. Similarly, to this study, other studies reported several chronic illnesses became more prevalent due to salinity intrusion in other coastal areas in Bangladesh. Jabed et al. (2020) found that villagers in the southeastern coastal region suffered from skin diseases, hair loss, diarrhoea, gastroenteritic conditions, and high blood pressure due to saline water consumption.Other studies like Khan et al. (2011, 2014), Grant et al. (2015), Scheelbeek et al. (2016), Dasgupta et al. (2016), and Talukder et al. (2016b) focused on the effects of salinity on maternal health and infant, impacts of maternal hypertension, infant mortality, and one explored the possibility of a relationship to cholera outbreaks. Chakraborty et al. (2019) found drinking water salinity to be associated with cardiovascular diseases (five times higher for those living in high drinking water salinity areas than in lower drinking water salinity areas), diarrhoea (more than twice as high), and abdominal pain (more than three times higher) in the south and southwest regions of Bangladesh. Abedin et al. (2014) found that due to scarcity of safe drinking water, southwestern coastal people suffered from diarrhoea, especially infants and children, indigestion, black spots on the skin, and so on. Furthermore, Shuvo et al. (2020) evaluated the impact of soil and water salinity on dietary behaviour and health risk. They revealed that consuming high-salt meals and drinking potable water were linked to hypertension, renal illness, and skin disease (Shuvo et al. 2020). The current study investigated three conditions (namely diarrhoea, arsenicosis, and tuberculosis) in relation to salinity and other indicators.

The present study also reported the adaptation and coping strategies adopted to cope with saline water problems in DacopeUpazila. Like this study findings, other studies such as those by Alam and Rahman (2014, 2019) reported that southern coastal people use safe drinking water from ponds with PSF, deep tube wells, saline purification, and rainwater harvesting plants. Abedin et al. (2014) reported that small units of ponds with PSF and rainwater harvesting plants were considered the most suitable safe drinking water options in the southwestern coastal region. Shammi et al. (2019) revealed PSF, rainwater harvesting plant, and regulated aquifer recharge are identified as significant treatments to decrease difficulties linked with drinking salty water. Solar-powered desalination facilities, reverse osmosis, and electro dialysis devices are among the advanced and expensive potable water alternatives that can provide saline-free water to Bangladesh's coastal residents.

Local government institutions in Bangladesh's coastal area have not had the opportunity to work with any more efficient levels of the government as mandates, and inadequate budgets hinder them from carrying out their functions and obligations (Islam et al. 2019). Moreover, the salinity issue has not to date received much attention in the government's disaster management policy. The 'Coastal Zone Policy 2005' did not consider the salinity issue. This policy stated that, in case of shortage of drinking water,

> Rainwater harvesting plant and conservation shall be promoted, ponds and tanks will be excavated for conservation of water and local technology for water treatment (such as, PSF) will be used for the supply of safe water; steps will be taken to ensure sustainable use and management of groundwater.

The National Adaptation Programme of Action (NAPA), on the other hand, briefly outlines the origins and effects of water salinity on the coastal economy, although climate change and SLR received a lot more attention (Islam et al. 2019). In this respect, it needs more attention by government as well as NGOs to overcome the salinity intrusion problem in the near future.

Conclusion

Many waterborne infections that affect persons living around the shore are linked to salinity. SLR is predicted to play a substantial impact in raising saline levels in coastal drinking water sources. Waterborne diseases are transmitted either through drinking or through increased contact with contaminated saline water or less than optimal alternate sources while salinity-induced diseases (e.g., hypertension-related conditions) are related to consumption of high-salt levels through diet, including salinity-affected drinking water. The local health facility, which has just 50 beds and is insufficient for a sub-district population, is where most people go. People seek treatment from a Kabiraj (traditional herbalist) and informal healthcare professionals (dubbed 'village physicians') due to the constraints of the healthcare system. Our findings suggest that the presence of a high level of salinity in drinking and cooking water in DacopeUpazila is a cause of major public health concern and a challenge for the government and NGOs. Therefore, it is imperative to develop and evaluate affordable approaches to providing safe drinking water with low-salt content in the entire coastal area and specifically in DacopeUpazila. This research will alert policymakers to the increased danger of consuming saline water on human health in coastal locations, allowing them to include suitable preventative and mitigation measures into their strategic plans. In coastal areas, special treatments should be created to address the negative health impacts of excessive salt intake.

References

Abedin MA, Habiba U, Shaw R (2014). Community perception and adaptation to safe drinking water scarcity: Salinity, arsenic, and drought risks in coastal Bangladesh. *International Journal of Disaster Risk Science* 5:110–124. DOI: 10.1007/s13753-014-0021-6

Ahmad SA, Khan MH, Haque M (2018). Arsenic contamination in groundwater in Bangladesh: implications and challenges for healthcare policy. *Risk Management and Healthcare Policy* 11:251.

Alam K, Rahman MH (2014). Women in natural disasters: A case study from southern coastal region of Bangladesh. *International Journal of Disaster Risk Reduction* 8:68–82. DOI: 10.1016/j.ijdrr.2014.01.003

Alam K, Rahman MH (2017). The role of women in disaster resilience. In: Madu CN and Kuei C (eds), *Handbook of disaster risk reduction and management*. New Jersey: World Scientific Press, pp. 697–719. DOI: 10.1142/9789813207950_0029

Alam K, Rahman MH (2019). Post-disaster recovery in the cyclone Aila affected coastline of Bangladesh: Women's role, challenges and opportunities. *Natural Hazards* 96:1067–1090. DOI: 10.1007/s11069-019-03591-7

Allen TR, Crawford T, Montz B, Whitehead J, Lovelace S, Hanks AD, Christensen AR, Kearney GD (2019). Linking water infrastructure, public health, and sea level rise: Integrated assessment of flood resilience in coastal cities. *Public Works Management & Policy* 24(1):110–139. DOI: 10.1177/1087724X18798380

Aston SJ (2017). Pneumonia in the developing world: Characteristic features and approach to management. Respiratory Infections in the Asia-Pacific Region. *Wiley Online Library* 22(7):1276–1287. DOI: 10.1111/resp.13112

BBS (Bangladesh Bureau of Statistics) (2011). *Population and housing census 2011: Socio-economic of demographic report*. Ministry of Planning, Dhaka.

Brecht H, Dasgupta S, Laplante B, Murray S, Wheeler D (2012). Sea-level rise and storm surges: High stakes for a small number of developing countries. *The Journal of Environment & Development* 21(1):120–138. DOI: 10.1177/1070496511433601

CEGIS (Center for Environmental and Geographic Information Services) and DoE (Department of Environment) (2011). *Final Report on Programmes Containing Measures to Facilitate Adaptation to Climate Change of the Second National Communication Project of Bangladesh*. Dhaka, Bangladesh.

Chakraborty R, Khan KM, Dibaba DT, Khan MA, Ahmed A, Islam MZ (2019). Health implications of drinking water salinity in coastal areas of Bangladesh. *International Journal of Environmental Research and Public Health* 16(19): 3746. DOI: 10.3390/ijerph16193746

Climate Change Cell (2016). *Assessment of sea level rise on Bangladesh coast through trend analysis*. Department of Environment, Dhaka.

Craig RK (2010). A public health perspective on sea-level rise: Starting points for climate change adaptation. *Widener Law Review* 15(2):521–540.

Dasgupta S, Huq M, Wheeler D (2016). Drinking water salinity and infant mortality in coastal Bangladesh. *Water Economics and Policy* 2(1):1650003. DOI: 10.1142/S2382624X1650003X

Dvorak AC, Solo-Gabriele HM, Galletti A, Benzecry B, Malone H, Boguszewski V, Bird J (2018). Possible impacts of sea level rise on disease transmission and potential adaptation strategies, a review. *Journal of Environmental Management* 217:951–968.

Ebi KL, Lindgren E, Suk JE, Semenza JC (2013). Adaptation to the infectious disease impacts of climate change. *Climatic Change* 118(2):355–365.

Ericson JP, Vörösmarty CJ, Dingman SL, Ward LG, Meybeck M (2006). Effective sea-level rise and deltas: Causes of change and human dimension implications. *Global Planetary Change* 50(1–2):63–82. DOI: 10.1016/j.gloplacha.2005.07.004

Ferdous F, Ahmed S, Das SK, et al. (2018). Pneumonia mortality and healthcare utilization in young children in rural Bangladesh: A prospective verbal autopsy study. *Trop Med Health* 46:17. https://doi.org/10.1186/s41182-018-0099-4

Flanagan S, Johnston R, Zheng Y (2012). Arsenic in tube well water in Bangladesh: Health and economic impacts and implications for arsenic mitigation. *Bulletin of the World Health Organization* 90(11):839–846. DOI:10.2471/BLT.11.101253

Grant SL, Tamason CC, Hoque BA, Jensen PKM (2015). Drinking cholera: Salinity levels and palatability of drinking water in coastal Bangladesh. *Tropical Medicine & International Health* 20(4):455–461. DOI: 10.1111/tmi.12455

Haider MZ, Hossain MZ (2013). Impact of salinity on livelihood strategies of farmers. *Journal of Soil Science and Plant Nutrition* 13(2):417–431.

Haque A, Rahman MH, Rahman D (2019). An evaluation of sea level rise vulnerability and resilience strategy to climate change in the coastline of Bangladesh. *International Journal of Environmental Science & Natural Resources* 18(1):555983. DOI:10.19080/IJESNR.2019.18.555983

Hoque MA, Saika U, Sarder BC, Biswas KK (2013). Environmental and socio-economic impacts of salinity intrusion in the coastal area: A case study on Munshigong Union, Shymnagor, Satkhira. *Jahangirnagar University Environment Bulletin* 2:41–49.

IPCC (Intergovernmental Panel on Climate Change). (1990). The IPCC scientific assessment. In Houghton JT, Jenkins G, Ephraums JJ (eds). https://www.ipcc.ch/site/assets/uploads/2018/03/ipcc_far_wg_I_full_report.pdf

IPCC (Intergovernmental Panel on Climate Change). (2013). *Climate change 2013: The physical science basis.* Contribution of Working Group I to the Fifth Assessment Report of the Intergovernmental Panel on Climate Change. New York: Cambridge University Press.

IPCC (Intergovernmental Panel on Climate Change). (2014). *Climate change 2014: Impacts, adaptation, and vulnerability.* Part A: Global and Sectoral Aspects. Contribution of Working Group II to the Fifth Assessment Report of the Intergovernmental Panel on Climate Change. Cambridge, New York: Cambridge University Press.

IPCC (Intergovernmental Panel on Climate Change). (2019). Summary for Policymakers. In: *Climate Change and Land: an IPCC special report on climate change, desertification, land degradation, sustainable land management, food security, and greenhouse gas fluxes in terrestrial ecosystems*. Cambridge, New York: Cambridge University Press.

Islam MR (2004). *Where land meets the sea: A profile of the coastal zone of Bangladesh*. Dhaka: The University Press.

Islam MS, Afrin S, Ahsan MN, Haider MZ, Mamun TM, Das DK (2019). Households' willingness to pay for disaster resilient safe drinking water sources in southwestern coastal Bangladesh. *International Journal of Disaster Risk Science* 10:544–556. https://doi.org/10.1007/s13753-019-00229-x

Jabed MA, Paul A, Nath TK (2020). Peoples' perception of the water salinity impacts on human health: A case study in south-eastern coastal region of Bangladesh. *Exposure and Health* 12:41–50. https://doi.org/10.1007/s12403-018-0283-0

Khan AE, Ireson A, Kovats S, Mojumder SK, Khusru A, Rahman A, Vineis P (2011). Drinking water salinity and maternal health in coastal Bangladesh: Implications of climate change. *Environmental Health Perspectives* 119(9):1328–1332. DOI: 10.1289/ehp.1002804

Khan AE, Scheelbeek PFD, Shilpi AB, Queenie Chan Q, Mojumder SK, Rahman A, Haines A, Vineis P (2014). Salinity in drinking water and the risk of (pre)eclampsia and gestational hypertension in coastal Bangladesh: A case-control study. *PLOS One* 9(9):e108715. DOI: 10.1371/journal.pone. 0108715.t001

Khan JR, Awan N, Archie RJ, Sultana N, Muurlink O (2020). The association between drinking water salinity and hypertension in coastal Bangladesh. *Global Health Journal* 4:153–158. https://doi.org/10.1016/j.glohj.2020.11.001

Li M, Chen Z (2019) An assessment of saltwater intrusion in the Changjiang (Yangtze) River estuary, China. In: Wolanski E, Day JW, Elliott M, Ramachandran R (eds), *Coasts and Estuaries*. Elsevier. DOI https://doi.org/10.1016/B978-0-12-814003-1.00002-2

Mustari S, Karim AHMZ (2014). Impact of salinity on the socio-environmental life of coastal people of Bangladesh. *Asian Journal of Social Sciences & Humanities* 3(1):12–18.

Nahian MA, Ahmed A, Lázár AN, Hutton CW, Salehin M, Streatfield PK (2018). Drinking water salinity associated health crisis in coastal Bangladesh. *Elementa Science of the Anthropocene* 6:2. https://doi.org/10.1525/elementa.143

Naser AM, Rahman M, Unicomb L, Doza S, Gazi MS, et al. (2019). Drinking water salinity, urinary macro-mineral excretions, and blood pressure in the southwest coastal population of Bangladesh. *Journal of the American Heart Association* 8:e012007.https://doi.org/10.1161/JAHA.119.012007

Nicholls RJ, Cazenave A (2010). Sea-level rise and its impact on coastal zones. *Science* 328(5985): 1517–1520. DOI: 10.1126/science.1185782

NRC (National Research Council) (2012). *Sea level rise for the coasts of California, Oregon, and Washington: Past, present, and future*. Washington, DC: The National Academies Press.

Parker CL (2014). Health impacts of sea-level rise. *Planning & Environmental Law: Issues and decisions that impact the built and natural environments* 66(5):8–12. DOI: 10.1080/15480755.2014.916166

Parvin GA, Takahashi F, Shaw R (2008). Coastal hazards and community-coping methods in Bangladesh. *Journal of Coastal Conservation* 12:181–193. https://doi.org/10.1007/s11852-009-0044-0

Rahman MTU, Rasheduzzaman M, Habib MA, Ahmed A, Tareq SM, Muniruzzaman SM (2017). Assessment of freshwater security in coastal Bangladesh: An insight from salinity, community perception and adaptation. *Ocean and Coastal Management* 137:68–81.

Ramasamy R, Surendran SN (2011). Possible impact of rising sea levels on vector-borne infectious diseases. *BMC Infectious Diseases* 11:18. www.biomedcentral.com/1471-2334/11/18

Scheelbeek PF, Khan AE, Mojumder S, Elliott P, Vineis P (2016). Drinking water sodium and elevated blood pressure of healthy pregnant women in salinity-affected coastal areas. *Hypertension* 68(2):464–470. DOI: 10.1161/HYPERTENSIONAHA.116.07743

Scheelbeek PFD, Chowdhury MAH, Haines A, Alam DS, Hoque MA, et al. (2017). Drinking water salinity and raised blood pressure: Evidence from a cohort study in coastal Bangladesh. *Environmental Health Perspectives* 125(5):057007. https://doi.org/10.1289/EHP659.

Shammi M, Rahman MM, Bondad SE, Bodrud-Doza M (2019). Impacts of salinity intrusion in community health: A review of experiences on drinking water sodium from coastal areas of Bangladesh. *Healthcare* 7:50. DOI:10.3390/healthcare7010050

Shuvo SD, Zahid MA, Rahman MM, Parvin R (2020). Exploring the impact of soil and water salinity on dietary behavior and health risk of coastal communities in Bangladesh. *Journal of Water and Health* 18(6):1110–1123. DOI: 10.2166/wh.2020.145

Sk Akhtar A, Manzurul HK, Mushfiqul H (2018). Arsenic contamination in groundwater in Bangladesh: Implications and challenges for healthcare policy. Risk Management and Healthcare Policy 11: 251–261. DOI: 10.2147/RMHP.S153188

Swaminathan A, Viennet E, McMichael AJ, Harley D (2017). Climate change and the geographical distribution of infectious diseases. In: Petersen E, Chen LH, Schlagenhauf-Lawlor P (eds), *Infectious Diseases: A Geographic Guide*. Chichester: John Wiley & Sons, Ltd.

Talukder MRR, Rutherford S, Phung D, Islam MZ, Chu C (2016a). The effect of drinking water salinity on blood pressure in young adults of coastal Bangladesh. *Environmental Pollution* 214:248–254. DOI: 10.1016/j.envpol.2016.03.074

Talukder MRR, Rutherford S, Phung D, Malek A, Khan S, Chu C (2016b). Drinking water contributes to high salt consumption in young adults in coastal Bangladesh. *Journal of Water and Health* 14(2):293–305. DOI: 10.2166/wh.2015.129

UNEP (United Nations Environment Programme). (2021). Emissions Gap Report 2021: The Heat Is On – A World of Climate Promises Not Yet Delivered. Nairobi. https://www.unep.org/emissions-gap-report-2021

Veenema TG, Thornton CP, Lavin RP, Bender AK, Seal S, Corley A (2017). Climate change–related water disasters' impact on population health. *Journal of Nursing Scholarship* 49(6):625–634. DOI: 10.1111/jnu.12328

Vineis P, Chan Q, Khan A (2011). Climate change impacts on water salinity and health. *Journal of Epidemiology and Global Health* 1:5–10. DOI: 10.1016/j.jegh.2011.09.001

WHO (2017). Guidelines for *d*rinking-*w*ater *q*uality: Fourth *e*dition *i*ncorporating the *f*irst *a*ddendum. Geneva: World Health Organization.

WHO (2021). *Climate change and health*. Geneva: World Health Organization. Available at: www.who.int/news-room/fact-sheets/detail/climate-change-and-health (accessed on January 8, 2022).

World Bank (2013). *4° Turn down the heat: Climate extremes, regional impacts, and the case for resilience*. New York: The World Bank.

4

'LIFE ON THE MOVE', IMPACTS OF RIVERBANK EROSION ON PEOPLE'S LIFE

A Case Study of an Eroding Village in Bangladesh

Nahid Rezwana and Saadi Islam

Introduction

Riverbank erosion is a natural phenomenon that displaces millions of people all over the world, and negatively affects their lives and well-beings. However, this disaster receives less attention despite its large impact on the affected society. There are an insufficient number of studies on this disaster, especially ones which focus on the displaced people's economic condition and social life. This chapter presents a study that involves research questions, such as how riverbank erosion affects the socio-economic conditions of the victims; what the prevailing plans and programmes that are managing this disaster, and if there is any gap between people's necessities and the disaster management plan. The study was located in Kalikanagar village, a small, beautiful village of Bangladesh, situated at the bank of mighty river Meghna. Natural resources (fish in the river and crops on the fields) offered a simple life to the inhabitants of this small village. Poverty and insufficiency were everyday's struggle of this village, still people were surviving on their own land with identity and pride. However, riverbank erosion brought a huge change to their lives. This chapter presents the impacts of riverbank erosion on the inhabitants of Kalikanagar village while emphasizing on the change in economic and social life of the victims who are living in the eroding village or adjacent villages. It also discusses the gaps in prevailing disaster management plans and programmes and mentions recommendations to improve disaster preparedness of the management authorities and hence, ensure better lives for the affected people.

Review of Literature

Geographic and geologic settings, soil composition and river morphology result in high risk of riverbank erosion in Bangladesh (Rana and Nessa 2017; Rahman 2010).

DOI: 10.4324/9781003253495-5

This country experienced land erosion of 105,928 acre during the year 2009–2014, which was 68.26% among the other disasters in the country. The damage and loss cost 36408.92 million BDT (BBS 2016). This is the topmost disaster concerning damage of land and loss of land productivity. Islam and his team revealed in their research that 56% of the resident living on the Brahmaputra floodplain areas in Mymensingh lost 0–5 acres and 33% lost 6–10 acres of their land due to riverbank erosion (Islam et al. 2017). Riverbank has a direct impact on the livelihoods of the rural society, downgrading the socio-economic status from solvent citizens to homeless extreme poor farmers or labourers. In Bangladesh, 15–20 million people are in danger of the effects of riverbank erosion (Hutton and Haque 2004; Rahman et al. 2015). Since riverbank affected regions' lack of ability to develop in industrial units, job opportunities are highly dependent on the agricultural sector, which reduces after erosion (Bhuiyan et al. 2017). Migration and displacement become a common coping strategy. Short-term temporary relocation to the adjacent villages and long-term permanent migration to urban areas are the types of movement mostly seen among the erosion victims (Islam et al. 2017). However, these migrants reach their destination with little savings, big loans for several movements and unsecured shelter or income sources (Rana and Nessa 2017; Bhuiyan et al. 2017; CEGIS 2009; Das 2011), which make their resettlement more difficult in a new culture and into traditions. Lack of skill, information and opportunities become constraints to get jobs whereas negative behaviours of the residents create harsh realities for the migrants. Sometimes a portion of these homeless people depend on illegal earnings for survival; different crimes are found to be increased in the society (Islam and Rashid 2011). Studies reveal that riverbank erosion brings significant amount of negative social consequences into the everyday life of victims as well as the most devastating aftermath compared to any other natural disasters (Elahi 1991; Rahman et al. 2015). Larger portion of erosion-hit families fall into the poverty cycle, from which they cannot break out in many cases. Extreme poor housing (Roy et al. 2017), lack of sanitation facilities, fresh water supply, education and healthcare facilities are the major problems in displaced people's lives. Many of them suffer from mental health problems (Rahman and Gain 2020), and aggression and hypertension increase within the members of the family, which results in conflicts and domestic violence (Akter et al. 2019). Rate of divorce or wife abandonment increases; men leave helpless families in the erosion-hit regions (Rahman et al. 2015). Women, girls and children are found more uncovered to insecurity such as sexual harassment, loss of privacy in disaster areas of Bangladesh (Islam and Rashid, 2011). However, most alarming is that the frequency of erosion attack is very high; families may face erosion and displacement up to 12 times in their lifetime (Islam et al. 2017). These families remain in constant risk of being homeless and acute poverty.

However, despite vast socio-economic losses and other severe impacts, cases of riverbank erosion do not get proper attention by the concerning authorities or the scientific society (Bhuiyan et al. 2017). There are disaster management plans at the broader level, but these are not sufficient and found to be ineffective at micro-level implementation. During the period of 2008–2018, the Ministry of Disaster Management and Relief constructed 653 shelters (cost 1965.772 million BDT or 25 million USD) in the flood and riverbank erosion-affected areas (MODMR 2018). The Government of Bangladesh and the Asian Development Bank approved USD $373.70 million for Flood and Riverbank Erosion Risk Management Investment Program (Loan from ADB USD $255.00 million, grant from the Government of the Netherlands USD $15.30 million and the Government of Bangladesh financed USD $103.4 million) in 2014 (ADB 2014). However, studies show that these efforts are insufficient and are not effective in helping the victims at personal level. Islam and his team (2017) revealed in their study that only 5% of the total assistance was received from the government and local Thana officials (Islam et al. 2017); a gap is easily noticeable (Islam and Rashid 2011). Erosion victims are widely unsuccessful in getting attention from the authority to receive help (Rana and Nessa 2017). So, this paper aims to focus on this gap by studying people's lives and their socio-economic conditions, under the constant threat of disaster being resident of the dynamic bank of River Meghna in Bangladesh.

Methodology, Methods and the Study Area

The study is located in the remote village Kalikanagar in Kachia Union of Bhola Sadar Upazila, Bhola (Figure 4.1). Kalikanagar village is situated at the active bank

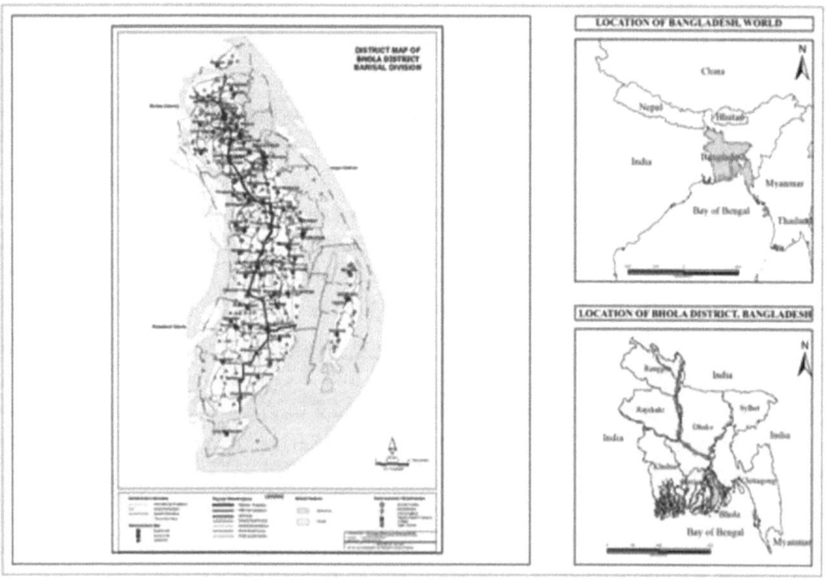

FIGURE 4.1 Location of Bhola District, Bangladesh

Source: Local Government Engineering Department (LGED 2020) (modified).

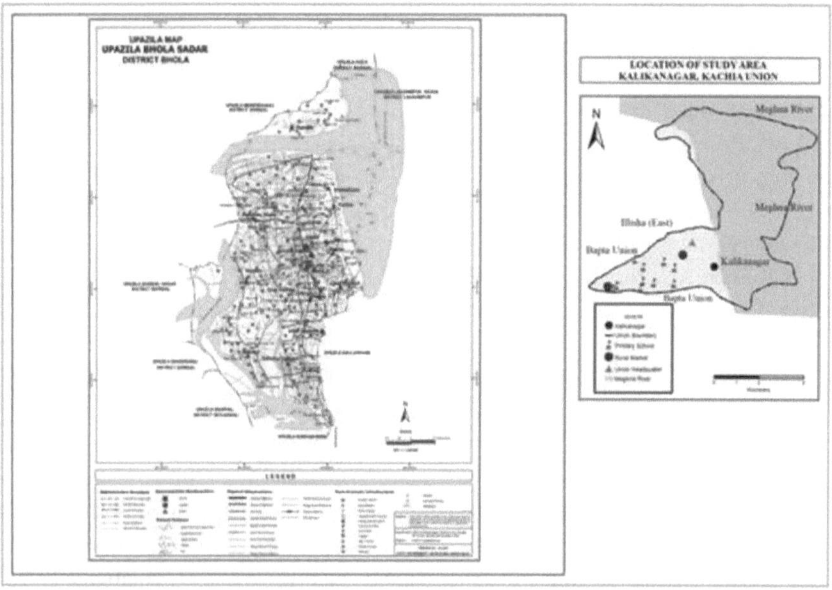

FIGURE 4.2 Location of the study area: Kalikanagar village

Source: Local Government Engineering Department (LGED 2020) (modified).

of river Meghna (Figure 4.2) and lost most of its land during the last 20–25 years. According to the Population Census 2011 there are 120 households in this village, though many of them are living in adjacent villages. There are 605 people in this village where 321 are male and 284 are female. Largest part of the population is in the 30–49 age groups. The average literacy rate (7+ years) is 12.2%. Most of the people earn their livelihood as fisherman. To understand the impacts of riverbank erosion on their lives, this study has followed mixed methodology and applied several methods: surveys at the households, key informant interviews (KII), focus group discussion (FGD) and observation. Seven KII were conducted among the local Upazila Chairman, member of the local government, officials of the Ministry of Disaster Management and Relief, local businessman, schoolteacher, student of postgraduate level and housewife. Ninety-three interviews were carried out at the household level. Sample size of the household survey was calculated applying statistical methods; sample size was 93 for 95% confidence level, 5% margin of error and 50% response rate (Raosoft 2019). Two FGDs were conducted with male and female local inhabitant groups. Quantitative data were analyzed by SPSS software and qualitative data were analyzed by thematic analysis methods. From the very beginning of the research, several visits to the study area helped to observe the inhabitants and their daily life closely. Observation was a strong method to understand the high impact of riverbank erosion on people's life.

Results Analysis

Riverbank Erosion in Kalikanagar Village and Displacement

Kalikanagar village has lost a large portion of its lands into the river Meghna. Most residents are severely affected by this natural event. They had to change their residence more than once, even up to 15 times to cope with this disaster. The study reveals that 56% of the families moved 1–5 times, while 35% had moved 6–10 times and 9% had moved about 11–15 times. Before even settling in one place, they have to move to another place. It shows the unstable and unsecured lives of the victims. They are on the move for years with their families, which include children (25.46%), older people (4.5%) and female members (48%).

Islam and Rashid (2011) also revealed similar results in their study area in Chapai Nawabganj and Rajshahi districts where erosion-hit families moved up to 12 times whereas Khan and his team found nine times in erosion-hit areas in Khulna (Khan et al. 2018). However, moving houses every time increases their social insecurity: theft, gendered harassments and child kidnapping. Displaced families work together and remain alert to ensure the security of their female members and children; however, some incidents happen in the period of instability (Islam and Rashid 2011).

Erosion-hit families of Kalikanagar village primarily move within the village when they lose their homestead. They try their best to stay within their village or move to the nearest village. The field study finds that at present about 85% of the families are living in the adjacent Bhabanipur village or left the area. Migration is a coping strategy for some families or for some members of the family; in this case, their destinations are mostly the nearest cities or mega cities like Dhaka or Chottogram. Islam et al. (2017) also revealed in their study that 73% of the people took to temporary migration and 27% migrated permanently to the town due to the Brahmaputra riverbank erosion (Islam et al. 2017). Destinations of their internal migration actually depend on their social connections, income sources, easy access and security (Islam and Rashid 2011; Bhuiyan et al. 2017; Mollah and Ferdaus 2015). However, changing residence needs extra money that increases the amount of loans, and has a high impact on their economic conditions.

Impacts of Riverbank Erosion on Household Economy and Poverty

Riverbank erosion deteriorates the local economy and increases poverty (Rahman et al. 2015). Inhabitants of the Kalikanagar village also experienced occupation change, increased loans and poverty along with losing agricultural lands into the river. Following sections present the situations created in the Kalikanagar village due to Meghna River erosion.

Changing Residence: Loans and Loanees

The present study reveals that most of the inhabitants (84%) of the Kalikanagar village are in debt (from 50,000 to 250,000 BDT or 624 to 3120.12 USD) due to the extra cost of changing residences: a huge amount of money is needed to change the living place. About 58% of the victims needed up to 500,000 BDT or 6240.25 USD, 27% needed 500,001–1,000,000 BDT or 6240.26 to 12480.5 USD and about 15% needed more than 1,000,000 BDT or 12480.5 USD. However, they have to take loan to meet this cost. About 85% of the victims took loan to change the places, which ranges from 20,000 to 700,000 BDT or 249.60 USD. They could not get loans from one source. About 61.29% of the residents took loans from NGOs, 58% from the relatives, *Mohajon* (locals who lend money with high interest), friends or neighbours and 42% from the banks or associations. However, it is alarming that 71% of loanees took loans with interest, which is very high in some cases. Loans for changing houses create double burden for the victims of riverbank erosion who lost agricultural lands and had to change their occupations.

Change in Occupation

The present study makes a comparison between the present and past occupations of Kalikanagar inhabitants. It reveals that people have changed their occupation from agriculture to other sectors after being affected by riverbank erosion. Nearly 25% of residents were farmers before the riverbank erosion whereas only 5.05% are employed in this sector now. Many wealthy farmers became marginal farmers or landless after the erosion. At present 46.46% of the residents work in the fisheries sector: mostly are employed in other boats (32.32%) whereas 36.29% worked in this sector before the erosion. Numbers of daily labourers, small businessmen and unemployment have also increased among the survivors (Table 4.1). Bhuiyan and his team revealed similar results from the Padma River erosion and Islam also revealed this change in occupation in their study in Mymenshing (Bhuiyan et al. 2017; Islam et al. 2017).

TABLE 4.1 Change in occupation

Occupation	*Present (Percentage)*	*Past (Percentage)*
Fishing (work on other's boat)	32.32	20.97
Fishing (work with own boat)	14.14	15.32
Business (small)	18.18	12.90
Business (medium)	7.07	9.68
Agriculture (work on others land)	4.04	9.68
Agriculture (work on own land)	1.01	15.32
Day labour/non-formal sector	16.17	8.87
Job	2.02	3.23
Unemployment	3.03	0.81
Others	2.02	3.22
Total	100	100

Source: Field study in 2019.

Increase in Poverty: Change in Landownership and Household Income

Riverbank erosion forced survivors to change their occupations several times to cope with the loss; however, these changes have strong negative impacts on household income (Islam and Rashid 2011). By comparing the present household monthly income and income before the riverbank erosion, the declination can be easily seen (Figure 4.3). Middle income group (10,000–20,000 BDT or 124.80 to 249.61 USD per month) have reduced at a mentionable rate: 38.54%–5.62%. Now, income of 86.52% of the families is below 10,000 BDT or 124.80 USD, which was 55.68% before the Meghna River erosion affected the residents of the village.

Poverty increased among the survivors of riverbank erosion. Per person income shows that 85.05% of the survivors earn less than 2,500 monthly (83.33 BDT or 1.04 USD per day) and 93.1% of them are living below the poverty line (Table 4.2) in Kalikanagar. National poverty rate was 29.5% in June 2020 (Dhaka Tribune 2020) and a total of 7,067,240 people were living in extreme poverty, which was 4% of the total population (6% live in rural and 2% live in urban areas) (World Poverty Clock 2020). Poverty has a direct impact on the land ownership status. At present, 66% families of Kalikanagar are living on rented or fellow lands (in their own village or in Bhabanipur village adjacent to the Kalikanagar village) and only 34% are living on their own lands.

Previous discussions reveal that riverbank erosion adversely affected the economic conditions of inhabitants of Kalikanagar village. Change in occupation, losing agricultural land, land and house rent costs and additionally big loans with

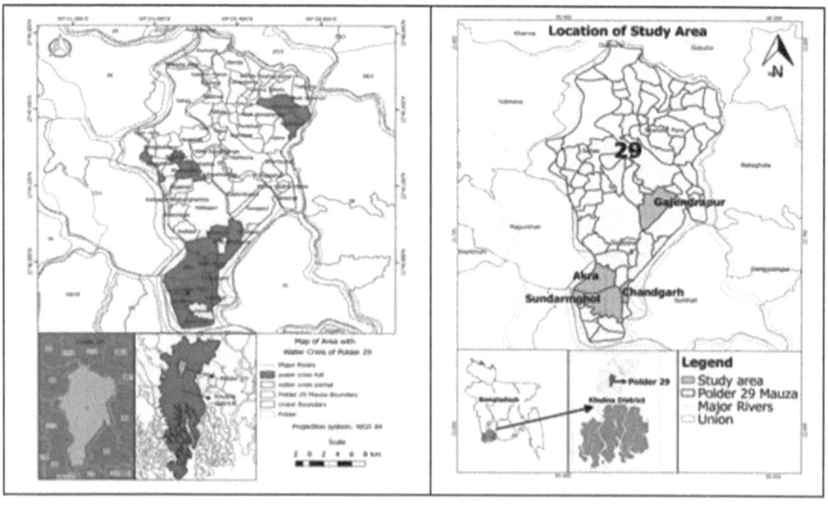

FIGURE 4.3 Change in household income

Source: Field study in 2019.

TABLE 4.2 Per person income and poverty level

Per month income per person in family (BDT)	*Number*	*Percentage*	*Poverty headcount ratio at $1.90 a day (PPP 2011)*
500–1,500	41	47.13	93.1% below poverty line
1,501–2,500	33	37.92	
2,501–3,500	7	8.05	
3,501+	6	6.90	
Total	87	100	

Source: Field study in 2019.

interest created huge economical pressure on the survivors of the riverbank erosion. This economic change highly affects the social status and mental health of the victims, which will be discussed in the following sections.

Social Impacts of Riverbank Erosion

Housing Condition

The housing condition of Kalikanagar village is very poor. About 93.54% families are living in katcha (made of wood, tin and bamboo) houses with very poor sanitation system. Beside poverty, due to repeated displacements people make their houses with light materials with low cost, so that they can move from places with low transport cost (Roy et al. 2017). However, many houses do not have toilets and especially no bathrooms, which is alarming. They take baths in the river. This is a common condition among the erosion-hit regions of Bangladesh (Islam et al. 2017; Rahman et al. 2015). However, housing without toilet and bathroom directly affects health conditions.

Education

Low family income, poverty, change in residence and unawareness create negative environment for education among the erosion-hit society. It has been revealed that 96.7% of the erosion survivors could not finish their primary education. Many of them (46.1%) could not attend school in their childhood. Only 3.3% of the residents could complete the Higher Secondary School Certificate level. However, many of the survivors mentioned that they did not want but had to leave their school to earn for the family and pay the loan. This low education level is highly related with unemployment, unsecured income and lower healthcare access (Islam et al. 2017).

Healthcare Access

Kalikanagar has a good transport system. It takes 20–25 minutes of automobile journey to reach the nearest city Bhola and government healthcare services. There are buses, *mahindro/auto* (engine-driven human hauler) available on this road. However, present study found that 77% inhabitants do not visit any healthcare centre for common health problems. They rather prefer to visit village doctors (who do not have MBBS degree), *Kabiraj* and also buy medicine over the counter from the local dispensary. Low service and transport cost, and close location, which allow them to walk to save travel cost, along with trust are the main reasons to choose these healthcare services. Only when in critical medical condition, they take patients to the government hospital, personal chamber and private hospital in Bhola city. This scenario shows the impact of poverty and unawareness among the residents. Again, most childbirth (93%) takes place at home in presence of untrained midwives and relatives. There were only five cases mentioned by the respondents during the fieldwork in 2019 where pregnant women were taken to Healthcare Centre or the village doctor visited.

Change in Address and Identities: Psychological Impacts

Riverbank erosion takes people's homesteads and their permanent address. However, losing their land means losing their address and identity. Survivors are collectively identified as the '*Nodi Bhanga Manush*' (people who lost everything in the river) or refugee by the locals. According to the erosion survivors, it is always very difficult to cope with the new location, occupations and environment. However, the hardest part is to face the social attitude towards the riverbank erosion victims. Following quotations present the social attitude towards victims after losing everything in the river.

> We are always treated as the floating people without any root, as if we are guilty for our fate ... losing home, losing everything.
>
> *Local Businessman, male, age 50*

> They always look at us with suspicion ... they do not believe us, do not want to employ us easily.
>
> *Student, male, age 20*

> It is difficult to arrange a good marriage for our daughters as we do not have a permanent address.
>
> *Housewife, female, age 40*

The above quotations illustrate the harsh reality when riverbank erosion victims take shelter in other villages or cities. They usually are not trusted by the local inhabitants where they resettle. Displaced people face relationship problems with the locals (Islam and Rashid 2011). It becomes difficult for them to get a job, use social facilities like shared ponds, toilets and open space as Bonano and team explained 'stress of disaster can erode sense of community' (Bonano et al. 2010). Inter-household conflicts increase (Islam and Rashid 2011). Present study reveals symptoms of psychological problems such as depression, anxiety, fear among the victims of riverbank erosion which are similar with Rahman and Gain's study located in the Koyra riverine area of Khulna district in Bangladesh (Rahman and Gain 2020). Amid these socio-economic and psychological problems, negative social attitudes can inflict more damage for the victims when surviving the disaster. As Harman (1997) mentions in her book, 'a supportive response from other people may mitigate the impact of the vent, while a hostile or negative response may compound the damage and aggravate the traumatic syndrome' (Harman 1997, p 61). Norris et al. (2008) and Bonano et al. 2010 also emphasize that 'community resilience', social relations during post-disaster period are important for success in coping and resilience. However, according to the present study, situations remain more difficult for the erosion-hit residents of the Kalikanagar village. It becomes a big challenge for them to cope with this harsh reality for the first few years. However, these conflicts slowly dissolve as time passes (Bonano et al. 2010). If they can improve their economic conditions, buy new lands and make a new address, situation improves. Nevertheless, this is impossible for most of the victims except very few families. Again, their identity as *Nadi Bhanga Manush* remains for the lifetime. Therefore, a disaster like riverbank erosion not only changes their economic conditions but also ironically changes their social status so drastically that it becomes impossible to make up the social loss. Many families expressed their worries, sadness and frustration in the present study whereas psychological effects of riverbank erosion rarely get attention in disaster management plans.

Recovery Plans and Resilience: Expectations of the Victims

The Government of Bangladesh declared riverbank erosion as a disaster in 1993. In May 2007, they planned strategies for allotment of general relief (food and cash) and financial help (cash) to build houses as relocation centres for the victims of cyclone, flood, fire, tidal wave, earthquake and riverbank erosion (Islam and Rashid 2011). At present, the Ministry of Disaster Management and Relief provides rice, cash money, blankets, dry food, children's food, animal food, metal roofing sheet and tents for temporary rehabilitation as relief for the riverbank erosion–affected people. Local government prepares a distribution list based on priority and reliefs are distributed accordingly. Widows, children, old people and extreme poor people get the priority.

However, according to the field study, no one has received any financial help from the government or any NGO in the Kalikanagar village. They did not receive corrugated iron (CI) sheet because their land is lost into the river, and they missed the general relief (cash) due to absence of fatality. According to government policy, one needs to have their personally owned land for house financing and metal roofing sheets, and they have to be living in the erosion-hit regions. Again, erosion victims often move to other administrative units, which create problems to locate them. Local administration of their origin cannot help them. Some of the victims in the Kalikanagar village received rice for the first few days after the disaster.

However, the amount is never enough to combat the severe impacts of riverbank erosion (Islam and Rashid 2011).

Again, the Ministry of Disaster Management and Relief does not allocate government land for riverbank erosion-affected people, the Ministry of Land does it. There is no loan facility from the government for riverbank erosion-affected people and there is no separate donor for riverbank erosion. NGOs like the Oxford Committee for Famine Relief (OXFAM), Rangpur Dinajpur Rural Service (RDRS), CARE-Bangladesh and Bangladesh Disaster Preparedness Centre (BDPC) work with displaced people to help the situation in certain regions of Bangladesh (Islam and Rashid 2011). This effort is insufficient for the large number of affected populations. Riverbank erosion also does not receive the proper attention of public, media and the government. Lack of information and awareness also become constraints for the erosion victims to get the help.

Present study makes some recommendations with the help of the survivors of the Kalikanagar village to improve this condition. The first thing Kalikanagar residents want from the government is to take necessary steps to build permanent accommodation for the erosion-hit families or allocate lands on the new river islands to the owners who lost lands. The process should be lawfully accountable and transparent (Roy et al. 2017; Mollah and Ferdaush 2015; Islam and Rashid 2011). Government or NGOs should plan for easy loans without interest and flexible time for the fishermen or farmers to help them recover their loss. Scholarships or grants could be arranged for the students of the displaced families. There should be programmes directed towards aiding them in healthcare facilities, education and vocational trainings to improve local disaster–coping capacity. Mass media can play an important role to create awareness on the severity of the riverbank erosion by broadcasting programmes and inspire the nation to participate in charity programmes. Above all, according to local residents, if the government takes structural measures like river training and riverbank strengthening with concrete blocks and sandbags or other appropriate measures to protect the riverbanks from erosion, they will not lose their homesteads, lands and address.

Conclusion

Riverbank erosion is a slow and silent disaster with a devastating nature, which damages lands, homes, other properties and changes people's life permanently. People become displaced from their origin. It changes people's identity along with changing socio-economic conditions. Solvent families are degraded to the poor group through losing everything into the river while the return remains uncertain and unsecured. Dependency on agricultural lands, lower livelihood opportunities in the region, lack of education and skills make recovery difficult for the erosion-hit families. They are collectively called *Nadi Bhanga Manush*, which has severe social and psychological impacts. Erosion victims need economic, health and mental support for recovery. However, it is a matter of concern that the response plan very insufficiently provides relief to retaliate this disaster. Survivors of riverbank erosion fight against this severe disaster on their own (Islam et al. 2017; Islam and Rashid 2011) and most of the time they cannot recover. Poor housings, lower education and healthcare accessibility with increasing insecurity and frustration among the erosion-hit residents in the studied Kalikanagar village show the alarming situations. These conditions raise questions, why these victims are fighting against disasters on their own. Present study reveals that there is a huge gap in understanding severe impacts of riverbank erosion on the victim's life, which is often misunderstood by comparing them with other disasters, even underrating them. As riverbank erosion does not occur with a large scale like cyclones and floods, fatality and injury are very low in this disaster, and there are less dramatic actions and uncertainty in the time of occurrence, which create ignorance among the disaster managers and the rest of the nation. This disaster does not catch the attention of international donors or charity organizations like the other disasters. While riverbank erosion silently changes socio-economic conditions and there is no help from the concerned authority and host community, feelings of isolation and frustration are created among the displaced people. To improve their lives, proper attention from the government, NGOs and voluntary organizations is essential. Mass media can play an important role in enhancing social awareness on this disaster. It is expected that along with effective disaster management plans and their proper implementation, change in social attitude and awareness can help in increasing community resilience and save people's lives and well-being from this devastative disaster.

References

ADB (2014). Proposed multi tranche financing facility, People's Republic of Bangladesh: Flood and Riverbank Erosion Risk Management Investment Program. Asian Development Bank. www.adb.org/sites/default/files/project-document/81254/44167-013-rrp.pdf, accessed 29/03/2020.

Akter, K., Dey S., Hasan, S. (2019). Riverbank erosion and its impact on rural women: Case study of Ulania village in Bangladesh. *Asian Journal of Women's Studies* 25(1):76–95. https://doi.org/10.1080/12259276.2019.1577343

BBS (2016). Bangladesh Disaster-related Statistics 2015: Climate Change and Natural Disaster Perspectives. Bangladesh Bureau of Statistics, Government of the People's Republic of Bangladesh.

Bhuiyan, M. A. H., Islam, S. D. U., Azam, G. (2017). Exploring impacts and livelihood vulnerability of riverbank erosion hazard among rural household along the river Padma of Bangladesh. *Environmental Systems Research* 6(1):25.

Bonanno, G. A., Brewin, C. R., Kaniasty, K., Greca, A. M. L. (2010). Weighing the costs of disaster: Consequences, risks, and resilience in individuals, families, and communities. *Psychological Science in the Public Interest* 11(1):1–49.

CEGIS (2009). Prediction of riverbank erosion along the Jamuna, the Ganges and the Padma Rivers. Center for Environment and Geographic Information Services. Dhaka, Bangladesh.

Das, B. (2011). Stakeholders' perception in identification of riverbank erosion hazard: a case study. *Natural Hazards* 58(3):905–928. https://doi.org/10.1007/s11069-010-9698-z

Dhaka Tribune (2020). Covid-19 impact: National poverty rate rise to 29.5 as of June. Dhaka Tribune, www.dhakatribune.com/health/coronavirus/2020/08/12/covid-19-impact-national-poverty-rate-rises-to-29-5-as-of-june, accessed on 11/08/2021.

Elahi, K. M. (1991). Riverbank erosion, food hazard and population displacement in Bangladesh: An overview. In Elahi KM, Ahmed KS and Mafzuddin M (eds.) *Riverbank erosion, flood and population displacement in Bangladesh*, Dhaka, Riverbank Erosion Impact Study (REIS), Jahangirnagar University, pp 364.

Herman, J. (1997). *Trauma and recovery: The aftermath of violence – from domestic abuse to political terror*. New York: Basic Books.

Hutton, D., Haque, C. E. (2004). Human vulnerability, dislocation and resettlement: adaptation processes of river-bank erosion-induced displaces in Bangladesh. *Disasters* 28(1): 41–62.

Islam, M. A., Parvin, S., Farukh, M. A. (2017). Impacts of riverbank erosion hazards in the Brahmaputra floodplain areas of Mymensingh in Bangladesh. *Progressive Agriculture* 28(2):73–83.

Islam, M. F., Rashid, A. B. (2011). Riverbank erosion displacees in Bangladesh: Need for institutional response and policy intervention. *Bangladesh Journal of Bioethics* 2(2):4–19.

Khan, B., Nabia, S. A., Rahman, M. A. (2018). The effect of riverbank erosion on lives and livelihoods of rural people: A study on Nolian village, Khulna, Bangladesh. *Journal of Science, Technology and Environment Informatics* 6(01):466–473.

LGED (2020). District/Upazila Map. Local Government Engineering Department. Ministry of Local Government, Rural Development and Co-operatives. Government of the People's Republic of Bangladesh. https://oldweb.lged.gov.bd/ViewMap.aspx, accessed on 20/01/2020.

MODMR (2018). Success story of last 10 years. Ministry of Disaster Management and Relief. Government of the People's Republic of Bangladesh. https://modmr.gov.bd/site/publications/d366f36d-27a7-496a-9d9e-fe1511ebfb49

Mollah, T. H., Ferdaush, J. (2015). Riverbank erosion, population migration and rural vulnerability in Bangladesh: A case study on Kazipur Upazila at Sirajgonj District. *Environment and Ecology Research* 3(5):125–131.

Norris, F. H., Stevens, S. P., Pfefferbaum, B., Wyche, K. F., Pfefferbaum, R. L. (2008). Community resilience as a metaphor, theory, set of capacities, and strategy for disaster readiness. *American Journal of Community Psychology* 41(1–2):127–150.

Rahman, M. R. (2010). Impact of riverbank erosion hazard in the Jamuna floodplain areas in Bangladesh. *Journal of Science Foundation* 8(1–2):55–65.

Rahman M. A. T. M. T., Islam, S., Rahman, S. H. (2015). Coping with flood and riverbank erosion caused by climate change using livelihood resources: A case study of Bangladesh. *Climate and Development* 7(2):185–191.

Rahman, M. S., Gain, A. (2020). Adaptation to riverbank erosion induced displacement in Koyra Upazila of Bangladesh. *Progress in Disaster Science* 5:100055. https://doi.org/10.1016/j.pdisas.2019.100055

Rana, M. S., Nessa, A. M. (2017). Impact of riverbank erosion on population migration and resettlement of Bangladesh. *Science Journal of Applied Mathematics and Statistics* 5(2):60–69.

Raosoft (2019). Sample size calculator. www.raosoft.com/samplesize.html, accessed on 02/02/2019.

Roy, D. K., Goswami, S., Ahmed, T., Saha, M. K., Emon, M. H., Rahim, M. A. (2017). Socio-economic impact of riverbank erosion on Durgapasha Union in Bakerganj Upazila, Bangladesh. *Barisal University Journal* 4(1):165–183.

World Poverty Clock (2020). People live in extreme poverty around the World. *World Data Lab.* https://worldpoverty.io/map, accessed on 29/03/2020.

PART II

Disaster, Water Insecurity and Management

5

WATER INSECURITY AND POVERTY

Changing Livelihood in Coastal Bangladesh

Monishankar Sarkar and Mahbuba Nasreen

Introduction

What causes poverty? It is still difficult to answer this century-old question in a single sentence. There is still no consensus about how poverty generates and ends. There is no single explanation of poverty because of its geographical and context-specific multidimensional nature. A single indicator is insufficient to explain the reason behind poverty. Poverty is instead a many-headed evil. It is hard to reach in consensus about the universal reasons for poverty that cause it across the countries. Poverty is very much context-specific, and therefore, the drivers differ significantly across countries. Rural livelihoods in developing countries are primarily agriculture-based. Good agricultural production is dependent on the availability of fresh irrigation water and humidity. Local agriculture practices often face water-borne challenges, which in turn amplifies poverty. Hence, it is crucial to understand the influence of poverty drivers.

It is hard to find households in a rural setting that are not connected with farming. Households usually practice farming as an essential strategy for their survival. The existing water-poverty connections can be understood when water-related hazards and challenges affect the household subsistence mechanism. Some households are engaged in various activities, but they still have an immediate or close connection with farming. Most of the households engage with agriculture for their subsistence. The occupations that are directly or indirectly related to agricultural production may be affected by waterborne challenges. The agriculture-based livelihood practices become affected by sudden-onset disasters (such as flood, embankment erosion, collapse and or breaching, and so on) or slow-onset disasters (salinity, gradual decrease in water and soil quality due to many waterborne factors

DOI: 10.4324/9781003253495-7

other than salinity). Both have a dramatic and negative impact on the agricultural fields and soil fertility, which threaten agriculture-based livelihoods.

Waterborne challenges and disasters caused by climatic alterations have created a powerful negative influence on the agricultural production and vulnerability context in many parts of South Asia (Orchard et al. 2016; Abdullah et al. 2017). The coastal communities in the South Asian region have been experiencing many challenges in recent decades due to various environmental transformations, increased aquaculture practices and industries, frequency and severity of environmental stresses and shocks (Abdullah et al., 2017; Orchard et al., 2016). Households often try to cope with these changes and reshape their livelihood choices based on available opportunities (Nasreen, 2012; 2019). Many factors (broadly covered biophysical, knowledge, financial, with less mentioned political, social and psychological barriers) interplay in such coping mechanisms and the households' coping capacity to keep pace with the changes (Shackleton et al., 2015).

This research judges the poverty situation with the water-related drivers and the other causes associated with poverty. This paper visits and explores the most frequent reasons that have been causing poverty for decades. Based on poverty literature and observed facts, this paper explores the drivers of poverty in coastal communities in Southern Bangladesh. This research is conducted with a view that this can help develop a broader understanding of how drivers of poverty are connected and influenced by waterborne challenges. Based on the empirical evidence and observed facts, this study explores how the drivers of poverty have been deeply affected by waterborne challenges over the years and how those challenges augment the poverty among poor inhabitants.

Study Area and Research Design

Study Area

The study location, as Figure 5.1 depicts, is in an 'active deltaic floodplain' situated at the lower reaches of the Ganga-Brahmaputra-Meghna catchment, and the exposure is high in pre- and post-monsoon phases. The study was conducted in three villages from Khulna, a coastal district in Southern Bangladesh, facing different waterborne challenges. Both Chandgarh and Baro Aria are highly exposed to waterborne difficulties, whereas the third village named Gajendrapur has successfully managed the waterborne challenges and experienced the waterborne challenges differently. The three villages studied in the research comprise 1,347 households (Statistics, 2011; Berman et al., 2012).

Data Collection and Analysis

The study was designed with a mixed-method approach that collected quantitative and qualitative empirical evidence. The quantitative part, composed of the household questionnaire survey, collected data from 384 households and tries to understand the linkage between water insecurity and livelihood decisions. The qualitative techniques included focus group discussions (FGDs), in-depth interviews (IDIs), key informant interviews (KIIs), semi-structured interviews (SSIs), case study and observation. The data collection was completed in several stages between January 2016 and May 2018.

This paper explores less quantitative data and more qualitative information on different drivers of poverty and water-poverty analysis. The households for the quantitative and qualitative survey were selected following a random walk sampling strategy, and consent was taken from each respondent before the survey. While conducting the survey, the heads of the households were the primary respondents and, on some occasions, other members were permitted to take part and discuss for the sake of drawing actual data and for intra-household verification. The survey questionnaire was concentrated on the household location, demography, water and sanitation, drinking and irrigation water source and accessibility, exposure, poverty condition, household power source, priority concerns, concerns regarding water and natural environment, land ownership and agriculture, income sources, crop production, homestead land livestock, aquaculture, hazard scenario, water-related risks and its impact on livelihoods, and regarding water management group.

Measuring Poverty through a Multidimensional Lens

This section explores an understanding of poverty in terms of the Multidimensional Poverty Index (MPI) – a measurement and analysis developed by Oxford Poverty and Human Development Initiative (OPHI) and United Nations Development Program (UNDP) jointly to calculate the Human Development Index (HDI). This section explores the condition of rich and poor, in terms of the HDI measurement scale, to better understand water-poverty relations in water-stressed and water-managed villages.

Education has a central role in poverty alleviation (Paraschiv, 2017; Janjua & Kamal, 2014; Kiani, 2011) and helps the members of the poor households to lift out of poverty. The collected information shows that the 'years of schooling' criteria are broadly fulfilled by the school-going children in most households. Years of schooling and school attendance, as Table 5.1 portraits, are powerful indicators of poverty situation of a given area. Higher school dropouts are significantly associated with the long-term poverty. There are very few cases where adults or parents fulfil the requirements. In most cases, the parents have little or no education. This is particularly true in both water-stressed and less water-stressed (or water-managed) villages. The change is not homogenous to both less and more water-stressed villages. The decrease in school attainment is persistent in water-stressed villages, but the situation is better or satisfactory in water-managed villages. Like the impacts, the reasons are also heterogeneous. The situation is changing at

TABLE 5.1 MPI indicators and their connection with field data (years of schooling and school attendance)

MPI measurement criteria	*Relevant findings from study*
1. Years of schooling: a household is deprived if no household member has completed six years of schooling	In both water-stressed and less stressed villages, this category is fulfilled by kids, most of the cases, and by parents also in few cases Stipend has changed the scenario (increased attendance in schools) and awareness level
2. school attendance: a household is deprived if any school-aged child is not attending school up to class 8	Water-stressed villages Large number of dropouts, usually starts after primary schooling Attendance of girls are remarkably low Water-managed village Attendance is satisfactory, as guardians are willing Less dropouts, attendance of girls is high

a slower pace in water-stressed villages, and the main reason is the introduction of the stipend in government schools. School attainment in water-managed villages is improving. The key reason includes parents' awareness regarding their children's future and the liberty of children from engaging in income-generating activities (IGAs) for family subsistence.

In the water-stressed villages, people tend to involve their school-aged children in various IGAs rather than sending them to schools. Parents see no benefit of sending them to schools except creating an extra burden of education expenses for the families. Parents see only immediate benefits by engaging themselves in various IGAs available within the areas (like collecting shrimp larvae, working in their own fields, helping parents by doing household chores when they are at work, etc.). Girls' school attendance is meagre in water-stressed villages. The girls are primarily engaged in household chores, helping their mothers while at work. Early marriages of girl children are notably high in stressed villages. The scenario is the opposite in less water-stressed (or water-managed) villages. The well-managed drinking and irrigation water security liberates them from focusing on areas other than subsistence farming. They endorse schooling for their kids, help them to pursue a college education. Consequentially, the rate of school dropouts is very low, and early marriage is very rare. The attendance is satisfactory for both boys and girls. Girls are likely to attend college similar to boys after successfully completing high school education.

Both child mortality and nutritional status indicate a community's health condition in an area. On the one hand, the death of a child below five years of age demonstrates severe health care inadequacy and support. On the other hand, the

TABLE 5.2 MPI indicators and their connection with field data (child mortality, nutrition, electricity and sanitation)

MPI measurement criteria	*Relevant findings from study*
3. Child mortality: deprived if any child has died in the family in the past 5 years	Qualitative data available/no official record
4. Nutrition: deprived if any adult or child, for whom there is nutritional information, is stunted	Qualitative data available/no official record

nutritional status of a household is difficult to measure and indicates an overall healthy arrangement of a family or household.

In the last five years, child mortality in both water-stressed and less water-stressed villages is not reported. This fact fills the third criteria of MPI, and at the same time, provides inadequate data on the nutritional status of the households surveyed. The objective of this research did not include the collection of broader input of quantitative data on the nutritional measurement information from families, and therefore the study lacks such information. The qualitative data are not adequate to fulfil this wider measurement field. Usually, the households facing acute poverty often fail to meet the basic standards of nutritional criteria.

As MPI considers, the power source of a household mentions the opportunities for improved living standards. There is more chance of improvement if a household enjoys an improved power source. If, for example, a household remains under electricity coverage, it can ensure services that require electric supply. The households are facing scarcity of electricity in the study villages. For water-stressed villages, nearly half the households of Chandgarh, and more than two-thirds of the villages of Baro Aria are beyond the electricity-covered area. The proportion is worse in the less water-stressed village, where only one-third of the households are under electricity coverage. Among these three villages, the coverage is partly disrupted in Chandgarh because of the breaching problem. The households deprived of electricity coverage depend on kerosene oil as their lighting source; leaves, straws, wood and dung as their cooking fuel source; hand- or peddle-driven technology as their irrigation source. They cannot afford or depend on any electric technologies required in everyday lives. This inability forces them to live in poor and substandard living.

The quality of sanitation falls under fundamental health issues in any measurement for the standard of living. In the MPI, both the questions were raised – whether sanitation is standard and whether it is shared with other households or not. Both are measured to identify the living standard in terms of sanitation

TABLE 5.3 MPI indicators and their connection with field data

MPI measurement criteria	*Relevant findings from study*
5. Electricity: deprived if the household has no electricity	Baro Aria: 67.74% HHs are poor (coverage: 32.26% HH) Chandgarh: 44.03% HHs are poor (coverage: 55.97% HH) Gajendrapur: 66.49% HHs are poor (coverage: 33.51% HH)
6. Sanitation: deprived if the household's sanitation facility is not improved (according to MDG guidelines), or it is improved but shared with other HHs	Baro Aria: 71.64% HHs are poor Chandgarh: 88.71% HHs are poor Gajendrapur: 87.23% HHs are poor
7. Drinking water: deprived if the household does not have access to safe drinking water (according to MDG guidelines) or safe drinking water is more than a 30-minute walk from home roundtrip	Water-stressed village Baro Aria: 100% HHs are poor/deprived Chandgarh: 100% HHs are poor /deprived Water-managed village Gajendrapur: 100% HHs are benefited

Source: This calculation was based on primary data collected within the 2016–2019 period.

services. If sanitation is not standard enough, it may cause various health problems. Toilet sharing is a prevalent scenario in study villages, and the less water-stressed villages are no longer different from the water-stressed village. Both the worst-stressed Chandgarh and least-stressed Gajendrapur are similar, where nearly 90 percent of households share their toilets with other families. The scenario is not different for different wealth classes, and the sharing is practiced highly across the classes. Very few extreme poor and wealthier families do not share toilets with other families.

Like many other parts of the world, rural Bangladesh's housing structure and housing conditions often portray a household's poverty and socio-economic status. In all the villages studied, clay houses are predominant, referred to as kachabari. Many affluent families have improved dwellings with a concrete structure, including the wealthier families in stressed villages. Corrugated tin is a common material for houses with a minimum standard of living. Most families, even some extreme poor households, also try to use corrugated tin in making their house roofs because the traditional material for preparing house roofs, usually straw and *golpata*[1] are no longer readily available. More than 80 percent of the houses of Baro Aria and Gajendrapur are made with an earthen floor. The dwellers of all these households could be termed deprived in accordance with the standards of MPI measurement.

Fuel material is also an important indicator of living standards in any household or community. If the fuel materials are not standard enough, it may invite serious health hazards, including various respiratory complications. It is observed that nearly all three villages' households use substandard raw materials as their cooking

TABLE 5.4 MPI indicators and their connection with field data

MPI measurement criteria	*Relevant findings from study*
8. Housing: deprived if the household has a dirt, sand or dung floor	Water-stressed village Baro Aria: 80.60% HHs are poor Chandgarh: 75.81% HHs are poor Water-managed village Gajendrapur: 80.85% HHs are poor – these HHs have earthen floor
9. Cooking fuel: deprived if the household cooks with dung, wood or charcoal	Water-stressed village Baro Aria: 98.51% HHs are poor/deprived Chandgarh: 96.77% HHs are poor /deprived Water-managed village Gajendrapur: 100% HHs are poor/deprived
10. Assets ownership: deprived if the household does not own more than one of: radio, TV, telephone, bike, motorbike or refrigerator and does not own a car or truck	This calculation is better understood in PCA, which is done for this research in order to determine the five wealth classes

fuel, including fuel wood, cow dung, straw, leaves, etc. Such materials may cause a health hazard for pregnant women, girls and children of different ages (Biswas & Lucas, 1997; Briscoe, 1979).

Asset allocation is one of the concerns of this study. Therefore, it applied to a range of questions related to assets to record them carefully to produce clear statements regarding each household's asset condition and asset poverty scenario. Based on the asset recorded in the household questionnaire survey, the present study plotted all the studied households into five wealth classes with the help of principal component analysis. The five wealth classes are rich, upper-middle, lower-middle, poor and extreme poor (as presented in Table 5.5). The livelihoods of all five classes have been recorded with particular attention to the trajectories of change, Table 5.4 shows. The agriculture-based livelihood was the key concentration for this study. It is observed how agriculture-based livelihood has been changing with significant environmental change in the area. The asset profile is primarily influenced by the key livelihood options of households, with the amelioration of traditional agriculture-based livelihood, households made significant advancement in terms of asset accumulation and other forms of improvement as part of better standards of living.

Livelihoods, Employment Opportunities and Precariousness

Unemployment, Partial Employment, Precarious Jobs

The water-related drivers, as Figure 5.2 portraits, are connected to poverty in multiple ways. Households may fall into poverty if they lose the key livelihood options (Olsson et al., 2014; Ellis & Mdoe, 2003; Duncan et al., 2019; Antwi-Agyei et al., 2014; Assan, 2014) and have less or no means to diversify their livelihoods (Oyinbo & Olaleye, 2016). Households that have more livelihood diversification capacity (Igwe, 2013), but the people who have less diversification options face difficulties for their subsistence and tend to fall into long-term poverty problems (Igwe, 2013).

The livelihood strategies vary, as Table 5.5 depicts, in line with asset ownership. The local farmers and agricultural labourers in the water-stressed villages face a

TABLE 5.5 Wealth indicators and livelihood strategies of different wealth classes in the three study villages

		Rich	*Upper middle*	*Lower middle*	*Poor*	*Extreme poor*
Wealth indicators and livelihood strategies of wealth classes in Baro Aria		n=10	n=14	n=23	n=47	n=40
Percentage of wealth class		7.46	10.45	17.16	35.07	29.85
Mean agricultural land (decimals)		959.5	701.07	255.04	93.5	99.6
Mean homestead land (decimals)		124.5	62.15	29.25	19.30	9.20
Mean pond area (decimals)		40.25	28.21	22.15	4.25	1.05
Education (High School Certificate/above)		43.15	25.12	20.5	13.3	3.15
Income from different sources within last 1 year	Agriculture	50.00	57.14	34.78	57.45	37.50
	Aquaculture/ pond	60.00	35.71	39.13	38.30	37.50
	Business	50.00	50.00	43.48	23.40	12.50
	Casual labour	0.00	0.00	21.74	61.70	80.00
	Livestock/poultry	50.00	35.71	13.04	25.53	25.00
	Property rent	60.00	57.14	39.13	6.38	12.50
Wealth indicators and livelihood strategies of wealth classes in Chandgarh		n=3	n=5	n=11	n=24	n=19
Percentage of wealth class		4.84	8.06	17.74	38.71	30.65
Mean agricultural land (decimals)		1299.7	720.2	239.4	190.5	42.8

TABLE 5.5 Cont.

		Rich	*Upper middle*	*Lower middle*	*Poor*	*Extreme poor*
Mean homestead land (decimals)		104.25	43.30	22.25	7.15	2.15
Mean pond area (decimals)		41.40	34.25	12.50	7.40	2.60
Education (High School Certificate/above)		43.50	38.50	33.60	11.25	3.50
Income from different sources within last 1 year	Agriculture	100.00	100.00	63.64	33.33	42.11
	Aquaculture/pond	66.67	60.00	54.55	25.00	5.26
	Business	0.00	0.00	18.18	8.33	10.53
	Casual labour	0.00	0.00	45.45	54.17	94.74
	Livestock/poultry	0.00	0.00	9.09	8.33	0.00
	Property rent	33.33	0.00	0.00	8.33	0.00
Wealth indicators and livelihood strategies of wealth classes in Gajendrapur		n=14	n=21	n=31	n=53	n=69
Percentage of wealth class		7.45	11.17	16.49	28.19	36.70
Mean agricultural land (decimals)		367.43	266.52	214.35	112.57	175.79
Mean homestead land (decimals)		145.6	125.5	85.6	29.5	12.6
Mean pond area (decimals)		43.5	28.6	22.6	15.2	2.05
Education (High School Certificate/above)		45.8	29.6	21.5	15.6	3.8
Income from different sources within last 1 year	Agriculture	50.00	52.38	77.42	83.02	57.97
	Aquaculture/pond	0.00	19.05	9.68	18.87	7.25
	Business	35.71	28.57	32.26	22.64	11.59
	Casual labour	0.00	0.00	32.26	54.72	71.01
	Livestock/poultry	50.00	61.90	67.74	58.49	60.87
	Property rent	64.29	61.90	25.81	13.21	5.80

Source: Household Survey, 2016–2018.

lack of employment opportunities several times of the year. The households fail to maintain savings; most of their earnings are expended for their subsistence. After the cultivation period, most of them face a dearth of employment opportunities. In the post-harvesting period, they are to face the same challenges. To cope with the working opportunity shortage, they often leave their villages and go to distant places for work. Many of them seasonally migrate to urban centres and do various manual odd jobs there. Some families also seasonally migrate to work in the brickfields.

The lack of employment opportunities has created many challenges for the members of poor and extreme poor households. Poor people often fail to ensure family consumption if they cannot guarantee livelihood opportunities. The situation is almost similar for both water-stressed villages. Failure to ensure employment opportunities enhances the chances of intergenerational poverty. It is necessary to raise the income of the poor households to eliminate their poverty, which will result from employment generation and the continuation of the existing livelihood opportunities.

Little or No Access to Livelihoods or Decent Jobs

It is pretty frustrating for households if they cannot continue or sudden inaccessibility to the livelihood options they usually follow. Households may fall into poverty if they do not have enough access to livelihood options or if they have less choice to do work to maintain their families; lack of access to livelihood options causes poverty (Mumuni & Oladele, 2016; Larson et al., 2007; Nasreen 2007; King, 2011; Barau et al., 2019; Barrett et al., 2001) and may create long-term livelihood insecurity (Geiser et al., 2011) and intergenerational poverty. Poor people who faced sudden livelihood failure and meagre income opportunities also faces long-term economic hardship (Sarkar & Kulkarni, 2022).

Ownership of land is often considered a defining criterion of poor and non-poor in the villages of South Asia. But this is not always the indicator, as the study villages bear the perfect examples for that. The landownership in Chandgarh, for example, does not necessarily mean that they are in a better position on the wealth indicator instead, some people have agricultural land but cannot cultivate, as saline river water visits recurrently their fields when the tide is high. Another reason is salinity. Both problems are exacerbated by recurrent embankment breaching in Baro Aria and Chandgarh villages.

Because of asset and income poverty, a section of Baro Aria and Gajendrapur does not have access to their traditionally agriculture-based livelihoods. They do not have the required capital, cannot manage any guarantor to get loan and therefore try to lease some land and invest there. To cope with the situation and for family maintenance, many poor households choose various non-farm jobs throughout the year. Many engage in agricultural employment for three to four months a year and are involved in non-farming activities for the rest of the year.

Non-farming jobs require additional knowledge and skills, which the members of poor families often fail to ensure. It is observed that the affluent families can afford better schooling for the children and make them competent for decent jobs (like teaching in local schools, having some medical orientation to become local doctors, working as development workers in local NGOs). Wealthy families often send their children to nearby cities after getting primary or high school education from the village. There are better job opportunities in the city than in villages.

Even the people (of water-stressed villages) suffer from a lack of information and are often misguided by wrong information. Sometimes, lack of information pushes them towards a devastating future. Students, getting no future hope or betterment, often lose motivation to carry on their studies. Consequently, school dropouts rise, parents become reluctant to support their children's studies. Parents often see benefits in recruiting their children in locally available IGAs from early childhood.

> I have completed my graduation from Govt. Brajamahan College (a renowned college in Khulna division[2]), and returned to the village, because I have heard from many of my friends and familiar ones that candidates who have relatives in high position, or have money to invest [as bribing] can get a job. I have neither relatives nor money, so I did not try for a job. I have returned village after my degree is completed.
>
> *A graduate student in Chandgarh, working as a day labourer*

The scenario described so far is the opposite in Gajendrapur. The villagers send their children to schools and support them to continue. Like their kids, the parents also feel competition with their neighbours' children. This competition is about doing better results and receiving higher degrees.

> Education is the topmost priority for our children. We try our best for the education of our children. The situation is not like this all the time. After we have managed to draw three or more crops from the same land in a year, parents have become serious about educating their children.
>
> *A farmer, interviewee, Gajendrapur*

Besides the employed members of households, the rest of the family members are just consumers or jobless consumers. What are the reasons for those cases? This can include the age barrier (under-aged child, over-aged); girls are not allowed to work to maintain their chastity or for security reasons, and parents usually do not permit their children to work outside.

Access to Agriculture, Food and Water

Food and Agricultural Issues

If poor households do not have access to agricultural practices, that may result in poverty (Osabohien et al., 2021), which creates food security. The access to food and fresh water improves the living condition of the household members, and the

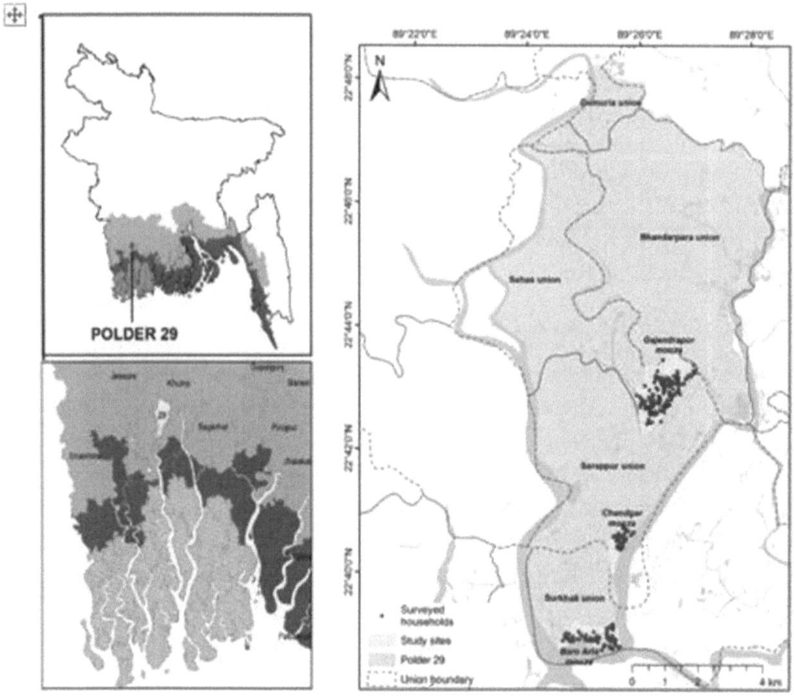

FIGURE 5.1 Country map and polder 29 location map

lack of those augments' poverty condition and suffering (Cohen & Sullivan, 2010; Nasreen, 1995, 2009). Both rich and middle farmers practice agriculture within the study villages. The wealthy farmers have enough capital, invest in farming and practice independently and lease in lands in their own and neighboring areas. Among the less wealthy category, people who have the capacity to pay the landowner can get land on lease. It is observed that most of the households of water-stressed villages fail to share a fair amount of nutritious food, but the situation is completely different in water-managed villages. Sufficient production of crops and vegetables lets people get more food for their families. An adequate amount of vegetables are grown within each homestead in water-managed villages, whereas the nearby local market of water-stressed villages lacks the provision of a sufficient amount of vegetables. The lack of provision may indicate a severe nutritional imbalance in the water-stressed villages.

Inadequate Access to Clean Water and Nutritious Food

These are the significant indicators of poverty measurement. People who fail to ensure fresh food and clean water for themselves face several challenges. Water poverty occurs when a household or a community does not have accessibility, availability and acceptable level of quality water (Grey & Sadoff, 2007). This water may be for drinking, household use or agriculture. When a community or household fails to ensure an acceptable level of quality water, they are considered water poor. According to this definition, the people of two of the water-stressed villages can be regarded as water poor, as they lack sufficient drinking water and water for household use and agriculture throughout the year. As discussed earlier, water is a significant concern for the water-stressed villages Chandgarh and Baro Aria. The people of Gajendrapur have adapted to the situation. Child mortality and nutrition data, mentioned in Table 5.2, are also significant indicators of multidimentional poverty but these indicators were beyond our concentration here. Table 5.3 depicts that the drinking water, sanitationand lifeline service like electricity also varies in a great scale between water stressed and water managed villages. Water poverty creates various challenges for the poor and insolvent families struggling for long to get out of poverty. They may fail to produce agricultural items, and insufficient or lack of quality drinking water may create various health problems.

Food poverty is associated with both poverty and healthy and nutritious food (Dowler & Connor, 2012; Molcho et al., 2007; Lang & Caraher, 1998). Such poverty may be created if a household cannot purchase food, afford food required for a healthy diet or if the household members suffer from a lack of required food. This involves both the quality and quantity of food necessary for a healthy life. The study findings suggest that the wealthier and solvent families in water-stressed villages can ensure sufficient food for their families, whereas the poor and extreme poor families often fail to do so. The more affluent families do not face any food poverty throughout the year, whereas the poor families face this several times, mainly when the working opportunities are less. Food poverty and associated nutritional deficit create a number of health challenges for the members of the poor households in the water-stressed villages. The poor households of Gajendrapur seldom face such challenges as the village produces a sufficient number of agricultural items throughout the year.

Socio-economic Inequality, Labour Market and Local Economy

Socio-economic Inequality and Discrimination

Inequality creates a possibility for less privileged individuals or groups to be deprived of their social rights, including the source of income, education, health care, political representation, participation and freedom of speech (Wade, 2014). It is not very unusual that social inequality will rise if resources of a society are distributed unevenly, or the allocation system is arranged in a way that deprives a segment of the population of their basic rights (Scheffer et al., 2017; Yifei, 2011; Niehof, 2004). Such a situation creates an unequal distribution of population and arrays them to

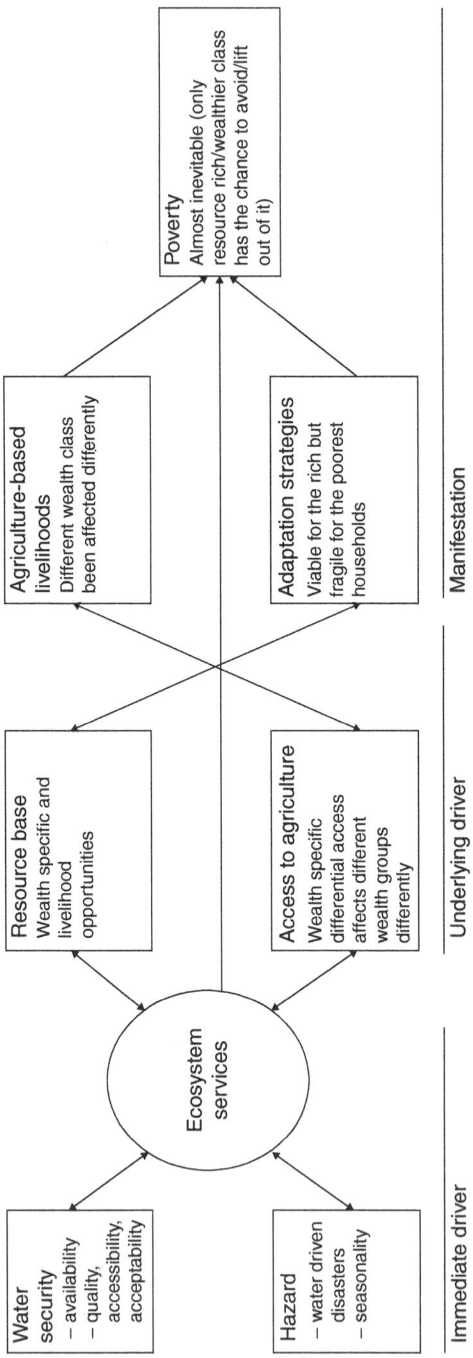

FIGURE 5.2 The way water-related driver (waterborne challenges) is connected to poverty

various socio-economic strata, that is, array population under different categories or wealth classes. In such contexts, the access to resources and social goods is controlled by power, status group, prestige, race-religion or ethnic identity, gender identity, age and various wealth categories.

Social inequality can be observed in every society, but it is necessary to examine how cultural affairs or discourses define the less privileged. It is crucial, for example, to see whether the poor are considered 'deserving' or 'under serving' (Walker et al., 2015) within a society. Households may face difficulty while living in a region or community where inequality is high, and the poor have less access to get or have no equal access to find jobs. Such disparities may lead them to further fall into poverty (Wan & Wang, 2019).

The most common type of inequality includes income disparity, social class, health care and gender inequality. It is also often observed in five specific types (Scott, et al., 2019) of inequality in many communities. Social inequality is closely connected with economic inequality –the unequal distribution of wealth and income. It is hardly possible to explore social inequality without proper investigation and interpretation of economic inequality. Therefore, as the Table 5.6 demonstrates, the present study has a clear investigation on economic profiles and wealth distribution for a clear understanding of the socio-economic context of the study areas. Such understanding helps to capture the nature of their livelihood choices and the reasons they choose various livelihoods to meet their necessities. This also helps us to understand why the agriculture-based livelihoods are so fragile in the face of various waterborne challenges and why poor and extreme poor feel the need to search for livelihood options other than their traditional agriculture-based options.

Poor Labour Market (Low Payment)

Households, when an area does not have a strong labour market, mean that the poor have low bargaining power for their labour, and the labour is not treated with better employment opportunities, they may compel to choose low-paid jobs (Tomlinson & Walker, 2012; Ural et al., 2007). The local economy does not offer viable compensation to the workers who finally get a chance to work within their own area and nearby villages. The payment system is not standard enough to maintain their families. It is variegated in different seasons within a year: the regular cropping season offers better compensation than the rest of the year.

> Working opportunities are unavailable throughout the year. Those are available in cultivation and harvesting periods. The rest of the time, one can observe that many workers are searching for jobs. Many of them seasonally migrate to distant villages, even cities. There is a wage discrepancy. We get better compensation during cultivation and harvesting, but the labour price

TABLE 5.6 Inequality and evidence from study areas

Type of social inequality	*Associated with*	*Evidences from study areas*
Income and wealth inequality	On the basis of what a person can earn daily	The wealthier families in three villages manage to secure income through livelihood diversification. Most of the poor and extreme poor households of water-stressed villages are engaged with subsistence farming. However, the poor of less water-stressed village is comparatively the earlier two villages.
Treatment and responsibility inequality	Ability to receive privilege than others	Some individuals have better access to health care and receive better professional care than others.
Political inequality	Ability to access to resources related to government	Individuals with political connection get benefits of social and political benefits. In case of relief distribution, for example, those people get benefits first.
Life inequality	Brought about by inequality of opportunities which, if any, improve the life, quality of a person	The people consider that either money or education can improve the living condition. The poor and extreme poor lack both categories. They cannot ensure the higher education for their children. Most of their children drop out from primary and secondary schools. Very few of them reach to local college.
Membership inequality	On the basis of number of members in a family or nation	The voice of the poor remains unconsidered. Even their participation in community groups cannot represent meaningfully. The poor member in local water management group, for example, often cannot create impact on community decision.

Source: Qualitative survey, 2016–2018.

> goes down the rest of the time. We cannot maintain our families properly with that limited job opportunities and decreased wage.
>
> *A male day labourer, male FGD participant, Baro Aria*

Moreover, there is wage discrimination between males and females, as there is a common belief that women are less capable of doing jobs that require physical effort (Tiwari & Joshi, 2014; Casey & Alach, 2004). Communities consider that as women have less physical strength than men, they must work less, and therefore they should be paid less. There were times when the demand for female workers rose, not because of women's empowerment but because of their lesser compensation money. When the crop fields are needed to be taken care of with hoes, the farm owners consider that the female workers can carry out this service with less payment.

> We are paid less. The scenario is the same for this and all nearby villages. In the cultivation and harvesting period, male day labourers are paid BDT 200 taka in a day, whereas female labourers are paid 160. In another period, the working opportunities become less available. There are plenty of workers, but the work availability becomes less. As a result, the wage goes down. Each male labourer is paid BDT 120, where as a female worker is paid 80. We do the same work, but the system always deprives the women.
>
> *A female day labourer, female FGD participant, Chandgarh*

Local Trade and Economy, Market Conditions

The local labour market is not well organized, and therefore, it fails to employ a vast number of surplus labourers. The labourers, therefore, choose a seasonal migration strategy in search of living and family maintenance. The local working opportunity includes jobs in agriculture and aquaculture sectors, which are not enough for family maintenance for most poor households. In addition, the local economy does not offer handsome compensation (as stated in previous section) to the workers who finally get chance to work within their own area and nearby villages.

There is a gender segregation of work as well. The gender role in agriculture is delineated based on job nature and the type of physical effort it demands. Farmers consider the jobs that require many physical efforts like ploughing, managing cows, levelling soil and sowing seedlings are the jobs for males; whereas the jobs that require less physical effort like caring for seedlings, applying hoes and weeding out unnecessary plants are jobs for female workers. The female workers are also working together with male workers during ploughing, sowing and harvesting. In most cases, the gender roles for women limit them to stay and work within family boundaries, as poverty compels them to work outside their households.

In case of subsistence farming, using family labour is very common in the study villages. Almost all eligible members, including children and aged, are seen working in the crop fields. Family labour reduces both labour and production cost. The families draw greater benefit in this case. However, such benefits are not homogenous, as production varies in water-stressed and water-managed villages due to water-borne challenges.

The study villages have not fully developed the market system and its mechanisms. Most of the exchanges are made within villages can be categorized as petty trade. The trade is limited to locally produced daily necessary items like vegetables, fishes and the likes. The people of Chandgarh cannot grow vegetables; people can produce vegetable in Baro Aria, and mostly for their own consumption. At times, the people of two water-stressed villages are engaged in petty business like buying locally produced vegetables (from nearby less water-stressed areas) and selling and taking them into nearby Khulna city for a better price. The people of Gajendrapur can practice agriculture throughout the year. They produce vegetables in both their homestead and crop fields. Most of them usually draw two to three crops per year. Large producers generally manage better transport, like large trucks and pick-up vans, to send the village products to Dhaka, the capital city of Bangladesh. In comparison, the people of Gajendrapur are in a better position in terms of local production and marketing. The poor are often left with little choice to get the actual share of their product as they are not the direct transporter of their products.

Household Vulnerability

Poor, Little or No Education

Education has a crucial role to play in poverty alleviation, and it opens a range of possibilities for the poor in particular (Paraschiv, 2017). And if the education process and schooling is disrupted, that can create intergenerational poverty. Education is one of the leading drivers that helps individuals and households lift out of poverty. In some families, the younger members, especially the children, often play a vital role here and improve their living conditions and family status through studying and getting jobs in the formal sector. The jobs liberate them from getting out from various malfunctions caused by poverty, including poor living standards. Education helps families by increasing household income and their living standards and the possibilities to ensure better schooling and training skill to later generations, etc. If any household member has basic formal education, there is a chance for improvement in a later stage and a hope of lifting out of poverty. The school enrolment means nothing if a child does not continue their primary education and move towards higher education. Many dropouts in less educated communities of two water-stressed villages cannot significantly improve the standard of living through education. The water-managed village has succeeded both in school enrolment and continuation through the past few years. The school attendance may indicate the continuation of schooling or active engagement in the academic process. Poor attendance rate often turns into detachment from the schooling process, leading to dropouts from schools.

Household Debt, National Debt

A growing literature has linked poverty with household and national debt (Sachs, 2002, 2004; Poku, 2002; Kemal, 2001; Peters, 2000). It is argued that poor people often fail to come out of poverty because the household debt works as a trap where the poor people cannot get out of it. To escape this trap, the households need a powerful economic support that household members fail to ensure in most cases. As a result, poor households continue to sink into poverty and poor living conditions over the years.

Many poor and extreme poor households fall into debt during the dry seasons. In water-stressed Chandgarh, dwellers make a seasonal migration to Khulna city searching for work in various non-farm sectors. Many of them work in the brickfields. When these people come back to the village during the cultivation and harvesting period, they take loan from brickfield owners to run their household over the next few months with a commitment that they will go back to the brick-field and work again to repay the loan.

> My husband took a loan from a brick-field manager when we migrated there in search of work. The loan was taken for a few months. We spent all the money there as the living cost is high in Khulna city. He retook a loan when returning home, with a commitment that he would work in the brick-field again when the work opportunity would be less in the village. We often would stay in debt. We would need to search for a new loan source immediately after we managed to reimburse a debt.
>
> *A female worker, FGD participant, Chandgarh*

No debt options and endeavours available in the study villages have the reputation to help the poor improve their livelihoods or assist them in lifting out of poverty. Debt does not help poor people to lift out of poverty, rather sometimes it increases poverty condition (Dent & Peters, 2019; Pressman & Scott, 2010; Soludo, 2003; Basu et al., 2000). The poor and extreme poor households, thus, engage in such a debt trap and cannot fight it out for years. They consider the debt trap to continue, which is one of the key reasons for their lingering poverty.

Lack of Resources and Poor Management of Resources

Poverty is a many-headed evil. The poor often fail to manage natural resources properly just because of their inaction and lack of sufficient knowledge on resource management. Lack of resources hinders the poor to invest and improve their living conditions. Even the poor cannot invest in their children's education properly (Kiernan & Mensah, 2011; Lacour & Tissington, 2011; Van der Berg, 2008; Nolan & Whelan, 1996). Lack of resources compels them to enter the vicious cycle of poverty and continue the poverty condition over generations.

Lack of Financial Means to Improve the Living Conditions

When poor households cannot invest in different sets of capital, the poor resource base and their incapacity to further investment may create long-term poverty and the vicious cycle of poverty (Elisha& Felix, 2021; Wang& Zhou, 2020). Poor resource base, along with water-related hazards, arguably, is one of the fundamental reasons for poverty in the water-stressed villages. The incapacity to invest compels poor households to enter the vicious cycle of poverty. Suppose fewer wealthy households have the capacity to invest more or have sufficient financial means to invest and improve their living condition, in that case, the overall situation can be changed significantly.

> All of us sitting here are very poor, and we live hand to mouth. Our families depend on the everyday income of the earning members. If there is no work opportunity, and if we stay without work for one week, all the family members will starve. We have no choice other than work. We have no savings, we have almost no support.
>
> *A female FGD participant, Chandgarh*

The situation is not the same at Gajendrapur. There are poor and extreme poor households, but they do not face starvation like water-stressed villages. The members of poor families in Gajendrapur have working opportunities within their own villages throughout a year.

Hazard and Environmental Condition

Climatic Factors, Extreme Weather, Vulnerability and Natural Disasters

Climate change-induced frequent disasters generate additional risks to smallholder households who practice farming, and such risks are often the cause for their poverty (Hansen et al., 2019; Anderson, 2011). Households may immediately fall into poverty do not have sufficient adaptation and coping mechanisms during and after disasters (Eriksen et al., 2007). In such contexts, widespread hunger and malnutrition might be unavoidable consequences in the long-run (Sarkar & Kulkarni, 2022).

As Figure 5.3 depicts, the people of different wealth classes experience the waterborne challenges differently. Most of the households in Chandgarh and Baro Aria are traditionally dependent on subsistence agriculture, and only a few of them are engaged in saline water-based aquaculture. People with poor economic profiles lack agricultural land, in both villages, work as day labourers for their family maintenance. People of Baro Aria and Chandgarh face high risk and exposure to waterborne challenges, including embankment erosion, tidal flooding, saline water intrusion, soil salinity and limited water logging. These drivers severely affect their traditional agricultural production. Therefore, very limited agricultural production forces subsistence farmers to search for new livelihood options for their family maintenance. The embankment breaching affects different wealth classes differently; the rich considered themselves the most affected as they either lost their cultivable lands or the production declined considerably, whereas the poor and extreme poor experienced the problem differently as they faced gradual diminishing opportunities as casual labour in agriculture. The rich households perceive salinity as a minor problem responsible for low agricultural production, whereas the poor and extreme poor see it as a significant problem as they cannot practice gardening in their homestead. Likewise, poor and extreme poor in Chandgarh regard breaching as a less severe problem as they have lost everything long ago and reside on the partially eroded embankment. In contrast, wealthy households consider this a significant threat to their homestead and agricultural fields.

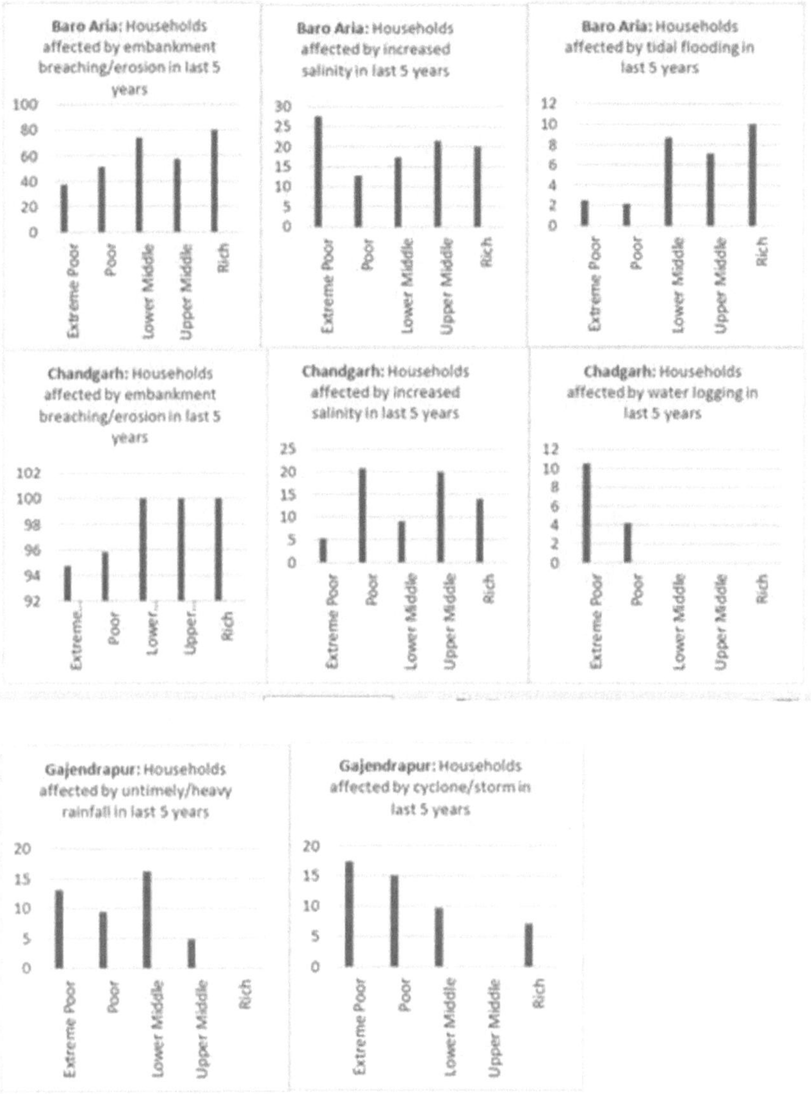

FIGURE 5.3 Major waterborne challenges experienced by the inhabitants of the studied villages

The breaching problem, which started a decade and a half ago, has reversed the situation. Because of breaching, a significant part of the village is already eroded, salinity has increased, soil quality is deteriorating gradually and vegetation and a good number of trees are decreasing. Many people, who were traditionally engaged with agriculture, are shifting their occupation. Many wealthy families with large land stakes are dependent on crop farming, migrated to nearby villages and cities.

The adult members of poor households are now working as agricultural day labourers and in many other non-farm activities within nearby villages or Khulna city. Seasonal migration is evident when there is a dearth of work opportunities. The migrant workers usually engage in odd jobs available in the cities and return to their villages when work is available.

The villagers also face drinking water problems as tubewell is unsuccessful in the village and nearby areas. The water in the nearest deep-water tubewell is salty and inedible. The villagers, therefore, depend on water vendors for fresh water in the dry season and on rainwater in the rainy season. Poor families pay a significant part of their income to buy freshwater for drinking purposes. Some low-income families do not purchase water, instead, they consume from unhealthy sources without proper treatment. Members of these families often face health problems, have to bear medical bills and thus a significant portion of their income runs out, which pushes the already insolvent families to further vulnerability.

Government and the Basic Social Support

Lack of Necessary Infrastructure

Infrastructure has a very crucial role to play in poverty alleviation, as necessary infrastructure helps communities and households to communicate and move to different places and gain vital advantages, which empower them to get better services and increase their income (Ogun, 2010; Seetanah et al., 2009; Setboonsarng, 2006; Cook, 2005; Ali & Pernia, 2003; Pouliquen, 2000). The lack of infrastructure (roads and other communication networks) is connected to the long-term and vicious cycle of poverty. Such lack hinders poor households from lifting out of poverty.

Like other rural communities, the infrastructure of study villages includes roads, bridges, wells and cables (for electricity, cell phones and the internet). These are necessary for keeping a community functional and connected with mainstream development. Without these support systems, the community becomes disoriented, feels isolated and detached from the mainstream development effort of the government. When individuals try to reach distant places for these services, they are to pay extra time and money, which pushes those families further into economic vulnerability. Living 'off the grid' (their inability to go to school, college, work or market to buy and sell products) often promotes people to fall into poverty and maintain the abject condition over generations.

The road and transport communication are not good enough, and therefore, the district-level administration cannot communicate with the community properly. Although the local government constructed the concrete pitch road years ago, those were not appropriately maintained by concerned authorities. Roads were damaged in some parts on the way to reach the community. Government and local efforts to remake the embankment are often futile by further breaching and tidal flooding. The breaching affects the communication negatively. One of the major reasons for the partial or fully destruction of roads lies in common problems like erosion and flooding. However, sufficient trees were needed to be planted on both sides of the concrete pitch road, which was not properly maintained. In places, illegal piped drains were made under the main road to draw river water in the agricultural fields for practicing aquaculture. The practice is frequent in some parts, and therefore, the roads and embankments have become vulnerable. Local government and community people often repair the embankment, which gets damaged again by the next breaching and tidal flooding. The number of bridges is not sufficient for the area. People in the locality mentioned that they are always anxious during the rainy or cyclonic season not due to the natural disaster itself but the damage or destructions of embankments/polders. They urged the authority for proper and timely maintenance and repair or reconstruction of those faulty or poorly maintained embankments (*beri bandh*).

The drinking water infrastructure is another crucial issue to be discussed, as this water-related driver also has a significant role in the poor inhabitants' everyday lives. The scenario of drinking water in water-stressed areas is disappointing for the community. The tube well is unsuccessful in both water-stressed Chandgarh and Baro Aria. However, the local people tried their best to install both shallow and deep tube wells but found that the groundwater was full of iron, salty and not drinkable. The people draw water from a long distance. A significant part of each day is invested in drawing a single pitcher of water. Local water vending system has been developed, which supports a tiny amount of the demand for local drinking water. In addition, only the solvent families buy water from vendors, not poor households. The water vendors cannot draw and sell water in the rainy season, as the roads are not safe to travel.

The drinking water problem is well managed by the less water-stressed village Gajendrapur. The villagers have managed to install an electricity-driven deep tube well, a large water tank and a piped drinking water supply for the households within the village. The drinking water is no longer a problem for the villagers, and these permit village women plenty of time to invest in their households and agriculture fields. Most of the vegetable and crop farming is managed by households using both family labour and hired labours.

The villages are under electricity coverage, but all residents do not have access to electricity services. Many families residing on the embankment in Chandgarh and broken part of Baro Aria remain out of the electricity facility. The households living in both areas are deprived of services powered by electricity.

The Limited Capacity of the Government

When poor households do not receive necessary essential services that the government usually offers, education and health care – for example, they do not get the welfare benefits of those services, and eventually, they are to spend more for those for what they cannot save money for future benefits (Rogers, 2014; Grindle, 2004; Francis & James, 2003; Omiti et al., 2002). The government capacity and willingness to reach poor households can make a big difference towards poverty alleviation (Slocum & Backman, 2011). Due to remoteness from the city growth centre, the villagers often cannot reach out to the services offered by the government in the nearest urban centres (usually the *Upazila sadars*). Most government services like better healthcare, schooling, legal and other social support are available in urban centres. As it is not offered at the local level, the villagers often are not willing or able to reach the urban centres to receive the benefits of those services. Government services are not available in remote areas always, and therefore the people residing there remain deprived of the state services.

> Local government should offer essential services based on the needs of people in local remote areas. If the government can do it, the root level served, political acceptance of government will increase.
>
> *A KII participant in Baro Aria*

These matters are often influenced by economic capability, required arrangement and political will of the government. But these are not enough. Attention is needed for curbing corruption, local power practice, nepotism, widespread inequality and lack of participation. This study has found no evidence in line with these, except some opinions of very few participants. The awareness and involvement of the villagers are too poor to get the welfare of the benefits offered by the government bodies and the various non-government, profit and non-profit organizations.

Social Services and Safety Nets

Social services, including widespread public works and services, are usually offered and supported by government, local and/or international non-government organizations, and other profit or non-profit agencies. The goal of social services is to build capacity with stronger communities and promote equal opportunities. The service includes promoting health care, education, food aid and agriculture subsidies, police, and legal support, fire service, training and employment, subsidized housing, adoption, community management, policy research and lobbying (Hills et al., 2007).

Both the health care and education services are available, though not sufficient, within and around the nearby areas. The government's health care services to the study villages are not sufficient to meet the requirements of local poor and extreme poor populations. The nearest hospitals are almost 14.5 kilometres far from the water-stressed areas. The community clinics are claimed to have good functionality by the local government. Still, those are not enough to support the densely populated villages (in terms of patient–doctor ratio and functionality). Though the community clinics offer equality and access, very few villagers are interested in going there to get service. Most of the maternity and child delivery-related issues are managed locally. The villagers seek hospital support in the cases of complexity during delivery, critical cases of surgery, etc. The educational institutions often fail to attract students to pursue higher studies. Though there are schools and colleges in the locality, the literacy rate is not improving and is being affected by dropouts, lacking family support, etc.

Government has several safety net programmes. Although those are not for poverty alleviation but mainly to provide support to the affected people in hazardous areas. The most common social safety net programmes of the government of Bangladesh include Employment Generation Program for the Poorest (EGPP), Vulnerable Group Development (VGD), Vulnerable Group Feeding (VGF), Old Age Allowance and Allowance for Widow and Destitute Women. Although the safety net programmes are creating notable changes in many parts of the country, they are not sufficient to reduce poverty in the study villages because of their low coverage, weak selection process, unaccountability in the selection process and lack of clear guidelines.

Overpopulation

Overpopulation is one of the significant reasons for poverty, but there is a considerable transformation of other dimensions of the population. It is evident that many people have migrated since the beginning of embankment breaching. Many villages (e.g. Jaliakhali) disappeared, and the displaced population started working for their living again in distant villages. Some of them migrated to urban poor settlements as workers. Many wealthy families migrated to urban centres leasing out their agricultural lands to other sharecroppers. The rest of the households, who did not move elsewhere, started staying in the government *khas* land, on partially broken embankments, in the homestead land permitted by relatives or neighbours whose land has not eroded yet.

Along with the internal migration feature, a notable change has been observed in the family structure. The traditional extended families have been going through a gradual change. The family members traditionally supported by the extended family have now become responsible for maintaining their costs. The newly created nuclear families have been initiated once the adult children get married and form new families. Breaking down of extended families influences the breakdown of their family property. The large lands then divided into smaller pieces eventually produce insufficient crops to support a family throughout the year. The nuclear family members thus are unable to practice subsistence farming, and finally, the shortage turns into poverty.

The gradual loss of essential livelihood items leads the household to poverty in the long run. The households, thus, maintain poverty as an intergenerational legacy. As they lack resources, capital and various inputs, they remain undernourished and bear this brunt over generations. Long-term food shortage or hunger creates intergenerational poverty and protracted crisis. The general strategy to address such a situation is providing food aid and associated reliefs to affected areas. The food aid and relief strategy are often criticized as they may propel the dependency of the recipient countries in the long run. There are examples that the recipient countries fall into long-term debt traps, which is often considered the key barrier of many developing countries for not getting rid of poverty. Many developing countries fail to repay their loans as the successive governments do not take responsibility and are willing to fill their regime successfully.

Conclusion

Poverty is a many-headed evil, and this paper explores how waterborne challenges influence those heads or drivers. It provides a detailed account of poverty through the lens of an MPI where poverty is seen in terms of three broad dimensions and ten specific indicators. All these indicators are capable enough to present the true nature of poverty among different wealth classes in the studied villages. The wealth class-specific employment scenario tells the type of jobs people do in terms of their socio-economic capability.

It shows that rich people can work at formal jobs, whereas poor people engage in precarious jobs. Small households are traditionally engaged in agriculture-based activities. As the agricultural practices are declining in water-stressed villages, the poor are losing opportunities and are compelled to choose various non-farm jobs. Many families migrate seasonally searching for jobs and often engage in a debt trap, which increases their vulnerability. The hazard and environmental conditions pose threats to the people who are engaged in agricultural practices. The considerable lack of basic social support and coverage is often insufficient, which plays a crucial role in augmenting poverty to the agriculture-dependent poor households facing difficulties due to waterborne challenges.

Notes

1 *Golpata* is long leaves of a mangrove palm (Nipa palm/Nypa palm), which grows abundantly in Sundarban (the largest mangrove forest in the world), and coastal people use that leaves to cover the roof of their houses. The leaves have a huge popularity among rich and poor residents.
2 There are ten districts in Khulna division.

References

Abdullah, A. N., Myers, B., Stacey, N., Zander, K. K., &Garnett, S. T. (2017). The impact of the expansion of shrimp aquaculture on livelihoods in coastal Bangladesh. *Environment, Development and Sustainability*, *19*(5), 2093–2114.

Ali, I., & Pernia, E. M. (2003). *Infrastructure and poverty reduction-what is the connection?* ERD POLICYBRIEF SERIES Economics and Research Department Number 13, Asian Development Bank.

Anderson, S. (2011). Climate change and poverty reduction. https://www.osti.gov/etdeweb/biblio/22073505

Antwi-Agyei, P., Stringer, L. C., & Dougill, A. J. (2014). Livelihood adaptations to climate variability: Insights from farming households in Ghana. *Regional Environmental Change*, *14*(4), 1615–1626.

Assan, J. K. (2014). Livelihood diversification and sustainability of rural non-farm enterprises in Ghana. *Journal of Management and Sustainability*, *4*(4). https://doi.org/10.5539/jms.v4n4p1

Barau, A. A., Afrad, M. S. I., Rahman, M. S., & Rahman, M. A. (2019). Indigenous peoples' livelihood practices in south-eastern Bangladesh and the question of sustainability. *Sustainable Development*, *27*(4), 809–816.

Barrett, C. B., Reardon, T., & Webb, P. (2001). Nonfarm income diversification and household livelihood strategies in rural Africa: concepts, dynamics, and policy implications. *Food Policy*, *26*(4), 315–331.

Basu, S., Mate, K., & Farmer, P. E. (2000). Debt and poverty turn a disease into an epidemic. *Nature*, *407*(6800), 13–13.

Becker, E., & Jahn, T. (2001). Social-ecological research-conceptual framework for a new funding policy: Synopsis of the Report for the German Federal Ministry of Education and Research. *Nature Sciences Sociétés*, *9*(1), 73–77.

Berman, R., Quinn, C., & Paavola, J. (2012). The role of institutions in the transformation of coping capacity to sustainable adaptive capacity. *Environmental Development*, *2*, 86–100.

Biswas, W. K., & Lucas, N. J. D. (1997). Energy consumption in the domestic sector in a Bangladesh village. *Energy*, *22*(8), 771–776.

Briscoe, J. (1979). Energy use and social structure in a Bangladesh village. *Population and Development Review*, 615–641.

Casey, C., & Alach, P. (2004). 'Just a Temp?' Women, temporary employment and lifestyle. *Work, Employment and Society*, *18*(3), 459–480.

Cook, C. C. (2005). *Assessing the impact of transport and energy infrastructure on poverty reduction*. Asian Development Bank.

Dent, M., & Peters, B. (2019). *The crisis of poverty and debt in the third world*. Routledge. https://www.amazon.com/Crisis-Poverty-Third-Routledge-Revivals/dp/1138393029

Dowler, E. A., & O'Connor, D. (2012). Rights-based approaches to addressing food poverty and food insecurity in Ireland and UK. *Social Science & Medicine*, *74*(1), 44–51.

Duncan, A. J., Öborn, I., Nziguheba, G., Temesgen, T., Muoni, T., Okeyo, I., ... & Vanlauwe, B. (2019). Supporting smallholder farmers' decisions on legume use in East Africa-The Legume CHOICE approach. *Aspects of Applied Biology*.

Elisha, O. D., & Felix, M. J. (2021). Destruction of coastal ecosystems and the vicious cycle of poverty in Niger Delta region. *Journal of Global Agriculture and Ecology*, *11*(2), 7–24.

Ellis, F., & Mdoe, N. (2003). Livelihoods and rural poverty reduction in Tanzania. *World development*, *31*(8), 1367–1384.

Eriksen, S. E., Klein, R. J., Ulsrud, K., Næss, L. O., & O'Brien, K. (2007). Climate change adaptation and poverty reduction: Key interactions and critical measures. *GECHS Report*, *1*, 1–44.

Francis, P., & James, R. (2003). Balancing rural poverty reduction and citizen participation: The contradictions of Uganda's decentralization program. *World Development*, *31*(2), 325–337.

Geiser, U., Bottazzi, P., Epprecht, M., Fokou, G., Fritschi, A., Ramakumar, R., ... & Strasser, B. (2011). Access to livelihood assets: inclusion, exclusion, and the reality of development interventions. *Research for sustainable development: Foundations, experiences, and perspectives*, 313–330.

Gilling, J., Jones, S., & Duncan, A. (2001). Sector approaches, sustainable livelihoods and rural poverty reduction. *Development Policy Review*, *19*(3), 303–319.

Grindle, M. S. (2004). Good enough governance: Poverty reduction and reform in developing countries. *Governance*, *17*(4), 525–548.

Grey, D., & Sadoff, C. W. (2007). Sink or swim? Water security for growth and development. *Water Policy*, *9*(6), 545–571.

Hansen, J., Hellin, J., Rosenstock, T., Fisher, E., Cairns, J., Stirling, C. & Campbell, B. (2019). Climate risk management and rural poverty reduction. *Agricultural Systems*, *172*, 28–46.
Hills, M., Mullett, J., & Carroll, S. (2007). Community-based participatory action research: transforming multidisciplinary practice in primary health care. *Revista Panamericana de Salud Publica*, *21*(2–3), 125–135.
Janjua, P. Z., & Kamal, U. A. (2014). The role of education and health in poverty alleviation a cross country analysis. *Journal of Economics, Management and Trade*, 896–924.
Kemal, A. R. (2001). Debt accumulation and its implications for growth and poverty. *The Pakistan Development Review*, *40*(4), 263–281.
Kiani, A. (2011). Role of university education in poverty alleviation in Pakistan. *Interdisciplinary Journal of Research and Business*, *1*(7) 30, 38.
Kiernan, K. E., & Mensah, F. K. (2011). Poverty, family resources and children's early educational attainment: the mediating role of parenting. *British Educational Research Journal*, *37*(2), 317–336.
King, B. (2011). Spatialising livelihoods: Resource access and livelihood spaces in South Africa. *Transactions of the Institute of British Geographers*, *36*(2), 297–313.
LaCour, M. M., & Tissington, L. D. (2011). Conflict resolution in the multicultural classroom. *Journal of Instructional Psychology*, *38*(1).
Lang, T., & Caraher, M. (1998). Access to healthy foods: part II. Food poverty and shopping deserts: what are the implications for health promotion policy and practice?. *Health Education Journal*, *57*(3), 202–211.
Larson, A. M., Pacheco, P., Toni, F., & Vallejo, M. (2007). The effects of forestry decentralization on access to livelihood assets. *The Journal of Environment & Development*, *16*(3), 251–268.
Molcho, M., Gabhainn, S. N., Kelly, C., Friel, S., & Kelleher, C. (2007). Food poverty and health among schoolchildren in Ireland: findings from the Health Behaviour in School-aged Children (HBSC) study. *Public Health Nutrition*, *10*(4), 364–370.
Mersha, A. A., & Van Laerhoven, F. (2016). A gender approach to understanding the differentiated impact of barriers to adaptation: responses to climate change in rural Ethiopia. *Regional Environmental Change*, *16*(6), 1701–1713.
Mumuni, E., & Oladele, O. I. (2016). Access to livelihood capitals and propensity for entrepreneurship amongst rice farmers in Ghana. *Agriculture & Food Security*, *5*(1), 1–11.
Nolan, B., & Whelan, C. T. (1996). *Resources, deprivation, and poverty*. OUP Catalogue.
Nasreen, M. (1995). *Coping with floods: The experience of rural women in Bangladesh*. Unpublished PhD Dissertation, New Zealand, Massey University.
Nasreen, M. (2007). *Social inclusion: Bangladesh case study* (Vol. 1). UNICEF Regional Office for South Asia.
Nasreen, M. (2009). *Violence against women during flood and post-flood situations in Bangladesh*. ActionAid Bangladesh.
Nasreen, M. (2012; 2019). *Women and girls: Vulnerable or resilient?* Institute of Disaster Management and Vulnerability Studies, University of Dhaka.
Niehof, A. (2004). The significance of diversification for rural livelihood systems. *Food Policy*, *29*(4), 321–338.
Ogun, T. P. (2010, August). Infrastructure and poverty reduction: Implications for urban development in Nigeria. *Urban Forum* (Vol. 21, No. 3, pp. 249–266). Netherlands, Springer.
Olsson, P., Galaz, V., & Boonstra, W. J. (2014). Sustainability transformations: a resilience perspective. *Ecology and Society*, *19*(4). https://www.jstor.org/stable/26269651?seq=10#metadata_info_tab_contents
Olsson, L., Opondo, M., Tschakert, P., Agrawal, A., & Eriksen, S. E. (2014). *Livelihoods and poverty*. https://nmbu.brage.unit.no/nmbu-xmlui/bitstream/handle/11250/2387830/WGIIAR5-Chap13_FINAL.pdf?sequence=5

Omiti, J., Owino, W., Otieno, W., & Odundo, P. (2002). *Poverty reduction efforts in Kenya: Institutions, capacity and policy*. Nairobi, Institute of Policy and Analysis.

Orchard, S. E., Stringer, L. C., & Quinn, C. H. (2016). Mangrove system dynamics in Southeast Asia: Linking livelihoods and ecosystem services in Vietnam. *Regional Environmental Change, 16*(3), 865–879.

Osabohien, R., Wiredu, A. N., Nguezet, P. M. D., Mignouna, D. B., Abdoulaye, T., Manyong, V., ... & Awotide, B. A. (2021). Youth participation in agriculture and poverty reduction in Nigeria. *Sustainability, 13*(14), 7795.

Oyinbo, O., & Olaleye, K. T. (2016). Farm households livelihood diversification and poverty alleviation in Giwa Local Government area of Kaduna State, Nigeria. *Consilience*, (15), 219–232.

Paraschiv, C. I. (2017). The role of education in poverty alleviation. *Theoretical & Applied Economics, 24*.

Peters, B. (2000). The Third World debt crisis—why a radical approach is essential. *The Round Table, 89*(354), 195–204.

Poku, N. K. (2002). Poverty, debt and Africa's HIV/AIDS crisis. *International Affairs, 78*(3), 531–546.

Pouliquen, L. (2000). Infrastructure and poverty. *Background paper to the World Bank's, 2001*. http://documents1.worldbank.org/curated/pt/358331468780319045/pdf/wdr27911.pdf

Pressman, S., & Scott, R. H. (2010). Consumer debt and poverty measurement. *Focus, 27*(1), 9–12.

Rogers, S. (2014). Betting on the strong: Local government resource allocation in China's poverty counties. *Journal of Rural Studies, 36*, 197–206.

Sachs, J. D. (2002). Resolving the debt crisis of low-income countries. *Brookings Papers on Economic Activity, 2002*(1), 257–286.

Sampath, R. K. (1992). Issues in irrigation pricing in developing countries. *World Development, 20*(7), 967–977. https://doi.org/10.1016/0305-750X(92)90124-E

Sarkar, M., & Kulkarni, K. G. (2022). Pandemic and global food insecurity: Experience from developing countries. *ANVESHAK-International Journal of Management, 11*(1), 56–88.

Scheffer, M., van Bavel, B., van de Leemput, I. A., & van Nes, E. H. (2017). Inequality in nature and society. *Proceedings of the National Academy of Sciences, 114*(50), 13154–13157.

Scott, C. A., Zhang, F., Mukherji, A., Immerzeel, W., Mustafa, D., & Bharati, L. (2019). Water in the Hindu Kush Himalaya. *The Hindu Kush Himalaya assessment*. https://doi.org/10.1007/978-3-319-92288-1_8

Seetanah, B., Ramessur, S., & Rojid, S. (2009). Does infrastructure alleviate poverty in developing countries. *International Journal of Applied Econometrics and Quantitative Studies, 6*(2), 31–36.

Setboonsarng, S. (2006). Transport Infrastructure and Poverty Reduction. https://think-asia.org/handle/11540/4064

Shackleton, S., Ziervogel, G., Sallu, S., Gill, T., & Tschakert, P. (2015). Why is socially-just climate change adaptation in sub-Saharan Africa so challenging? A review of barriers identified from empirical cases. *Wiley Interdisciplinary Reviews: Climate Change, 6*(3), 321–344.

Slocum, S. L., & Backman, K. F. (2011). Understanding government capacity in tourism development as a poverty alleviation tool: A case study of Tanzanian policy-makers. *Tourism Planning & Development, 8*(3), 281–296.

Soludo, C. C. (2003). Debt, poverty and inequality. *The Debt Trap in Nigeria*, 23–74.

Statistics, B. B. O. (2011). Statistical Yearbook of Bangladesh. *Statistics Division, Ministry of Planning, Dhaka, Government of the People's Republic of Bangladesh.*

Tiwari, P. C., & Joshi, B. (2014). Environmental changes and their impact on rural water, food, livelihood, and health security in Kumaon Himalaya. *International Journal of Urban and Regional Studies on Contemporary India, 1*, 1–12.

Tomlinson, M., & Walker, R. (2012). Labor market disadvantage and the experience of recurrent poverty. *The age of dualization. The changing face of inequality in deindustrializing societies*, 52–70.

Tse-ring, K., Sharma, E., Chettri, N., & Shrestha, A. B. (2010). *Climate change vulnerability of mountain ecosystems in the Eastern Himalayas*. International Centre for Integrated Mountain Development (ICIMOD).

Ural, B. P., Mitra, D., & Hasan, R. (2007, September). Trade liberalization, labor-market institutions, and poverty reduction: Evidence from Indian states. *India Policy Forum 2006* (Vol. 7, No. 3, pp. 71–122).

Van der Berg, S. (2008). How effective are poor schools? Poverty and educational outcomes in South Africa. *Studies in Educational Evaluation, 34*(3), 145–154.

Wade, R. H. (2014). The Piketty phenomenon and the future of inequality. *Real-world Economics Review, 69*(4), 2–17.

Walingo, M. K., Liwenga, E. T., Kangalawe, R. Y. M., Madulu, N. F., & Kabumbuli, R. (2009). Perceived impact of land use changes and livelihood diversification strategies of communities in the Lake Victoria Basin of Kenya. https://www.academia.edu/download/68553728/Perceived_impact_of_land_use_changes_and20210803-28018-15i49zy.pdf

Walker, E. T., McQuarrie, M., & Lee, C. W. (2015). Rising participation and declining democracy. *Democratizing Inequalities: Dilemmas of the New Public Participation*, 3–23.

Wan, G., & Wang, C. (2019). Poverty and inequality. *Asian Transformations*, 240–266.

Wang, Y. P., & Zhou, X. N. (2020). The year 2020, a milestone in breaking the vicious cycle of poverty and illness in China. *Infectious Diseases of Poverty, 9*(01), 1–8.

Yifei, S. (2011). China in the "Post-Patriarchal Era" changes in the power relationships in urban households and an analysis of the course of gender inequality in society. *Chinese Sociology & Anthropology, 43*(4), 5–23.

6

CHALLENGES OF SAFE WATER MANAGEMENT IN SOUTHWEST COASTAL REGIONS OF BANGLADESH

Syed Ashik-E-Elahi and Khondoker Mokaddem Hossain

Introduction

It is anticipated that in Bangladesh, the highest number of people are already facing the impact of frequent disasters and that the country will confront a significant disaster threat shortly. As a result, catastrophe specialists have classified Bangladesh as one of the most disaster-prone countries. It's worth mentioning that Bangladesh's coastal regions are more prone to natural disasters than the rest. A considerable number of people live in coastal belts, and they are struggling much due to frequent disasters. Such disasters are causing loss of shelter, loss of greenery, loss of occupation, loss of domestic animals, loss of preferred cultivation, and so on. Among others, this chapter has focused on one of the most fundamental crises in the coastal regions of Bangladesh. The loss of safe water sources, a lack of adequate management of available sources, and the risks that come with it in Bangladesh's southwest coastal districts. Figure 6.1 shows the study regions of Bangladesh.

Bangladeshi people have always had enough of water – in fact, have never had access to safe drinking water (Pal et al., 2011). Two-thirds of the world's population suffer from a severe lack of safe drinking water for at least one month each year (Hadi, 2019a). Though significant progress has been made toward fulfilling the MDGs, many of the targets, notably those relating to water, have yet to be met. For instance, one out of every ten people in the world lacks access to safe water as of 2015 (Chan, Roy, & Chaffin, 2016). The National Water Policy was established in 1992 to guide public and private efforts to guarantee optimal water resource management and development for the benefit of individuals and society. The strategy outlined six major objectives for ensuring long-term water management by acknowledging that the current water problem is the result of ineffective institutional management, regulatory mechanisms, and unsustainable practices (Hadi, 2019).

DOI: 10.4324/9781003253495-8

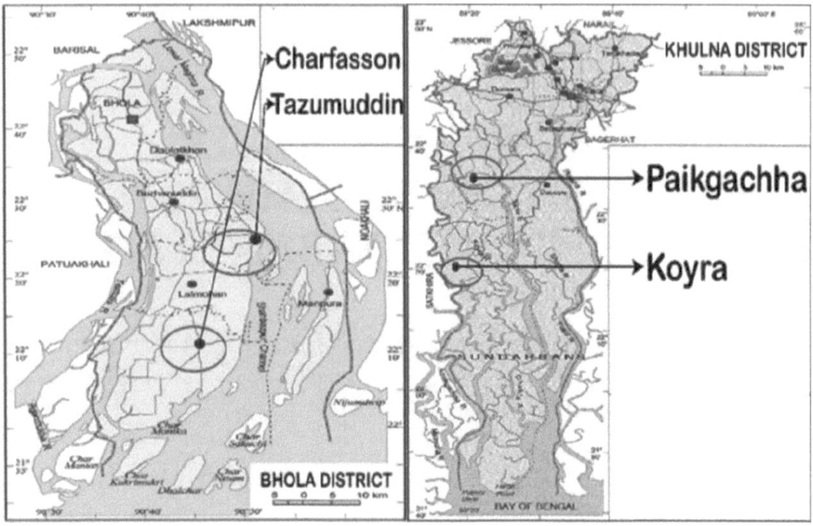

FIGURE 6.1 Bangladesh, highlighting study areas
Source: Self modification (Primarily taken from google map).

The world population is increasing by around 80 million people each year. This growing number is implying a 64 billion cubic meter rise in annual freshwater consumption. The majority of population growth will occur in developing countries, particularly in areas that are already facing water scarcity and lack access to safe drinking water and adequate sanitation (Neto, 2018). Bangladesh's National Adaptation Plan 2005 was developed in response to five climate change challenges: freshwater scarcity, drainage congestion, riverbank erosion, frequent floods, and widespread drought and salinity (Chan, Roy, & Chaffin, 2016). When the Prophet advised, "Do not squander water even if performing ablution on the edge of a fast-flowing (big) river," he meant even in seemingly abundant circumstances (Faruqui, Biswas, & Bino, 2001). Water scarcity affects roughly 2 billion people now, with that number estimated to climb to 4 billion by 2025. According to the World Water Assessment Program, more than 2.2 million people died in 2000 from diseases caused by contaminated water or drought. All of these numbers have come to pass as the water sector faces depletion, contamination, mismanagement, and even being hijacked by multinational corporations (Nurul & Mohammad, 2014).

The United Nations Water Meeting, held in Mar del Plata, Argentina in 1977, was the first worldwide conference dedicated solely to water. The International Drinking Water Supply and Sanitation Decade (1981–1990), for example, was sparked by the Mar del Plata Action Plan. The UN General Assembly declared this decade at the end of 1980 with the primary goal of providing full access to water and sanitation to all people in developing nations (Unesco, 2002). Goal 6 of the Sustainable Development Goals has given special concern to universal access to safe drinking water and sustainable use of water resources (Chan, Roy, & Chaffin, 2016).

Bangladesh is placed 130th out of 139 countries in the World Economic Forum's 2010 Global Competitiveness Report for its network of roads, power, and ports due to poverty and geography. In terms of certainty, urgency, and severity of impact, climate change poses major hazards to Bangladesh's infrastructural weaknesses, notably to its water and coastline resources (Earle et al., 2015).

In 1980, 32 percent of the world's estimated people (1.49 billion) living in rural regions had better access to water. According to recent estimates of rural water availability in "least developed nations," basic rural water access has increased from 43 percent in 2000 to 52 percent in 2015 (Hope et al., 2019). Water is the most essential element for the survival of living being. It affects many aspects of life, including agriculture, industry, domestic life, and all economic operations in general. So, it is highly recommended that water-related field activities have to be included in the management policy (Abou & Djebedjian, 2004). Since 1980, the usage of treated wastewater has increased dramatically in developing countries. This is mainly due to rising population and per capita water use, which has resulted in an increase in wastewater production in both urban and rural areas of emerging countries (Al Khateeb, 2001). In the international debate over water resource allocation priorities, there is a growing consensus. The provision of water for essential human needs should take precedence (Unesco, 2002). Water resources management involves a wide variety of operations, from infrastructure development and water resource distribution to financing arrangements and incentive implementation for efficient water use and protection (Figure 6.2) (Lenton & Muller, 2012).

"Rural people are not fully able to locate and provide answers to all of their problems due to low literacy rates and little exposure to the outside world" (Chadwick & Datta, 2003). So far too many of us cannot apply what we've learned about our existing paradigms. If we imagine ourselves 150 years in the future and then look back, it seems likely that practitioners of that era will smile at our belief systems once more (Mercer, Christesen, & Buxton, 2007). Ma', the Arabic term for water, appears in the Quran 63 times. The throne of God is said to be seated on water, and Paradise is depicted as "Gardens beneath which rivers flow." According to Caponera (this volume), water appears to be the most valuable creation in the Quran after people. The verse "And Allah has poured down the water from the heavens and therewith provides life to the land after its death" reflects the life-giving aspect of water. Water not only gives life, but it is also the source of all

1. Water is a limited, vulnerable, and necessary resource for life, development, and environmental sustainability.
2. Water resource development and management should be based on a participatory approach, involving users, planners, and policymakers at all levels.
3. Water provision, management, and conservation are all dominated by women.
4. Water has a monetary worth in all of its competing uses, and it should be treated as such.

FIGURE 6.2 The Dublin Principles

Source: Loucks & Beek, 2017.

life: "We fashioned every living thing from water" (Faruqui, Biswas, & Bino, 2001). Water governance is defined as "the political, economic, and societal processes and structures through which authorities, civil society, and the private industry make choices about how to best use, grow, and manage water resources," by the United Nations Development Program (Chan, Roy, & Chaffin, 2016).

Method

Table 6.1 shows the issues emphasized in coastal water policy summary of Bangladesh. The work is based on an empirical study to assess the prevailing challenges and threats in access to safe drinking water in southwest coastal regions of Bangladesh. Different social, cultural, political, and behavioral issues have been taken into account to understand the loopholes in management failure. The study's primary data were collected by administering mixed methods, and questionnaire survey and key informant interview (KII) techniques have been applied as data collection techniques. After data collection, the researcher processed quantitative data using SPSS software and qualitative data were analyzed as per specification.

TABLE 6.1 Coastal water policy summary of Bangladesh

Country	*Policy*	*Focal points*
Bangladesh	**Coastal zone policy – 2005**	The coastal zone policy was developed by the Ministry of Water Resources of the Government of the People's Republic of Bangladesh in 2005. It mainly discussed multisectoral aspects of coastal zone management, such as economic growth, basic needs of livelihoods, disaster management, empowering women and coastal communities, and ecosystem conservation in coastal areas. One of the components, sustainable management of natural resources, has focused on water management. It has emphasized rainwater gathering and conservation for daily use: ponds, tanks, and local water technology excavation. The policy also focuses on establishing pond sand filtering (PSF) plants in coastal areas where the saltwater intrusion has already occurred. It also focuses on storing freshwater, and the authority will construct an appropriate water management system based on specific areas. It will also ensure that groundwater in coastal areas is used sustainably. Initiatives will be taken by the concerned authority to gather tidal water for irrigation by constructing reservoirs.

Source: Authors' compilation 2021.

Areas Covered in This Study

Out of 64 districts of Bangladesh, 02 southwestern coastal districts (Namely Bhola and Khulna) have been chosen purposely by considering data availability as per objectives of the study. Again, following purposive sampling, 02 coastal Upazila have been selected from each district. Finally, 02 coastal Unions were selected from each Upazila, considering the severity of access to safe drinking water. At last, the researcher collected necessary data following stratified sampling techniques as respondents are diverse.

Data Analysis

Initially, collected data were sorted and scrutinized for data appropriation. After selecting appropriate data, data coding issues are processed carefully. The researcher then followed a specific cluster data entry procedure completed in SPSS software. After cross-checking and finalizing, data entry necessary statistical equations are administered to validate the study's quantitative data. Besides, the KII technique was used to meet the necessary qualitative data for further study validation.

The quantitative data were then analyzed as per the following equations:

Cross tabulation: This equation is applied to compare two different variables as per their ratio in a table. It gives a clear state of the variables and is easy to understand in a nutshell.

Linear regression: This equation demonstrates the significance of the variables on each other.

Cronbach's alpha: This equation shows the reliability of the variable.

Chi-square test: This equation displays the significance of the nominal variable.

One-sample test: This equation confirms the significance of specific variables.

ANOVA: Analysis of variance shows the significance of a group of variables.

Correlations: This equation indicates the significance of numerical variables.

The qualitative data is described as per the following technique:

KII: This technique administered on a respondent who is expert or knowledgeable on a certain issue.

Result and Discussion

Challenges of Safe Water Management in Southwest Coastal Regions of Bangladesh

Bangladesh will be home to around 27 million climate refugees by 2050. Bangladesh government has invested more than 10 billion USD to solve the problem and assure future adaptation of coastal communities (BRAC, 2016). Most of these victims will come from the country's coastal regions. Bangladesh has been recognized as a role model in disaster risk management by the United Nations. Despite that, there are regulatory flaws in assuring a secure future for coastal residents. The country's coastal residents suffer greatly from daily home issues, and agriculture requires freshwater. According to field experience, negligence and mismanagement are the main reasons for the coastal community's current suffering.

Several studies have highlighted Bangladesh's water governance shortcomings, including (i) a lack of major water users' participation in planning and implementation; (ii) a miscommunication between legislators and interested parties; (iii) an absence of political will to create an enabling environment for effective water governance; and (iv) citizens' exclusion and failure to participate (Chan, Roy, & Chaffin, 2016). During a discussion at a workshop on "People's Participation in the Water Sector: Lessons Learned from Experience" in 1997, respondents confirmed that there had been no real participation in water sector operations in Bangladesh, reflecting the arising consensus on the failure of prior participatory approaches. They suggested that local government and community leaders, as well as people from all walks of life, be involved at every stage of the process, from problem recognition to needs assessment, viability testing, project plan, execution, and operation and maintenance, to ensure authentic stakeholder engagement. According to the text, the following three factors should be considered when determining meaningful involvement: What is the expected level of participation? Who is involved in the procedure? What is the method for participation? (Chadwick & Datta, 2003). Geographical circumstances, excessive sedimentation owing to river morphology, climate-induced SLR, and other variables all contribute to a rise in salt intrusion, notably in Bangladesh's coastal region. Climate warming has hastened the salinization process (Hadi, 2019).

Local pond, tubewell, river, and deep tubewell are the four examined regions' reliant sources of safe drinking water, as shown in Table 6.2. According to the table value, the maximum number of respondents in Goroikhali union were reliant on a local pond for safe water collection, accounting for 50 percent of all respondents in this category, while the second highest respondents in Uttar Bedkashi union are in the same category. Furthermore, it is seen that tubewell usage is high in Char Kukri Mukri and Sonapur unions, with 60 percent and 29 percent, respectively. Furthermore, the Char Kukri Mukri and Sonapur unions have the most reliance on river water, with 43 percent and 36 percent, respectively. The only exception is the use of deep tubewells, which are used exclusively in Sonapur union and account for 100 percent of this category. We may deduce from this data that coastal people in Khulna district rely more on small ponds, whereas people in Bhola district rely more on tubewells, deep tubewells, and rivers.

Due to the blatant exclusion of political opposition from a rightful role in democratic debate, undue influence of state institutions to pursue political goals, and the general politicization of bureaucracy, the current political system is basically a "winner take all" society. Administrative operations hampered by bureaucratic challenges are common in Bangladesh. Therefore, few effective check and

TABLE 6.2 Drinking water source before installation of safe water plant in the study areas

Drinking water before * Union (Cross tabulation)

			Union				*Total*
			Uttar Bedkashi	*Goroikhali*	*Char Kukri Mukri*	*Sonapur*	
Drinking water source before safe water plant installation	Local pond	Count	85	101	3	13	202
		% within drinking water before	42.1%	50.0%	1.5%	6.4%	100.0%
	Tubewell	Count	14	1	93	45	153
		% within drinking water before	9.2%	0.7%	60.8%	29.4%	100.0%
	River	Count	3	0	6	5	14
		% within drinking water before	21.4%	0.0%	42.9%	35.7%	100.0%
	Deep tubewell	Count	0	0	0	39	39
		% within drinking water before	0.0%	0.0%	0.0%	100.0%	100.0%
Total		Count	102	102	102	102	408
		% within drinking water before	25.0%	25.0%	25.0%	25.0%	100.0%

Source: Field survey 2020.

balance practices result petty corruption and crime across the water management sector (Chan, Roy, & Chaffin, 2016). Human activities raise benefit from rivers and floodplains. But several destructive activities causing physical, economic, environmental, and social costs and harms to these bodies. Such loss arises as a result of a misalignment between what humans expect or demand and what nature provides (Loucks & Beek, 2017). Changing the social and institutional components of water resource systems is frequently the most difficult since it requires individuals to change their thinking and behavior. Any reform process will necessitate a shift in our institutions and the rules by which we operate as a society. Individuals are the ones who are primarily accountable for and adjust to changing political and social circumstances. Sustainability necessitates public and private organizations to adapt to the changing requirements of individuals and society across time (Loucks & Beek, 2017).

According to the results of the linear regression, the researcher discovers that plant damage has a considerable impact on water irregularity (Table 6.3). The value

TABLE 6.3 Water abnormality and plant damage: a linear regression analysis

Model summary[b]

Model	*R*	*R square*	*Adjusted R square*	*Std. error of the estimate*	*Durbin-Watson*
1	.157[a]	.025	.022	1.370	1.303

[a] Predictors: (Constant), kind of damage
[b] Dependentvariable: Kind of abnormality

ANOVA[a]

Model		*Sum of squares*	*df*	*Mean square*	*F*	*Sig.*
1	Regression	19.338	1	19.338	10.296	.001[b]
	Residual	762.535	406	1.878		
	Total	781.873	407			

[a] Dependentvariable: Kind of abnormality
[b] Predictors: (Constant), kind of damage

Coefficients[a]

Model		*Unstandardized coefficients*		*Standardized coefficients*	*t*	*Sig.*
		B	*Std. error*	*Beta*		
1	(Constant)	1.302	.121		10.777	.000
	Kind of damage	.086	.027	.157	3.209	.001

[a] Dependentvariable: Kind of abnormality

Source: Field survey 2020.

of R square, in this case, is around.157 (the "**R Square**" column), indicating that the independent variable explains 15.7 percent of the total variation in the dependent variable. The dependent variable is significantly predicted by the ANOVA table regression model. Because the significant value here is.001 (the "**Sig.**" column), the result is significant. Finally, the coefficients table shows that the B value of the type of damage is.086 (the "**B**" column) and the P-value is.000 (the "**Sig.**" column), indicating that damage has a highly positive significant impact on water abnormality.

Large river systems have already been classified as being in a state of severe water stress in many regions of the world, implying that extracting further water from these systems may cause irreparable environmental harm (Lenton & Muller, 2012). Since surface water availability varies seasonally and regionally, the use of tube wells to elevate shallow groundwater was advocated in rural and urban areas to enable intensive agriculture and provide safe drinking water. Groundwater is used to provide about 95 percent of drinking water and 68.5 percent of agricultural water (Chowdhury, 2010). Mismanagement of river resources, on the other hand (for example, haphazard infrastructure development), has exacerbated the harmful consequences of seasonal flooding, compounding already widespread pollution issues. As a result, Bangladesh has relied heavily on groundwater for irrigation, drinking, and manufacturing (Chan, Roy, & Chaffin, 2016).

Table 6.4 demonstrates that the most common cause of illness is fungal contamination, with 95 percent of males and 5 percent of women suffering. The second greatest group of responders addressed water plant maintenance carelessness, with women suffering the most from this problem, accounting for 91 percent of the total. Point source pollution, salinity, and fecal from domestic animals are among the other notable issues raised by the responders.

Though Bangladesh's Cabinet approved a national water strategy more than a decade ago, it is remarkable that the policy has yet to be implemented. This must be done as a first step in correcting the current state of misunderstanding in the country's water resource management (Pal et al., 2011). According to the study, there is a shortage of resources and planning to ensure that coastal residents have a secure livelihood. Furthermore, the lack of appropriate policy and strategy puts the minimum investment at risk. Though there are difficulties in combating coastal crises due to a lack of enough investment opportunities, it is possible to alleviate coastal people's suffering by ensuring proper planning and management authority.

Problems Prevailing in Coastal Regions of Bangladesh

In Bangladesh's coastal regions, salinity in the surface water is a widespread issue. Due to the constant presence of saline water in the surface water of coastal regions, such suffering occurs daily. People are mostly living hand to mouth in Bangladesh's coastal regions. Because of their poverty, most residents cannot afford potable water. As a result, many suffer from various health difficulties due to a lack of freshwater. In addition, they are unable to perform their agricultural operations due to a lack

TABLE 6.4 Gender-sensible suffering to different surface water contamination

			Gender suffering			*Total*
			Female	*Male*	*None of them*	
The problem in surface water	Fungal contamination	Count	184	10	0	194
		% within problem in surface water	94.8%	5.2%	0.0%	100.0%
	Point source pollution	Count	54	7	0	61
		% within problem in surface water	88.5%	11.5%	0.0%	100.0%
	Maintenance problem	Count	88	9	0	97
		% within problem in surface water	90.7%	9.3%	0.0%	100.0%
	Domestic animal nursing	Count	5	0	0	5
		% within problem in surface water	100.0%	0.0%	0.0%	100.0%
	Salinity	Count	10	0	0	10
		% within problem in surface water	100.0%	0.0%	0.0%	100.0%
	Oyster grows and rotten	Count	3	0	0	3
		% within problem in surface water	100.0%	0.0%	0.0%	100.0%
	Frog in the water body	Count	1	0	0	1
		% within problem in surface water	100.0%	0.0%	0.0%	100.0%
	Death animal body	Count	2	0	0	2
		% within problem in surface water	100.0%	0.0%	0.0%	100.0%
	Iron	Count	3	0	0	3
		% within problem in surface water	100.0%	0.0%	0.0%	100.0%
	Do not face any problem	Count	27	3	2	32
		% within problem in surface water	84.4%	9.4%	6.2%	100.0%
Total		Count	377	29	2	408
		% within problem in surface water	92.4%	7.1%	0.5%	100.0%

Source: Field survey 2020.

of fresh water supplies, as evidenced by the current situation. Moreover, people are fighting for survival.

Climate change, land erosion, arsenic-contaminated aquifers, and water contamination are among the natural and man-made risks that have harmed the coastal population's livelihood (Hadi, 2019). To undertake a project, improved water resource planning, management review process quality, a focus on outcomes rather than output should be addressed. A better understanding of some of the linkages between diverse human actions and the hydrology and ecology of the basin, estuary, or coastal zone should be among the benefits (Loucks & Beek, 2017). Even though the Bangladeshi government has developed initiatives to mitigate water governance issues, little attention is given to how these regulations are carried out on a local level. According to Mandal, most policies are based on political objectives that lack empirical evaluation and a vision for large-scale implementation. A fundamental hurdle to promoting sustainable development and holistic water governance changes is the lack of real execution and evaluation of results (Chan, Roy, & Chaffin, 2016). According to the Fourth Assessment Report of the Intergovernmental Panel on Climate Change, the water sector would be one of the most vulnerable to climate change. Climate change manifests itself in extreme weather events, rising sea levels, and salty intrusion into aquifers and rivers (Hadi, 2019).

Table 6.5 shows that Cronbach's alpha acceptance values range from .7 to .8. The table value is.770 in this case. As a result, we can confidently state that variables (drinking water sources before and after plant installation) are widely recognized. These study variables are highly significant to the research objectives in terms of the outcome. For the past 40 years, excessive groundwater withdrawal has resulted from a lack of effective tubewell planning. The absence of effective water resource management has resulted in a variety of stressors that have the potential to cause conflict among different groups of rural society. The government's failure to ensure rural people's participation in the formulation and execution of water resource management policies (Safiuddin et al., 2003). WARPO was created by the GoB to oversee the country's comprehensive water sector planning. WARPO's objective is to develop and implement national water policies and strategies to ensure that water resources are managed efficiently by a variety of users, including agriculture, fisheries, navigation, public health, and industry. WARPO is also in charge of collating information from other water sector agencies. WARPO, on the other hand, is widely acknowledged to have struggled to meet its sectoral planning and coordination tasks, and individual line ministries frequently rely on their own planning offices, which sometimes work in isolation from other sectors (Chadwick & Datta, 2003).

The P-value is .000, and the Pearson Chi-square value is 408.000, as seen in Table 6.6. As a result, statistical evidence suggests that respondents' disease and the condition of water sources are highly reliant on one another.

TABLE 6.5 Reliability test of the source of drinking water before and after plant installation

Reliability statistics		
Cronbach's alpha	*Cronbach's alpha based on standardized items*	*No. of items*
.770	.778	2

Source: Field survey 2020.

TABLE 6.6 Significant test result of illness and reasons

Chi-square tests

	Value	*df*	*Asymp. sig. (2-sided)*
Pearson Chi-square	408.000[a]	3	.000
Likelihood ratio	322.661	3	.000
Linear-by-linear association	163.985	1	.000
N of valid cases	408		

a 0 cells (0.0%) have an expected count less than 5. The minimum expected count is 7.41.

Source: Field survey 2020.

TABLE 6.7 The impact of political influence on plant installation

One-sample test

	Test value = 0					
	t	*df*	*Sig. (2-tailed)*	*Mean difference*	*95% confidence interval of the difference*	
					Lower	*Upper*
Political influence	40.850	407	.000	.804	.77	.84
Kind of impact	27.903	407	.000	1.784	1.66	1.91

Source: Field su-rvey 2020.

By 2010, the government has promised to provide everyone with access to clean water and sanitation. Currently, 74 percent of Bangladeshis have access to safe drinking water, while just 33 percent have access to sanitation (Chowdhury, 2010). Several water-related policy instruments in Bangladesh have become obsolete due to a lack of integration of climate change issues. To ensure long-term water resource management and IWRM implementation, climate change policy must be included in surface water-groundwater planning (Chan, Roy, & Chaffin, 2016). Climate change, according to World Bank forecasts (2000), will cause widespread inundation in low-lying areas, adding to an increase in the rate of saline intrusion (Hadi, 2019).

Table 6.7 shows that the t-value is 40.850 and the P-value is.000, both of which are highly significant. As a result, it can be inferred that political influence has a major impact on plant installation in the research areas, such as plant location, maintenance, and monitoring.

Bangladesh's water management is disjointed, and citizen engagement in water provision is uncommon, if at all. A continuous conflict to overcome basic hurdles to reaching water-related MDG targets, such as considerably inadequate quality of public water services and inefficient practices for operations and maintenance of existing facilities, can be attributed to the lack of focus on IWRM and effective water governance. Regional socioeconomic gaps, access discrepancies, and poor institutional ladders exacerbate these flaws (Chan, Roy, & Chaffin, 2016). Water utilities' dominance has had a significant impact on the form of this organization and its impact on governance. Water utility resource planning techniques have been studied, and it has been discovered that they tend to use low or no risk strategies, preferring to stay "invisible" socially and politically rather than maximizing their approach on estimated risks (Walker, 2014). Analysts' models and simulations of water resource systems are frequently far more complex than they can handle. The issue is that we don't fully know the myriad interconnected physical, physiological, ecological, social, legal, and political (human) elements that control the behavior of such water resource systems (Loucks & Beek, 2017).

The P-value for a one-tailed test is.000, as seen in Table 6.8. As a result, the observed differences among the category means can be attributed to random ability to influence various disorders. As a result, we can conclude that after the establishment of a safe water plant, diseases such as diarrhea, hypertension, eye irritation, and diabetes are greatly reduced in the study unions.

Determining what the notion of the right to water means in practice could be the most helpful step in seeking to solve the problem's core issues. Rather than disputing whether or not the water right is a human right, it is more important to get an agreement on who has a right to water and who has priority access to water. Is there a right to water for a single citizen, a community, a state, or a corporation, and if so, what kind of right does it have, and what is the sequence of precedence? Is it required to own property to have water access? (Hukka, Castro, & Pietila, 2010). The information and knowledge that actors use to interpret a situation are referred to as cognition. Infrastructure-focused bureaucracy, a lack of employees and talents, a faulty monitoring system, and an erroneous poverty classification – all led to the implementation gap (Hueso & Bell, 2013). The water industry is no exception to the world's fluctuating and difficult-to-predict social, economic, and environmental situations. This uncertainty creates a difficult climate in which to construct policies and institutions (Funke, Claassen, & Nienaber, 2013).

TABLE 6.8 Evidence of disease reduction following the establishment of a clean water plant

ANOVA					
Union					
	Sum of squares	*df*	*Mean square*	*F*	*Sig.*
Between groups	254.835	8	31.854	49.810	.000
Within groups	255.165	399	.640		
Total	510.000	407			

Source: Field survey 2020.

KII – 01
MostainUddin
Sonapur, Tazumuddin, Bhola
October 10, 2020

Age: 75
Teacher (Retired)
Educational Status: Honors

The outcome of KII:

Rainwater and pond water, according to the respondent, are the main sources of drinking water for his union's general population. He added that few people purify water by boiling it or processing alum. However, most people drink water directly from the pond without filtering it. During the rainy season, usually, people collect and store rainwater for drinking. But, he said during the wet session, it's more difficult to find fresh water. He further added people were habituated to collecting drinking water from a neighboring government pond. Since a lease mechanism has been imposed on this pond, fish food and medicine are imposed for fish farming. As a result, pond water becomes contaminated and unsafe for drinking.

There are few ponds available in the village to provide fresh water for the majority of the villagers. He mentioned that such private pond owners have expressed privacy concerns and imposed barriers for others to collect water to some extent. However, a few small size ponds at the community level are dry up during the winter season. As a result, during the crisis, the poor residents had to rely on privately owned large ponds to meet their daily needs. Deep tube wells are not available in this area, he added. Instead, local people depend on three to four shallow tube wells. However, saline water contamination is making these available water sources unbearable. Furthermore, due to the frequency of natural disasters, the sufferings of local coastal people are escalating. The pond's water became contaminated after sudden storms or flooding and further worsened the crises.

Usually, they store drinking water in pitchers to survive in emergency situations. He mentioned a pond sand filter (PSF) had been installed beside the government pond at Union Sadar with the help of Oxfam and BRAC to excel pure drinking water supply for the locals. But those who reside far away from Union Sadar are deprived of the facility. He further added in November 2020 the government agency declared a "drinking water reservoir Project" to meet the water crisis. Another provision he has noticed is a water tank supply for the poor household to preserve and use rainwater for certain months of the year. During the wet season, they can store rainwater in this tank. The capacity of each tank is 3,000 liters. He stated that installing a new water supply plant would benefit the middle and lower classes of his community.

	Before purification plant installation	*After purification plant installation*
Health	People were more susceptible to waterborne diseases in the past such as diarrhea, dysentery, typhoid fever, and so on. Several infections wreaked havoc on children in particular.	The number of waterborne diseases has significantly decreased. The health of those living near the purifying facility, in particular, has improved.
Livelihood	1. Treatment costs would have been higher.	1. As a result of the decrease in waterborne diseases, medical costs have decreased.
	2. Boiling water used to take longer in the past. Furthermore, it would have cost more keroseneat that time.	2. People in the neighborhood have saved time and gasoline since the installation of the purifying plant.

Source: Field study 2020.

The association between initial ignorance and types of challenges in using new technology among respondents is shown in the above statistical result. The correlation is.974, indicating that there is a positive and extremely significant association between respondents' knowledge and the induced technology, as seen in Table 6.9.

Groundwater contamination can occur as a direct result of waste discharge through wells or because of surface material filtration into aquifers with no upper containment. In many aquifers accessible to surface drainage, irrigation is the source of the most contamination. Because large volumes of salty water naturally occur near fresh water supplies, a fall in aquifer level caused by pumping may cause the brine to infiltrate a previously fresh supply. This is a particularly common problem

TABLE 6.9 Relation between difficulties of using new technology and kind of troubles

Correlations

		Difficulties in using technology	*Kind of difficulty*
Difficulties in using technology	Pearson correlation	1	.974**
	Sig. (2-tailed)		.000
	N	408	408
Kind of difficulty	Pearson correlation	.974**	1
	Sig. (2-tailed)	.000	
	N	408	408

** Correlation is significant at the 0.01 level (2-tailed).

Source: Field survey 2020.

at the coast, where rising sea levels or extraction of freshwater from an aquifer adjacent to saltwater can result in saltwater intrusion. Chlorides and dissolved sediments find their way into what was once a freshwater aquifer (Wescoat, White, & White, 2003). It is still unclear whether authorities should ensure water service availability in sparsely inhabited areas or whether governments in developing economies should spend significantly on the water industry (Hukka, Castro, & Pietila, 2010).

KII – 02
Md. NasirHawaladar
Uttar Bedkashi, Koyra, Khulna
October 10, 2020

Age: 29
Business
Educational Status: SSC

The outcome of KII:

According to the respondent, tube wells are the only way for people to get drinking water. Since it is low land and there are no dams to safeguard the area, other water sources (such as ponds) are quickly polluted by river water intrusion. Due to the tainted river water and the tube well mixing up, collecting water from the available sources becomes more difficult during the rainy season. He noted that drinking water contains mud and other contaminants, frequently resulting in a horrible odor. Another problem is that they frequently can't find a nearby tube well that drains floodwaters, exacerbating the situation.

When queried about the "purification" procedure, he said he had never heard of it before. He states that just a few people are familiar with the notion of "filtration," and most residents have never heard of it. People frequently use a piece of cloth as a filter in front of the tube well's mouth to collect clean water. Despite their ignorance of the water purification process, individuals routinely use alum to cleanse water during the rainy season since their neighbors' practice this mechanism.

> As per the respondent's experience, seven or eight homes, with on an average 30 to 40 family members, rely on a single tube well as their sole source of drinking water throughout the year.

> Key issues:
> 1. Most people are unaware of water purification.
> 2. No one has ever shown up to help with the drinking water issues during the wet season.
> 3. They are unaware of their rights, as seen by their failure to notify the authorities of their difficulties.
> 4. They can't afford to set up new tube wells on their own because they are financially needy.

They are doubtful, however, whether this procedure can filter the water. They used to drink water that has boiled for two to three minutes on occasions. No governmental or non-governmental organizations have attempted to examine their circumstances to address the current challenges, and locals must be concerned about the situation. As a result, the purification system remains in the dark, and installing a water purification project at the union level remains a mystery	

Source: Field study 2020.

The concept of environmental and ecological sustainability originated as a result of growing concern about the long-term health of our planet. There is rising evidence that our current resource management and usage activities and actions, even at the local level, can have a considerable future impact on the welfare of those living in considerably larger regions. Water resource management concerns at the river basin level are rarely technical, and they are only of concern to those who live in the river basins. They're getting entwined with societal institutions, expectations, and goals (Loucks & Beek, 2017). For Muslims, the Prophet was the ideal human role model and leader, and the society he constructed around him is a paradigm of compassion and equity. In Islam, the Prophet is held in such high regard that it may not be found in other religions. If the Quran serves as a compass for Muslims, the Sunnah serves as a more detailed map of humanity's path on this planet (Faruqui, Biswas, & Bino, 2001). Droughts, floods, and other natural calamities substantially impact (i) the availability of healthy drinking water and irrigation water in Bangladesh, along with (ii) an alarming increase in pollution due to uncontrolled industrialization (Chan, Roy, & Chaffin, 2016). Forecasts for sea-level rise vary from 30 to 50 centimeters in 2050 and 105 centimeters in 2100, indicating a potentially significant impact on the social and economic issues of the coastal community (Hadi, 2019).

According to the research, women and children in Bangladesh's coastal regions have more difficult survival issues. They can't go far to achieve their necessities since they can't travel far. Furthermore, this population cannot meet their basic needs due to a lack of financial resources. Moreover, due to the regular mega-disasters, people's misery worsens day by day. Residents in KoyraUpazila, for example, are still dealing with the aftermath of storm Ayla in 2009. Salinity in surface water and land fertility loss is the most vulnerable outcomes of this cyclone.

Possible Way Out

Bangladesh stands as an example of disaster risk reduction. Now it is time to set another example in coastal management. As a delta country, it has ample opportunities in a blue economy. It has to set up a holistic approach to exploring every possible way out in doing so. In doing so, it is time to set proper planning and management to grab the future flourishment.

Culture, especially religion, has a significant impact on how people view and manage a resource like water. Even though local culture and values were previously overlooked in development initiatives, development agencies are increasingly recognizing the importance of local culture and values in their policies (Faruqui, Biswas, & Bino, 2001). Regardless of how powerful or well-coordinated the administrative machinery of commissions, boards, committees, or executive directors is, it cannot prevent or address the water problem. Understanding of the hydrologic environment, within which their power is exercised, and an informed public whose goals and demands are known are both essential for such officers to be effective (Leopold, 1959). To reduce the number of people without adequate water access, international and national water law would have to favor water use over other uses, as many countries now do. Indeed, there is a clear need to prioritize water consumption, with community water supply being the main concern (Hukka, Castro, & Pietila, 2010). Transparent and responsive political, organizational, and administrative systems must also result in consensus on effective and efficient, equitable, and inclusive techniques and practices for protecting and sharing our shared legacy, which includes both national and international water resources (Hukka, Castro, & Pietila, 2010). A dynamic and adaptive strategy for water resource development, management, and usage is a critical necessity for environmental sustainability in the face of certain changes with unknown consequences (Loucks & Beek, 2017). Imams are better at reaching out to the general people than water experts. Although they are normally well-versed in fiqh, sunnah, and sharia. Their understanding of water resources and conservation measures are frequently insufficient for them to serve as instructors on the issue. As a result, water specialists must not only teach, educate, and advise imams about water shortages, water conservation measures, and the need to involve the public, but also about audiovisual methods and resources to assist them in reaching out to the public (Faruqui, Biswas, & Bino, 2001).

In the study, two major religious responses are shown in Table 6.10. The Char Kukri Mukri union has the highest percentage of Muslim respondents, accounting for around 95 percent of all respondents. On the other hand, the biggest proportion of Hindu respondents is found in the Uttar Bedkashi union, which accounts for only approximately 18 percent of all respondents. In general, the study's respondents are 90 percent Muslim and 10 percent Hindu.

BDP 2100 is a long-term strategic plan for land and water resource management for the next few decades. "Ensure maximum water and food security, economic expansion, and environmental sustainability while effectively managing natural catastrophes, climate change, and other delta concerns through rigorous, adaptive,

TABLE 6.10 Religious ratio of respondents as per study unions

*Union * Religion (Crosstabulation)*

			Religion		*Total*
			Muslim	*Hindu*	
Union	Uttar Bedkashi	Count	84	18	102
		% within union	82.4%	17.6%	100.0%
	Goroikhali	count	92	10	102
		% within union	90.2%	9.8%	100.0%
	Char Kukri Mukri	count	97	5	102
		% within union	95.1%	4.9%	100.0%
	Sonapur	count	95	7	102
		% within union	93.1%	6.9%	100.0%
Total		count	368	40	408
		% within union count	90.2%	9.8%	100.0%

Source: Field survey 2020.

integrated, and inclusive water governance," according to the delta plan's goal (Hadi, 2019). In Bangladesh, 79 percent of rural people consume groundwater drawn from shallow tube wells. It is difficult for them to make the transition to using surface water and/or rainwater swiftly. As a result, the use of groundwater from shallow tube wells with permitted arsenic concentrations should be continued until alternate water sources with acceptable levels of acceptability are found. Parallel to this, arsenic-free groundwater sources must be sought (Safiuddin et al., 2003). Interfering as little as possible with natural life cycles in river basins, estuaries, and coastal zones is one of the most successful ways to accomplish future generations' desires (Loucks & Beek, 2017). Many of the concerns and issues that water resource planners and managers face now are identical to those that they faced previously. There are a few exceptions to this rule. Two trends are responsible for the majority of the new ones: (1) an increasing recognition of the need for a bottom-up, "grassroots" participatory approach to planning, management, and decision-making, and (2) a growing concern for natural ecosystem sustainability (Loucks & Beek, 2017). The section on water and the environment discussed policies for restoring, conserving, protecting, and preserving water resources through regulatory, planning, and management measures. None of these themes included climate change (Hadi, 2019).

Table 6.11 demonstrates that, except Goroikhali union, respondents often collect water from tube wells. On the other hand, the pond sand filter, which is only found in the Uttar Bedkashi union, is the least accessible water source. The table data reveal that the nature of clean water sources in Bhola and Khulna districts is notably diverse.

"The basic dignity to water is essential for living a dignified existence." It's a requirement for the exercise of other rights. The Declaration of the United Nations (General Commentary No. 15 of the United Nations Committee on Economic,

TABLE 6.11 State of drinking water sources among unions following plant installation

Drinking water after * Union (Crosstabulation)

			Union				*Total*
			Uttar Bedkashi	*Goroikhali*	*Char Kukri Mukri*	*Sonapur*	
Drinking water after	Pond sand filter	Count	3	0	0	0	3
		% within drinking water after	100.0%	0.0%	0.0%	0.0%	100.0%
	Rainwater harvesting (personal)	Count	0	90	0	5	95
		% within drinking water after	0.0%	94.7%	0.0%	5.3%	100.0%
	Rainwater harvesting (charity)	Count	0	12	0	0	12
		% within drinking water after	0.0%	100.0%	0.0%	0.0%	100.0%
	Tubewell	Count	99	0	52	28	179
		% within drinking water after	55.3%	0.0%	29.1%	15.6%	100.0%
	Deep tube well	Count	0	0	50	69	119
		% within drinking water after	0.0%	0.0%	42.0%	58.0%	100.0%
Total		Count	102	102	102	102	408
		% within drinking water after	25.0%	25.0%	25.0%	25.0%	100.0%

Source: Field survey 2020.

Social, and Cultural Rights) added a new dimension to the global debate by recognizing access to drinking water and sanitation as a fundamental right (Neto, 2018). It is also critical to recognize that variables both inside and outside the water sector play a role and that the water sector does not develop in a vacuum, but rather as an integral and crucial element of a larger development process (Hukka, Castro, & Pietila, 2010).

The policy aims to create a robust framework for distributing human water rights, providing water for the environment, and developing implementation and operational plans. These procedures are crucial because they define the characteristics of consumer and environmental entitlements, determine when and how water can be returned to the environment, and provide the parameters for water markets. These are the following: (i) Identify the types and behaviors of surface and groundwater resources, resource availability under various climate scenarios, and environmental values and their water requirements within a specific water system, watershed, or aquifer. (ii) Participate in the process with important stakeholders and communities. (iii) Defining and clarifying the rights of existing consumers, particularly about water availability. (iv) Determine the agreed-upon share of water for environmental reasons; establish implementation plans for seasonal water allocation for consumption and environmental purposes, system operation, ecological obligations, and, if appropriate, market procedure and financing (Doolan & Hart, 2017).Experts believe that converting a consolidated database into a management tool is one of the most effective approaches to solving water scarcity, water stress, and diminishing quality (Tundisi, 2008).

Bottom-up planning must seek for all stakeholders to have a clear or "shared" vision of goals and priorities. It must be aware of all applicable laws and regulations and follow them. It should attempt to discover and analyze a variety of alternatives and performance criteria, including sustainability criteria, while avoiding creating a wish list of everything each stakeholder wants (Loucks & Beek, 2017). Bangladesh was one of the first LDCs to develop a long-term climate change strategy with a ten-year timeframe (2009–2018). To be successful, a policy must first be properly aligned with official and informal power systems (to avoid political impediments). Second, the strategy's design and implementation must be sufficient to achieve the targeted result (suitable instruments) while keeping any unwanted consequences (spillovers) of the policy's implementation within acceptable bounds (Martin & Williams, 2013).

Bangladesh's seashore is noted for being a zone of both vulnerabilities and opportunity. Natural calamities such as cyclones, storm surges, and floods are common. As a result, the government has developed this coastal zone policy (CZP), which will provide general guidelines to all parties involved in the management and development of the coastal zone in such a way that coastal residents can live and work in a safe and conducive environment (Islam, 2015). Agenda 21 was approved at the United Nations Conference on Environment and Development in Rio de Janeiro in 1992. Chapter 18 on freshwater is very interesting for water managers. To properly utilize limited water resources and protect those resources from pollution, innovative technologies, including the upgrading of indigenous technology are required. It then goes on to say – The widespread shortage, slow degradation, and worsening contamination of freshwater resources in many parts of the world, as well as the progressive encroachment of incompatible activities, necessitate integrated water resource planning and management. Agenda 21 is a list of activities that all nations will focus on in the following decades, as the name suggests (Unesco, 2002). When you tell someone about a water management problem and ask how they would address it, you will most likely hear one of the two responses. One, consider the sources of the problem, as well as who is involved; consider the problem's effects on others; consider several different solutions; evaluate the effectiveness, costs, and benefits of each solution; and finally, adopt the best solution. Two, invite all stakeholders and ask them what they think about the problem, allow them to provide solutions, and look for compromises that all parties can agree on. The first method emphasizes the problem's scientific examination, whereas the second emphasizes the process that should lead to a solution. Water resource management is a combination of the two (Unesco, 2002).

Indicators and indices have been employed to provide a meaningful summary of factors such as economic return on water utilized, number of species per area, and so on. More developed indices, such as the Water Poverty Index and the more extensive Climate Vulnerability Index, provide a clearer understanding of the relationship between water availability, abstraction ease, and welfare, and they integrate numerous data to identify five basic components (resources, access, capacity, use, and environment). It is feasible to aggregate and characterize the water resource situation using this information (McDonnell, 2008). Area-specific experiments will occasionally survive to tell the tale of the past discursive debate on the basics of water management; then, the tensions between these "new" and "old" policy arrangements will become key issues. Is the paradigmatic scenario implying a more inclusive and far-reaching shift in discourse hegemony in favor of welcoming water? (Backes, 1999). Experience suggests that the primary cause of apparent policy failure, or community dissatisfaction with proposed changes, is a misinterpretation of the policy problem. Hydrologists and economists can help policymakers be more consistent by defining supply and demand configurations that fit within system features (Loch, Adamson, & Mallawaarachchi, 2014). Researchers propose three periods of rural water policy in Africa and Asia between 1980 and 2030 to (i) identify key principles of rural water policy formation, (ii) analyze how they have changed, and (iii) offer progress priorities to ensure that everyone has access to at least basic drinking water by 2030 (Hope et al., 2019).

Even though Bangladesh has been independent for nearly 50 years, the government is still working to develop a comprehensive strategy to ensure the safety of coastal residents. The coastal territory is home to around 40 million people. As a result, the government must immediately begin a strategic plan to achieve a long-term outcome from this large workforce. Along with development preparation, stakeholders must ensure mutual respect and regularity with the beneficiary groups.

Conclusion

From the study, it is evident that drinking water sources in two coastal districts are significantly different. Classified data show before the safe water plant installation, people of Khulna district used to collect water from local ponds, whereas inhabitants of Bhola district mostly depended on shallow tube well. Another analyzed data show a significant relationship between the damage of water plants and the illness of the coastal inhabitants. Data show that respondents of the study areas have noticed nine different problems. Remarkably in coastal regions of Bangladesh, women counterparts are suffering more as they have limited scope to go far away from home. Another result shows that the condition of water sources is highly responsible for waterborne diseases of coastal inhabitants.

Significant sample test shows in coastal regions of Bangladesh; political influence plays a remarkable role in selecting the place of water plant installation and its maintenance work. Another significant test result shows that the implementation of drinking water plants remarkably reduces several waterborne diseases among residents of coastal regions. The outcome of KII 01 shows that waterborne diseases are remarkably reduced after installing a water plant. In addition, it is observed that coastal residents' expenditure on medical and gasoline was reduced due to availing of freshwater. Moreover, people enjoy spare time and get relief from managing safe water for daily living. On the other hand, KII 02 shows a scarcity of many tube wells compared to the number of residents in some coastal areas. In addition, an acute shortage of safe water arises during the dry season. Moreover, ignorance is a vital issue as many coastal people are not aware of safe drinking water even though they do not get enough support from the concerned authority.

Another result shows that imposed technology made trouble for the inhabitants. People welcome new technologies that are socially, culturally, and economically suited. Religious value has an impact on using surface water even if it is contaminated. Finally, tube well installation is on top of other water purification plants installed among coastal regions of the study regions. But remarkably, there are few deep tube wells installed in coastal regions of the Bhola district, but none of it noticed in the coastal regions of Khulna district. On the other hand, rainwater harvesting systems are mostly noticed in the coastal regions of the Khulna district. However, in the end, it is noticeable that by ensuring proper management of remaining installed plants and ensuring further necessary plant installation, it is possible to boost the healthy livelihood of coastal residents. Therefore, the concerned authority should address the above-mentioned issues to ensure a sustainable coastal economy. Though Bangladesh has a large coastal region, the nature of its coastal resources is diverse. Considering this fact, stakeholders must assess individual places' unique culture, habits, and socioeconomic realities to ensure policy success. However, it is possible to tackle current challenges in water management in Bangladesh's coastal zones by considering all the factors mentioned above.

References

AbouRayan, M., & Djebedjian, B. (2004, May). Egypt's water demand, supply and management policies. *International Water Demand Management Conference, Dead Sea, Jordon* (Vol. 30).

Al Khateeb, N. (2001). Sociocultural acceptability of wastewater reuse in Palestine. *Water management in Islam*, 79–84.

Backes, C.W. (1999). Towards international water boards and cross-border spatial plans. *The State of Water. Koninklijke Vermande, Lelystad*, 179–194.

BRAC. (2016). 27m Bangladeshis at risk of sea-level rise by 2050. *Dhaka Tribune*, 1.

Chadwick, M., & Datta, A. (2003). Water resource management in Bangladesh. *Policy Review Paper, 1.*

Chan, N. W., Roy, R., & Chaffin, B. C. (2016). Water governance in Bangladesh: An evaluation of institutional and political context. *Water, 8*(9), 403.

Chowdhury, N. T. (2010). Water management in Bangladesh: An analytical review. *Water Policy*, *12*(1), 32–51.

Coastal Zone Policy. (2005). Ministry of Water Resources. *The People's Republic of Bangladesh*, 1–14.

Doolan, J., & Hart, B. T. (2017). Water resource policy, planning and management in Australia—an overview. *Decision Making in Water Resources Policy and Management*, 3–19.

Earle, A., Cascão, A. E., Hansson, S., Jägerskog, A., Swain, A., & Öjendal, J. (2015). *Transboundary water management and the climate change debate*. Routledge.

Faruqui, N. I., Biswas, A. K., & Bino, M. J. (2001). *Water management in Islam*. Ottawa, IDRC.

Funke, N., Claassen, M., & Nienaber, S. (2013). Development and uptake of scenarios to support water resources planning, development, and management: Examples from South Africa. *Water Resources Planning, Development and Management*, 1–27.

Hadi, T. (2019). An analysis of water policies and strategies of Bangladesh in the context of climate change. *Asia-Pacific Journal of Rural Development*, *29*(1), 111–123.

Hope, R., Foster, T., Koehler, J., & Thomson, P. (2019). Rural water policy in Africa and Asia. *Water Science, Policy, and Management: A Global Challenge*, 159–179.

Hueso, A., & Bell, B. (2013). An untold story of policy failure: The Total Sanitation Campaign in India. *Water Policy*, *15*(6), 1001–1017.

Hukka, J. J., Castro, J. E., & Pietilä, P. E. (2010). Water, policy and governance. *Environment and History*, *16*(2), 235–251.

Islam, Mazharul. (2015). Water resourse management and policy in Bangladesh. *ACADEMIA*, 1–21.

Lenton, R., & Muller, M. (2012). *Integrated water resources management in practice: Better water management for development*. Routledge.

Leopold, L. B. (1959). Water resource development and management. *Journal American Water Works Association*, *51*(7), 821–827.

Loch, A., Adamson, D., & Mallawaarachchi, T. (2014). Role of hydrology and economics in water management policy under increasing uncertainty. *Journal of Hydrology*, *518*, 5–16.

Loucks, D. P., & van Beek, E. (2017). Water resources planning and management: An overview. *Water Resource Systems Planning and Management*, 1–49.

Martin, P., & Williams, J. (2013). Water governance: A policy risk perspective. *Water Resources Management*, 7(171), 73.

McDonnell, R. A. (2008). Challenges for integrated water resources management: How do we provide the knowledge to support truly integrated thinking? *International Journal of Water Resources Development*, *24*(1), 131–143.

Mercer, D., Christesen, L., & Buxton, M. (2007). Squandering the future—Climate change, policy failure and the water crisis in Australia. *Futures*, *39*(2–3), 272–287.

Neto, S. (2018). Territorial integration of water management in the city. *Water challenges of an urbanizing world*. IntechOpen.

Nurul, S., & Mohammad, I. (2014). State of water governance in Dhaka Metropolitan city of Bangladesh: Evidence from three selected slums. *International Journal of Interdisciplinary and Multidisciplinary Studies*, *1*(2), 19–38.

Pal, S. K., Adeloye, A. J., Babel, M. S., & Das Gupta, A. (2011). Evaluation of the effectiveness of water management policies in Bangladesh. *Water Resources Development*, *27*(02), 401–417.

Safiuddin, M. D., Karim, M. M., Murphy, T., & Guo, J. (2003). Water resources management in the remediation of groundwater arsenic contamination in Bangladesh. *Aquatic Arsenic Toxicity and Treatment*, 1–17.

Tundisi, J. G. (2008). Water resources in the future: Problems and solutions. *Estudosavançados, 22*(63), 7–16.
Unesco, Y. (2002). *Knowledge for sustainable development-an insight into the encyclopedia of life support systems*. Oxford: EOLSS Publishers/UNESCO.
Walker, G. (2014). Water scarcity in England and Wales as a failure of (meta) governance. *Water Alternatives*, 7(2).
WescoatJr, J. L., White, G. E., & White, G. F. (2003). *Water for life: Water management and environmental policy*. Cambridge University Press.

7

WATER MANAGEMENT ISSUES OF COASTAL BANGLADESH

Syed Hafizur Rahman and Prabal Barua

Introduction

The coastal environment of the worldwide nations is repeatedly changing due to the interactions among the ocean, winds, currents, waves, and anthropocentric variables. The sustainability of coastal ecosystems is contingent on the degree of use of the coast by the stakeholders and the external impacts of natural disasters. However, the extent of coastal use and external impacts may underwrite global warming and increase it significantly. Coastal areas supply grave ecological services like nutrient cycling, flood control, shoreline stability, beach replacement, and genetic resources. The Ocean and coastal systems added 63% of the total value of the world's ecosystem services. The increasing population is a significant apprehension for the coastal areas, with more than 50% of the world's inhabitants concerted within 60 km of the coast. The continued growth of world population and per capita consumption has the consequences of unsustainable exploitation of the world's biological diversity, worsening climate change, ocean acidification, and other anthropogenic environmental effects. The effective conservation of biodiversity is necessary for the survival of humans and the maintenance of ecosystem processes (Barua et al., 2017; Barua et al., 2020: Eslamian et al., 2020).

At present, Bangladesh has been called the 'ground zero' of climate change. However, the impact of 'climate change' is causing an increase in the frequency and severity of these disasters adversely. The coastal area of Bangladesh is outlined in different ways, and significant vulnerabilities of the coastal zone are tidal fluctuations, salinities, and cyclone and storm surge risk. According to these characteristics, the coastal zone of Bangladesh covered 19 districts, where approximately 42 million inhabitants live in the zone, which is 26% of the total population and 32% of the total land area of the country (Figure 7.1). According to the different geographical characteristics and locations, the coastal zone of Bangladesh is divided into three

DOI: 10.4324/9781003253495-9

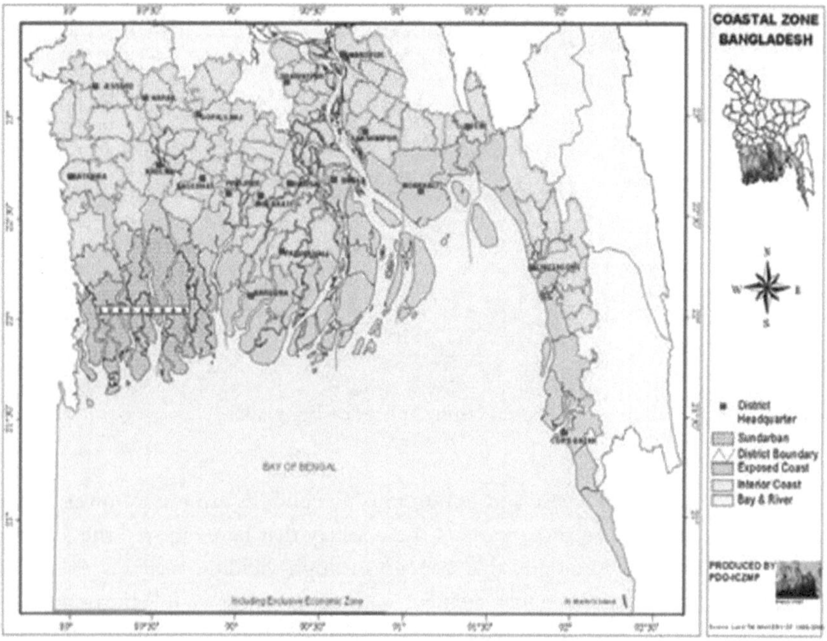

FIGURE 7.1 Geographical location of the coastal zone of Bangladesh

distinct zones according to the different geographical characteristics and locations, which are (1) Eastern zone, (2) Central zone, and (3) Western zone. Most coastal communities are living in hardcore poverty, experience suffering a massive loss due to natural disasters and man-made hazards (Akter, 2020; Biswas et al., 2017; Barua et al., 2021).

Coastal areas of Bangladesh are stressed by the significant impact of climate change–induced natural hazards and anthropogenic causes. So, this immense part of the coastal area requires sustainable management and a comprehensive scaffold for integrated, comprehensive development, coastal ecosystems conservation, and protection for the coastal communities (Figure 7.2). Coastal resource management and development programs of Bangladesh started in the early 1960s through starting a different approach like coastal embankment construction, developing coastal greenbelt, reclamation of new land, community-based management, and development of new delta establishment and management (Haque, 2017; Hossain et al., 2020).

Approximately 30% of the fertile land in Bangladesh is available in the coastal zone, which is cultivable, and different salinity levels influence approximately 5 million hectares of land. Agriculture farmers are producing mostly low-yielding, traditional rice varieties at the moment of the wet season. Maximum lands remain uncultivated in the summer season because of salinity intrusion in the sediments and the short of good-quality irrigation water. It was found that cropping intensity,

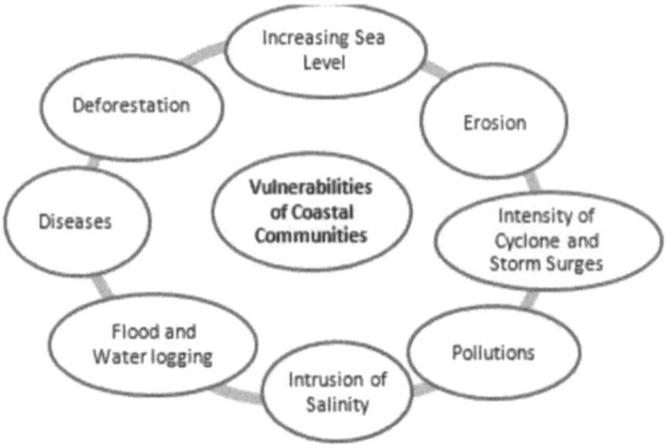

FIGURE 7.2 Vulnerabilities of coastal communities of Bangladesh

crop yields, production levels, and community livelihoods are much lower in the coastal areas compared to other parts of the country that have enjoyed the fruits of recent agricultural technologies that depend on high-yielding varieties, enhanced levels of fertilizer, and water management and improved pest and disease control actions (Islam et al., 2015; Barua and Rahman, 2021).

Bangladesh's water management is significant because of the growing requirement and increasing divergence between its substitute uses in various sectors. Water demand is increasing speedily in agriculture (for irrigation), urban and industrial sectors, inland navigation, fisheries sectors, and control of salinity intrusion (Chowdhury, 2010). Water specialists, policy planners, and international donors usually agreed that improving water veracity is one of the excellent solutions to water insecurity in Bangladesh (Araral and Yu, 2012).

Supplying agriculture farmers with alternative crop production systems with high land and water productivity is decisive for food security, increasing farmers' livelihood and environmental sustaining of the coastal zone. The employment of short-duration, high-yielding varieties of rice, the effectual application of rainwater, and proper crop arrangement linking crop water requirements with the water supply and eminence dynamics could enhance the cropping pattern, production, and productivity of paddy fields in the coastal zone (Mahmuduzzaman et al., 2014; Rahman et al., 2016; Rahaman et al., 2018).

According to the report of the World Bank, by 2050, climate change will create enormous alteration for the salinity levels in rivers of coastal Bangladesh at the moments of the dry season. This problem becomes the consequence of the drinking and irrigation water supply shortage with changes in aquatic ecosystems. To solve the drinking water in coastal Bangladesh, it has spawned a new business: selling potable water. A crowd in front of a water shop is a common sight nowadays. The shops are usually set up at points where relatively uncontaminated deep aquifers have been found. The water is pumped up, treated at a reverse osmosis (RO) plant at the back of the shop to get rid of the salt, and sold. The present study has assessed that the cost of potable running water in the coastal areas is far higher than in the urban city of Bangladesh (Saha et al., 2017; Rabbani et al., 2018; Salauddin et al., 2018; Taher and Rahman, 2018). Rahman et al. (2017) developed the DPSIR (Driver-Pressure-State-Impact-Response) framework for Integrated Water Resource Management (IWRM) that is possibly applied for sustainable planning and preparedness under the probable risks and susceptibilities of climate change on water resources of Bangladesh.

This chapter describes the different policies and strategies related to water management and administrative institution involving water management for Bangladesh's coastal zone. Besides, this paper seeks to assess salinity intrusion on water bodies and crop fields in the coastal inhabitants' areas and resilient practices through different agricultural adaptation mechanisms in response to climate change to develop efficient, equitable, and sustainable water management.

Water Management Policies and Strategies for Bangladesh

Bangladesh has different national policies and strategies for various prime sectors to accelerate the balanced way of development. Several policies, strategies, and acts establish the rules and regulations on the general utilization of water for different sectors. The significant policies and acts are as follows:

1. Ports Act (1908)
2. Territorial Waters and Maritime Zones Act (1974)
3. Bangladesh Irrigation Water Rate Ordinance (1983)
4. The Marine Fisheries Ordinance (1983)
5. Ground Water Management Ordinance (1985)
6. Water Resource Planning Act (1992)
7. National Water Policy (1999)
8. Coastal Zone Policy (2005)
9. Coastal Development Strategy (2006)
10. The Environment Conservation Rules (1997)
11. National Policy for Arsenic Mitigation (2004)
12. National Water Management Plan (2011)
13. National Strategy for Water and Sanitation for Hard to Reach Areas of Bangladesh (2011)
14. The Ship Breaking and Recycling Rules (2011)
15. Bangladesh Water Act (2013)
16. National Strategy for Water Supply and Sanitation (2014)
17. Bangladesh Water Rule (2018)
18. Ground Water Management Act (2018)
19. Bangladesh Ship Recycling Act (2018)
20. Bangladesh Delta Plan (2100)

The Ports Act was developed in 1908to merge the performances related to the ports and port charges. The various chapters of this act elaborate on the government's influences, port administrators and their duties and responsibilities, safety rules for shipping and port protection and conservation, raising signals for shipping vessel protection from natural disasters. This act offers the elimination or devastation of anything hindering the channel leading to any port in Bangladesh. This act has highlighted water safety, pollution control, and the environmental management process for sustainable management of the port of Bangladesh.

Territorial Waters and Maritime Zones Act (1974) declares Bangladesh's territorial waters and maritime zones. Clause-4 stated that the planning and preparation for any type of emergency condition would integrate with the water services and security of the country. The government has the vision of safeguarding the production of aquatic biodiversity of the sea as in clause-4 and clause-6 by setting up conservation zones in such places neighboring the territorial waters. According to the clause-8, the Bangladesh Government might, with a vision to protecting the seawater and coastal area from marine pollution and safeguarding the superiority and ecological balance in the coastal and marine environment in the deep seas contiguous to the territorial waters, take such actions as it may estimate suitable for the purpose. The ordinance should be updated within the present demand as the climate change perspective should be integrated for Bangladesh's standard coastal water resource management.

The Bangladesh Irrigation Water Rate Ordinance was formulated in 1983tocombine and adjust the law related to the obligation of water rate for delivering storage or regulation of water for drainage and irrigation purpose. The ordinance offers for charges of water applied to occupants or landowners profiting from the supplies of water on behalf of the Government, Bangladesh Agricultural Development Corporation (BADC), and Bangladesh Water Development Board. The water channel is all over the land, and any activities undertaken to build and sustain the waterworks will be no fee for their landowners and tenants. According to Article 9, there is a provision for diversion of normal water flow as a crime subject to a fine up to ten times the charge for illegal water use for that financial year.

The Marine Fisheries Ordinance was adopted in1983 to strengthen the proper management, protection, conservation, and increase marine fisheries production in Bangladesh. This ordinance explores the categories, classes, and amounts of fishing vessels, substances for the validation of license, responsibility to give information concerning harvesting, catches, license issue, local fishing ships, and vessel to grip valid of inspection certificate, utilization of explosives, fishing purpose, dredging and other issues forbidden in coastal and marine reserves, authority to stop the vessel, etc.

Ground Water Management Ordinance is a rule formulated in 1985 to manage and utilize groundwater sources for agricultural production. In this ordinance, it is mentioned in clause-5 that a license should be provided for the installation of tube well near the agriculture field with fairness, competence, and sustainability that explain the impartial service provision. In this ordinance, precedence is specified to necessary services for water and sanitation in all the plans, policies, budgets, and relevant expenditures. According to the ordinance, there will be Upazila irrigation that will be responsible for necessary permission to install tube wells in the respective areas by providing license by monitoring the location of the installation places whether it will be necessary or not. In the ordinance, there are no options for groundwater and surface water, and monitoring systems.

The Water Resource Planning Act has been established to secure the development and impartial utilization of water resources in January 1992, and this assists the strategy execution, but this has a few missing points in the integrity issue. Although the planning for the Water Resource Planning Act was prepared to secure the progress and balanced utilization of water resources, rural and urban accessibility to overhaul and sufficient planning and arrangements have not been stated in any clause of this act. Community participation, gender context, the voice of local people, and their preference not expectant due to this act have any clause connected to these matters, but current water management requires these issues to be integrated. Environmental management is optimistic in this act because the Bangladesh Government's initiative might think significant, so to do, at any moment, ask for a story and statement on any issue from the organization. The institution would be spring to furnish such documents and statements to the Bangladesh Government.

The National Water Policy was developed in1999 as the main rule of the water applied to the coast, drought, hilly areas, water management, and the Ministry of Water Resources responsible for conducting all the activities that coordinate with this policy for securing the appropriate application of the inadequate water resources. This policy has precise goals that are generally the major achievement of the segmented achievement as elements of sustainable water management.

In this policy, few distinguishing characteristics of accessible water mechanisms and few significant extreme challenges are identified, which are as follows:

1. Individual consumers of water are considered as the significant negotiator for development.
2. Requirement of the vigorous endorsement of the private sector to the improvement of sustainable water management.
3. Advancement of the water sector development with the support of lumpy and significant levels of capital investment.
4. Economics of scale validates the involvement of the public sector.
5. Necessity to gratify multi-sector water requires with the inadequate resources.
6. Defining responsibilities on public and private sectors for water management.

The Bangladesh Government formulated the Coastal Zone Policy in 2005 to provide a common direction to coastal zone management and development issues to pursue their lives and livelihoods within a secure and conducive environment. The Coastal Zone Policy was developed through a different stakeholder level discussion procedure over three years. Coastal development progression aspires to convene, on an overall basis, increase the rate of economic growth, reduce poverty level, social harmony and development, special coding conduct for fisheries management, special conservation approach for wetland management as well as other international treaties and conventions to obtain the target of Millennium Development Goals (MDGs).

The Coastal Development Strategy was developed in 2006 with the aim of coordination between different sector policies for the coastal region of Bangladesh. The development of coastal zones progression intends to achieve the National Goal for Accelerated Poverty Reduction (PRSP); the MDGs, special coding conduct for fisheries management, special conservation approach for wetland management, and other international conventions. The key objective of the Coastal Development Strategy is to generate circumstances in which the poverty decline, creation of secure life and livelihoods, and the incorporation of the coastal zone into the national progression could take place for accessing resources.

The Environment Conservation Rules were formulated in 1997, and the last amendment was completed in 2003 that assists policies and strategy execution in organizational authorization where regulation attains the fairness, competence, and sustainability in allotment and management of water resources. According to clause-3 and clause-5 of this rule, the ecologically critical area and application are connected to pollution or environmental degradation. The main concern is provided in the rules and policies and based on clause 12, the water standard is resolute in harmony with the principles practiced in schedule-3 for inland surface water, the purpose of drinking water, and the industrial waste effluent. According to schedule-12, monitoring and scheming the ground and surface water pollution, different water standards were assessed, and the values of the release or discharge of garbages of different industrial companies will be assessed according to the values specified.

The National Water Management Plan was adopted in 2011to provide a framework plan for the national and regional levels. Local stakeholders, civil societies, local Government administration, and Government water sector organizations should take the proper plan and execute their programs and projects. It also gives instructions for protecting and liberating water and water-related various services and developing sustainable water resources in Bangladesh.

National Policy for Arsenic Mitigation adopted in2004 has been started to extenuate the impact of arsenic on community people and the environment. This policy mentioned the characteristics and degree of the problem and mitigation process of arsenic, organizational arrangement, and partnership among them. This policy suggests applying suitable options and reasonable technologies as shallow tube wells could no longer supply drinking safe water and cooking in arsenic-affected areas.

The Ship Breaking and Recycling Rules were adopted in 2011, which the Ministry of Industry manages. The Ship Breaking Industry is operating in Chittagong coastal areas and is the main steel production sources of Bangladesh. According to Rule 13, gaining beaching permission, if this is obtained, the meticulous ship is not assembled and operating cargo vessel, domestic or ocean, departing to be declared an illegal offense by the yard owner or company to whom beaching permission was granted. However, according to Rule 18, this is mentioned that the ship recyclers are necessary to severely provide environmental compliance in line with coastal soil, water, air under the Environment Conservation Act 1995 (Act 1 of 1995 amended in 2010), and other related national environmental acts or laws. So these rules are essential for governing the ship breaking industries of Bangladesh.

National Strategy for Water and Sanitation for Hard to Reach Areas of Bangladesh was adopted in 2011; this strategy highlights the progress of the water and sanitation facilities in the hard-to-reach hill areas, river islands, haor areas, urban slums, and coastal areas. Good service supplies have been affirmed in clause-3 of this strategy so that the poor, vulnerable, and marginalized communities could obtain evenhanded admittance to the sector services. Besides, sufficient proper planning and arrangements have been prepared for emergencies that integrate water and sanitation facilities for the susceptible communities. In general, the WatSan strategy highlights the local communities' involvement in the conventional decision creation process for sustainable service provision. Besides, proper planning, financing, transparency, accountability, and promotion of gender-friendly, context-specific water and sanitation facilities are addressed in the strategy. Besides, climate change adaptation matters incorporated in the development plan of coastal areas of Bangladesh also added as necessary perceptively in this strategy.

Bangladesh Water Act was initiated in2013to create provision for the development, appropriate management, distribution, conception, utilization, proper protection, and national water resource conservation in an integrated approach for the different water sectors in Bangladesh. This Water Act is appropriate for groundwater, surface water, seawater, rainy water, and atmospheric water in Bangladesh. Beneath this Water Act, various swap could be potential at government level for water resource development that are (1) review and exchange for whichever general water resource-related information; (2) economies of levels justify the engagement of public sector; (3) necessity to satisfy multi-sector water requires with inadequate resources; (4) describe the responsibilities of public and private sectors.

National Strategy for Water Supply and Sanitation was developed in 2014is an essential element of the Sector Development Plan (SDP) 2011–2025 for the sustainable water and sanitation sector in Bangladesh.This national strategy provides the sector situation, aim and objectives, steering principles, appropriate framework, direction, organizational understanding, and execution plan for sustainable freshwater supply and sanitation at local, regional, and national levels. This strategy recommends attaining the sector purpose, and in agreement with the steering principles, a set of 17 different strategies has been developed. This strategy is generally categorized into three themes: (a) interventions of WASH – raising the exposure and advancing the quality of interventions for water, sanitation, and hygiene activities, (b) promising challenges – addressing the budding confront in the sector, and (c) sector governance– intensification of the segment governance.

Bangladesh Water Rule was formulated in August 2018 following the Bangladesh Water Act, 2013. The government decided to take a leading role to The Water Resources Planning Organization (WARPO) to implement this rule. In this rule, WARPO has been authorized to provide groundwater abstraction license together in groundwater-strained areas. According to this rule, all kinds of water resource development programs and projects have a clearance certificate from the respective union, Upazila committee of water resources management. Under this committee, any program like the establishment of the coastal embankment, installation of deep tube well, pond sand filter, conservation of pond and other water sources, water resource storage, water resources for industrial purposes, hydraulic infrastructure establishment, wetland conservation, river dredging, flood management, or groundwater resources for agriculture purposes, a No Objection Certificate should be taken from union and Upazila level water resource management committee to national technical committee for water resource management. They also give power to WARPO for declared Water Vulnerability Area according to the criteria of community consultation, reasons of vulnerability, crisis, report of the technical committee, and analysis of socioeconomic and environmental reports based on scientific assessments. Involvement of the communities and civil society is highly encouraged in this water rule. There is a provision for representatives of the CSO members in the committee of (1) the district water resource management committee, (2) the Upazila water resource management committee, and (3) the Union water resource management committee. The rule is mentioned for developing the social web portal for applying, processing, and tracking these No Objection Certificates as described in Chapter 10 of Bangladesh Water Rule 2018.

Ground Water Management Act for Agriculture purpose was formulated in 2018 to decline the wastewater and secure, well-integrated planning for utilization of groundwater in a high yield of agriculture production all over the country. Besides, this act is prepared to secure the best potential utilization of water and its well-planned management for enduring the development drift in the agriculture sector. Upazila administration has obtained full empowerment for implementing the Ground Water Management Act, and this administrative body is responsible for forming and supervising Upazila Irrigation Committee (UIC). After conducting field inspections, the UIC will provide the permitted license for installing tube wells and any other water sources by the concerned and experienced crop farmers in exchange for a fee. Once the law comes into effect, the penalty for the use of unlicensed groundwater will be a maximum Tk10,000 fine and seven days of imprisonment in case of default. In the rule, it is mentioned that the Upazila administration will monitor the utilization of water in agriculture in all the locations of Bangladesh and provide the list to the national level for water crisis area. According to the crisis, the administration will provide necessary permission to establish the deep, shallow tube wells and other water sources in their areas with certain application process. The Ground Water Management Act (2018) was established after closing the Ground Water Management Ordinance (1985).

Bangladesh Ship Recycling Act was formulated in 2018 for an environment-friendly ship recycling industry to operate because this industry provides an enormous negative impact of coastal pollution in Bangladesh. Section 4 of the Act has mentioned creating a zone for the ship breaking industry where Section 25 identifies that whoever establishes any ship breaking yard without the government's permission will be a punishable offense. According to the law, Bangladesh Government should establish the Treatment Storage and Disposal Facility (TSDF) within three years for proper management of wastage produced from ship breaking for protection of coastal pollution, aquatic environment along with coastal communities.

Bangladesh Delta Plan 2100 is a long-term unique strategic plan for the nation's sustainable water and land management plan until 2100. The vision of this Delta Plan is 'Secure long-term food and water safety, environment and economic sustainability that efficiently resilience and adopting with natural disasters, climate change adaptation and various delta related matters through the vigorous, adaptive, coordinated, integrated and evenhanded water sector governance.' The Bangladesh Delta Plan recommends three national policies integrated with national sustainable plans and goals of six delta-based themes that would add to that ambition. The Delta Plan has recognized six hotspot locations with water sector-based challenges, according to the ease of developing strategies and addressing them at the national levels. Among the challenges, river flooding is the most significant confront in all coastal regions. The potential actions suggested to address the existing troubles are as follows: (a) river management, (b) coastal zone management and protection, (c) flood risk management, (d) water availability and irrigation management during the dry season, (e) freshwater supply, and (f) wetland protection.

Administration of Coastal Water Management

Bangladesh Government established different regulatory administration institutes to implement and manage different activities based on various policies and acts on coastal water management, which are as follows: Ministry of Water Resources, National Water Resources Council (NWRC), Bangladesh Water Development Board (BWDB), Water Resources Planning Organization (WARPO), Institute of Water Modeling, River Research Institute (RRI), Joint Rivers Commission (JRC), Center for Environmental and Geographic Information Services (CEGIS), Local Government Engineering Department (LGED), Department of Public Health Engineering (DPHE), Department of Environment (DoE), Bangladesh Agriculture Development Corporation, Ministry of Industry, and Water and Sewerage Authority (WASA).

Ministry of Water Resources is the top administration of the Bangladesh Government for the expansion and management of the entire water properties of the nation. This ministry develops the policies, acts, plans, strategies, regulations, guidelines, and instructions for the country's inhabitants. This ministry is responsible for developing the water sector and management together with an extension of the irrigated areas, water conservation, surface- and groundwater utilization, and river management.

NWRC was founded under the Bangladesh Water Act 2013, an inter-ministerial uppermost policy-making and decision-making administrative body. It is authorized to decision- and policy-making, the direction of policy, harmonization of effective water resources planning, and appreciative and securing execution of the National Water Resources Plan.

BWDB was established in 1959 as the water division of the preceding East Pakistan Water and Power Development Authority. As the major organization of the government for water resource management, BWDB was provided the accountability of completing the errands of implementing the flood control, irrigation, and drainage programs to raise agricultural production and fisheries resources.

WARPO was established in 1992, which is an autonomous national institution responsible for the implementation of water resource planning for the country. This organization has further been reinforced through the implementation of thc National Water Policy in 1999. WARPO has full responsibilities for developing and implementing the National Water Resources Management Plan, a scaffold plan to translate the National Water Policy into the different activities under the scaffold of Integrated Water Resources Management (IWRM).

RRI was established in 1973, a national institute working as a public legislative authority as an autonomous administrative body of the Ministry of Water Resources. This research institute consists of three divisions: Hydraulic Research, Geotechnical Research, and Administration& Finance. RRI helps out the Bangladesh Government agencies for managing the rivers and assessing the impact of development of the rivers.

JRC was established in 1972 whose functions are (1) to create the sustainable liaison among the neighboring nations in order to secure the most effectual dual attempts in exploiting the more benefits from the general river systems to the nations; (2) to eloquent flood control activities and to suggest the execution of dual programs; (3) to develop the complete proposals on early warnings of flood, flood and cyclone forecasting; (4) to conduct the research investigation on the control of floods and agriculture irrigation programs.

CEGIS is a trust founded by the Government of Bangladesh as a public trust. CEGIS aspires to function as a center of excellence for assisting the research on coastal and river erosion, flood signals, analysis of environmental quality, remote sensing, and geographic information systems for the coastal and mainland of Bangladesh. CEGIS is responsible for conducting all Environmental Impact Assessment connected to the water sector as a trustee of the Ministry of Water Resources.

Flood Forecasting and Warning Centre (FFWC) is an institute founded under the Bangladesh Water Development Board accountable for an emergency response to flood conditions to reduce the loss of life and damage of property. This center aims to attain this by increasing the community capacity and the government to administer the flood-related disasters of the nation efficiently.

DPHE is the leading national organization under the Ministry of Local Government, Rural Development and Cooperatives (MoLGRD&C) responsible for providing water supply and sanitation services in Bangladesh except for Dhaka, Chittagong, Khulna, Rajshahi, and Narayanganj, where WASA operates. DPHE is working in the coastal areas to provide a safe water supply, install water sources, mitigate arsenic problems, and waste management activities to improve the coastal environment.

Institute of Water Modelling (IWM) is an autonomous research organization funded by the Bangladesh Government that deals with statistical modeling of Bangladesh river systems, coastal environment, river and coastal erosion, Geographical Information Systems, environment, and surface, along with groundwater modeling.

LGED is a department under MoLGR&D that aims to integrate community participation in the planning and implementation of water resource development initiatives. This department executes many programs and projects in the water sector, frequently to strike an equilibrium among the water utilized for the small-scale and large-scale water-related projects. Smaller-scale projects are executed mainly by LGED. LGED is a highly decentralized department, with most of its human resources is deployed at the district and sub-district levels. It is one of the significant public sector institutions in the country commended for proper planning and execution of local level rural–urban and infrastructure development of small-scale water resources programs.

WASAs were founded by the WASA Act 1996 under MoLGR&D in five cities of Dhaka, Narayganj, Khulna, Chittagong, and Rajshahi. The WASAs are accountable for the sustainable planning, management, development, construction, and operations and maintenance of the safe water supply and sewerage facilities, stipulations within their authority. WASAs also have the power to provide permission licenses for the installation of deep tube wells in their influence.

DoE is the administrative department mandated under the Environment Conservation Act 1995 to monitor activities impacting environmental quality (water, air, land pollution, etc.) and implement conservation and management standards of environmental perspectives. This department is under the Ministry of Environment, Forest and Climate Change and is authorized to control and implement environmental management, including water pollution. This department is also responsible for securing the sufficiency of the Environmental Impact Assessments.

Ministry of Industry manages the progression and procedures engaged in the foundation of industries. This industry is pertinent from creating industries that might be noteworthy water utilizers or factors impacting the water quality through the effluent discharge in water bodies. This ministry has the sole responsibility of managing the shipbreaking industry in Bangladesh to control environmental pollution.

BADC is a department under the Ministry of Agriculture that is responsible for giving support to farmers for irrigation facilities through ground- and surface water sources, assist in providing climate resilience crops to the farmers, and innovating modern technology for production and harvesting of crops as well as irrigation mechanisms.

Water and Livelihood of Coastal Communities

Coastal areas of Bangladesh are recognized as the formerly affluent hub for fisheries and agricultural production. Sweet freshwater was accessible; saline water was useful due to natural flow; the forest was resourceful to provide food and shelter for local people; and the surrounding community villages were wealthy with an abundance of commercially important trees; families have domestic and pet animals. In general, the coastal ecosystem benefits were highly abundant at a stage that provided a livelihood for coastal communities. Because of the sufficient natural situation, there were adequate agriculture crops; shared cropping practices that offer the marginalized landless and small crop farmers to produce adequate crops for their yearly stock of food. Highly potential coastal and aquatic biodiversity and surrounding natural environment carried the production of livestock. Farmers were relied on each other for agricultural activities, which reserved them to constrict in collective approaches. Thus, coastal communities lived in social harmony; enriched biodiversity and high accessibility of available natural food sources permissible for the marginalized poor and deprived people to stay away from divergence with landlords or big farmers in the coastal areas (Rahman et al., 2019).

Researchers identified that unplanned expansion of immense coastal aquaculture predisposed water and land-use changes that changing agricultural lands into coastal aquaculture farms bringing a salinity level into freshwater-fed agriculture lands. Since the 1980s, coastal aquaculture was initiated in the shrimp farms surrounded by the earthen dykes, situated by riversides. Two main factors jointly provided a catalyst to the progression of coastal aquaculture farming: brawny international market requirements and large prices for shrimp and prawn products; and this was no longer economically viable to rice production as coastal polders were waterlogged because of the poor drainage.

Rahman and Islam (2019) mentioned that coastal communities facing a regional reduction of natural resources like safe drinking water resist to continue sustainable life and livelihoods. Natural disasters and the effects of climate change put raising pressure on the coastal region, hampering livelihoods. Over the last 50 years, development done through controlling the tidal water and estuarine rivers has altered the natural environment. However, communities botched to control cyclone and storm surge that is a pouring force of ecosystem obliterations. The unplanned and increasing rate of coastal aquaculture and salt farming shifting agricultural crop production, sponsoring aquaculture in paddy fields that centralized the scheming of coastal resources in the hands of the influential people; establishment of engineering infrastructures shaped huge congestion of the drainage system of the rivers, canals, and channels (Figure 7.3).

For drinking water supply in the coastal areas, deep tube wells and pond sand filter (PSF) technology are now using for solving the saline water intrusion in the

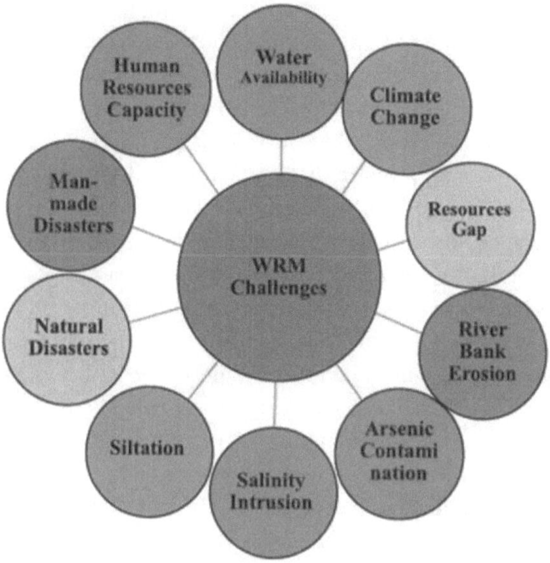

FIGURE 7.3 Water resource management problem for coastal communities of Bangladesh

water sources by shifting the thousands of years of utilizing ponds as a dependable source of drinking water. These modifications have caused widespread environmental spoil where coastal communities live with the threat of water scarcity from groundwater-level decline. The livelihoods of coastal communities are under significant threats, and the natural environment is tremendously delicate and under growing pressure.

Domestic livestock creates fundamental contributions in the marginalized coastal peoples' livelihoods regarding nutrition and income-generating activities. Livestock generally suffers two types of susceptibilities because of increasing coastal aquaculture and salt production: declined sources of fodder and augmented mortality rates because of salinity intrusion in water sources of livestock. Saltwater coastal aquaculture activities practiced in the state-owned lands (*Khas* land) that coastal people grazed their livestock and declined the fodder quantity and other domestic animal feeds. The present amount of livestock had declined enormously in contrast to the cattle number before shrimp and salt farming started. As a result, poor and marginalized coastal farmers were forced to sell their cattle or took them exterior of the coastal aquaculture and salt shrimp farming areas.

The coastal community practice of differing livelihoods that the contestation among the saline water and freshwater in the coastal areas of Bangladesh could be traced in history in the approach water has been directed and the manner political-economic services influenced the coastal water systems. The exclusive tidal wetlands of the coastal areas have always sustained some salinity level, yet the soil stayed fertile, and rice production was significantly high. This was not awaiting the introduction of the embankment system and succeeding, endorsement, and proliferation of coastal aquaculture and salt production that salinity intrusion turn out to be a somber problematic issue. In recent years, the southwest and southeastern coastal areas suffer development-induced disasters as salinity penetrates the soil and water tables, threatens the crops, and kills vegetation. Coastal aquaculture farming effects and amplifies salinity levels in the coastal region, declining livelihood diversification options, and day-laboring occasions. Coastal inhabitants are often forced to displace and migrate to urban cities for work and livelihood opportunities.

In the past, crop farmers utilized canal water for irrigation purpose, which was freshwater. However, since 2000, canal water could no longer be utilized for irrigation purposes due to high levels of salinity intrusion, the input of saline water in coastal aquaculture farms. The situation of pond water is also in the same condition, where areas of the pond and the canals are engaged by the owners of coastal aquaculture farms for economic beneficiaries, they merged with shrimp farming. This activity indicated the absence of good governance in the coastal areas and the refutation of rights to utilize water resources and sustainable water management for different purposes of the coastal communities.

This is a general circumstance concerning the freshwater availability for irrigation purpose in cropland. Coastal aquaculture, water logging, cyclone, storm surge, and salinity in groundwater created water crises for agriculture-related activities such as tilling and plowing in the cropland, lifting paddy seedlings, etc. Crop farmers were familiar with applying low value and insufficient water for the purpose of irrigation, which declined crop production to the extent that the crop farmers lost attention in crop cultivation as they could not afford it. This is also a situation that the increase of rice plants reduces to raise the salinity level in the irrigation water. The groundwater is extremely impacted by salinity intrusion and sodium and the use of such water for irrigation creates high level of sodium soils, breaks soil structure, and declines soil aeration and water infiltration.

In the coastal areas, rainwater is the mere source for irrigation of Aman rice cultivation for most farmers. This is necessary for heavy rainfall to remove the soil salinity during the start of the rainy season. However, in current times, rainfall pattern in coastal areas has changed completely and become erratic, and there is a declining rate of rain in the early monsoon period that is unfavorable to agriculture production in the coastal zone of Bangladesh. The amount of rainfall is reduced, especially in the pre-monsoon and monsoon seasons in the country.

The Resilience of Coastal Communities

Coastal communities are unsurprisingly resilient to natural hazards. They have indigenous knowledge for disaster warning and coping practices for resilience on economic losses through the disaster. They have higher information on climate resilience crop practices about the coastal context up to a higher level. The semi-diurnal tidal rivers of the coastal areas support the creating ability of coastal communities living beside the polders to enter water during the time of high tide and drain excess amounts of water at the time of low tide.

A high level of salinity intrusion straightway affects the lives and livelihoods, fresh and safe sources of drinking water, human health, and the coastal region's surrounding environment. Though, coastal communities, Bangladesh Government, and various national and international non-government organizations (NGOs) and donors are attempting through their great work to endure these unenthusiastic circumstances. Coastal communities are involved in using various coping and adaptation mechanisms to reduce the effects of a high level of salinity intrusion. Community-based adaptation practices are generally carried out for family requirements. Local communities are familiar with rainwater harvesting, conserving sweet pond water, and collecting drinking water from the constructed ring wells to solve the catastrophe of safe drinking water. As rainwater harvesting is a relatively costlier option and relatively time-consuming, sweet water pond conservation and pond water are the most popular options by the coastal communities. Besides, excavation of artificial pond and use of PSF are additional drinking water sources that are commonly applicable for many coastal communities. Different government organizations and NGOs contribute to providing sweet drinking water facilities all over the country. National and international NGOs are conducting various programs like excavating the pond and installing deep tube well to combat drinking water scarcity in the study areas.

In the coastal areas, the local water management committee starts the operation of sluice gates that allows fresh or non-saline water from the river during the rainy season to the nurseries. During the wet season, the sluice gate is moreover kept open, or a few times managed on an ad hoc basis for the flow of water or process of drainage in case of a high rate of rain in the wet and dry season.

Most of the agriculture farmers near the coastal polders produce tall traditional crop varieties that can tolerate 0.5–0.8-m-height water depth and no utilization of inorganic fertilizer. The local water management committee does not take adequate measures in managing exchange of water and operation of sluice gates. However, the dwarf high-yielding varieties (HYVs) of paddy crop need advanced water management with the decline of standing water level and to drain water before the initiation of panicle formation and at the direction to endorse high productivity of crop. Coastal communities that face such challenges of excess water limit the adoption of HYV paddy crop in the polder ecosystem (Rabbani et al., 2018; Rahman et al., 2017).

Although utilization of land for crop production is declining steadily, some alternative methods are adopted for agriculture production that reduces salinity intrusion in water and soils around the coastal areas. The major salinity-tolerant practices in the coastal areas are (a) salinity-tolerant rice varieties, (b) cage fishing culture, (c) cultivation of mele (reed), (d) floating dhap cultivation, (e) changing the crops planting time, (f) increasing short-duration rice production, (g) integrated cultivation, (h) cultivation of crabs, (i) platform shed goat, duck, and hen rearing, (j) net fishing, (k) farming in dyke, (l) production of salt-tolerant wheat, (m) salt-tolerant potato production, (n) production of salt-tolerant sweet potato, (o) cultivation of salt-tolerant pulses, (p) short-duration and salt-tolerant oilseeds production (q) increase the production of salt-resistant jute varieties, and (r) raising the production of salt-tolerant sugarcane varieties. Besides, farmers adopt enhanced agricultural technologies and all-year basis cropping because of existing hydrology in the polders of the coastal zone. This technique and technologies need a harmonized cropping system between the cultivators within the community-based water management units and a coordinated approach by the coastal communities at a large level from within the adjacent locations of sluice gate within the polder area. Without these recognized processes, coastal agriculture farmers of Bangladesh might not be able to cope and benefit from the advanced and modern agricultural technologies in response to climate change (Rabbani et al., 2013; Rahman, 2016).

Conclusion and Recommendation

Impact of climate change, salinity intrusion, and soil productivity status are the imperative factors for the production of crops in the coastal zone of Bangladesh. In general, the coastal area of Bangladesh is a low-lying topography, has enormous tidal creeks, and is relatively low in soil fertility. Salinity intrusion in freshwater sources and agriculture is now a global problem, while Bangladesh has no exemption. In Bangladesh, salinity intrusion is one of the significant natural hazards that disrupt versatile crop productions, particularly Rabi crops or winter crops are impacted because of the various levels of salinity. The salinization process changes the paddy crop rotation and discourages farmers from cultivating crops in the coastal areas of Bangladesh. Presently, rapid population growth increased national food demand; climate change, increase salinization process, and loss of soil fertility are the leading causes that place tremendous pressure on the nation. So, time to come applying integrated coastal zone management (ICZM) that may reduce saline water intrusion and colonization process, including increased soil fertility status, can ensure the nation's food security and the next generation.

In this chapter, the authors mentioned that although salinity intrusion crisis fetches more frightful instances with respect to safe drinking water, accessible important stakeholders and legitimate management operators do not pay any heed to it. Few aspects related to the socioeconomic circumstances and environmental issues are rising vulnerabilities for the coastal communities to solve the catastrophe of protected drinking water, although they are stressing to decline their susceptibilities by executing various adaptation options. Despite different coping practices such as harvesting of rainwater, conservation of pond water, and PSF have been utilized at the community levels in remote coastal areas and financially supported by various government organizations and NGOs, salinity intrusion problem till now has not been specified significance in the Bangladesh Government's disaster management policy. The Government of Bangladesh and different national and international NGOs should provide appropriate concentration to conquer the salinity intrusion crisis for the near future.

The current adaptation practices by the farmers are insufficient to adapt to raise the concentration of salinity particularly due to the severe measures. The level of poverty, low-down resilience capacity, lack of alternative livelihoods, and such climate-induced hazards cause huge losses for study communities and the people of the whole coast. Increasingly, people are moving from the coast, mainly because of loss of livelihood opportunities. This internal migration (rural–urban, coastal–central) will intensify as the sea levels continue to rise, as extreme weather events become more frequent, and if adaptation options remain inadequate. The ultimate destination of these low-income families is in the urban slum. Various climatic factors, including salinity, cause this migration. The intrusion will further put the capital, and other nearby cities and towns, in a challenging situation as they attempt to meet the additional needs for adequate utility and other services for old and new city dwellers.

This book chapter has explored the experiences of vast numbers of coastal inhabitants that affect lives and livelihood-related problems due to salinity intrusion that is grounded to a large level through the tropical cyclone, storm surges, tidal flood, and water logging coastal erosion. Increase of salinity level in sediments over the last decades has severe effects on crop production, increase unproductive land, food insecurity, and livelihood options. This is now imperative to reorganize long-term integrated coordinating planning for crop production in the coastal areas, not only in areas presently impacted through the salinity intrusion but also in these locations which could potentially be impacted, according to the forecasted climate variability scenarios. This is found that low-lying agricultural land requires to be steadily protected, even at the same moment, inventive and circumstance precise practices, together with salinity resilience crops and vegetable production, should be researched and applied all over the coastal zone of Bangladesh.

The authors recommended developing the national coastal management policy where government, NGOs, and coastal communities should work together with solid coordination for sustainable water management programs in the coastal zone of Bangladesh. This process would help out and encourage the practical actions that will simplify the incorporation through expressive and accomplishing coastal community's perception, resilience, and coping practices and prospects of community-based adaptation practices and resolve the safe drinking water crisis. The straight voice of coastal communities and traditional indigenous knowledge must be integrated into the formulation of integrated water-linked policies and approaches, which could be through the government consultation or the platform of multi-stakeholder water governance for the coastal zone of Bangladesh. There is the massive possibility of bringing into line with the various national and international development/donor agencies behind a combined apparition for strong coastal water governance. This donor assistance could be reinforced if the robust method for lucidity and insurability are in place.

Therefore, the Bangladesh Government must develop policies, incentives, and instruction to private and public programs in the direction of escalation of adaptation practices for the sustainable development of the water resources sector. There is also a requirement for IWRM to allow for the effects of climate change on Bangladesh's water resources sector. The coastal communities' effective participation should make sure in susceptibility and assessment of impacts for enhanced preparedness and integrated planning of climate change–induced water-associated risks and hazards. Social and environmental indicators in risk and susceptibility and impact measurement should be painstaking appropriately to recognize the related risks and vulnerabilities for the integrated water resources sector. Besides, transboundary water sharing matter under the changing climatic circumstance is also an essential apprehension as the four principal rivers of Bangladesh are collective with bordering countries like India and Myanmar. Incomplete sharing of two-pronged knowledge and documents about multi-stakeholder's water resources, be deficient of information on international documents (policy) on the water resources availability and dominance and its multi-dimensional applications in the river upstream and downstream segments should also be in use into deliberation conversation with the water sharing between India and Myanmar. The combined watershed management conception should be a sustainable approach to decline the vulnerabilities and impacts of climate change on common water resources of Bangladesh.

References

Akter, S. 2020. Social cohesion and willingness to pay for cyclone risk reduction: The case for the Coastal Embankment Improvement Project in Bangladesh. *International Journal of Disaster Risk Reduction*, 48(2), 101–120.

Araral, E., and Yu, D. 2012. Water governance: Critique, theory, and evidence from Asia. *Asian Journal of Water Management*, 10(3), 50–70.

Barua, P., Rahman, S.H., and Morshed, M. H. 2017. Sustainable adaptation for resolving climate displacement issues of southeastern islands in Bangladesh. *International Journal of Climate Change Strategies and Management*, 15(2), 440–465.

Barua, P. and Rahman, S. H. 2018. Community-based rehabilitation attempt for the solution of climate displacement crisis in the coastal area of Bangladesh. *International Journal of Migration and Residential Mobility*, 1(4), 358–378.

Barua, P., and Rahman, S. H. 2019. Impact of river erosion on livelihood and coping strategies of displaced people in South-Eastern Bangladesh. *International Journal of Migration and Residential Mobility*, 2(1), 34–55.

Barua, P., Rahman, S.H., Barua, S., and Ismail, M.M.R. 2020. Climate change vulnerability and responses of fisherfolk communities in the South-Eastern coast of Bangladesh. *Water Conservation and Management*, 4(1), 20–31.

Barua, P., Rahman, S. H., and Eslamian, S. 2021. Global climate change and inequalities, reduced inequalities. In Leal Filho, W., Azul, A.M., Brandli, L., Lange Salvia, A., Özuyar, P.G., and Wall T. (eds.), *Encyclopedia of the UN sustainable development goals*. Cham: Springer.

Biswas, P., Chaudhuri, R., Tanmay, D., Pallavi, R., Manju, D., Zaman, S., and Abhijit Mitra. 2017. Footprints of climate change on fish community of Mangrove dominated lower Gangetic Delta. *Journal of Fisheries and Aquaculture Development*, 5(1), 1–3.

Chowdhury, N.T. 2010. Water management in Bangladesh: An analytical review. *Water Policy*, 12, 32–51.

Chakma, N. 2014. Inter island relocation due to natural hazards and socio-economic challenges: A case study on Hugli Estuary Islands. *Golden Research Thoughts*, 4(2), 80–90.

Eslamian, S., Parvizi, S., Ostad-Ali-Askari, K., Talebmorad, H. 2020. Water. In P. Bobrowsky and B. Marker (eds), *Encyclopedia of Engineering Geology. Encyclopedia of Earth Sciences Series*. Springer, Cham.

Haque, M. 2017. Community-based adaptation to climate change: Experience of the coastline of Bangladesh. In Z. Islam, H. Shafie, and R. Mahmood (eds.), *Culture, adaptation and resilience: Essays on climate change regime in South Asia*. Bangladesh Climate Change Trust and Department of Anthropology, University of Dhaka, Bangladesh.

Hossain, B., Crispin, M., and Sohel, M. S. 2020. Climate change induced human displacement in Bangladesh: A case study of flood in 2017 in Char in Gaibandha District. *Asian Research Journal of Arts & Social Sciences*, 10(1), 47–60.

Islam, A. M., Hossain, M.S., and Murshed, S. 2015. Assessment of coastal vulnerability due to sea level change at Bhola Island, Bangladesh: Using geospatial techniques. *Journal of Indian Society for Remote Sensing*, 20(4), 80–95.

Islam, S.N., Reinstädtler, S., and Gnauck, A. 2016. Coastal environmental degradation and ecosystem management in the Ganges deltaic region in Bangladesh. *International Journal of Ecological Economics*, 37(4), 59–81.

Islam, M.M., Rahman, M.S., Kabir, M.A., Islam, M.N., and Chowdhury, R.M. 2019. Predictive assessment on landscape and coastal erosion of Bangladesh using geospatial techniques. *Remote Sensing Applications: Society and Environment*, 17(2), 120–145.

Mahmuduzzaman, M., Ahmed, Z. U., Nuruzzaman, A. and Ahmed, F. R. S. 2014. Causes of salinity intrusion in coastal belt of Bangladesh. *International Journal of Plant Research*, 4(2), 8–13.

Rabbani, G. M., Rahman, S. H. and Lucy, F. 2013, Impacts of climatic hazards on the small wetland ecosystems (ponds): Evidence from some selected areas of coastal Bangladesh. *Sustainability*, 5(4), 1510–1521.

Rabbani, G. M., Rahman, S. H., and Sirazoom, M. 2018. Prospects of pond ecosystems as resource base towards community-based adaptation (CBA) to climate change in coastal region of Bangladesh. *Journal of Water and Climate Change*, 9(1), 223–231.

Rahman, S. H. 2016. Stocktaking and assessment of adopted technologies related to water use in coastal areas of Bangladesh. Assessment Report, Christian Commission for Development in Bangladesh (CCDB), Dhaka.

Rahman, S. H., Faisal, B. M. R., and Islam, S.T. 2017. Water integrity in Bangladesh: An analytical review of legal documents. *SUB Journal of Sustainable Environment and Development*, 2(2), 216–229.

Rahman, S. H., Faisal, B. M.H., and Islam, S.T. 2019. A critical review on water governance gaps in Bangladesh. *SUB Journal of Sustainable Environment and Development*, 4(3), 40–58.

Rahman, S.H. and Islam, S. T. 2019. Country baseline assessment: Water sector integrity. Water Integrity Network and Transparency International Bangladesh, Dhaka.

Rahman, S. H., Islam, S.M., and Rume, T. 2016. Seawater intrusion in coastal region of Bangladesh: Causes, impacts and adaptation. *Journal of SUB*, 6(2), 44–63.

Rahaman, M. A., M. Rahman, and S. Hossain. 2018. Climate-resilient agricultural practices in different agro-ecological zones of Bangladesh. In Walter Leal Filho (ed), *Handbook of climate change resilience*. Switzerland, Springer Nature.

Saha, A., Nabonita, N., Amin, G., Zaman, S., and Abhijit, M. 2017. Response of mangroves to climate change: A case study from the World Heritage Site of Indian Sundarbans. *International Journal for Innovative Research in Multidisciplinary Field*, 3(9), 70–75.

Salauddin, M., Hossain, K.T., Tanim, I.A., Kabir, M.A., Saddam, M.H. 2018. Modelingspatio-temporal shoreline shifting of a coastal island in Bangladesh using geospatial techniques and DSAS extension. Annals of Valahia University of Targoviste. *Geographical Series*, 18(1), 1–13.

Taher T.B. and Rahman S.H. 2018. Impact of cyclone Sidr on rural livelihood using Participatory Rural Appraisal (PRA) tools: A case study of a coastal unit in Bangladesh. *Journal of Sustainability Science and Management*, 13(1), 91–102.

PART III

Gender, Disaster and Social Inclusion

8

WHY IS WOMEN'S LEADERSHIP IMPORTANT FOR ENHANCING DISASTER RESILIENCE TO NATURAL PERTURBATIONS?

Important Insights from Coastal Communities of Bangladesh

Mohammed Moniruzzaman Khan, M. Abul Kalam Azad, Abu-Hena Mostofa Kamal, Dilara Zahid and Md. Abdul Malak

Introduction

Because of its deltaic shape, land characteristics, multiplicity of rivers and the monsoon climate, Bangladesh is extremely vulnerable to natural hazards (Ahmed, 2006; Asaduzzaman, 2015). Climate change has added a new dimension to community risk and vulnerability (Rahman et al., 2018; NPDM, 2007–2015). It is predicted that, by 2030, an additional 14% of Bangladesh will become extremely susceptible to floods and currently vulnerable areas will experience higher levels of monsoon and pre-monsoon flooding (Faisal & Parveen, 2004; Karim & Mimura, 2008). Climate change-induced disasters (e.g., storm surges, extreme cyclone) frequently affect marginalized groups, especially pregnant mother, disabled, and women. Women are highly vulnerable and exposed to these climate change-triggered hazards in comparison to their counterparts because of social status, position, socio-cultural discrimination role in income earning resources and the decision-making process (Asaduzzaman, 2015). Women felt insecure in terms of their rights both at homes and in cyclone shelters (Alam & Rahman, 2014; Enarson, 2002; Goldsworthy, 2010; Rahman, 2013). Women are encountered violence and other gender-related problems (e.g., lack of access to institutional initiatives of disaster management) in different phases of a disaster (Nasreen, 2010; Alam & Rahaman, 2014). However, to cope with a disaster, women bear double physical burden than men because of gender-specific tasks, such as procuring and cooking food, collecting and storing

DOI: 10.4324/9781003253495-11

safe drinking water for family members and fuelwood, and rearing and caring for children and elders (Nasreen, 2008; Alam& Rahman, 2014). Despite performing critical roles during disastrous situation, the lack of access to information and technology, the dearth of access and entitlements to resources augment their vulnerability to natural hazards (Kumbetoglu & User, 2010; Asaduzzaman, 2015). Additionally, women become innocent victim of sexual violence, psychological harassment, and physical torture by male counterparts during and post-disaster. These factors are limitedly addressed in most scholarship (Nasreen, 2008; Kumbetoglu& User, 2010; Parkinson& Zara, 2013).

However, women have been the agents of change, and have enormous potentials to survive, cope, recover from, and adapt to nature-triggered disasters (Nasreen, 2012, 2008). For instance, based on their lived experiences, women adopt preparedness initiatives (e.g., gathering necessary food for family members, protecting the plinth of houses after anticipating a disaster During a disaster, women, in addition to their daily routine activities, take a leading role in caring for injured persons and elders, and they provide emotional support to family members, collect relief items food and firewood (Malak et al., 2020). These roles unquestionably imply that women are not only victims of climate change and natural disasters, but they are also the stewards of the protection of family and community at large. They are proactively involved in disaster emergencies and take necessary actions for the survival of family members (Malak et al., 2020).

In this research, we argue that women's stewardship is essential in disaster risk reduction and sustainable management resources. Taking this into consideration, the primary objectives of the study are to (i) identify the roles of women as leaders in disaster risk reduction (DRR) and disaster management both at the community as well as the family level, (ii) explore risks encountered by women due to nature-triggered disasters (NTD), (iii) understand challenges experienced by women as leaders at the institutional level.

Theoretical Consideration and Analytical Framework

In order to examine the nature of women's vulnerability, women leadership role in DRR, especially in coastal Bangladesh, we have applied a number of theoretical lenses – including access theory (Blaikie et al., 1994; Wisner et al., 2004), entitlement theory (Sen, 1981), and conservation of resources (COR) theory (Hobfoll, 1988, 2001). The theoretical standpoints suggest that mainstreaming of women in the disaster management process certainly fosters agency and stewardship capacity of women for disaster risk management.

Blaikie et al. (1994) and Wisner et al. (2004) argue in access theory that the vulnerability of women is disproportionate in society because of societal mechanisms such as gender, class, income, employment opportunities, unequal distribution of resources and unequal access to resources. Lack of entitlement and access by marginalised and underprivileged individuals and groups such as women to foods is denied or reduced via the operation of social, economic and political process

(Blaikie et al., 1994), which is identified by Sen (1981) as the legal means (normal and legitimate) of society.

Individuals, who have rights and command over, and access to many income opportunities, have low risks and wide choice (Blaikie et al., 1994) in securing a sustainable livelihood under unfavourable conditions. However, individuals who have limited, and no access to entitlements have few opportunities in living and have the least flexibility in adverse conditions. According to the access theory (Blaikie et al., 1994), the access to resources over time can improve and release individuals or households from being at risk and disproportionate vulnerability.

However, in terms of entitlement to resources, Sen (1981) argues that resource scarcity is not resulted from drought and inadequate food supply but by the incapacity of the people to keep decree for food through the authorised means, including the use of production opportunities, scope of trade and other methods that are available in the society. Starvation is caused by failure to have entitlement and rights, including the availability of food in the society. It indicates that the inability to adjust and recover from adverse situations in society is the outputs of the existing mechanisms of that society.

Sen (1981) argues that people's access to resources relies on "entitlement bundles" (e.g. a subsistence farmer depends on a "direct entitlement" whereby food is directly produced by the household). As per Sen (1981), problems arise when one food entitlement strategy is disrupted and the individual is unable to switch, for example, to an "indirect entitlement" (i.e. working for a wage) or to a "transfer entitlement" (where someone transfers an entitlement in the form of charity and food aid). Sen's (1981) focus on constraints that prevent people from changing entitlement strategies. Although the entitlement approach of Sen (1981) is an explanation in terms of the causes of famine, it has drawn the attention and interest of scholars of disaster in assessing that the assets make people able to use when confronting problems such as disasters (Sen, 1981). With these lenses, this study aims to apply how the leadership role of women in DRR in the coastal area of Bangladesh for ensuring access to a bundle of entitlements.

Conservation resource (COR) theory (Hobfoll, 1988, 2001) focuses on analysing vulnerabilities and adaptation practices of the community people during natural disasters in terms of individuals' effort towards conserving resources. The COR theory (Hobfoll, 1988, 2001) asserts that individuals strive to obtain, retain and protect that what they value. People have an innate and a learned desire to conserve the quality and quantity of their resources and limit any state that may jeopardise the security of resources. When (a) an individual's entitlements are threatened with loss, (b) resources are lost or (c) individuals fail to gain resources after investments of other resources, it leads to mental or physical stresses. In the short run, these may result in negative emotions, such as anger, frustration or fear, and dire mental and physical consequences (e.g., burnout, depression or coronary heart disease), which generally occur over the period.

Recognizing the theoretical perspective, our analytical figure (Figure 8.1) underpins that women's stewardship is a keystone pre-, during, and post-disaster

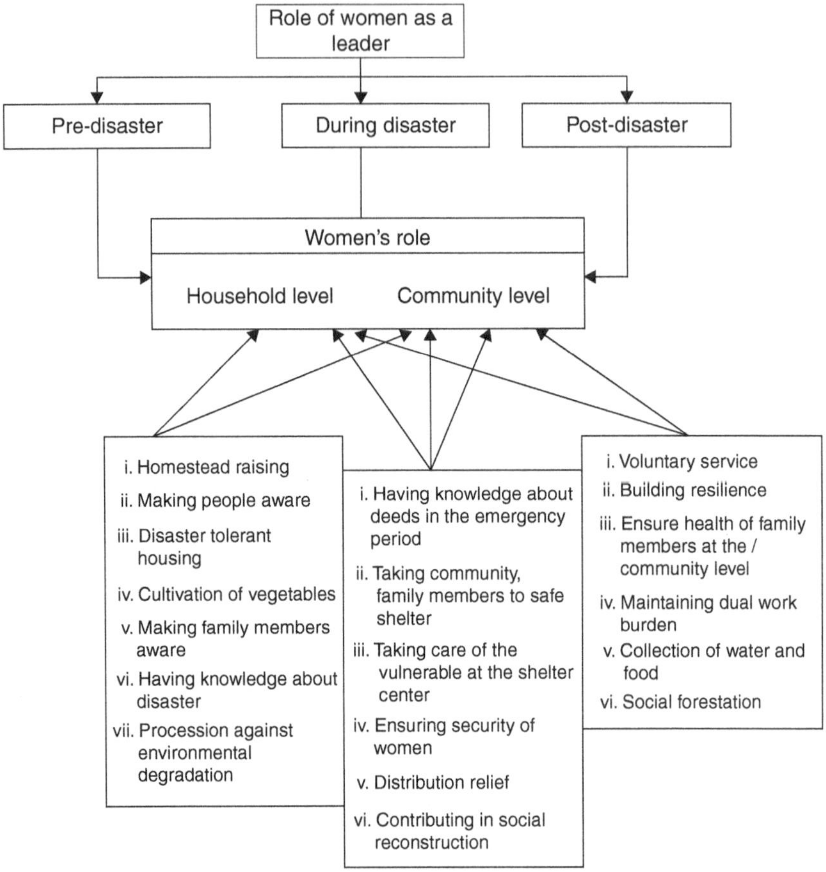

FIGURE 8.1 Engendering leadership in disaster risk mitigation

period as it works as an instrument to organize resources at the household as well as the community level. For instance, despite challenging socio-cultural roles, women perform custodian roles for protecting household by heightening plinth of houses, assuring food security by cultivating homestead vegetables, and applying lived experience to protect environment. These functions are not limited only at the household level; at the community level, women engage in disaster management, social forestation, and social reconstruction programs (e.g., distributing relief, and safety net programs). The involvement in diverse issues fosters their leadership capability to disasters. While socially-created constraints (e.g., lack of access to information and technology, resources, and institution) destabilise their disaster resilience building process, women's proactive involvement from the household to the community level in a disastrous state advance disaster risk reduction apparatus, undermine vulnerability of all groups, and create a route for sustainable disaster management.

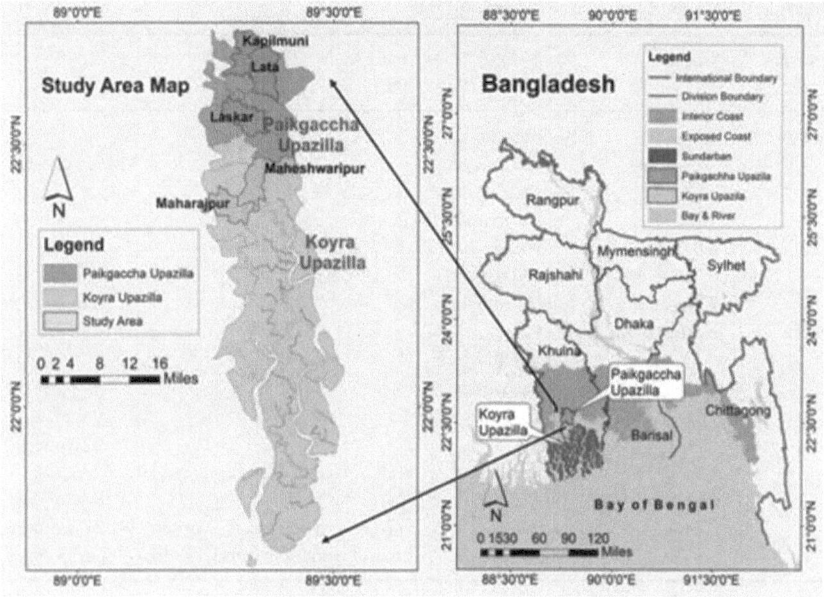

FIGURE 8.2 The location of the study area

Research Study Area

This study was undertaken in coastal Bangladesh, especially in Khulna district within the framework of qualitative research. Five unions were chosen from two sub-districts of Khulna district (Figure 8.2). Data were procured from different cross-sectoral groups, including women, chairpersons and members of union disaster management committee, disabled, teachers, and political elite). For an in-depth understanding, a total of 11 villages were selected (see Table 8.1). The villages were chosen based on severity and exposure to Cyclone Aila and SIDR as they are located nearby the Bay of Bengal. Due to geographical position, the 2007 Cyclone SIDR and the 2009 Cyclone Aila adversely impacted livelihood and subsistence of locality. Subsequently, community members have experienced a number of coastal floods and storm surges in a recent decade. Eventually, community members had liveable memory on how to deal uncertain risks and perturbations, which empowered us to examine how marginalized groups, particularly women, perform pivotal role as major agents or leader to withstand environmental stresses.

Research Method and Data Collection Tools

A qualitative research approach to in-depth understanding of women's vulnerability, local political system in disaster management, and forms of deprivation of women at the community as well as the household level. To achieve the objectives, participants were selected from multiple levels – from the household level, community level,

TABLE 8.1 Sample distribution of respondents

Upazila	*Union*	*Village*	*Case study interview*	*KII*	*FGD*
Paikgacha	**Lata**	Tentultala	2	1	
		Shamukpota	2	1	1 (9 women)
		Haldarchok	2		
		Katthamari	2	1	
	Loskor	Paikgacha	1	1	
		Loskor	1	1	
	Kapilmuni	Voirobghata	1		1 (8 women)
Koyra	**Maheshwaripur**	Gilabari	1	1	
		Shakharipur	1	1	
		Sathaliya	1	1	
		Amtoli	1	1	1 (8 men and 2 women)
		Kalikapur	1	1	
	Moharajpur	Deyara	1	1	1 (8 women)
	Total		17 (15 women and 2 men)	11 (7 women and 4 men)	4 (27 women and 8 men)

to the institutional level (e.g., the members of the Union Disaster Management Committee [UDMC], Upazila Disaster Management Committee [UzDMC] and NGO officials). We conducted five focus group discussion (FGD), and 28 semi-structure interviews (17 case study interviews and 11 key informants' interviews). Of these, 15 case study interviews were with women and 2 were with men; seven KIIs were with women who were the members of the UDMC and four were the chairpersons of the Union Parishad; three FGDs were marginalized women from farming communities, and two FGDs were with men (see Table 8.1).In order to procure in-depth data, we used an open-ended checklist for each group.

Data Processing and Analysis

Data analyses were engaged in a number of steps. Data were primarily transcribed from audio recordings. As the qualitative techniques produced voluminous data, transcriptions were edited and cleaned for analysis.The major goal was to summarize considerable amount of data for understanding the pattern of data in a meaningful way. For this, we applied a data reduction approach to identify, summarize, and simplify key themes that emerge in the collected data.We identified *priori*(etic) key themes and emerging (emic) themes applying indexing and coding methods. Data were indexed on the basis of prior themes (e.g., nature of risks)deductively, whereas emerging themes (emic that evolved from data) were coded through running the texts (Strauss & Corbin, 1998). The collected data were analysed and interpreted "thick description", in which an in-depth interpretation was performed to elucidate naturally situated meanings and information.

Results and Discussion

Disaster and Risk in Everyday Life

The funnel shape of the Bay of Bengal regularly leads to the formation of tropical cyclones, tidal surges and riverbank erosion in the southern part of Bangladesh. The study sites of Paikgacha and Koyraare highly disaster-prone sub-districts, in which the lives and livelihoods of the local community are at multiple risks resulted from the intersection of natural hazards and social discrepancies.. For example, the intersection of geographical exposure, low-income options, limited knowledge on reproductive rights, and the dearth of access to health facilities augmented their vulnerability to diseases and precarious livelihood. Extreme natural events, such as cyclone and coastal flooding were more considerable in the locality and severely affected health of locals. A majority of participants reported that natural disasters first impacted the livelihood of local people and then made them more vulnerable to multifarious diseases, such as diarrhoea, dysentery, fever, and other waterborne diseases. Moreover, during menstruation, women become extremely vulnerable due to using saline water. The most direct impacts of a disaster on health were overwhelmingly adverse, especially for on women and pregnant mother. For instance, the impact of a disaster includes facing childbirth with no privacy and minimum care as well as risk of abortion and miscarriage (Ahmed et al., 2008; Alston, 2015). A woman's menstrual cycle is an extremely private matter in Bangladeshi culture; it becomes a real crisis and challenge after puberty. The use of saline water was the underlying factor behind these problems. Accordingly, one case study participant asserted:

> There is salt in the water to such a degree leading to ulcers in our body. Since, we are poor, we have nothing to drink and except saline water in our daily life.

In coastal Bangladesh, women, primarily at the household level take responsibility of the protection of household resources, ensuring safety of livestock, taking care of children, assuring food security for family members, and repairing and building the damaged houses. This leadership role is a steppingstone for the survival during and post-disaster situation. Similar results are also noted by Alston (2015). His research asserts that women perform imperative roles from childrearing to family subsistence. This role is fostering their disaster resilience and leadership to tackle future uncertainty (Alston, 2015). However, for substantial reduction of income sources after *Sidr* and *Aila*, the burden for women increased to meet household requirements. For example, the daily income of the people of the study area reduced after disasters (e.g., *Sidr* or *Aila)*. Additionally, the diseases during that time made their livelihood more difficult as the diseases increased their daily expenditure. A participant disclosed:

> How can we ensure our savings? The half (50%) of our total income need to expend for treating different water borne diseases.

Disasters destroyed normal lifestyle of community members; for instance, Cyclone *Sidr* and *Aila* damaged the houses , submerged roads and destroyed normal communication system, increased drop out of school going children due to remaining closed schools and colleges, and caused mental trauma among children for long-run. Acknowledging post-disaster mental shocks, a study participant noted:

> In earlier time, my son was quiet in nature. Aftermath of Aila, he became extremely scared; frequently, he woke up from sleep at midnight, cried, expressed his fear at water. He could not take a bath because of his fear of water. For the reduction of his mental trauma, I took him to a psychiatrist for counselling.

Nature-induced disasters have been appearing to be gender-neutral events that affect people and their livelihoods regardless of class or gender; the impacts are greater on poor communities and mostly on women (Akhter, 1992; Nasreen, 2008, 2012). During a disaster, gender-based disparities impact to a higher degree than normal because of the breakdown of social and family order, institutional security and protection. Furthermore, women are mostly at risk when hazardous conditions unfold during a disaster such as cyclones (Ikeda, 1995) and gender; a pervasive gender division taking access to social and economic resources away from women. Our study unveils that a disaster is an event that opens out natural practices, traditions, genders, relationships and power relations in a community and even within intimate relationships. The empirical findings are also confirmed by other research (Kamal et al., 2018); this study asserts that salinity intrusion is a major factor that leads to considerable health, sexual and reproductive problems among women and girls.

Disaster, Livelihood Options, and Women

Employment opportunities for women are extremely limited to certain types of work; they are mostly paid in kind rather than cash, which is excluded from the market (Nasreen, 2010. A disaster breaks down the whole economic structure of the society. Our empirical findings implies that agriculture is the primary source of income for most of the coastal people; it has subsequently shifted to shrimp farming and other petty business related to shrimp farming. In the past (~20 years ago), the local people could produce 15–20 mound (1 mound = 37.32 kg) of rice in one *bigha* (1 bigha = 61.98 decimal) of land; however, because of extremely saline water, now hardly five mounds of rice is obtained in the same amount of land. Other seasonal crops such as gourd, bitter gourd, eggplants and tomato are no longer available in the area. The total economic structure of the society has shifted from a rice cultivation-based economy to a shrimp cultivation-based economy. The results of this study indicate that people of this locality either earn from their own or leased shrimp farm or other small business developed based on shrimp or crab farming. Therefore, marginal people, especially women have to work as wage labourer in shrimp farms or in shrimp processing units of the local market to lead to their family.

However, shrimp cultivation is no longer profitable because of the lack of demand abroad; farmers often lose their shrimp in the pond because of unknown viruses. Now, they neither can produce rice nor have enough profit from shrimp farming. However, who does not have their own land or cannot afford to lease any others land either work in other's firm or join to a newly created earning, that is, opportunities such as driving share motorcycle as rent. Certain participants report that disasters ruin shrimp farms and impact other businesses. When they do not get enough profit from their shrimp firms, they then reduce their daily purchases. Therefore, the people who depend on other small business have a low demand of their services or products. This creates a vicious cycle of poverty in the local economy. Accordingly, a participant reported that people from lower income group received support from NGOs and GOs; however, the middle-income group people neither received help from any organisation nor did they seek help from others.

Certain participants report that NGOs are working for community development in coastal regions. NGOs are more likely to be a part of the recovery phase of a disaster and played an important role in reducing the impacts of hazards on villages (Alston, 2015). During a disaster, NGOs provide different support, including microcredit, relief and different training on poultry and fisheries, sewing, vegetable gardening, livestock rearing and disaster preparedness. Most programmes conducted by NGOs are seasonal; however, the people in the community focused on sustainable rural development for their region as because of salinity intrusion all around the area, most efforts of these NGOs were not properly working. Nevertheless, the participation of women in different training created income opportunities for women in the study area, which increased the capability of women, for example, expenditure in the family. This indicates that increasing the income is a tool/component for women to be equipped to encounter poverty or for reducing the poverty level because it has improved the material quality of life such as food habits, amount of food intake in a day and education of children.

Disaster acts as a "push factor" in the decision to migrate; it can be attributed to a destruction of livelihood options and houses, landlessness, poverty, lack of optimism for improvement and insecurity of their children, as has been reported by the respondents and reflected in other research (e.g., Kitkiri, 2011, Alston (2015). Better employment (seasonal jobs and non-seasonal jobs) at the destination and lack of employment opportunities in the place of origin are the primary reasons for migration (Nguyen and Wodon, 2015). Our results show that aftermath of disasters people do not find their previous income source to be useful, and therefore they are forced to move to a new place. Consequently, a disaster forces them to migrate to other places to earn money for sustaining their livelihood. When examining the strength of cause and effect in the relationship between disaster and migration, cyclone *Aila* emerges as the central predictor of migration. Furthermore, migration is fundamentally influenced by socio-cultural contexts in which local community people are embedded and living. However, the pattern of migration is both temporary and permanent depending on the extent nature-triggered disasters. On this note, a primary school teacher, expressed:

> People had to migrated as there was no jobs. Most of local people moved to India to their relatives and rest of the people migrated to Shatkhira, Gopalgong, Dhaka, and other cities to maintain their livelihood as daily labourers.

Women as Leaders in Disaster Risk Management

Women are primarily responsible to manage almost everything at their household during and after disaster. They are responsible to collect water, foods and firewood, all of which are responsible to take care of cattle, children and elderly family members. Women are even responsible to save money, valuable assets and dry foods before a disaster. The findings note that women save valuable assets before a disaster. As most male members work outside their locality during the early stage of a disaster, it becomes the responsibility of women to collect their assets and dry food for future uncertainty. Considering future perturbations, women's stewardship are engaged in collecting dry firewood, local vegetables from nearby garden or river bank as most of the families do not have sufficient money to buy them from market at a high price as well as fresh drinking water for their family. Moreover, they are looking after the children, elderly and cattle. After a disaster, rebuilding the houses and arranging the family assets also become a woman's role. These results confer that women are disaster stewards at the household when males are busy with their primary earnings and working outside of the locality.

The role and position of women outside of the household are imperative, especially in both personal and public sphere for disaster management (Kabeer, 2011). The results report that women stewardship performs substantial role in disaster management, especially at the institutional level. For instance, female members of the Union Disaster Management Committee play pivotal role in the decision-making process, particularly for arranging delivery system for pregnant mother at shelter centers, providing recommendations for managing hygienic kits for girls and pregnant mother. Cyclone *Aila* was a milestone for coastal people in terms of changing roles and disaster stewardship of women as it unlocked learning opportunities for women how to involve in community affairs. This also fosters a sense of self-worth and social identity, ability to question on critical issues, and capacity to exercise and take control over community-based resources and entitlements. However, we found that women's leadership is rarely valued as they are the constituent members of the union parishad. The dearth of institutional recognition has been a major challenge to establish them as a agents of change maker.

Disaster and Women in Decision-Making Process

Both women and men play an important role in disaster management; however, most of the time, women cannot play a role in the decision-making process as the results of this study suggest. Participants reported that women in the locality have less involvement and in some cases no involvement with the decision-making process at both family and community level though they contributed to rebuilding the community. The lack of participation of women in the decision-making process is more socio-cultural than the problem with legislation; moreover, they are created by the society. For example, poor women are at risk because they lack status, resources and education (Fothergill, 1998; Kabeer, 2011). The limited participation of women in the decision-making processes is related to a belief that women are less competent, lack of funds, religiously more affectionate, keen to family members (Kabeer, 2011; Panday, 2016). Eventually, men appear as the primary decision-makers in households, especially for earnings and the decision-making process. Accordingly, a vulnerable women noted:

> In most families, females are not allowed to go outside of the house to participate in socio-economic activities; women's knowledge and experience are considered to be extremely limited regarding disaster and they rarely have the capacity to contribute to the decision-making process, except the management of household chores.

However, conventional perceptions regarding women's capability are gradually changing; after the cyclone *Aila*, women are now allowed to work outside their home because families have limited income sources. The NGOs are working to empower women. women are now much more aware about their role in the family, which is fostering their participation in the decision-making process. The decision-making of a woman in the family depends on certain factors, such as level of education of women, demands of extra hands for total family income, work opportunity and women's self-esteem. Often, women do not get sufficient respect from their superior co-workers. As one of the NGO officials denoted:

> Trainings of different sectors are being provided in my office, but my opinion is not properly evaluated during and after the training programs.

The involvement of women with the decision-making process increases the capacity of individuals to transform their choice into actions and enable then to independently work (Panday, 2016). Many participants and respondents feel that whatever projects and plans are in place for the coastal community, they will not be effective until the voices of women are listened, and they are taking effective initiatives and measures to meet the women's requirements, by considering their experiences. This increases bargaining capacity, adaptation capacity and potentiality of coastal women. Kabeer (2011) argues that allowing women to be part of projects and project plans allows changes that may be uneven because there is a greater desire for change in certain domains. For example, Panday (2016) reveals that women's participation in community activities ensures equality in decision-making processes because they claim to have an improved understanding of women's requirements. Participation and sharing experiences with decision-making strengthen the capacity of society and family to recognise what is just and what is unjust to determine actions to be taken for a better society.

Women in Disaster Risk Management at Family and Community/Local Level

The results of this study suggest that women are important disaster managers in a family. In some cases, even women perform very well compared to men in a family setting. In case of community, the women participate at a community level in disaster management such as repairing roads, dams, embankments and canals or schools. Because women maintain household chores, they are aware of the problems of a family as well as the solutions. The results showed that women manage households and household-related activities; however, they cannot share and contribute for the greater interest of the community because their voices always remain unheard and ignored by the community. After a disaster and during a disaster, only women collect drinking water, vegetables and cook food for family to save lives of family members. In terms of decision-making, women require to take consent from their husband or other male members of the family whether they will participate.

In addition to family arena, the results show that women participate in the community-level DRR activities. Currently, NGOs work in coastal areas to empower women by involving them with different activities taken for the community during a disaster. For example, women are active members of different committees, such as UDMC, Cyclone Preparedness Program (CPP) and Project Implementation Committee (PIC), all of which are coordinated by both GO and NGOs. In terms of ensuring participation, NGOs have trained women on DRR and women are trained to disseminate their knowledge to people in the community. Kabeer (2011) argues that allowing women to be part of projects and project plans allows changes that may be uneven because there is a greater desire for change in certain domains. For example, Panday (2016) reveals that women's participation in community activities ensures equality in decision-making processes because they claim to have an improved understanding of women's requirements. The results show that women do participate in reconstruction activities in post-disaster period. They help to reconstruct dam, embankment, road and culvert. For example, a study participant replied that she did not feel any trouble to work as a member of the UzDMC; she rather enjoyed working there. Moreover, there seems to be the games of power practice at institutional level based on political party and ideology. Female UDMC members are often deprived for being a member of the opposition political party. As one of the member noted: "She gets less aid from the local member of the parliament and Upazila Chairman as well as from the UpazilaNirbahi Officer (UNO)".

It is unquestionably asserted by the results of the study that woman leaders face massive challenges to perform disaster risk reduction actions. . Women feel trouble to face disaster and to mitigate vulnerabilities by not having proper training and knowledge about a particular disaster and worst situation. Even, local female leaders do not properly get early warnings of a disaster. A local female leader expressed:

> We, female , do not get disaster warnings early properly because we cannot go outside the home; the male members of our house get information from others and share with us. Hence, it is necessary to disseminate disaster warnings early properly. Moreover, we feel the problem is caused by a patriarchal societal system.
>
> To overcome this situation, women take or get the help of NGOs. NGOs provide basic training to build up capacity of the women as women are the best disaster managers i n coastal areas of Bangladesh. Females try to adopt with the changed environment and teach others of their community about coping with the changed environment. Women participate in making the community members educated such that superstitions might be eradicated about women's role in terms of disaster mitigation.

Women in Protecting Environment

Alston (2015) reports that women and girls are important agents of change in circumstances where climate change is creating major challenges and are important to the global battle to preserve our planet. This study shows that women as household managers not only manage household activities during a disaster, they take measures to protect the environment too. In terms of protecting the environment, the respondents report that women in the study area earn money by performing homestead gardening, which meets their household requirements; sometimes, they sell the produce in the local market. Moreover, women perform forestation around their house, plant wood trees and fuel wood trees in the coastal area, which are not only for earning money, but it also protects them from storms and strong winds during a disaster.

The respondents argue that additional homestead gardening and forestation produces oxygen and reduces carbon emission. It is argued by respondents that women are important activists against environmental degradation. Women participate in the human chain against and protest movement against shrimp cultivation, the man-made source of causing saline water in the community. They make people in the community aware of environmental pollution and its impacts, as well as inspire others to cultivate vegetables around the residence. The respondents report that women sometimes organise themselves to visit local political leaders and raise their demand, including additional aid and assistance and build a strong dam on the bank of the river, thus forming a cooperative society by community members to solve the sudden crisis such as a disaster. Furthermore, certain participants report that they cannot take any major decision because of a traditional social structure and particularly because of the patriarchal social structure. Patriarchy and masculinities are indicators of gender inequality, which are created by social settings. Oppression and exploitation are constant at a certain time and place as well as within a specific socio-economic structure of society. Women are overlooked and subordinated in the Bangladeshi coastal area by patriarchy (Mies & Shiva, 1993; Rocheleau et al., 1996). The study suggests changing these notions to enhance women's capability to mitigate disaster-related risks. The results of this section are like the fourth proposition of a constructionist perspective where it is argued that the importance of a disaster is the conjunction of historical conditions and social definitions of physical harm and social disruption at the community or higher levels of analysis (Kreps, 1995a, 1995b).

The result of the study shows that women of the study area undertake many measures to cope with climate change and mitigate disaster-related vulnerabilities, for example, rearing domestic animals such as cows and goats, ducks and hens, planting trees and constructing a small dam near the household. One of the respondents report:

> It may not be possible to prevent a disaster, but the loss caused by a disaster can be reduced by taking certain steps such as making the floor of the house high, tying the house with a large tree, planting trees close to the household, rearing domestic animals, ducks and hens, and preserving dry food. Moreover, disaster-related risks can be reduced in most parts if the public can be informed about the ruin of a disaster and the usefulness of agriculture.

Similarly, participants reported that change of people's attitudes towards women is necessary to play a potential role in the adaptation mechanism; moreover, governmental and non-governmental patronisation is required for them to go ahead. The results of the study of Enarson and Chakraborti (2009) are like those of this study where it is said that:

> women's roles in securing water, food and shelter, during and after disasters, have been documented well. However, women continue to be marginalized in decision-making on disaster-related issues, even if they often hold important social knowledge and vast untapped capacity for reducing risk at a community level.

Conclusion

This study provides an understanding of the vulnerability of coastal women in the context of gender and gender relations in their everyday lives. It shows that the coastal society is a highly gender-segmented society because of social customs and practices where women are discriminated against, as well as controlled and restricted in their everyday work, starting from the household. Often, natural disasters and climate change exacerbate existing inequalities and discrimination, including gender-based ones, and can lead to new forms of discrimination. The study results reveal that the existing socio-cultural practices leave women in a more vulnerable condition during a disaster in the coastal area in Bangladesh.

Moreover, there is greater marginalisation of women in the context of physical and mental health issues, access to information, greater restriction on following religious practices and additional obligations to collect drinking water and fuel wood. Women are restricted, discriminated against and controlled in the coastal area but women have a greater role in caring for the sick, injured children and the elderly, thus helping male household/family members that create a double burden on the shoulders of the women. Women in the coastal area are distressed because of traditional practices by society and family, as well as migration and absence of male members.

In this study, "gender" is the socially constructed roles, behaviours, activities and attributes that a society considers appropriate for a person based on his or her assigned sex at birth. Understanding the gender implications and facets of natural disasters and climate change is important for effective disaster risk management practices that enable communities and countries to be disaster resilient. Women play important roles during all stages of disaster and climate risk management; they are often at the frontline as responders and bring valuable resources to disaster and climate risk reduction and recovery. However, the important roles or potential roles women take on are often not recognised; moreover, women themselves "are largely marginalized in the development of DRR policy and decision-making processes and their voices go unheard."

This study examines certain gender-related vulnerabilities and capacities in natural disasters, why it is important to adopt a gender-based strategy for planning and response, what a gender-based approach to disaster management looks like and recommendations to relevant actors. DRR that delivers gender equality is a cost-effective win–win option for reducing vulnerability and sustaining the livelihoods of whole communities. Policies that address social safety, as well as training and awareness programmes have led women out of the home, thus ultimately reducing their vulnerability. Changes in societal, patriarchal, religious attitudes and economic hardship forced women to get involved with natural resource management, outdoor work, homestead gardening, poultry and livestock, thus improving access to resources for women as reported by the access theory and entitlement approach.

The evidence shows that the increased access to available resources in the coastal area such as ensuring the integration of gender-sensitive approaches, active engagement of civil society particularly human rights groups and women's associations to monitor both the development and implementation of policies will lead to a positive outcome. National human rights commissions, for example, can play a role in identifying areas where women have been excluded from planning processes. Women's groups can develop mechanisms by which women can come together to identify their common concerns and develop strategies to voice these concerns to those in policy-making positions. International organisations should require reporting by their field staff on the concrete results of women's participation in relevant programmes. For their part, donors should ensure that the organisations they fund integrate a gender perspective in all of their work. Adopting a gender-sensitive approach to disaster risk management is an issue of basic human rights and effective at the practical level. Simply, policies that ensure that women and men are completely involved in planning DRR strategies and are full participants in recovery efforts are more likely to succeed. Disaster response strategies that protect and assist women and men are better for the overall community. A gender-sensitive approach is a smart policy in that it enables resources of all members of an affected community to be completely utilised.

References

Ahmed, A. U. (2006). *Bangladesh Climate Change Impacts and Vulnerability: A Synthesis*. Climate Change Cell, Department of Environment.

Ahmed, A. U., Neelormi, S., Adri, N., Alam, S., & Nuruzzaman, K. (2008). *Climate Change, Gender and Vulnerable Groups in Bangladesh*. Dhaka: Climate Change Cell, MoFDM.

Alam, K., & Rahman, M. H. (2014). Women in natural disasters: A case study from southern coastal region of Bangladesh. *International Journal of Disaster Risk Reduction, 8*, 68–82.

Alston, M. (2015). *Women and Climate Change in Bangladesh*. New York: Routledge.

Akhter, F. (1992). Women are not only victims. In H. Hameeda, D. Cole, & F. H. Abel (Eds.), *From Crisis to Development: Coping with Disasters in Bangladesh* (pp. 59–66). Dhaka: University Press.

Asaduzzaman, M. (2015). *Livelihood Vulnerability of Women in the Context of Climate Change Impacts: Insights From Coastal Bangladesh*. The University of Newcastle. http://hdl.handle.net/1959.13/1311968

Blaikie, P., Cannon, T., Davis, I., & Wisner, B. (1994). *At Risk: Natural Hazards, People's Vulnerability and Disasters*. London: Routledge.

Corbin, J., & Strauss, A. (2008). *Basics of Qualitative Research: Techniques and Procedures for Developing Grounded Theory*. Thousand Oaks, CA: Sage.

Enarson, E. (2002). Gender issues in natural disasters: Talking points and research needs. In: International Labour Organization, *Crisis, Women and Other Gender Concerns* (pp. 5–12). Geneva: ILO Recovery and Reconstruction Department.

Enarson, E., & Chakrabarti, P. D. (Eds.). (2009). *Women, gender and disaster: Global issues and initiatives*. SAGE Publications India.

Faisal, I. M., & Parveen, S. (2004). Food security in the face of climate change, population growth, and resource constraints: implications for Bangladesh. *Environmental Management, 34*(4), 487–498.

Fothergill, A. (1999). An exploratory study of woman battering in the Grand Forks flood disaster: Implications for community responses and policies. *International Journal of Mass Emergencies and Disasters, 17*(1), 79–98.

Fothergill, A., (1998). The neglect of gender in disaster work: An overview of the literature. In E. Enarson & B. H. Morrow (Eds.), *The Gendered Terrain of Disaster: Through Women's Eye* (pp. 11–25). Westport, Connecticut: Praeger.

Goldsworthy, H. (2010). Women, global environmental change, and human security. In R.A. Matthew, J. Barnett, B. Mcdonald, & K.L. O'Brien (Eds.), *Global Environmental Change and Human Security* (pp. 215–236). Cambridge: The MIT Press.

Hobfoll, S. E. (1988). *The Ecology of Stress*. Washington, DC: Hemisphere.

Hobfoll, S. E. (2001). The influence of culture, community, and the nested-self in the stress process: Advancing conservation of resources theory. *Applied Psychology*, 50(3), 337–421.

Ikeda, K., (1995). Gender differences in human loss and vulnerability in natural disasters: A case study from Bangladesh. *Indian Journal of Gender Studies*, 2(2), 171–193.

Kabeer, N. (2011). Between Affiliation and autonomy: Navigating pathways of women's empowerment and gender justice in rural Bangladesh. *Development and Change*. 42(2), 499–528.

Kamal, A. H. M., Umama, U., Roman, S., &Khan, M. M. (2018). Impact of flood on women's sexual and reproductive health: An empirical evidence from Northern Bangladesh. *Global Journal of Medical Research:K Interdisciplinary, 18*(5), 55–64.

Karim, M. F., &Mimura, N. (2008). Impacts of climate change and sea-level rise on cyclonic storm surge floods in Bangladesh. *Global Environmental Change, 18*(3), 490–500.

Kitkiri, K. (2011). Climate change and migration: A case study from rural Bangladesh. *Gender & Development*, 19(1), 23–38.

Kreps, G. A. (1995a). Disaster as systemic event and social catalyst: A clarification of subject matter. *International Journal of Mass Emergencies and Disasters*, 13(3), 255–284.

Kreps, G. A. (1995b). Excluded perspectives in the social construction of disaster: a reply to Hewitt's critique. *International Journal of Mass Emergencies and Disasters*, 13(3), 349–351.

Kumbetoglu, Fatma Belkis & User, Inci (2010).Gender aspects of the disaster recovery process, In S. Dasgupta, I. Siriner, & P. Sarathi De (Eds.), *Women's Encounter with Disaster*. London: Frontpage Publications Limited.

Malak, M. A., Sajib, A. M., Quader, M. A., & Anjum, H. (2020). "We are feeling older than our age": Vulnerability and adaptive strategies of aging people to cyclones in coastal Bangladesh. *International Journal of Disaster Risk Reduction, 48*, 101595.

Mies, M., & Shiva, V. (1993). *Ecofeminism*. New Delhi; London: Kali for Women and Zed Books.

Nasreen, M. (2008). *Violence Against Women: During Flood and Post-flood Situations in Bangladesh*. Women's Rights & Gender Equality Sector, Action Aid Bangladesh.

Nasreen, M. (2010). Rethinking disaster management:Violence against women during floods in Bangladesh. In S. Dasgupta, I. Siriner, & P. Sarathi De (Eds.), *Women's Encounter with Disaster*. London: Frontpage Publications Limited.

Nasreen, M. (2012). *Women and Girls Vulnerable or Resilient?* Dhaka: Institute of Disaster Management and Vulnerability Studies, University of Dhaka.

National Plan for Disaster Management 2007–2015 (2007). Government of the People's Republic of Bangladesh.

Nguyen, M. C., &Wodon, Q. (2015). Temporary and permanent migration. In: A. O'Donnell & Q. Wodon (Eds.), *Climate Change Adaptation and Social Resilience in the Sundarbans* (pp. 123–150). Hoboken: Taylor and Francis.

Panday, P. (2016). *Women's Empowerment in South Asia: NGO Interventions and Agency Building in Bangladesh*. London & New York: Routledge.

Parkinson, D., & Zara, C. (2013). The hidden disaster: Domestic violence in the aftermath of natural disaster. *The Australian Journal of Emergency Management, 28*(2), 28.

Rahman, M. S. (2013). Climate change, disaster and gender vulnerability: A study on two divisions of Bangladesh. *American Journal of Human Ecology, 2*(2), 72–82.

Rahman, H. T., Hickey, G. M., Ford, J. D., & Egan, M. A. (2018). Climate change research in Bangladesh: Research gaps and implications for adaptation-related decision-making. *Regional Environmental Change, 18*(5), 1535–1553.

Rocheleau, D., Thomas-Slayter, B., & Wangari, E. (1996). *Feminist Political Ecology: Global Issues and Local Experiences.* New York: Routledge.

Saha, C. K. (2015). Dynamics of disaster-induced risk in southwestern coastal Bangladesh: An analysis on tropical Cyclone Aila 2009. *Natural Hazards, 75*(1), 727–754.

Sen, A. (1981). *Poverty and Famines: An Essay on Entitlement and Deprivation*. Oxford: Clarendon Press.

Sen, A. (1999). *Poverty and Famines: An Essay on Entitlement and Deprivation*. Oxford University Press.

Strauss, A., & Corbin, J. (1998). *Basics of Qualitative Research: Techniques and Procedures for Developing Grounded Theory*. New Delhi: Sage Publications.

Wisner, B., Blaikie, P., Canon, T., & Davis, I. (2004). *At Risk: Natural Hazards, People's Vulnerability and Disaster*. London and New York: Routledge.

9

GENDER CULTURE IN WATER SECURITY OF COASTAL BANGLADESH

A Maxim or Redress?

Sabrina Zaman, Mahbuba Nasreen, and Fahreen Hossain

Introduction

Water insecurity have been one of the global threats, especially for water-stress countries. It poses dire consequences for populations (Vaughan et al., 2016). According to the WHO, about 4000 children die every day from shortage of pure drinking water and water borne disease and around 2.4 billion people are short of access to sanitation due to unhealthy water and unhygienic sanitation system (WHO, 2015), while nearly 663 million people still do not have access to safe water sources (Carver, 2016). This trend implies that water crisis is undermining sustainable livelihood of populations in water stress countries, like Bangladesh. In Bangladesh, over 5 million people do not have access to safe water and 85 million people have lack improved sanitation (Water.org, 2019).

Bangladesh is one of the most vulnerable countries in the world; it encounters various unfriendly geomorphological and hydrological stressors such as cyclone, floods, salinity intrusion, drought, unpredictable patterns of rainfall, tidal surge, etc. as a result of climatic variations and variability. In the 'National Water Security Index' of 2016, Bangladesh scored 35.3 out of 100, which is very meager. The indicators were water security of households' level, the economic perspective of water security, urban and ecological water security, and access to water (ADB, 2016). This report predicts that the people living in coastal areas of Bangladesh are continuously suffering from water scarcity due to the shortage of sweet water. Due to not having any alternative options for sweet and safe drinking water, people are forced to drink saline water that causes serious health complications such as high blood pressure, gastric problems, and preeclampsia in women and others (REACH, 2015; Tuongetal., 2014; Tusar and Moumita, 2013). Approximately, 40 million people are vulnerable to cyclonic squall, water and soil salinity, and river flooding in

DOI: 10.4324/9781003253495-12

the coastal zone of Bangladesh and affected 70 percent of the agricultural area and their livelihoods (REACH, 2015).

Water risks are associated with gender in terms of fetching water, distance of water point, crop failure, washing clothes, cleaning utensils, psychological stress, unhygienic latrine, and food storage (Stevenson, 2012). Gender dimension determines how different groups are affected during disasters and how they confront the problem (Nasreen, 2008, 2012; Dasgupta et al., 2010). Though, water relates to feminine power (Lawuyi, 1998; Coles and Wallace, 2005), the gender dimension is invisible to most of the analysis, and the role of women is deniable in terms of water security. The concept of water fetching depends on the women, girls, and children in the literature (Nasreen, 1995,2012; Jambrina, 2013; Earle and Bazilli, 2013; Sorenson et al., 2011) as well as in the pragmatic life. Women are also managers of domestic/household water, responsible for household health and well-being, and water users. They are positioned in these roles by their gender, age, and household status (Nasreen, 2008; Hemson, 2002). In the water sector, gender tends to be equated with women and, beyond the notion that improved water access will reduce women's workload and therefore is good for gender equity; this analysis has tended to be superficial for Polder 29. For many years, women's productive and community management roles in the water sector were hidden. However, their direct contribution to agriculture, homestead gardening, securing food and water for household level is still unrecognized (NCS, 2016).

Though women are traditional members of the household for domestic water management within the family (Khosla and Pearl, 2003; UNICEF-WHO, 2011), there are significant gender differences in use, access, and behavior of water management. Gender discrimination is surrounded over the world and conniving the development of the society where patriarchal domination is seen in every aspect of social life. Gender discrimination constrained the water usage behavior of men and women in society as well as their access to water resources (Cap-Net, 2006). Unequal access to safe water has increased the burden of women, which in turn causes social insecurity due to the unavailability of safe drinking water points. The collection of water from distant places often exposes women to sexual harassment and reduces their productive and economic functionalities (Sarwar, 2005; Nasreen, 2008; Mahmuduzzaman et al., 2014). Hence, the contribution of women to water security is hindered by the shade of patriarchy cultural barriers, norms, and values.

Despite the fact, Sustainable Development Goal (SDG) 5 also emphasized gender equality and empowerment of all women and girls by declining all forms of gender discrimination and violence against women; to achieve the SDG goals there is a need for equal distribution of economic resources and unpaid work between women and men (UN Women, 2017). Other than SDG 6, which indicates safe drinking water for all, Millennium Development Goals (MDGs) also aims to achieve access to improved water sources for more than 2 billion people.

Moreover, water crisis causes detriment to the women and children more due to climate-induced hazards, water logging, salinity intrusion, floods, riverbank erosion, and agricultural droughts, which causes higher death, personal insecurity, dropout of girls from school and early marriage, unhygienic menstrual system, reproductive problems, disabilities, physical and sexual violence, etc. (Geere et al., 2010; REACH, 2015). Different survey data (Nasreen, 2008; Shamsuddoha and Chowdhury, 2007) showed that the living pattern, individual life expectancy, per capita agricultural land, education, health, and other basic services, social security, etc. in the coastal and disaster-affected areas are failing to meet the expectation level in comparison to national average. Due to the geo-socio-economic characteristics, the recurrent climatic variability and natural disasters put marginal communities entirely at risk, especially women who face much higher risks compared to men. To adapt with the calamities, women had to bear more physical burden than men due to gender-specific tasks (Nasreen, 2008). Women are also faced with grave poverty, increasing instances of violence and abuse after the disaster, which are often poorly taken into consideration (Nasreen, 2005, 2009). Generally, disasters affect all segments of the population, but the burden of impact has been observed on persons with disabilities, elderly, women, and children due to their gender and disability status (Nasreen, 2008, 2012) during any disasters and post-disasters situation (Nasreen, 1995, 2008, 2012; Reyes, 2002). Thus, gender identity leads to a vulnerable condition for women and children due to water-related disasters and it is extremely convoluted issues (Nasreen, 2003; Wisner et al., 2004). Furthermore, poor women, children, and elderly person are conceded as a carrier of 'vulnerability bundles' and place them at the pinnacle of risk (Wisner et al., 2004, p. 1).

Moreover, considering this view, the authors tried to explore the impact of water risks on gender in water-secure and water-insecure areas of the coastal polder. The study aims to show the core comparison between water-secure and water-insecure areas in terms of spatial differences, gender division of labor, and water collection challenges. This paper also examines gender disparities at the household level and at times the community level.

Methods and Materials

The aim of this article has explicated an interplay between gender and water and an insight into how gendered-different behavior intensifies water insecurity in coastal Bangladesh. To carry out this study, intensive fieldwork was conducted between 2017 and 2018. A mixed method (i.e. both qualitative and quantitative method) was administered to analyze water security and its' associated risks from a gender perspective. The study was dominated in qualitative nature, especially to explore the gender dimension with water insecurity. The qualitative nature was more conducive to producing in the intrinsic nature gender dimension with water insecurity in coastal Bangladesh.

The Polder 29 located in Khulna district of the coastal region of Bangladesh was selected as the study location for this research. The total area of Polder 29 is about 7930 hectors[1] (MoWR: EIA report, 2016). Four mouzas of Polder 29: Akra, Chandgarh, Gajendrapur, and Sundar Mahal (Figure 9.1), were selected purposively from Khulna district following the risk profile of REACH (2016/2017). To capture

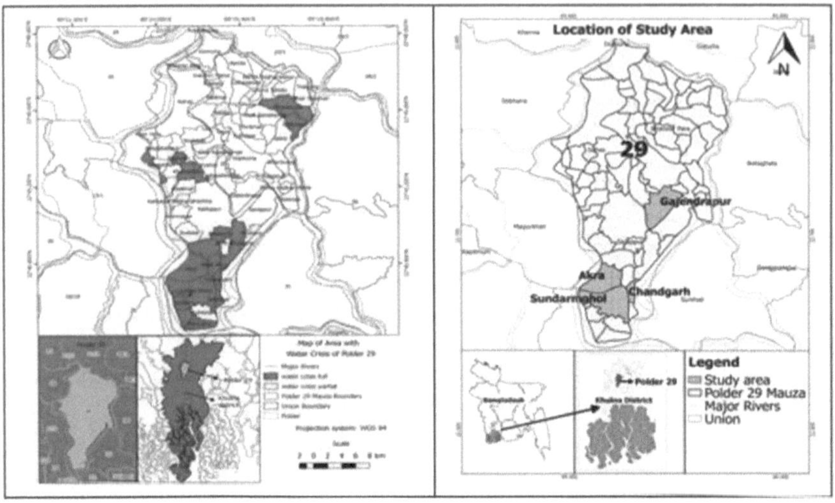

FIGURE 9.1 Polder 29 indicating water crisis and study area of Polder 29

Source: Ashroy Foundation and BUET.

a holistic aspect of water insecurity pertinent to gender issues, the participants of the study were categorized into three groups – men, women, and mixed groups. The reasons behind the selection of these mouzas were the prevalence of salinity intrusion, riverbank erosion, and unsafe water stress (Figure 9.1). However, all mouzas were not in similar conditions, for instance, water stress was lower in Gajendrapur compared to the other three areas.

The survey was conducted among 2103 households of Polder 29 in two phases based on purposive selections of mouzas[2] within the unions, and random selection of households based on the risk profiles of differential water security (REACH, 2017). Phase-1 focused on communities with spatial differences with different risk profiles and phase-2 was persistent with the communities who have relatively higher risk profiles in terms of water services (REACH, 2017). The REACH Khulna observatory team consisting of the Bangladesh University of Engineering and Technology, the University of Dhaka, and the University of Oxford provided support to collect data from the field.

The total surveyed households of Polder 29 were 2103; for this present study, four mouzas were selected purposively based on the risk profile of REACH. A total of 515 sample households (Table 9.1) were selected; of them, 142 households were from the first phase and 373 were from the second phase. Furthermore, qualitative data were garnered by conducting 12 focus group discussions (FGDs), 12 case studies, and8 in-depth interviews to identify a comprehensive view of water-related risks and its impacts on the vulnerable communities. Majority of the households of the four mouzas were headed by males; only a few households were run by females. The participants of FGDs were chosen randomly in the selected areas of

TABLE 9.1 List of mouzas and sample size

Union	*Mouza*	*Total households*	*Number of sample household in phase 1*	*Number of sample household in phase 2*	*Number of female-headed households*	*Number of male-headed households*	*Risk profile of water insecurity*
Sarappur	**Akra**	193	19	39	3	55	**High**
Sarappur	**Chandgarh**	195	20	42	2	60	**High**
Surkhali	**Sundar Mahal**	659	68	139	6	201	**High**
Sahas	**Gajendrapur**	726	35	153	10	178	**Moderate**
Total		1773	142	373	21	494	

Source: Field Survey, Khulna Observatory Research Team of REACH (2017).

coastal Bangladesh. At least 10–12 participants were present in each FGD. The goal of the interviews was to formulate a better understanding about type and causes of water-related risks, differential impacts based on gender, water-dependent livelihood and use of water, gender-based water conservation behavior, gender-based challenges while collecting water, gender-based alternative adaptation options, and decision-making in household level as well as outside in coastal polder. Mainly the questions were selected to identify the impacts of water insecurity on both women and men and their livelihoods, spatial variations of water-secure and water-insecure areas, diversification of jobs in different water-secure and water-insecure areas. For research purposes, the majority of the respondents were female and, in some cases, both female and male respondents have been selected from the same households. To identify the water insecurity situation in the coastal polder, the study focuses on three water-insecure areas namely Akra, Chandgarh, and Sundar Mahal with a glimpse of the water-secure area of Gajendrapur. Both female- and male-headed households were included in the study to examine the overall impact of water insecurity in households.

To collect the qualitative data, an audio recorder was used and then translated from Bengali to English. Ensuring the privacy and confidentiality of the participants', written consent was also collected from all respondents. Additionally, secondary data were used to explain the gender and water context in the coastal polder.

Analysis of the Data

Analyzing the intensive fieldwork in coastal Bangladesh, several issues emerged from the research associated with the gender dimension of water security. To collect quantitative data a structured questionnaire was developed by the researcher and electronic software of ONA, a software used for collecting and visualizing quantitative data, was applied for this household survey. From ONA the data was downloaded to the Microsoft Excel and was analyzed using Statistical Package for Social Sciences (SPSS) version 20. The REACH Khulna observatory team (BUET, University of Dhaka and University of Oxford) supported the programming process of the software and collecting the survey data. REACH provided the required number of tablets installed with ONA software for data collection purposes. The survey data were checked by a data analyst and REACH team. The data were transformed from ONA to Excel format. For further analysis, SPSS program was used. To analyze the qualitative data, three broad themes were categorized for this present article such as gender water collection challenges, gender division of labor, and spatial differences of water-insecure and water-secure areas.

Spatial Variations: Dialectical Relation with Water and Gender

The study findings indicated that most respondents in water-insecure areas explained that riverbank erosion and salinity were the major problems in their locality. In Chandgarh and Sundar Mahal respondents mentioned riverbank erosion was 96.8 percent and 92.8 percent, respectively, while 26.1 percent and 12.9 percent of respondents indicated salinity intrusion (Figure 9.2). Conversely, the participants of Gajendrapur reported the iron problem in water and unanticipated environmental shocks, especially cyclone. While people of the Gajendrapur mouza were economically solvent, the predominant problem of iron in water caused reduction in crop production.

In water-insecure areas, the respondents reported that they lost all they received through their different occupations due to erosion and salinity of the coastal polder (Figure 9.3). Male respondents indicated that they suffered more, since they were key earners and had to worry the most regarding the income sources for the family. The ensuing river erosion mostly seized their houses because of the surge of river water, and they lived with a sense of danger all the time for this. The river water increases every year, bringing up the salinity level significantly, creating problems in the cultivation of fish and paddy. After the river erosion, all the tube wells in the locality were affected by salinity and became completely undrinkable. All women and men had stomach ailments after drinking the water and the tube wells' water is still unsuitable for drinking.

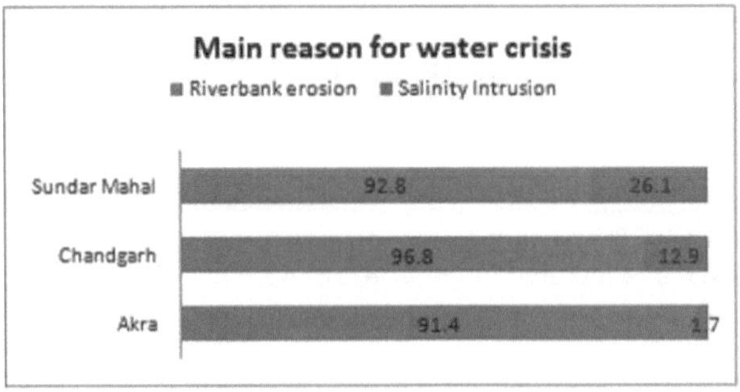

FIGURE 9.2 Main reason for the water crisis

Source: Survey data, 2018 (multiple responses).

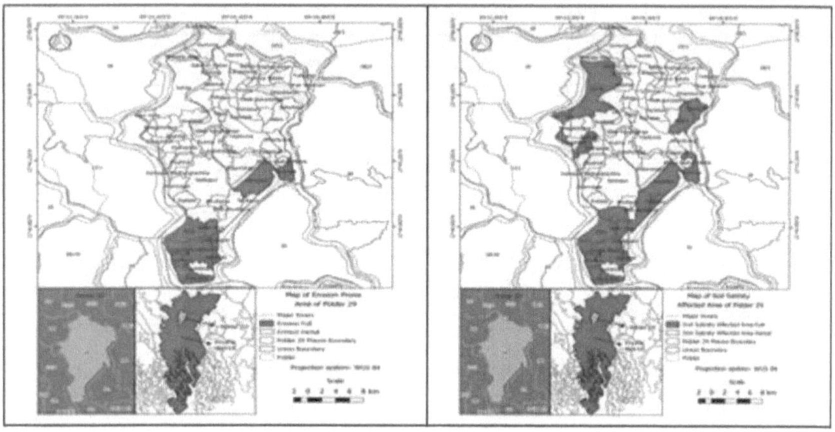

FIGURE 9.3 River erosion in Polder and soil salinity in Polder 29
Source: Ashroy Foundation and BUET.

When asked about financial/environmental losses, the participants revealed that the fish and paddy cultivation was damaged after erosion. For two years they could not cultivate paddy, however, they added that from the last year (from 2018) they could do paddy cultivation. The FGD participants revealed that after erosion their houses were destroyed, and they took shelter next to the roadside; there was no gender-friendly shelters and no prevalence of toilet facilities for both men and women. Moreover, female-headed households also exposed that they did not get any benefit due to their social identity. The leaders only provided facilities for those families who they are close/familiar with. Thus, a dialectical relation was observed from empirical results (Table 9.2).

In water-insecure areas, women's resilience building capacity is undermined by several factors, for example, water collection conflict, religion, and ownership of land and water sources.. The qualitative data also illustrate that the resilience capacity of men decreases due to water collection conflict and ownership of land and water sources at a moderate level. In contrast, a number of contributory factors – for example, women's decision-making power in installing tubewell at the household level, workshop and training on water security, women's engagement in income earning sources outside of their locality – foster women's resilience. 'Women take the decision of establishing tube-well through physical labor and selling their domestic animals'– said one of the women participants from FGD.

As, both women and men faced more health issues in Akra, Sundar Mahal, and Chandgarh (Zaman, 2018), they perceived salinity to be the key driver of health hazards. However, to reduce the impact of salinity, women kept water in the washroom for three to four days and used it later. They assumed that it would decrease the salinity level of water and reduce their health problems.

In the study, it was also found (Figure 9.4) that more water insecurity leads to more involvement of women in the workplace in Akra (3.57 percent), Chandgarh

TABLE 9.2 Dialectical relation between water and gender (FGDs and case studies' findings)

Dialectical relation between water security and gender dimensions				
Thematic variables and reason of conflict	*Women resilient capacity*		*Men resilient capacity*	
	Water-insecure areas	*Water-secure area*	*Water-insecure areas*	*Water-secure areas*
Land ownership	- -	++	- -	+ +
Ownership of water resources	- - -	++	- -	+++
Religion	- - -	- - -	+++	+++
Shrimp cultivation	- -	++	- - -	+++
Funding from donor agencies	- - -	++	++	++
Water collection conflict	- - -	++	- -	++
Divorce rate	- - -	+	- - -	+
Wage conflict	- - -	- - -	+ + +	+ + +
Workshop and education, trainings	++	++	++	++
Migration	- -		---	
Diversification of job	++		++	++
Health impact	- - -		- - -	++
Installation of water sources	++	++	++	++
Involvement of water management group	++	++	- - -	- -
Involvement of women and men in outside village	++	- -	++	-

Note: Increase in high level= +++ ↔ Decrease in high level= - - -
Increase in moderate level= ++ ↔ Decrease in moderate level= - -
Increase in low level= + ↔ Decrease in low level= -

(3.57 percent), Sundar Mahal (32.14 percent) compared to Gajendrapur (0.0 percent). On the other hand, in Gajendrapur women were more involved in work inside the village compared to water-insecure mouzas. The qualitative findings indicated that long-term and short-term rainfall hampered the overall crop production that created a shortage of food for both humans and animals in water-insecure areas. Some of the villagers practiced combined crop production like fishing and paddy cultivation in the same land. They had no coping strategies except storing some fertilizers. Women of the village also played an important role in agriculture by working as a helping hand for the males in both water-secure and water-insecure areas. Besides agriculture they also helped in fishing activity. They reared various domestic animals and cultivated vegetables in their courtyards. Women were more conscious of water because they were involved in more water-centric chores. The participants of the FGD used rainwater for agricultural purposes in water-insecure areas. Sometimes the coastal people used deep tube well for agriculture purposes, since shallow tube well water is not usable. Due to high level of iron and salinity in the soil, the people of Akra, Chandgarh, and Sundar Mahal were unable

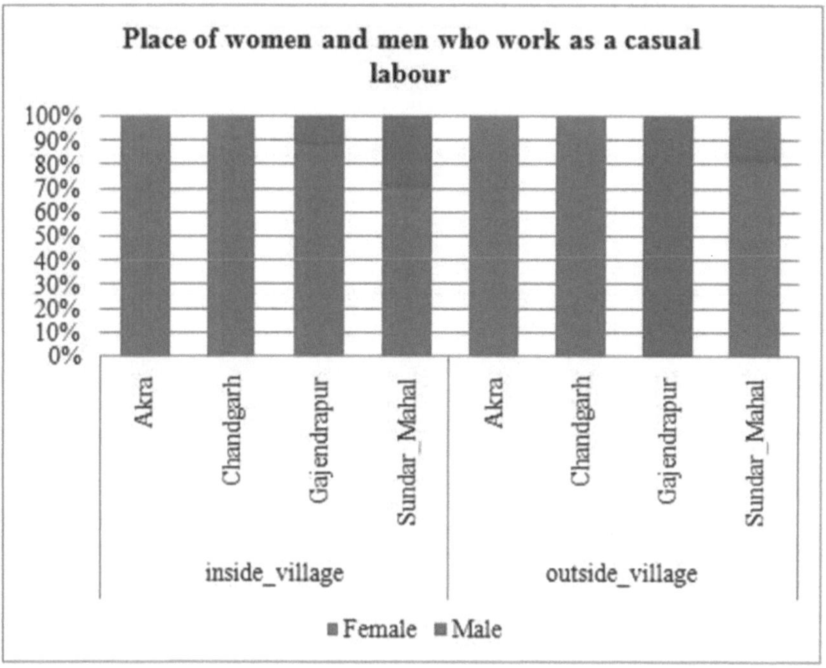

FIGURE 9.4 Location of work as casual labor based on gender

Source: Survey data, 2018.

to grow paddy. People used raw sugar in soil and water with the hope of decreasing the iron and salinity. One of the male respondents in Chandgarh shared his own experience regarding this matter, where he used 4 kg of sugar, first in water then mixed it with soil. The few respondents of water-insecure areas assumed that these practices will increase the production of the crops. Thus, dialectical changes were seen between the water-insecure and water-secure areas.

Paradoxical Challenges towards Gender and Water

The survey results indicated (Table 9.3) that most of the women were the collector of water for daily life chores in the family in terms of water-secure and water-insecure areas, while men played a supportive role in water-secure areas. In Sundar Mahal men were involved in fetching water, but the number of male respondents (Table 9.3) was less (5.5 percent). The interviewees explained that the reason behind it was that social norms demonstrated water fetching as women's responsibility. When fetching water from long distances, girls and boys of the family accompanied their mothers to help reduce the physical burden. Men also collected water from far away locations and used their vehicle as a mode of transport. Gender-inclusive

TABLE 9.3 Water collectors of the mouzas

Name of the mouza	*Collector of water (N=515)*			
	*Women (n=413)**	*Men (n=144)**	*Girls(<16 years) (n=47)**	*Boys(<16 years) (n=44)**
Akra	4.6	2.2	0	0
Chandgarh	7.2	4.6	0.2	0.9
Gajendrapur	36.9	19.2	6.3	7.4
Sundar Mahal	41.5	5.5	3.7	1.3

* Multiple responses (Household Survey data, 2018).

TABLE 9.4 Transportation system of water collection

Transport using to collect water	*Name of the mouza and percentage of the respondents (N = 515)*			
	Akra	*Chandgarh*	*Gajendrapur*	*Sundar Mahal*
Byindividual carried	69.23	63.04	95.00	98.48
Nossimon	30.77	6.52	0.00	1.02
Bicycle	0.00	30.43	3.89	0.51
Trawler	0.00	0.00	1.11	0.00

* Multiple responses (Survey data, 2018).

dimensions (GIDs) in organizational and institutional level are recognized in the coastal zone (Polder 29) and these gender dimensional vulnerabilities will be exemplified in connection with the water security model.

The study findings showed that water risks are related to gender in terms of fetching water, distance of water points, washing clothes, cleaning utensils, psychological stress, unhygienic latrines, and food storage. In terms of the availability of water in insecure areas, most of the FGD participants revealed that for about eight months they had easy access to water from different sources, however, for the remaining months they had to purchase water or drink dirty pond water or contaminated tubewell water in water-insecure areas. They also added that, they mostly used the shallow tube well water for their daily needs such as washing clothes, cleaning dishes, taking bath, and sometimes cooking too.

It was a tedious job for both men and women to fetch water from distant places. It is apparent from Table 9.4 that the nearest 70 percent, 64 percent, 95 percent, and 99 percent people of Akra, Chandgarh, Gajendrapur, and Sundar Mahal, respectively, collected water by walking to cover the long distance. People in Akra mostly used *nossimon*[3] as their water transportation (30.77 percent), and bicycle was used more in Chandgarh (30.43 percent) compared to other mouzas.

In the water-insecure areas, the female respondents complained of health-related problems such as itching, skin diseases, menstrual cramps, vomiting, abdominal pain, and so on. Most of the male respondents had gastric problems from drinking saline water in water-insecure areas. In contrast, people of the water-secure area also indicated some health problems due to excess iron in their tube well water.

Both female- and male-headed households have faced a similar situation while collecting water from a far distance. From time to time, they had to pay to the water vendors to transport the water that was brought in on the vehicle from 7–8 km away. They gave the water pot to the water providers and then they would deliver the water in the evening. It costs BDT 20–30 per jar, which was hard to afford for vulnerable population.

Water body was approximately 6–7 km away from Akra and Chandgarh and 3–4 km from Sundar Mahal (Qualitative Field data, 2018). It took 30–60 minutes (Figure 9.5) to collect water in Chandgarh. Most respondents collected water from the dwelling yard. In the four mouzas, the women respondents reported on gender-based challenges while collecting water such as becoming sick with back pain, sometimes they could not stand on their feet, and so on. During this time family members such as their children and husbands fetched water.

The distance of water points and quality of water amplify the tensions between different groups in coastal Polder 29. The women respondent added that when pond water was limited or thicker, they increased pressure on tubewell water. It was also regarded as one of the reasons for breaking serial to collect water and quarrelling

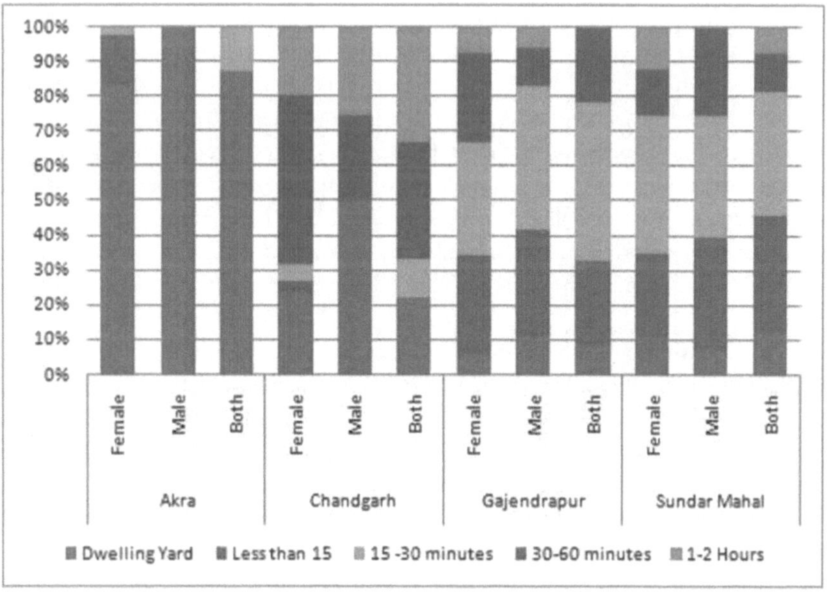

FIGURE 9.5 Time to collect water from the sources

* Measures the following questions: How much time does it take to go the source, get water, and come back?

** Multiple responses (N=515), Household Survey data, 2018.

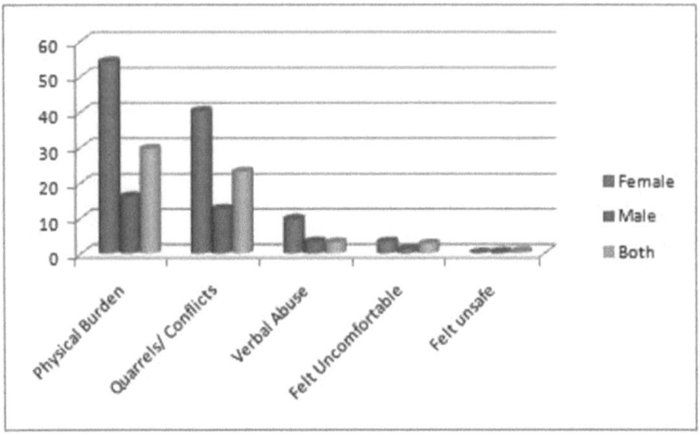

FIGURE 9.6 Challenges of women in water collection (N–515)
Source: Survey data, 2018 (multiple responses).

with each other. Sometimes they have to wait until midnight to collect water or to get an early serial. The FGD participants pointed out that if they did not have to spend so much time to fetch water, they might have been able to spend more time with their family members or their children, and they could also complete more household activities. Some families also faced divorce problems due to unsafe water in Sundar Mahal and Chandgarh. Some people left their paternal house and are currently working in the city. Some wives even divorced their husbands and they never went back to the husband's family. Around 146 respondents from the four mouzas mentioned physical burden as the major problem of collecting water from a distance, 45 respondents (Figure 9.6) mentioned verbal quarrelling and verbal abuse due to the long serials for water collection. Most of the female FGD participants (111 out of 120) strongly reported the same challenges were faced while collecting water in their locality. 'First come, first serve' was the rule for collecting water of the villagers.

> Wife beating by husband often happens in our locality. One of the reasons is that she was not able to prepare food timely due to the paucity of water. However, this ill practice is not present on a massive scale. If I talk about myself that I have to do all the household activities on my own. I bring water for my family. The experience of collecting water is not very good. During the fajar azan (call for prayer time at dawn) we (the women) wake up and go out for collecting water. Some of us have to go even during easha azan (call for prayers at night). We always go in a group to fetch water. It gives us sometimes to talk to each other, gossip and chat about different things. Such meetings and storytelling help us to forget about the heavy load of the water bucket while travelling a long distance.
>
> *A female FGD participant, Chandgarh*

When women go to collect water from other tube wells or ponds, sometimes the owner of that pond or tube well make them deliberately inaccessible, so women are unable to fetch water. It was humiliating for them to collect water from other people's pond or tube well, and would often at times lead to social unrest. The women would have to end up going to water points far away, the extra distance was about half a kilometre from Chandgarh, which exacerbated the suffering of women in that area. Moreover, if the owner asked BDT 10 (USD 0.118021) to collect one pitcher of water from that reservoir, people did not agree to do it due to their meager economic conditions. Most respondents (106 out of 120) mentioned that they had tube wells nearby, but the water quality was not good. During the rainy season, women collected rainwater with the help of men by using polythene/plastic sheets to store on the roof or the house premises in Sundar Mahal and Chandgarh. Sometimes unknowingly drinking water was misused by men mostly, for example, washing hands with drinking water after eating. One female respondent mentioned, 'When there is not enough water in the house, we mix safe water with unsafe water and then drink it' (FGD participants, 2018). When there was a shortage of water, they would use deep tube well.

Gender Division of Labour, a Jeopardizing Dimension

High levels of salinity and erosion in the southern parts of the Polder 29 in Khulna district (Chandgarh, Sundar Mahal, and Akra) affect agricultural production and predominantly drive women and men to search for work outside of the locality, being susceptible to multiple gendered discriminations at their workplace. Gender division of labour and wage discriminations are found everywhere in the coastal polder. Women who worked outside their family farms encounter various types of gender-based discrimination in their workplace in the form of lower payment, an unequal division of labour (e.g. water collection is only women's work), and working longer hours than men. Women living in the water-stressed areas earned half the wage compared to men, when working for agriculture, aquaculture, and as casual labors (Figure 9.7). Despite the discriminations, women laborers were still easier to find for agricultural activities, due to their poor economic conditions. In the distinctively water-secure Gajendrapur village, women were capable of self-sustaining themselves without having to search work outside the locality. Still, the recognition of women's work is limited at the household level and they are not valued by the patriarchal social norms. Marginalized women (poor, lack of access to rights, and treated as less important) are recognized as less valuable and available as casual laborers in the agriculture sector. Men make most decisions regarding income-generating activities such as agricultural activities and investment in farming, whereas women make decisions about daily household chores. More women are members of community organizations such as NGOs and other private institutions, but decisions are articulated by men at the community and household level. Men spoke on behalf of the community and their decisions have remained. Women also spoke but received less attention to their voices (Qualitative

FIGURE 9.7 Woman pushing the van in Chandgarh (as casual labor)

data, 2018). Some male members in the family do not want to hear the voice of women; this turns the situation grave. Though women work with equal pressure as men do, their contribution is not considered as valuable in water-insecure areas such as Akra, Chandgarh, and Sundar Mahal. In contrast, in the water-secure area as Gajendrapur, women help their husbands to work on their land instead of going outside to work. They revealed that the production in their locality is good because of the lack of saltwater in their soil.

In water-insecure areas, the female FGD participants said that they plant, nurse, and cut the paddy, then they bring it home and harvest it through machines, etc. They additionally cultivate paddy, potatoes, brinjal, and many others on the fields. Whereas they revealed that, female members received BDT 150 (USD 1.77043) per day and the male member received BDT 250 (USD 2.95072) per day. However, women claimed that they received less money though they worked as much as their counterparts and for the same length of time. They also added that

> We receive less money only because we are women, there is no other reason. The women work in the enclosure (*Gher*) from 8:00 am to 2:00 pm, they get BDT 150 (USD 1.77043) per day. The women who work in Khulna get BDT 450 (USD 5.31215), in contrast, men get BDT650 (USD 7.67311) per day.

Additionally, both women and men are involved in different types of work in coastal Polder 29, such as casual labor, business, agricultural and aquaculture activities, income from property rent, poultry, and so on (Figure 9.8). Majority of the women are involved in the agricultural sector (62.7 percent) and as casual labor (48.7 percent). When comparing with men the situation is same, 69.9 percent and

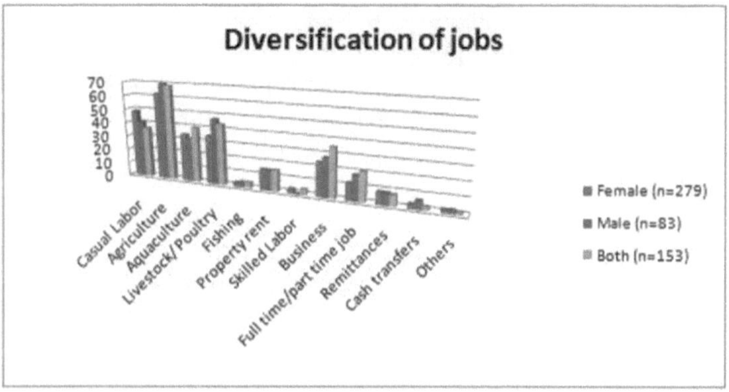

FIGURE 9.8 Diversification of job in coastal polder

Source: Survey data, 2018 (multiple responses).

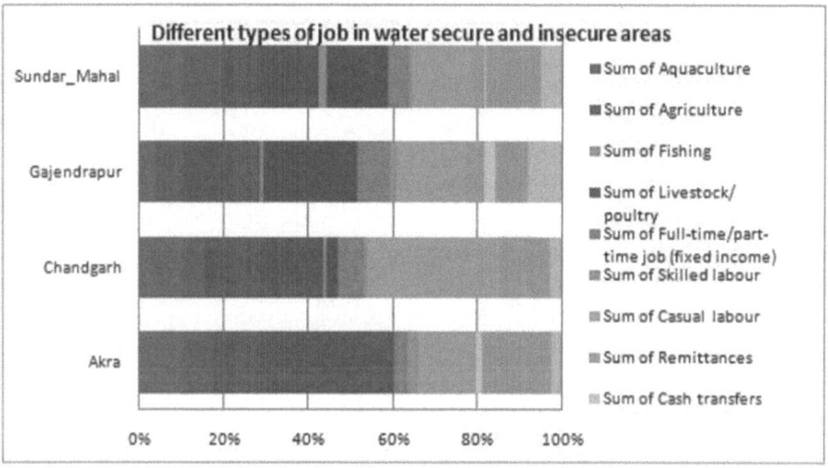

FIGURE 9.9 Different types of jobs in the four mouzas depending on water-secure and water-insecure areas

Source: Survey data, 2018 (multiple responses).

41 percent are, respectively, involved in the agricultural sector as well as casual labor. Some of the households in the survey also indicated that both women and men are involved in the same job. Livestock is another popular job sector where both women (34.8 percent) and men (47 percent) are involved in Sundar Mahal and Gajendrapur (Figure 9.9). Most of the households of Akra and Sundar Mahal are additionally involved with business because of their loss in the agricultural sector due to salinity and river erosion.

Women are earning half the amount that men earn in agriculture and aquaculture, and as casual labors, though they have equal work pressure.

> It is not that women work any less than men, rather, in some instances, they work more. But a woman will get BDT 200 (USD-2.36) and a man will get BDT 500 (USD-5.9). But the workload is equal. Men bear the load on their head; women bear it on their waist. It is just that you are a woman, and men want to portray you as the weaker gender. Women are becoming more aware. In the near future, this discrimination will go away.
>
> *Chandgarh, KII, Female participant*

Discussions

Our findings showed water insecurity leads to differentiated impact on the gendered culture of the coastal polder. Salinity intrusion and river erosion were the main concerns highlighted by respondents in the southern part of the polder, in Chandgarh and Akra. In contrast, respondents in Gajendrapur only highlighted the issue of iron and salinity. These issues are especially severe during the dry season (January–April) when cultivable land becomes more shriveled and the soil becomes more brackish. As such, seasonal drought was also identified as an important stress. Some farmers used sugar in agricultural fields to reduce salinity levels in the soil for their crops. However, marginalized and economically disadvantaged groups did not have this opportunity.

In addition, the southern parts of the polder faced higher risks from the cyclonic storm, damaging people's land and properties. This also contributed to men and women leaving the area to seek better livelihoods, often leaving their families behind. A significant number of women reported that they sold their valuable assets like jewellery, land, livestock, and household utensils to sustain themselves. Higher levels of hazards in the southern part of the polder led to more negative impacts on agricultural production, drinking water, and overall well-being and health.

High levels of salinity and erosion in the southern parts of the polder (Chandgarh, Sundar Mahal, and Akra) affected agricultural production; so many women had to work outside of the area to sustain their livelihoods. There, they faced various types of gender discrimination: women were often paid less than men, even for the same workload, and worked long hours (in addition to their household chores, such as collecting water). More women also reported to have increased workload in planting food crops compared to their male partners (Nelson and Stathers, 2009), but men get different/higher income for the same production (Bynoe, 2009; Zaman, 2018). The women, who worked in the agriculture and aquaculture sectors, and as casual labor workers, earned only half the amount men earned. In contrast, Gajendrapur village, which suffers from fewer water risks, women were, for the most part, able to self-sustain themselves without needing to find work in other areas. The women work in other's farm/plots and they face fewer difficulties such as getting a lower payment and longer working hours. The theory was also supported by the findings of the study that women have a greater physical burden than men in the event of calamities because of their gender-specific tasks. Nasreen (2008, 2012) explained the situation in her gender-disaster resilient theory that gender discrimination is a major factor leading to increased vulnerability for women and the risk of water in accessing and controlling information and services, technology, lack of access to or control of reproductive resources.

It was noted in the literature that in 44 countries across the world, women are the most common water collectors: 58.6 percent on average, and men 30.4 percent, children 9.1 percent, respectively (Sorenson et al., 2011). The situation is also similar in the context of coastal Bangladesh. In developing countries, performing as the collectors of water, women constitute only 17 percent of the formal and paid water workers, whereas they constitute 80 percent of the casual and unpaid water collectors or workers (IWA, 2014). In some countries, this percentage is more than 90 percent (UNICEF-WHO, 2011). Due to the low quality of water in Chandgarh and Akra, women walked on average around 3–4 km and 6–7 km, respectively, to fetch water from shallow tube wells. However, the quality from those shallow tube wells remains low, and their only alternative was to purchase it from water vendors by paying BDT 20 (USD-0.24) for 20 liters of water. In situations where they were not able to buy sufficient water for their households, women stored the safer water (from water vendors) for their children and husbands and they drank the unsafe water from shallow tube wells. Men will only collect water when there is no woman at home or women of the households were suffering from illness and unable to collect water from a distance. Nasreen (2008) and Hemson (2002) also argued to explain the situation of women that was similar to the study findings that women are positioned in these roles by their gender, age, and household status.

The distance of water points and quality of water amplify the tensions between different groups in coastal Polder 29. The women respondents added that when pond water is limited or went thicker, it raised pressure on tubewell water. It is also regarded as one of the reasons for serial breaking to collect water and quarrel with each other. Sometimes, they have to wait till midnight to collect water and get early serials (Qualitative data, 2018). The FGD participant reveals that if they hadn't spent so much time collecting water, they might have given more time to their family members or children as well as pursued more household activities. Some families also face divorce problems due to unsafe water in Sundar Mahal and Chandgarh (Qualitative data, 2018). Some people leave their paternal houses and move to the town or cities to work. Some wives divorced their husbands and never went back to the husband's family.

Recognition of women's works and decisions is not valued by the men of the coastal society due to patriarchal social norms. Brym and Lie (2012) argued that male domination is rooted highly in the relationships of patriarchal authority relations, family structures, and socialization and cultural patterns that exist in nearly all societies. In the line of the present study, it is observed that marginalized women (poor, lack of access to rights, and treated with less importance) are considered less valuable and usually available as casual laborers in the agriculture sector. Men are taking most of the decisions regarding income-generating activities such as agricultural activities and investments in farming, whereas women make decisions about daily household chores. The number of women in the community is higher, but decisions are articulated by the men in both, the community as well as household levels. Men speak on behalf of collectives in the community and their decisions remain. Women also speak but get less attention to their voices (Qualitative data, 2018). It is evident that women also face grave poverty, increasing instances of violence and abuse after disasters, which are often poorly taken into consideration (Nasreen, 2008; Kumbetoglu and User, 2010). However, some literature provides evidence that poverty and scarcity of women are affected in different ways compared to men (FAO, 2011) and shocks also differ from each individual (Cohen and Young, 2007). Women have less access to assets, capital, and resources than men (Antonopoulos and Floro, 2005; Deere and Doss, 2006; Quisumbing, 2009; Nasreen, 2012) and they have less decision-making power in the household, community, and national level.

Discriminating roles of organizations in the distribution of water facilities in coastal Polder 29 (Chandgarh, Sundar Mahal) are also gender inhospitable. The allocation of water by the organizations working in the community was not based on the needs of coastal people. A group of people with beneficial connections gets most facilities related to water and the poor have no access to safe water. These people get more support during a disaster or water crisis by the local leaders. However, female-headed households do not receive such support in times of crisis or risk reduction activities. In Sundar Mahal as well as Chandgarh there is no allocation or regulation system of water sources such as tube well.

It was revealed from the findings that a certain group of people dominate in the water management sector whereas the marginalized group has less participation in water-related decision-making. The chairman of the locality does not monitor these activities, instead the work is carried out by the local inhabitants who are comparatively in better economic conditions and have a close relationship with the chairman (mostly by fawning him). The respondents have informed that any aid from the government or non-government organizations is handed over to the chairman and no further monitoring of the system is done to ensure that the people in need benefit from them. The Chairman usually hands over the aid materials to their relatives/close person (Qualitative findings, 2018) who are usually backscratchers (local leaders).

Securing water management and gendered behavior are juxtaposed in the concurrent literature and realities of the coastal area, but both issues are not simultaneously working for the development of water security. This correlation also depended on the time of water collection as well as the distance of the water point. Literature showed that if the societal burden of collecting water was less, the role of collecting water would also shift to men and children (Sorenson et al., 2011). Thus, fetching water, time, distance affected the health, opportunity, and economic condition of the household as well as amelioration (Brouwer et al., 2007; Sorenson et al., 2011) of the situation of coastal women.

The struggle for water on a daily basis creates the painful physical and social burden of women and girls who are spending most of their valuable time to fetch water over long distances (UN Water, 2013). Millennium Development Goals (MDGs) have already been met and Sustainable Development Goals (SDGs) have yet to address its goal. But, the water conditions of coastal people are worsening day by day due to climate change and gender behavioral approach. From the evidence of the literature, it was found that women arranged food, reconstructed houses with their resources, collected water, and made it safe to drink (Nasreen, 2012). Indeed, understanding the gender dimension is the prerequisite to mitigate risks among the coastal community because they are not only the consumers but the safeguard of water.

Conclusion

The empirical results state that vulnerability to water-related risks and other climate effects are closely linked to gendered culture, sustainable development, and livelihood. Though Bangladesh has achieved remarkable success in managing disaster risk reduction and given equal opportunity for both women and men at all level, the issue of water security has not been adequately addressed to mitigate water-related vulnerabilities in coastal areas. As coastal communities are more vulnerable to hazards and disasters compared to other parts of the country, more attention needs to be given for policy implementation. Despite the rhetoric on water collection, women's participation is less in water management. Additionally, male's domination is the barrier for women's involvement in development pathways. Over the past few decades, women's empowerment is embryonic in every aspect of life. However, still, they have limited access to organizational, institutional, or household-level decision-making processes in a country like Bangladesh. The roles of women hindered by the blockade of socio-cultural values of patriarchy and accelerated the gap in the gendered culture as a maxim in terms of water security. Women's participation in decision-making roles in the water management sector needs to be extensively addressed by the government, community level, institutions, and organizations. The study also finds that the accountability of sharing the water sources materials of the community is less and confined with the power structure in Polder 29. The water resources distribution process should be at the community level, placing women's strategic interests and practical needs at the center. A 'people-centred' participatory inclusive system is required to conciliate water risks and to establish a sustainable water management process by reducing the discrimination of payment for work and valuing women's work in the coastal polder. By giving women equal opportunity as their male counterparts as well as increasing wages, their work will be valued more. Institutional capacities of government need to be strengthened and enforced for effective equal distribution of water sources across all categories. Empowering women and increasing their education level for mitigating water risks of the community is crucial for ensuring water security. Gender-specific access to land and market is vital for social recognition and encourage decision-makers to build a water-resilient society at the community level. In broad, a balanced representation of both men and women is required for water security in the coastal zone. As revealed in the study, coastal zone in Bangladesh is highly gender sensitive; women's participation in decision-making roles in securing water sources will create vital space to ensure gender equality. The gender balance world is requisite for the development of the water-poor community and therefore would support minimizing the water risks of the coastal zone.

Acknowledgment

This article is part of a study funded by the REACH program led by the University of Oxford, UK in collaboration with Bangladesh University of Engineering and Technology (BUET) in Bangladesh. REACH program aims to work on ways to improve water security for the poor especially in developing countries, and this article contributes to the REACH Coastal Water Security observatory by exploring the impacts of water insecurity through the gender lens. The immense support from the 'REACH Khulna Observatory' team comprising Bangladesh University of Engineering and Technology (BUET), University of Dhaka, and University of Oxford is immensely appreciated.

Notes

1 1 Hector= 10,000 square metres.
2 Mouza= Synonymous with the gram or village in Bangladesh.
3 A form of local transportation in the villages.

References

ADB. (2016). *Gender equality and food Security: Women's empowerment as a tool against hunger.* Asian Development Bank, Mandaluyong City, Philippines.

Antonopoulos, R., & Floro, M.S. (2005). *Asset ownership along gender lines: Evidence from Thailand.* Gender, Equality, and the Economy. Working Paper, No: 418. www.levyinstitute.org/pubs/wp418.pdf

Brouwer, R., Akter, S., Brander, L., & Haque, E. (2007). Socioeconomic vulnerability and adaptation to environmental risk: a case study of climate change and flooding in Bangladesh. *Risk Analysis: An International Journal, 27*(2), 313–326.

Brym, R., &Lie, J. (2012). *Sociology: Pop culture to social structure.* Nelson Education.

Bynoe, P. (2009). Case study on the impact of climate change on the agriculture of an indigenous community in Guyana.

Cap-Net, G. W. A. (2006). Why gender matters: A tutorial for water managers. *Multimedia CD and booklet. CAP-NET International network for Capacity Building in Integrated Water Resources Management, Delft.* www.thewaterchannel.tv

Carver, C. (2016). Everything to know about Global Goal 6: Clean Water and Sanitation. Global Citizen.

Cohen, M., & Young, P. (2007). *Using micro insurance and financial education to protect and accumulate assets. Reducing global poverty: The case for asset accumulation.* Washington, DC: Brookings Institution Press.

Coles, A. & Wallace, T. (Eds.). (2005). *Gender, water and development.* Oxford and New York: BERG.

Dasgupta, S., Şiriner, İ., & De, P. S. (Eds.). (2010). *Women's encounter with disaster.* Ismail Siriner.

Deere, C.D., & Doss, C.R. (2006). *Gender and the distribution of wealth in developing countries.* Research Paper No: 2006/115. UNU World Institute for Development Economics Research (UNU-WIDER). www.wider.unu.edu/stc/repec/pdfs/rp2006/rp2006-115.pdf

Earle, A., & Bazilli, S. (2013). A gendered critique of transboundary water management. *Feminist Review, 103*(1), 99–119.

FAO (Food and Agricultural Organization). (2011). *The state of food and agriculture 2010–2011-Women in agriculture: Closing the gender gap for development.* Rome: Food and Agriculture Organization of the United Nations. www.fao.org/docrep/013/i2050e/i2050e00.htm

Geere, J. A. L., Hunter, P. R., & Jagals, P. (2010). Domestic water carrying and its implications for health: A review and mixed methods pilot study in Limpopo Province, South Africa. *Environmental Health, 9*(1), 52.

Hemson, D. (2002). *'Women are weak when they are amongst men': The participation of women in Rural Water Committees in South Africa* (Vol. 3). HSRC Press.

IWA, (2014). *An avoidable crisis: Wash human resources capacity gaps in developing economies.* London: International Water Association (IWA). https://iwa-network.org

Jambrina, F. (2013). *Nicaragua: Final evaluation report.* New York, Millennium Development Goals Achievement Fund (MDG-F).

Khosla, P., &Pearl, R. (2003). *Untapped connections. Gender water and poverty: Key issues government commitments and actions for sustainable development.* New York: Women's Environment and Development Organization (WEDO). https://wedo.org/wp-content/uploads/untapped_eng.pdf.

Kumbetoglu, F.B., and User, I. (2010). Gender aspects of the disaster recovery process. In: *Women's encounter with disaster*, edited by S. Dasgupta, I. Siriner, & P. Sarathi De, 22–51. London: Frontpage Publications Limited.

Lawuyi, O. B. (1998). Water, healing, gender and pace in African cosmology. *South African Journal of Ethnology, 21*(4), 185–190.

Mahmuduzzaman, M., Ahmed, Z. U., Nuruzzaman, A. K. M., &Ahmed, F. R. S. (2014). Causes of salinity intrusion in coastal belt of Bangladesh. *International Journal of Plant Research, 4*(4A), 8–13.

MoWR. (2016). Environmental impact assessment (EIA) on rehabilitation of Polder 29: Final report. Ministry of water Resources, Government of the People's Republic of Bangladesh.

Nasreen, M. (1995). *Coping with floods: The experience of rural women in Bangladesh.* Unpublished PhD Dissertation, Massey University, New Zealand.

Nasreen, M. (2003). Gender and sustainable development in Bangladesh: Myths and realities. *Journal of Social Studies-Dhaka*, 84–98.

Nasreen, M. (2005). *Gender inequality, poverty, water and sanitation in Bangladesh.* Baseline survey of Sustainable Environmental Health (ASEH) programme of Water Aid Bangladesh, Dhaka: Pathway International, July.

Nasreen, M. (2008). Impact of climate induced disasters and adaptation strategies: A gender perspective. *Climate change impacts and adaptation strategies for Bangladesh. ITN Centre and Dept. of Civil Engineering-BUET*, 256.

Nasreen, M. (2009). *Violence against women during flood and post-flood situations in Bangladesh.* ActionAid Bangladesh.

Nasreen, M. (2012). *Women and girls vulnerable or resilient.* Institute of Disaster Management and Vulnerability Studies, University of Dhaka, Dhaka-1000.

NCS. (2016). Bangladesh National Conservation Strategy (2016–2031). Ministry of Environment and Forest Government of the People's Republic of Bangladesh.

Nelson, V., & Stathers, T. (2009). Resilience, power, culture, and climate: A case study from semi-arid Tanzania, and new research directions. *Gender & Development, 17*(1), 81–94.

Quisumbing, A.R. (2009). *Do men and women accumulate assets in different ways? Evidence from rural Bangladesh*. Washington, DC: International Food Policy Research.

REACH. (2015). *Country Diagnostic Report Bangladesh. Working paper 1.* Oxford: University of Oxford. https://reachwater.org.uk

REACH. (2016/2017). *Report*. Oxford: Smith School of Enterprise and the Environment, University of Oxford.

REACH. (October, 2017). *REACH Bangladesh Coastal Observatory Household Survey*. Oxford, University of Oxford.

Reyes, R.R. (2002). Gendering responses to El Nino in Rural Peru. *Gender and Development, 10*(2), 60–69.

Sarwar, M. G. M. (2005). Impacts of sea level rise on the coastal zone of Bangladesh. http://static.weadaptorg/placemarks/files/225/golam_sarwar.pdf

Shamsuddoha, M., &Chowdhury, R. K. (2007). *Climate change impact and disaster vulnerabilities in the coastal areas of Bangladesh*. COAST Trust, Dhaka.

Sorenson, S. B., Morssink, C., &Campos, P.A. (2011). Safe access to safe water in low income countries: Water fetching in current times. *Social Science & Medicine*, *72*(9), 1522–1526.

Stevenson, E. G., Greene, L. E., Maes, K. C., Ambelu, A., Tesfaye, Y. A., Rheingans, R., & Hadley, C. (2012). Water insecurity in 3 dimensions: An anthropological perspective on water and women's psychosocial distress in Ethiopia. *Social Science & Medicine*, *75*(2), 392–400.

Tuong, T. P., Humphreys, E., Khan, Z. H., Nelson, A., Mondal, M. K., Buisson, M. C., & George, P. (2014). Messages from the Ganges Basin development challenge: Unlocking the production potential of the polders of the coastal zone of Bangladesh through water management investment and reform.

Tusar, K. D., & Moumita, C. (2013). Climate change influence water use pattern in south-west coastal belt of Bangladesh. *Journal of Environmental Science and Natural Resources*, *6*(2), 217–225.

UN Water. (March, 2013). *Water security: Experts propose a UN definition on which much depends*. Hamilton: United Nations University. Water related news. UN-Water website.

UN Women. (2017). In focus: Women and the sustainable development goals: SDG 5: Gender equality.

UNICEF-WHO. (October, 2011). *Drinking water equity, safety and sustainability: thematic report on drinking water.* New York: UNICEF-WHO. https://data.unicef.org

Vaughan, S., Roy, D., & Miletto, M. (May, 2016). *Building transboundary water security*. UN World Water Assessment Programme (WWAP), UNESCO.

Water.org. (2019). *Bangladesh's water and sanitation crisis*. Bangladesh. https://water.org/our-impact/bangladesh/

WHO. (2015). World Health Statistics 2009-World Health Organization. www.who.int>gho>> publications>world_health_statistics> 2015

Wisner, B., Blaikie, P., Canon, T., & Davis, I. (2004). *At risk: Natural hazards, people's vulnerability and disaster*. London & New York: Routledge.

Zaman, S. (10 July, 2018). *How women bear the brunt of water-related risks in coastal Bangladesh*. Newsletter. MPhil Student, Institute of Disaster Management and Vulnerability Studies, University of Dhaka. reachwater.org.uk. https://reachwater.org.uk/reach-early-career-researcher-feature-how-women-bear-the-brunt-of-water-related-risks-in-coastal-bangladesh/

10

WHERE LIES THE PROBLEM WITH ELDERLY POPULATION BENEATH THE NON-RESPONSE TO EVACUATION ORDER?

A Study on Cyclone Bulbul in Cyclone-Affected Islands of Bangladesh

Zawad Ibn Farid and Mahbuba Nasreen

Introduction

Amid the numerous natural disasters that affect humans and other species, cyclones are one of the most commons. It possesses the ability to cause an upward bulge in the ocean, resulting in widespread flooding, heavy rainfall, and destructive fierce gusts that are beyond the coastal communities' ability to prevent. According to historical records, Bangladesh experiences at least one devastating cyclone every four and a half years (Shamsuddoha & Chowdhury, 2007;Paul & Dutt, 2010). Three cataclysmic cyclones killed approximately 5,00,000, 1,38,000, and 3,460 people, respectively, in 1970, 1991, and 2007. Several consecutive catastrophic cyclones have struck Bangladesh's coast recently, including cyclone Fani (in May 2019), cyclone Bulbul (in November 2019), and cyclone Amphan (Occurred in June 2020). Between the two cyclones that struck Bangladesh in six months in 2019, cyclone Bulbul had the greatest impact on the country's islands. Bulbul was classified as a "category 2" tropical cyclone because it made landfall in the Indian state of West Bengal and Bangladesh in November 2019, causing heavy rain, flash flooding, and storm surge in the affected areas (IMD, 2019). According to multiple sources, authorities assessed and confirmed that a cyclone struck the Bangladesh coast, killing 24 people (Cyclone Bulbul, 2019). Owing to the fact that Bangladesh conducted one of the largest evacuation drives in its history during this cyclone, moving approximately 2.1 million people to cyclone shelters specifically designed to minimize casualties from such storms, which can claim thousands of lives, more than 30 million coastal residents were not evacuated. However, this population faced an increased risk (Ahmad, 2019). This is to emphasize that, while the number

DOI: 10.4324/9781003253495-13

of fatalities caused by tropical cyclones has steadily decreased in the country, economic loss and damage have increased (Hussain, 2017). As a result, this state of severe damage to agriculture and fisheries sectors and government policies has a profound negative impact on the rural coastal communities of Bangladesh.

Earlier studies have revealed that the death toll, injury toll, and damage caused by previous devastating cyclones in Bangladesh, such as cyclone Sidr, were significantly lower than expected (Paul, 2009). In comparison to Gorkey (1991) and the great Bhola cyclone (1970), this was possible due to timely and proper dissemination of cyclone warnings and evacuation of vulnerable coastal residents (Heath, 2007;Hossain et al., 2008). Now, as cyclone early warning is a process of direct communication between recipients and issuing organizations (Sorensen & Sorensen, 2006), the effectiveness of a warning is contingent upon a number of factors, including proper hazard detection, information dissemination, and human response (Haque, 1997). Additionally, various socioeconomic, psychological, and cultural factors influence how vulnerable people respond to early warnings of cyclones (Drabek, 2004).

Various studies demonstrate that significant factors influencing proactive evacuation decisions include the warning system itself, the reliability of the warning message, an individual's perceived risk from hazards, and evacuation logistics (Paul, Rashid, Islam, & Hunt, 2010). Individual response characteristics and sociodemographic variables such as age, gender, household income, education, and ethnic origin all play a critical role in the decision-making process for human evacuation (Blanchard-Boehm & Cook, 2004;Paul et al., 2010). Human response to hazard warnings is a social cognitive process involving receiving warning messages, assessing personal risk based on available information, and acting proactively. Additionally, a review of the existing literature reveals that the reliability of a hazard warning is contingent on a variety of factors, including the message's content and characteristics, its source, the message's compatibility with the recipient's pre-existing beliefs and experience with similar hazard warnings. Furthermore, social networks, the enforcement of evacuation orders, logistics, and evacuation infrastructure (i.e., evacuation routes, transportation, and evacuation locations) all play a significant role in determining the effectiveness of an evacuation during a disaster (Hyndman & Hyndman, 2006). While the Bangladesh government has made significant strides in disaster management, the current cyclone warning system has been criticized for its lack of simplicity, ambiguity in the warning message, and difficulty identifying the cyclone's precise location and landfall timing (Paul & Routray, 2013).

Although a number of studies have been conducted to ascertain vulnerable peoples' responses to hazard warnings at the individual and organizational level in Bangladesh and elsewhere (Lindell & Perry, 1992; Mileti & Sorensen, 1990; Sorensen, 2000; Milelti & Peek, 2000), very little research has been conducted on vulnerable coastal population's responses to warnings to seek shelter (Sorensen, 2000). Additionally, no study has been conducted specifically on cyclone Bulbul and its significance for evacuation decision-making to date. On a different note, not focusing specifically on an endangered population, such as elderly population of both genders, during an evacuation in response to a warning signal is a significant gap in comparable studies. Coastal regions of Bangladesh with rapidly ageing populations, vulnerable older people need assistance in the event of a cyclone for being evacuated at the right time (HelpAge International, 2003).

Context and Issue

Different contexts of past cyclones have gotten disaster researchers' attention in the current patterns of disaster research in Bangladesh, such as reasons for increased or decreased deaths and injuries; adapting response and aiding procedures to cyclones; effects of cyclones on actual coastal regions and social conditions; cyclone early warning, non-response to evacuation order, food security status of marginal farmers, and the rest (Ahmad, 2019; Karim & Mimura, 2008). Having said that, the lack of explicitly lensing on the elderly coastal population during the emergency evacuation is a significant gap in similar studies. While the aged population is generally regarded as a vulnerable or possibly weak group, almost nothing is conducted to meet their specific needs or recognize their unique limitations and commitments. People beyond the age of 60 in Bangladesh are often referred to as "old" or "aged" (Barikdar, Ahmed, & Lasker, 2016). Due to age-related difficulties, such as physical disability or frailty, these aged persons find it challenging to move to different places. A lack of financial resources further hampers their mobility. With deteriorating health and dwindling resources, they become more and more reliant on others. In addition, the number of people who can be relied upon in an emergency for support among the elderly is considerably smaller (Gibson & Hayunga, 2006). Elderly persons may be unable or unwilling to evacuate prior to a cyclone's landfall due to factors such as their psychological condition (which may include depression, and greater acknowledgement of disaster as "fate" or "Divine Will" but rather something to be battled), as well as other undefined factors (Hartog, 2014). As a result, the elderlies are highly vulnerable to cyclones.

Compatible studies often disregard elder individuals' unique needs, which undermined their ability to help themselves in emergencies, such as evacuation prior to a devastating cyclone (Paul, 2010). Moreover, women, being in the elderly age cohort, on the other hand, are more powerless against tropical cyclones than men due to socially imposed roles and obligations, physical weakness, as well as their less fortunate and monetarily weak position, particularly in Bangladesh's remote rural coastal islands. According to Nasreen (2012), gender disparities in social, political, and economic status, exposure to violence, and health, particularly reproductive health, make women more vulnerable, particularly during a disaster. As a result, in contrast to previous studies, it is necessary to concentrate on how various factors influence the evacuation responses of both male and female elderly residents. This chapter attempts to fill the research gap by identifying factors contributing to elderly island residents' non-compliance with evacuation efforts during cyclone Bulbul. These factors are compared to previous research to determine common influencing factors concerning tropical cyclones.

The respective book chapter attempts to investigate the responses of elderly male and female residents at risk in Bangladesh's coastal islands to cyclone warnings and evacuation orders. To accomplish this, it investigated plausible determinants related to cyclone shelters, cyclone warning messages and evacuation orders, as well as evacuation perception factors that influenced their evacuation decisions during cyclone Bulbul. The chapter also examines the influence of selected socioeconomic and demographic variables on the evacuation response behavior of elderlies in response to cyclone warnings. It identifies significant issues based on empirical findings that may assist stakeholders in advising necessary interventions for effective and efficient evacuation processes in order to make the coastal at-risk elderly population more resilient to devastating cyclone shocks. Similarly, the findings should be used to guide future interventions to improve the effectiveness of the country's existing cyclone early warning system. However, this chapter's primary focus on a recent tropical cyclone incident reduces the likelihood that respondents would be influenced by recall bias.

Study Area and Research Methods

Bangladesh's only island district, Bhola, is a 15-kilometer by 150-kilometer flat expanse of land bordered on both sides by massive rivers and on the south by the Bay of Bengal. Rising sea levels are engulfing coastal areas like Bhola. Climate change is expected to increase the frequency and severity of cyclones' landfall on Bhola, making the island more vulnerable. As a result, Bhola's location puts it squarely in the middle of the storms when they take place. On the other hand, Bhola is also home to roughly 2 million people. Thousands of coastal households in Bangladesh have experienced abysmal cyclones repeatedly as the effects of global warming continue to transform their environment. Already, the toll is being felt. Massive tidal waves inundate thousands of houses, causing banks to break, flooding and demolishing livelihoods. This study considered 200 elderly individuals (both men and women) as corresponding samples from rural and exposed coastal communities from two exposed coastal unions of Borhanuddin upazila (sub-district) of Bhola district. For the study, Bara Manika and Gangapur Union Parishad have been chosen as the study area (Figure 10.1), from which the anticipated number of participants of the study has been selected randomly.

The core quantitative data were gathered using a survey method from older men and women respondents living in rural and exposed coastal settlements in Bangladesh's sole island district, Bhola. The sample was drawn from the population aged 60 years and above per Bangladesh's elderly population age range (Barikdar et al., 2016). The data collection instrument was a standardized questionnaire administered during the interview. The questionnaire was constructed and designed with care to ascertain the elements that contribute to non-compliance with evacuation response decisions based on a review of relevant prior national and international literature in order to classify respondents' evacuation response behavior. The

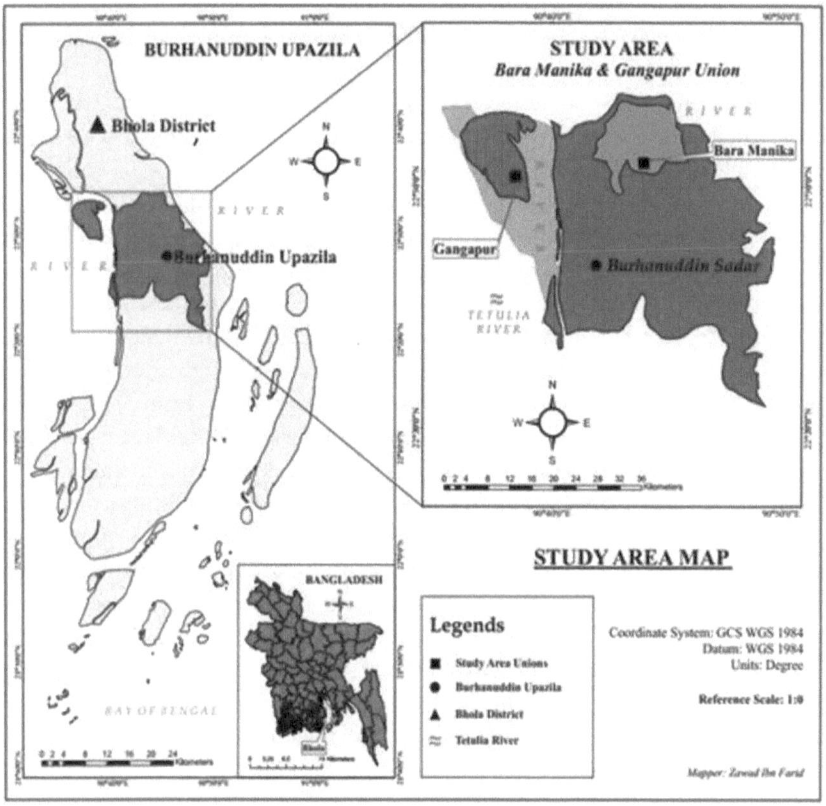

FIGURE 10.1 Map of the study area

questionnaire's determinants were chosen based on their relevance to the study area and target group. The instrument has a total of 33 measurement items. It is divided into two significant portions. "Section-1" contains respondents' personal information and primary information about cyclone risk. The "section-2" portion has 33 separate assertions that are quantifiable by a Likert-type scale ranging from 1 to 5 (Strongly Disagree to Strongly Agree). This section contains three major types of determinants of evacuation decision-making: cyclone shelter-related determinants (the first nine questions), cyclone warning message-related determinants (the following six items), and non-evacuation perception-related determinants (the final 18 items). The instrument concludes with a free-response question. Also, before the final data werecollected, a thorough pretest was conducted to ensure the scale's validity and reliability. In the reliability test, Cronbach's alpha value for the overall score for all the scales of the 33 items within three determinants in the questionnaire demonstrated inter-item consistency and was found reliable. In the validity (Pearson's correlation coefficient) test, all the total acquired correlation values for each scale of all the items were statistically significant and more than the specified value. After the device had been verified and shown dependable, the final data were collected.

However, due to the distant coastline and cyclone-prone portions of the island district Bhola and given the local dialect of the area, it was necessary to recruit trained indigenous enumerators to collect data. Thus, two teams of young and enthusiastic students from Borhanuddin Mohila College and Abdul Jabbar College were led and supervised by senior peers who had the experience necessary to conduct such an endeavor. Two appropriately gender-balanced teams were explicitly constituted for the data collection, as about one-third of the study sample was female (elderly women). Prior experience conducting relevant fieldwork in the coastal zone, educational qualifications, compatibility with the local dialect, ability to operate in a team atmosphere, mutual respect among team members, and gender sensitization were all considered criteria for selecting field team members. A rigorous one-day training session was conducted for degree college students who would be data collectors and for BDRCS volunteers who would provide orientation on data collection tools. Following mock interview sessions immediately following the training sessions, the final applicants for the quantitative survey were chosen based on their level of understanding of the questionnaire items. To assure the quality of data, all data collection team members, including the team leaders, were monitored on a continuous basis throughout the entire process (both piloting and final data collecting) via a simultaneous communication line. The study's entire course of action required ethical assurance, including considering respondents' advantages and drawbacks.

Nonetheless, each study respondent was clearly told of the study's goal, scope of information, interview time limit, confidentiality, and risk factors prior to beginning an interview. Additionally, all respondents had the right to decline participation in the interview because they were all well aware of the subject matter. They were informed throughout the study that participation was voluntary and that any personal information or information pertaining to their identity would be treated with absolute confidentiality. Before performing any given interview, verbal consent was obtained from each respondent willing to participate in the study.

Findings

At least a third of the 200 elderly people surveyed believe they are at significant risk from cyclones. A sizable percentage (32.5 percent) do not know whether or not a cyclone-related evacuation order has been issued during cyclone Bulbul. During cyclone Bulbul, 30.5 percent elderlies were not evacuated to cyclone shelters. Upon asking the primary reason for not being evacuated, almost half of the elderly islanders opined that they felt safe at home. Other primary reasons included family did not take them to the shelter (23 percent) and it was not possible to follow evacuation order because of the old age (29.5 percent) (Table 10.1).

TABLE 10.1 Perception of cyclone risk of elderly respondents

SL	*Cyclone risk perception*	*Frequency*	*Percent*
1	Level of risk of living area during a cyclone		
	Moderate	15	7.5
	Major	70	35
	Severe	115	57.5
	Total	200	100
2	Whether an emergency evacuation order issued or not during cyclone Bulbul		
	Yes	123	61.5
	No	12	6
	Do not know	65	32.5
	Total	200	100
3	Evacuated to shelter or not during cyclone Bulbul		
	Yes	139	69.5
	No	61	30.5
	Total	200	100
4	If not, the reason for not being evacuated/not following evacuation order		
	Family did not take to the shelter	14	23
	Felt safe at home	29	47.5
	It was not possible to evacuate because of old age	18	29.5
	Total	61	100

Source: Field Survey.

An evaluation of cyclone shelter-related drivers of evacuation decisions by elderly respondents is shown in Table 10.2 (first nine elements). There is a strong correlation between older residents' non-evacuation decisions and elements connected to their cyclone shelter behavior, as seen by the section's overall average mean value (3.95).

Elderly coastal residents emphasized highly the issue of women in the respective area not going to the cyclone shelter as shelter centers are filled with male members during a cyclone (M = 4.48, SD = .76). Moreover, cyclone shelters not having the least adequate accommodation capacity (M = 4.14, SD = .74) and not having enough (M = 4.19, SD = .81) or separate (M = 4.20, SD = .63) sanitary latrines for women received higher mean scores. Feeling insecure in staying in a cyclone shelter in terms of social crime was pointed out with due importance by the respondents. Overcrowded cyclone shelters with unhygienic surroundings received high mean scores as cyclone shelter-related determinants behind non-evacuation behavior. Likewise, the participants pointed out that evacuation routes being unsuitable for structure acquires a major impact behind demotivation against following an evacuation order.

The average mean score of elderly respondents' overall evaluation on cyclone warning message and evacuation order-related determinants is 3.45, which is comparatively lower than the shelter-related determinants' score. Regardless, the mean

TABLE 10.2 Mean scores of elderly respondents' evaluation on cyclone shelter-related determinants behind non-evacuation behavior

Cyclone shelter-related determinants	*Mean*	*SD*	*N*
There are not enough cyclone shelters in this area compared to the cyclone-affected population.	2.77	.610	200
The evacuation route to cyclone shelter from my home is unusable in terms of structure.	3.73	.701	200
The cyclone shelter nearest to my residence is usually overcrowded during a cyclone.	3.99	.626	200
The internal environment of the cyclone shelter is too unhygienic to stay with the family members.	3.99	.734	200
I feel insecure staying in a cyclone shelter in terms of social crime.	4.11	.760	200
The cyclone shelter nearest to my house does not have enough sanitary latrines.	4.14	.741	200
The cyclone shelters nearest to my house do not have separate sanitary latrines for women.	4.19	.806	200
The cyclone shelters in my area do not have the least adequate capacity for cyclone-affected people.	4.20	.626	200
Women in this area do not go to the cyclone shelter as shelter centers are filled with male members during cyclone.	4.48	.757	200

Source: Field Survey.

score reveals warning message and evacuation order-associated elements having a considerable impact on following an evacuation order.

Not having adequate lead time[1] to follow and execute evacuation order received the highest mean score (M = 4.39, SD = .73). Elderlies also elucidated facing difficulties understanding the meaning or language of officially issued cyclone warning (M = 3.95, SD = .76) and finding fluctuations and sudden changes in cyclone warning signals (M = 3.75, SD = .93). Nevertheless, the participants uttered their non-belief or less trustworthiness toward the early warnings issued for cyclones (M = 3.56, SD = 1.01). Concerns about not hearing or receiving any cyclone warning and even receiving evacuation order prematurely received more than moderate scores (Table 10.3).

Table 10.4 depicts the item-wise mean values (out of 5-point Likert scale) of elderly residents' evaluation on emergency evacuation perception-related determinants underlying evacuation decision. Given the age factor, elderly strata of the coastal population find it very difficult to go to cyclone shelters because of their age, which received the highest mean score (M = 4.50, SD = .87). The other determinants identified as the extreme catalyst per the Likert-type scale (M = 4.00–5.00) in this section were difficulties of families with small children to evacuate during cyclone (M = 4.44, SD = .77), businessmen not being able to evacuate leaving possession behind (M = 4.18, SD = .68), and having the fear of burglary/loot/theft (M = 4.04, SD = .92). The elderly residents opined strongly that women in this area do not

TABLE 10.3 Level of elderly respondents' evaluation on cyclone warning message and evacuation order-related determinants behind non-evacuation behavior

Cyclone warning and evacuation order-related determinants	*Mean*	*SD*	*N*
I did not hear/receive any cyclone warning in my area during the cyclone Bulbul.	2.40	.669	200
I received the evacuation order as per the cyclone warning prematurely.	2.57	.992	200
I do not believe/trust in the early warnings issued for cyclones in this area.	3.56	1.01	200
Most of the time, I do not understand the meaning/language of officially issued cyclone warning.	3.75	.929	200
I find many fluctuations and sudden changes in cyclone warning signals at a time.	3.95	.762	200
I do not get adequate lead time to evacuate.	4.39	.732	200

Source: Field Survey.

evacuate their homes if there are no adult male members in the home. On the other hand, they perceive that female household members should not be allowed to evacuate to cyclone shelter without male members of the family (M = 3.90, SD = .65) and that it is tough for women to evacuate if the father/husband is absent at home (M = 3.95, SD = .73). Similarly, perception of security and safety at own home rather than being in the cyclone shelter received a very high-level evaluation from the elderly perspective (M = 3.94, SD = .77).

Nonetheless, the elderlies particularly consider strongly that cyclones' landfallin other times than a particular season should not be too severe to evacuate. They frequently struggle to comprehend the risk of an impending cyclone that plays a crucial function beneath non-evacuation decisions. The statement related to not relying on the evacuation order following the warning with past experiences of the "false alarm," which received the lowest score (3.01) amid this section, also indicated a considerable impact on the non-evacuation decision. However, the total average mean score in this area (3.67) indicates that elderly persons' perceptions of emergency evacuation significantly impact their decision to comply with evacuation orders.

An independent sample t-test was used to compare the mean score differences between male and female elderly respondents' (n=200) evaluations of the determinants of evacuation decision-making. In all three determinant categories examined as continuous variables, there were substantial differences between male and female elderly respondents, as shown in Table 10.5. The t-test revealed statistically significant differences in mean scores for cyclone-related determinants ($t_{198} = 1.476, p < .05$), cyclone warning ($t_{198} = 3.132, p < .05$), and evacuation perception ($t_{198} = 2.810, p < .05$).

An open-ended question was asked to the participants through the query of steps to make people aware and motivated to follow emergency evacuation orders

TABLE 10.4 Level of elderly respondents' evaluation on emergency evacuation perception-related determinants behind the non-evacuation decision

Evacuation perception-related determinants	*Mean*	*SD*	*N*
With the experience of the "false alarm," I do not rely on the warning to evacuate.	3.01	.701	200
I take shelter in coastal earthen embankments and behind large trees during cyclone rather than evacuating.	3.06	.802	200
Mahajan/Arotdar (market/shop owner) pressurize the fishermen to go fishing even during cyclone warnings.	3.10	.828	200
I believe in indigenous knowledge of predictions more than official weather forecasting/cyclone warnings.	3.20	.804	200
I fear that cyclone shelter would collapse by the storm, so I prefer not to evacuate.	3.27	.675	200
I do not feel the necessity of evacuation, as I often believe that the approaching cyclone would not strike in this area.	3.30	.964	200
Cyclone is God's will, and I have nothing to do with this.	3.38	.895	200
I think that cyclones that appear in other times than a particular season would not be too severe to evacuate.	3.62	.770	200
I often do not understand the danger of the approaching cyclone toward our lives for deciding to evacuate.	3.64	.851	200
I intend not to leave my domestic cattle in the home while I evacuate to cyclone shelter.	3.69	.972	200
Women in this area do not evacuate their homes if there are no adult male members in the home.	3.84	.775	200
I think that female members should not be allowed to evacuate to cyclone shelters without male family members.	3.90	.648	200
I feel secure and safe at my own home rather than the cyclone shelter.	3.94	.771	200
It is very hard for women to evacuate if the father/husband is absent at home.	3.95	.733	200
Those families have small children face difficulties evacuating during a cyclone.	4.04	.924	200
People in this area, who have business/shops, must protect that shop by themselves and thus cannot evacuate, leaving that possession behind.	4.18	.677	200
I have a fear of burglary/loot/theft of my house in my absence to evacuate to cyclone shelter.	4.44	.786	200
I think old people like me find it very difficult to follow the evacuation order to go to cyclone shelters because of their age.	4.50	.866	200

Source: Field Survey.

in a coastal island area like Bhola. Most of the responses (25.90 percent) denoted that government should build more cyclone shelters in the risk-prone localities of the coastal region. They also articulated that the government should build a coastal community cyclone-resistant home (25.88 percent). Given the age cohort of the respondents, the necessity of special attention for the elderly was stressed upon

TABLE 10.5 Independent sample t-test differences in evaluation on determinants behind non-evacuation decision of elderly population (n=200) according to their sex (male/female)

Determinants groups	*F*	*SD*	*df*	*Calculated t*	*P*
Cyclone shelter	1.14	0.27,.25 (M,F)	198	1.476	.014*
Cyclone warning	20.49	0.52,0.32 (M,F)	198	3.132	.002**
Evacuation perception	17.66	0.30,0.20 (M,F)	198	2.810	.005**

**P<.01 and *P<.05 (2-tailed); df, degree of freedom; SD, standard deviation.

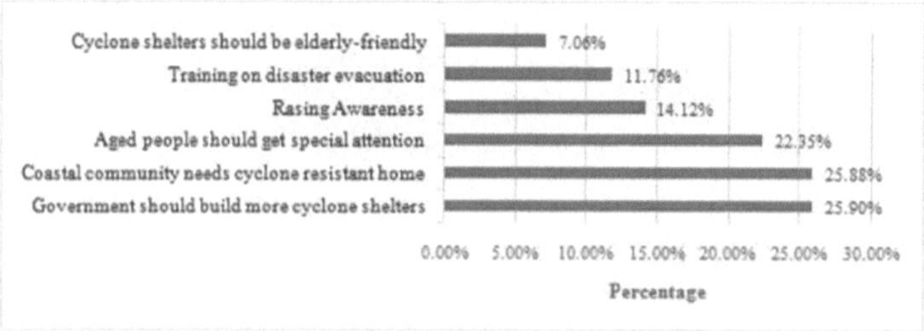

FIGURE 10.2 Elderly residents' top recommendation on making island people aware and motivated to follow the emergency evacuation order

Source: Field Survey.

(22.35 percent) in terms of providing an advantage for evacuation during cyclones. Besides these recommendations, a significant number of responses came regarding some specific factors, for example, elderly friendly cyclone shelters, awareness-raising activities, and the necessity of training on disaster evacuation in cyclone risk-prone areas (Figure 10.2).

Discussion

In terms of exposure and mortality, tropical cyclones are the most severe and common natural catastrophes in Bangladesh (Saha & James, 2017). According to Shamsuddoha and Chowdhury (2007), Bangladesh is hit by strong cyclones every four and a half years. As per Nasreen (2009), cyclones are a common occurrence for coastal people in Bangladesh. Around 6–10 percent of all tropical cyclones that occur globally begin in the Bay of Bengal, which is bounded by Sri Lanka, India, Myanmar, and Bangladesh's whole southern border. Additionally, these tropical cyclones generate strong winds, resulting in storm surge. These cyclones result in widespread death and property damage due to extreme wind speeds and colossal storm surges (Paul, 2014). Since 1877, when records began, cyclones have killed over a million people in Bangladesh (Paul, 2010). According to Haque et al. (2012), about 7,18,000 deaths in Bangladesh have been attributable to cyclones during the last 50 years. The most recent cyclone to make landfall in Bangladesh was the cyclone Amphan, the country's first super cyclone and the strongest cyclonic storm in history (Abbas, 2020). Amphan claimed the lives of 20persons, evacuated over a million people along the coast, and caused damage of US$130 million (Nagchoudhary, Paul, & Roy, 2020; Amphan: Kolkata devastated as cyclone kills scores, 2020). Furthermore, cyclone Amphan strikes during the Covid-19 pandemic, necessitating a new level of preparedness (Nasreen, 2020).

Compared to earlier research on statistics and destruction patterns, the outcomes of this study seem to be very consistent. Bhola district, which is the only wholly exposed coastal district in Bangladesh, has the highest percentage of elderly residents who consider themselves exposed to severe risk and 35 percent who consider themselves at major risk of cyclones (Uddin & Kaudstaal, 2003). Also, Bhola is located in the country's most dynamic coastal region and likewise is subject to constant erosion and accretion (Thomas, Wratten, & Sotherton, 1992). Climate change–induced events like tropical cyclones, sea-level rise, coastal inundation, storm surge, land erosion, and salinity intrusion frequently affect the elderly population of Bhola (Ahmad, 2019; Iftekhar, 2006; MoWR, 1999).

Cyclone Shelter-Related Determinants

In the context of cyclone shelter-related determinants of non-evacuation behavior of elderly residents, the issues found in this study, such as susceptibility for being elderly, space unavailability, distant location, unhygienic and unseparated sanitation, feeling insecure in shelters, availability of *killas*[2] or *Mujib Killas*[3] near shelters, etc., are relevant in this context. These items that received higher mean values than the average range are consistent with the findings of studies on the similar field (Ikeda, 1995; Paul & Routray, 2013; Paul & Dutt, 2010; Paul et al., 2010; Ahsan et al., 2016; Matsuda, 1993; Islam et al., 2004; Saha & James, 2017). The issue of women in the relevant area not traveling to the cyclone shelter because shelter centers are crowded with male members during a cyclone obtained a considerably higher score in this regard. This implies that the adequacy, crowd situation, and housing capacity of cyclone shelters also have an essential impact on women's decision to heed evacuation orders, if they can decide on their own (Ahsan et al., 2016; Islam et al., 2004). Additionally, it provides an opportunity to emphasize the critical lack of access to services such as contraception, safe water during pregnancy and childbirth, and good sanitation in cyclone-prone communities. Besides that, overcrowded shelters heighten women's vulnerability to theft, rape, harassment, and other forms of physical, emotional, and sexual violence.

However, given the status quo of the cyclone shelters in Bangladesh, officially, 13,097 cyclone shelters have been established for them in 19 coastal districts across the country. A total of 59,67,677 people can seek shelter in these cyclone shelters in the event of a cyclone (GoB, 2020). In these shelters, there is a severe lack of space, inadequate equipment, insufficient lighting, poor sanitation arrangements, a shortage of potable water and clean drinking water, and the dearth of separate restrooms for men and women, to name a few (Paul et al., 2010). According to

Wahra (2020), a significant percentage of cyclone shelters is not used due to insufficient maintenance and management efforts. The field observation of the respective study likewise found that cyclone shelter administration standards are lacking, and the public is not receiving relevant information from local governments and non-governmental organizations (NGOs) on cyclone shelters. As a result, some of the essential cyclone shelters were locked even during cyclones, preventing many people from evacuating and seeking refuge due to a lack of keys to the locks. This finding is quite similar to Miyaji et al. (2020). Another issue worthy of mentioning is the excess of too old to move or persons with disabilities in the coastal areas who are at risk of being impacted by tropical cyclones yet are unable to use these shelters due to the lack of ramps in many of these cyclone shelters (Molla et al., 2019). Thus, the conclusions that cyclone shelters lack the least adequate accommodation capacity and lack sufficient or separate hygienic latrines for women are well supported by the cited literature.

Cyclone Warning Message and Evacuation Order

The findings from elderly respondents' evaluation on cyclone warning message and evacuation order-related determinants indicate that the respective factors play a major impact on the decision to follow an evacuation order. Some factors are exclusively associated with the elder coastal population in this context. Not getting adequate lead time to follow and execute evacuation orders and not hearing or receiving any cyclone warning and even receiving evacuation orders prematurely were opined to be the most crucial background factor behind the non-evacuation decision. These determinants were similarly mentioned by Paul et al. (2010); Paul and Dutt (2010); Paul and Routray (2013); Saha and James (2017) as having no time or less time to evacuate. Elderlies also strongly elucidated facing difficulties understanding the meaning or language of officially issued cyclone warning. Considering this, Wilson and Tiefenbacher (2012) implied that response to the received warning signals or messages depends on how the recipients interpret the information in the message. Hence, elderly coastal residents, after all, are simulated by different social and environmental signals such as sounds, sights, neighbors' and peers' behavior (Lindell& Perry, 2012). This is also highly associated with aged people finding complexity in warning signals and their level of perception and interpretation of received warning signals (Paul & Routray, 2013).

Another significant determinative factor of cyclone warning messages and evacuation orders for the elderly population identified in this study was a lack of trust and the presence of doubt. The statement "With the experience of the 'false alarm,' I no longer rely on the evacuation warning" was ranked as a significant factor in non-evacuation decisions, as was similarly found in both some older and more recent studies (Chowdhury et al., 1993; Matsuda, 1993; Haque & Blair, 1992; Paul et al., 2010; Saha & James, 2017). Nonetheless, based on personal experience, the phenomenon of elderly people frequently disregarding hazard warnings when such warnings have been proven to be false in the past is referred to in disaster literature as the "cry wolf effect" in addition to the "false alarm" (Paul & Dutt, 2010). In this context, Paul (2012) discovered that approximately 19 percent of people did not believe the cyclone warning they received during cyclone Sidr, and one of the primary reasons for this mistrust was a false Tsunami alert issued two months prior to Sidr's landfall in Bangladesh's coastal zone. While this portion may appear small, it demonstrates the importance of determining the disconnect between risk perception and the reliability of warning messages for the at-risk coastal population. According to Bandura's (1986) Social Cognitive Theory (SCT), individuals whom disasters have directly impacted are more receptive to predictions and warnings of impending disasters than persons who have never been impacted by a disaster. Nevertheless, the sample of this respective study, the elderly coastal residents, opined on their non-belief or less trustworthiness toward the early warnings issued for cyclones based on previous experience as one of the dominant factors behind the non-evacuation decision. This was an interesting finding from the study as these elderly participants reportedly experienced every cyclone to date from the 1960s.

Apart from these, the other elements emphasized in this study's findings—premature issuing of the evacuation order, insufficient or missing information in a warning message, and quick and unexpected alternation of warning signals/information—are backed up by several disaster literature sources (Chowdhury et al., 1993; Paul & Dutt, 2010; Ikeda, 1995). Overall, in general, these findings claim that the existence of cyclone-warning systems and evacuation orders are not enough to ensure that elderly island residents comply with both warnings and evacuation orders (Saha & James, 2017).

Emergency Evacuation Perception

In consideration of the age cohort of the study's participants, it is unsurprising that the top-ranked response on "I believe that elderly people find it extremely difficult to follow evacuation orders to go to cyclone shelter due to their age." Ikeda (1995), and later Paul and Routray (2013), identified age or being too old as a barrier to moving toward public cyclone shelters for the elderly population strata. Participants' perceptions of people who own businesses/shops being unable to evacuate, leaving possessions behind, fear of burglary/loot/theft of their own house in their absence, a sense of security in own home, and the difficulties faced by families with small children in rushing to cyclone shelter received immediate attention in the emergency evacuation perception context. These findings suggest that when it comes to cyclones, prospective cyclone-prone elderly people frequently disregard cyclone warnings and opt to stay home out of fear of their belongings being looted or stolen (Ikeda, 1995; Dow & Cutter, 1998). To put it another way, this fear of the elderlies has a particular root cause, which is that the elderly coastal residents face a severely vulnerable predicament to cyclones because they have a limited supply of resources and are heavily dependent on others (Malak et al., 2020). Besides, there are several concerns, which also are found in previous studies, for example, the well-being of domestic cattle or pets, being unable to return to home after cyclone passes, and job responsibilities also provoked individuals to stay home disregarding the evacuation order.

We found that there exist other personal characteristics from the participants that may lead to cyclone warning failure and eventually inability to evacuate to a safe place because of the sole beliefs that "God alone will protect us" or "it will not happen to us/to me" (Biddle, 2007). Statements like "I do not feel the necessity of evacuation, as I often believe that the approaching cyclone would not strike in this area" and "Cyclone is God's will and I have nothing to do with this" appeared as stimulating factors behind how participants perceive evacuation. A distinct response was received in this context, which is very much related to the local sociopolitical context that *Mahajon/Arotdar* (market/shop owner) pressurizes the fishermen on the island to go for fishing even when a cyclone warning is already issued. This can be seen from the vulnerability and political-ecological paradigm, which acknowledges that marginal groups support the dominant class's interests during a natural disaster, such as a tropical cyclone (Bolin & Stanford, 1999; Wisner et al., 2003). However, Paul and Routray (2013) reported on this issue of pressure from fish *Aratdar* (merchant/stocker) to go fishing in their research.

On a different note, we discovered that participants rated the statement "I believe in indigenous knowledge of forecasting more than official weather forecasting/cyclone warnings" as a significant predictor of non-evacuation behavior. This means that elderly people can also reassure themselves about the warning message or decide to comply with the evacuation order by observing natural signals (e.g., listening to the wind, observing the sky, witnessing the alignment of the stars, observing the animal movements) and by checking the alternative source of warning message (e.g., radio and television channels). Sorensen and Mileti (1987a) addressed similar issues relating to environmental determinants.

Gendered Perspective in Decision to Follow Evacuation Order

To be more specific, the findings of this study indicate that elderly coastal residents place a high position on the issue of women in their area not attending cyclone shelters, as these facilities are typically overcrowded with male residents during a cyclone. When the context is considered, the issue's paramount importance becomes clear, as the coastal population is 35 million within a landmass of 19 districts, and 49 percent of this population is female (BBS, 2014). Distinctly high importance on the issue can be indicated as the absence of gender inclusivity and obligation of maintaining *parda* (veil) of women to this respective context, which is evident from previous research. Findings on cyclone Gorky (Ikeda, 1995; Haque, 1995; 1997) showed that excessive members from a specific gender (particularly filled with males) and gender discrimination were the subjects of complaints from cyclone-affected population against not wanting to go to the shelter. Again, violating *parda* norms works as a dispiriting agent for women against the movement as per the situational context (Nasreen, 1995; 1998). Nonetheless, Nasreen (2012) established that gender disparities in social, political, and economic status, exposure to violence, and health, particularly reproductive health, make women more vulnerable, particularly during a disaster. At the same time, evacuation to emergency cyclone shelter emerges, which forces women to compromise with the cultural norms of *parda* (veil). In line with the findings of this study, Paul (2012) and Ikeda (1995) considered the *parda* as a socio-religious customary practice amid adult Muslim women to be a crucial factor either delaying or dissuading evacuation of household members at risk.

Furthermore, it was found that statements like "women in this area do not evacuate their home, if there are no adult male members in home," "I think that female members should not be allowed to evacuate to cyclone shelter without male members of the family" and "It is very hard for women to evacuate if father/husband is absent at home" were remarked as critical factors in women's compliance with the evacuation order. These socially ascribed gender roles during the emergence of a cyclone can be implied by the fact that when communicating risk, a gendered perspective is blatantly applied to marginalized groups such as women, children, the poor, the elderly, people with disabilities, and ethnic minorities (Blanchard-Boehm & Cook, 2004; Sorensen & Mileti, 1987b; Wisner et al., 2003).

Observational data from the field study also revealed that the most visible and overwhelming effect of cyclones occurs when the issue of the evacuation of women and girls to cyclone shelters comes forth. The burden unavoidably increases where women are aged. Typically, these shelters are overcrowded, with little personal space, privacy, and separate toilets. As a result, the study's findings pertain to the intergenerational cycle of imposing patriarchal-rigid gender standards, limiting financial independence, and limiting access to decision-making options, all of which impose additional burdens on already aged women. These findings corroborate the result from the study's independent sample t-test, which found significant differences between male and female elderly respondents in all three determinant groups analyzed as continuous variables, with female scores being higher than male scores. It perfectly exemplifies how gender roles in a gendered power system affect evacuation decision-making by encouraging male and female behaviors during a disaster (Ayeb-Karlsson, 2020).

Conclusions

Based on the experience and perception related to cyclone Bulbul, this study tried to examine the factors steering vulnerable elderly residents of the coastal region to decide whether to evacuate or not and select the evacuation destination, that is, cyclone shelter. Despite the passage of time, cyclone Bulbul left an indelible impact on the psyche of those who witnessed it. The residents of island district Bhola's coastal community appeared to remember most of what we asked them throughout our field investigation. The study indicated that a variety of factors impact and steer elderly people's compliance or non-compliance with emergency evacuation orders based on their evaluations of the determinants behind the decision-making process. To examine the strength of the selected influencing determinants, this study attempted to develop a prototype Likert-type tool to analyze elderly people's actions and attitudes toward evacuation, categorizing the plausible factors into three different groups, that is, cyclone shelters, a cyclone warning message and evacuation order, and evacuation perception-related determinants. Among the three types of determinants, female elderly residents were shown to have a higher level of non-compliance with specific reasoning than male elderly residents. It likewise identified that this reflects the inevitability of putting much concern on female elderlies as the more susceptible group for reducing the non-compliance behind evacuation and decreasing the heavy toll by the tropical cyclones.

Analyzing the perceptions, experience, and actions of two different strata of vulnerable people, that is, elderly female and male, the study explored the extent of cyclone warning and emergency evacuation order, timeliness of evacuation order, availability of lead time, time gap for evacuation preparation, individual risk perception of people, indigenous knowledge, condition of roads, availability of evacuation route, the distance of cyclone shelter from residing place, condition and capacity of cyclone shelter, and tendency to save the valued belongings. Gender and age sensitization are also found to be the substantial factors to influence and guide the target population's evacuation or non-evacuation decision. In fact, per the recommendations, distinct strata of vulnerable coastal aged communities at risk should be treated by ensuing distinguish approach on designing and imposing cyclone evacuation awareness programs, drills, and hands-on training based on their evacuation behavior and serving further policy actions. Also, it is time to reassess the cyclone warning language and the system for disseminating the warnings while keeping in mind the practicality of the coastal localities.

Bangladesh must make tactical measures to reduce the potential death toll from cyclones in the future to meet the Sendai Framework global target of considerably reducing disaster mortality by 2030 and lowering the global average mortality rate. In order to keep the total death toll by a cyclone to a minimum, efforts must be made to motivate potential victims to evacuate per responding to cyclone warnings and evacuation orders in time. Therefore, it is substantial that the cyclone warning language and system, adequate cyclone shelters, and proper shelter management should be incorporated and operationalized with the people's responses to warning and evacuation orders.

Notes

1 The time it takes for individuals to arrive to a safe location (i.e., cyclone shelter) after receiving a warning or becoming aware of a particular disaster is known as lead time.

2 A *killa* is an earthen platform that offers shelter to about 400 livestock animals, primarily cattle and poultry. To protect cattle from cyclonic wind and storm surges, 196 *killas* have been constructed in cyclone-prone regions, out of a total of 872 *killas* constructed across Bangladesh till now.

3 The "*Mujib Killa*" was the name given to these earthen mounds, which were built in various parts of the country following a cyclone, which made a landfall in the southern region of the country in 1970. The construction of such earthen mounds was in higher

ground level than the food level. The initiative was taken under the leadership of Bangabandhu Sheikh Mujibur Rahman, the founding father of Bangladesh, to safeguard the resources, notably livestock, from cyclones, storm surges, and floods, and per the name of the Sheikh Mujibor Rahman, it became known as "*Mujib Killas*," However, *killas* which already have been constructed or are in the process of being constructed are referred to as "*Mujib Killas.*"

References

Abbas, M. (2020, May 20). Strongest storm ever recorded in Bay of Bengal amid a pandemic. *The Daily Star*. Retrieved September 2, from www.thedailystar.net/cyclone-amphan-in-bangladesh-strongest-storm-ever-recorded-in-bay-bengal-amid-pandemic-1904173

Ahmad, H. (2019). Bangladesh coastal zone management status and future trends. *Journal of Coastal Zone Management, 22*, 1–7. https://doi.org/10.24105/2473-3350.22.466

Ahsan, M. N., Takeuchi, K., Vink, K., & Warner, J. (2016). Factors affecting the evacuation decisions of coastal households during Cyclone Aila in Bangladesh. *Environmental Hazards, 15*(1), 16–42. https://doi.org/10.1080/17477891.2015.1114912

Ayeb-Karlsson, S. (2020). When the disaster strikes: Gendered (im)mobility in Bangladesh. *Climate Risk Management, 29*, 100237.

Bandura, A. (1986). *Social foundations of thought and action: A social cognitive theory*. Englewood Cliffs, NJ: Prentice-Hall.

Barikdar, A., Ahmed, T., & Lasker, S. P. (2016). The situation of the elderly in Bangladesh. *Bangladesh Journal of Bioethics*, 7(1), 27–36.

BBS. (2014). *Population & Housing Census-2011*. Bangladesh Bureau of Statistics. https://doi.org/10.1017/CBO9781107415324.004

Biddle, M. D. (2007). *Warning reception, response, and risk behavior in the 3 May 1999 Oklahoma City long-track violent tornado* (Doctoral Dissertation).

Blanchard-Boehm, R. D., & Cook, M. J. (2004). Risk communication and public education in Edmonton, Alberta, Canada on the 10th anniversary of the 'Black Friday' tornado. *International Research in Geographical & Environmental Education, 13*(1), 38–54.

Bolin, R., & L. Stanford. (1999). Constructing vulnerability in the First World: The Northridge Earthquake in Southern California, 1994. In *The angry earth: Disaster in anthropological perspective*, edited by A. Oliver-Smith & S. M. Hoffman (eds.), 89–112. New York: Routledge.

Central Intelligence Agency. (2019). *The World Factbook*. The federal government of the United States.

Chowdhury, A. M. R., Bhuyia, A. U., Choudhury, A. Y., & Sen, R. (1993). The Bangladesh Cyclone of 1991: Why so many people died. *Disasters*. https://doi.org/10.1111/j.1467-7717.1993.tb00503.x

Cyclone Bulbul. (2019, November 11). Cyclone 'Bulbul' Destroys 9,455 Homesteads in Khulna. *The Independent*. Retrieved February 1, from http://www.theindependentbd.com/post/223210

Dow, K., & Cutter, S. L. (1998). Crying wolf: Repeat responses to hurricane evacuation orders. *Coastal Management, 26*, 238–252.

Drabek, T. E. (2004). *Social dimensions of disaster: Instructor guide*, 2nd ed. Emmitsburg, MD: Emergency Management Institute, Federal Emergency Management Agency.

Gibson, M., & Hayunga, M. (2006). We Can Do Better: Lessons Learned for Protecting Older Persons in Disasters. *Undefined*. Retrieved from https://www.semanticscholar.org/paper/We-Can-Do-Better%3A-Lessons-Learned-for-Protecting-in-Gibson-Hayunga/600c11dcf1d6dd851be49e6a4cf971490e01b34d

GoB. (2020). *Special report on cyclone Amphan (05).* Ministry of Disaster Management and Relief, Dhaka, Bangladesh: Government of Peoples Republic of Bangladesh, Sher-e-Bangla Nagar, Dhaka 1207, Bangladesh: Government of Peoples Republic of Bangladesh. Page 02. Retrieved from https://modmr.gov.bd/sites/default/files/files/modmr.portal.gov.bd/situationreport/10e3ee64_8265_4781_b8cb_0df2200ea0b3/Cyclone%20Amphan%20Special%20Report%20-05.pdf

Haque, C. E. (1995). Climatic hazards warning process in experience of, and lessons from, the April Cyclone Bangladesh: 1991. *Environmental Management, 19*(5), 719–734.

Haque, C. E. (1997). Climatic hazards warning process in experience of, and lessons from, the April Cyclone Bangladesh: 1991. *Environmental Management, 19*(5), 719–734.

Haque, C.E., & Blair, D. (1992). Vulnerability to tropical cyclones: Evidence from the April 1991 cyclone in coastal Bangladesh. *Disasters, 16*(3), 217–229.

Haque, U., Hashizume, M., Kolivras, K. N., Overgaard, H. J., Das, B., & Yamamoto, T. (2012). Reduced death rates from cyclones in Bangladesh: What more needs to be done? *Bull World Health Organization, 90*(2), 150–156. doi:10.2471/BLT.11.088302

Hartog, J. (2014). *Disaster resilience in an ageing world: How to make policies and programmes inclusive of older people.* HelpAge International.

Heath, M. (2007). UN grants aid to Bangladesh to assist 3 million hit by cyclone. Retrieved from www.bloomberg.com

HelpAge International. Secretariat. Communications Team. (2003). *Older people in disasters and humanitarian crises: guidelines for best practice.* HelpAge International.

Hossain, M.Z., Islam, M.T., Sakai, T., & Ishida, M. (2008). Impact of tropical cyclones on rural infrastructures in Bangladesh. *Agricultural Engineering International, 10*(2), 1–13.

Hussain, A. (2017, April 30). Cyclone fatalities fall, economic losses on rise. *Dhaka Tribune.* Retrieved from www.dhakatribune.com/bangladesh/2017/04/30/cyclone-fatalities-fall-economic-losses-rise

Hyndman, D. & Hyndman, D. (2006). *Natural hazards and disasters.* Northwoods, MN: Cole Publishing.

IDM. (2019). *Cyclonic storm 'Bulbul' (pronounced as Bulbul) over coastal Bangladesh.* New Delhi, India: Cyclone Warning Division, Office of the Director General of Meteorology, India Meteorological Department, Ministry of Earth Sciences. Retrieved from https://mausam.imd.gov.in

Iftekhar, M. S. (2006). Conservation and management of the Bangladesh coastal ecosystem: Overview of an integrated approach. *Natural Resources Forum.* https://doi.org/10.1111/j.1477-8947.2006.00111.x

Ikeda, K. (1995). Gender differences in human loss and vulnerability in natural disasters: A case study from Bangladesh. *Indian Journal of Gender Studies, 2*(2), 171–193. https://doi.org/10.1177/097152159500200202

Islam, M.S., Ullah, M.S., & Paul, A. (2004). Community response to broadcast media for cyclone warning and disaster mitigation: A perception study of coastal people with special reference to Meghna Estuary in Bangladesh. *Asian Journal of Water, Environment and Pollution, 1*(1–2), 55–64.

Karim, M., & Mimura, N. (2008). Impacts of climate change and sea-level rise on cyclonic storm surge floods in Bangladesh. *Global Environmental Change, 18*(3), 490–500. https://doi.org/10.1016/j.gloenvcha.2008.05.002

Lindell, M., & Perry, R., (1992). *Behavioral foundations of community emergency planning.* Washington: Hemisphere Publishing Co.

Lindell, M. K., & Perry, R. W. (2012). The protective action decision model: Theoretical modifications and additional evidence. *Risk Analysis, 32*(4), 616–632. https://doi.org/10.1111/j.1539-6924.2011.01647.x

Malak, M. A., Sajib, A. M., Quader, M. A., & Anjum, H. (2020). "We are feeling older than our age": Vulnerability and adaptive strategies of aging people to cyclones in coastal Bangladesh. *International Journal of Disaster Risk Reduction*, *48*, 101595. https://doi.org/10.1016/j.ijdrr.2020.101595

Matsuda, I. (1993). Loss of human lives induced by the Cyclone of 29–30 April, 1991 in Bangladesh. *GeoJournal*, *31*(4), 319–325.

Mileti, D., & Sorensen, J. (1990). Communication of emergency public warnings: A social perspective and State-of-the-art assessment. *Landslides*. https://doi.org/10.2172/6137387

Mileti, D.S., & Peek, L. (2000). The social psychology of public response to warnings of a nuclear power plant accident. *Journal of Hazardous Materials*, *75*, 181–194.

Miyaji, M., Okazaki, K., & Ochiai, C. (2020). A study on the use of cyclone shelters in Bangladesh. *Japan Architectural Review*, *3*(4), 590–600.

Molla, M. H., Shahjahan, M., Barua, P., & Samia, R. (2019). Inclusion of persons with disabilities in the climate change and disaster management related rules and policies of Bangladesh: An empirical study. *Journal of Disability Studies*, *5*(1), 11–15.

MoWR. (1999). Integrated coastal zone management: Concepts and issues; a Government of Bangladesh Policy. *The Ministry of Water Resources, GoB.*

Nagchoudhary, S., & Paul, R. (2020, May 20). Cyclone kills 14 in India, Bangladesh leaving trail of destruction. *REUTERS*. Retrieved September 2, from www.reuters.com/article/us-asia-storm-india/cyclone-kills-14-in-india-bangladesh-leaving-trail-of-destruction-idUSKBN22W0MT

Nasreen, M. (1995). *Coping with floods: The experience of rural women in Bangladesh*. Unpublished Ph.D. dissertation. New Zealand: Messey University.

Nasreen, M. (1998). *Women in disaster: An analysis of division of labour and survival strategies in disaster*. Dhaka: Academic Publishers.

Nasreen, M. (2009). Disasters and disaster research: A case study of floods in Bangladesh. In *Understanding the Global Environment* edited by S. Dasgupta. Delhi: Pearson-Longman.

Nasreen, M. (2012). *Women and girls vulnerable or resilient?* Bangladesh: Institute of Disaster Management and Vulnerability Studies, University of Dhaka.

Nasreen, M. (2020). 'Gender and WASH in emergency: What happens when a Super Cyclone hits at the Height of a Pandemic?', 7 September. REACH, Oxford University https://reachwater.org.uk/gender-and-wash-in-emergency-what-happens-when-a-super-cyclone-hits-at-the-height-of-a-pandemic/

Paul, B. K. (2009). Why relatively fewer people died? The case of Bangladesh's Cyclone Sidr. *Natural Hazards*, *50*(2), 289–304. doi:10.1007/s11069-008-9340-5

Paul, B. K. (2010). Human injuries caused by Bangladesh's cyclone Sidr: An empirical study. *Natural Hazards*, *54*(2):483–495.

Paul, B. K. (2012). Factors affecting evacuation behavior: The case of 2007 Cyclone Sidr, Bangladesh. *Professional Geographer*, *64*(3), 401–414. https://doi.org/10.1080/00330124.2011.609780

Paul, B. K., & S. Dutt. (2010). Applications of participatory research methods in a post-disaster environment: The case of Cyclone Sidr, Bangladesh. In *Participatory research methodologies: Development and post-disaster/conflict reconstruction*, edited by A. Ozerdem & R. Bowd (eds.), 85–98. Burlington, VT: Ashgate.

Paul, B. K., Rashid, H., Islam, M. S., & Hunt, L. M. (2010). Cyclone evacuation in Bangladesh: Tropical Cyclones Gorky (1991) vs. Sidr (2007). *Environmental Hazards*, *9*(1), 89–101. https://doi.org/10.3763/ehaz.2010.SI04

Paul, S. K. (2014). Determinants of evacuation response to cyclone warning in coastal areas of Bangladesh: A comparative study. *Oriental Geographer*, *55*(1–2), 57–84.

Paul, S. K., & Routray, J. K. (2013). An analysis of the causes of non-responses to cyclone warnings and the use of indigenous knowledge for cyclone forecasting in Bangladesh. In *Climate change and disaster risk management*, edited by W. Leal Filho (ed.), (15–39). Berlin: Springer Berlin Heidelberg.

Perry, R. W., Lindell, M. K., & Greene, M. R. (1982). Crisis communications: Ethnic differentials in interpreting and acting on disaster warnings. *Social Behavior and Personality: An International Journal, 10*(1), 97–104.

Saha, S. K., & James, H. (2017). Reasons for non-compliance with cyclone evacuation orders in Bangladesh. *International Journal of Disaster Risk Reduction, 21*, 196–204. https://doi.org/10.1016/j.ijdrr.2016.12.009

Shamsuddoha, M., & Chowdhury, R. K. (2007). Climate change impact and disaster vulnerabilities in the coastal areas of Bangladesh. *COAST Trust*.

Sorensen, J. H. (2000). Hazard warning systems: Review of 20 years of progress. *Natural Hazards Review, 1*(2), 119–125.

Sorensen, J. H., & Sorensen, B. V. (2006). Community processes: Warning and evacuation. In *Handbook of disaster research*, edited by H. Rodríguez, E. L. Quarantelli, & R. R. Dynes (eds.), (183–199). New York: Springer.

Sorensen, J. H., & Mileti, D. S. (1987a). Decision-making uncertainties in emergency warning system organizations. *International Journal of Mass Emergencies and Disasters, 5*(1), 33–61.

Sorensen, J. H., & Mileti, D. S. (1987b). *Public response to emergency warnings*. Reston, VA: US Geological Survey.

Thomas, M. B., Wratten, S. D., & Sotherton, N. W. (1992). Creation of 'Island' habitats in farmland to manipulate populations of beneficial arthropods: predator densities and species composition. *The Journal of Applied Ecology, 29*(2), 524–531.

Uddin, A. M. K., & Kaudstaal, R. (2003). Delineation of the coastal zone. *Integrated Coastal Zone Management Plan (PDO-ICZMP)*. Ministry of Water Resources, Government of Bangladesh. Retrieved from http://warpo.portal.gov.bd/sites/default/files/files/warpo.portal.gov.bd/page/aa04373f_0ca3_49a5_b77e_5108186638dc/wp005.PDF

Wahra, G. N. (2020, November 11). Cyclone-resistant houses are better option than shelters. *The Daily Prothom Alo*. Retrieved from https://cutt.ly/GmvO6nJ

Wilson, S. N., & Tiefenbacher, J. P. (2012). The barriers impeding precautionary behaviors by undocumented immigrants in emergencies: The Hurricane Ike experience in Houston, Texas, USA. *Environmental Hazards, 11*(3), 194–212.

Wisner, B., Blaikie, P., Cannon, T., & Davis, I. (2003). At risk: Natural hazards, people's vulnerability and disasters, 2nd edition. *Human Ecology*.

11

EMPOWERING WOMEN TO ENHANCE SOCIAL EQUITY AND DISASTER RESILIENCE IN COASTAL BANGLADESH THROUGH CLIMATE CHANGE ADAPTATION KNOWLEDGE AND TECHNOLOGIES

Mahbuba Nasreen, Dwijen Mallick and Sharmind Neelormi

Introduction

Bangladesh is one of the most vulnerable countries to the impacts of climate change and natural disasters. Climate change accelerates the intensity and frequency of occurrences of cyclone, storm surge, long-duration floods and erosion, flash flood, salinity, waterlogging and drought. The agriculture and water are the most vulnerable sectors due to climate change events. The adoption of socially transformative adaptation technologies may support pursuing the goals of accelerating climate action and reducing gender vulnerability for achieving social inequity and empowering women and disadvantaged, impoverished people. This article is based on the findings of a research project "Scaling climate change adaptation knowledge and technologies for empowering women, and to enhance social equity and disaster resilience in Bangladesh" (SAKTEE). Its' aims were to promote pro-poor and gender-responsive adaptation technologies in water and agriculture in the two most climate-affected regions in Bangladesh: Satkhira and Sunamganj districts. The present paper will provide the findings of the coastal region Satkhira and adjacent areas only. Satkhira is a climate disaster-prone coastal area. Frequent cyclones, tidal surges, salinity and waterlogging are affecting water, agriculture and livelihoods of the common people.

The research intended to identify best practices in adaptation to climate change for resilience building and evaluate innovative climate change adaptation technologies in water and agriculture for scaling and implementation through multi-level institutional integration and adoption. Besides literature review and expert

DOI: 10.4324/9781003253495-14

consultation, the project further organized local-level stakeholder workshops to understand the current adaptation technology matrix and needs and priorities of people in this regard. The aim of the local workshops was to focus on women and identify women-friendly technologies in water, sanitation and small agriculture.

Agriculture is one of the major livelihoods in the coastal Bangladesh and future change climate is likely to affect the agricultural activities. Different adaptation measures have been taken in response to climate change in coastal Bangladesh (Saha et al., 2016). The coastal areas of Bangladesh are different from rest of the country because of their unique geophysical characteristics. Coastal region is the biggest victim of natural disasters and is highly affected by climate change hazards including cyclone, tidal surge, salinity intrusion, erratic rainfall, heat wave, drought, thunderstorm, soil erosion and flooding.[1] Most recent cyclone "Amphan" of 13 May 2020 left a trail of devastation in the coastal districts of Bangladesh including Satkhira with wind speed of up to 180 kph and over 10 feet tidal surges. It inflicted heavy damage to lives, livelihood, agriculture, infrastructures and environment in Satkhira region of the country.[2]

According to IPCC 5th Assessment Report, global climate change is likely to further aggravate the flooding and salinity situations in Bangladesh. Under a high-emission scenario, total sea-level rise may reach up to 98 cm by 2100. Erratic weather pattern and increased salinity intrusion will hamper the agricultural activities as well as the lives of the coastal inhabitants who have an agriculturally based livelihood. Under these circumstances, adaptation is the only feasible choice to manage the climate change impacts for a country like Bangladesh having high population density and limited resources (IPCC, 2014).

Climatic hazards are the greatest threats to human lives and livelihoods in coastal regions especially in Satkhira[3] district of Bangladesh. Furthermore, climate change and agriculture possess an inverse relationship as climate impacts hamper agriculture production. Among other sectors, agriculture faces deadliest experiences from climate change-related natural disasters as it is the main livelihood option for the coastal poor men and women. People, particularly the poor, women and marginal communities mainly depend on agriculture and they have local-level experience to adapt with climate changing situation. Hence, agricultural adaptation approach is one of the key aspects that are helpful to reduce agricultural crop vulnerability in the coastal zone of Bangladesh.

The hazards of natural disasters cannot completely be prevented, but the risk and vulnerability can be reduced through adaptation and increased capacity of the people. In this connection, the main objective of this review is to find out suitable and viable agriculture adaptation technologies that have the potential to help poor and marginalized farmers especially women farmers to adapt climate change impacts. The adaptation technologies include identification of suitable cropping pattern, choice of seed, irrigation and water management, soil management, crop intensification, suitable transplanting and so on.[4] The study depicts that the integration of local knowledge and wisdom with modern science and technology may produce better result toward community preparedness and building resilience at community level (Nasreen, 2018).

One of the key objectives of SAKTEE is to identify, evaluate and demonstrate innovative climate change adaptation knowledge and technologies (CCAT) in water and agricultural sectors for multi-level institutional adoption as well as to empower women and poverty-stricken disadvantaged groups in the project districts. The research has to identify and demonstrate CCAT and best practices that are proven and have potentials for benefiting vulnerable women and disadvantaged groups in the project areas. These were accomplished through literature and documents review for preparing an inventory on CCAT, consolations with sectoral experts and relevant government officials and engagement of stakeholders at regional and local levels. In this backdrop, the research has intended to prepare a technology inventory report based on relevant published and unpublished reports and documents, scientific papers as well as in consultations with sectoral experts.

Objectives and Steps of the Literature Review

The main purpose of the review and technology inventory is to improve the understanding of the project team about the available technologies in water and agriculture as well as to identify the potential adaptation technologies. Information for this review was collected through desk review of existing reports and publications, web search, meetings and consultation with colleagues, sectoral experts and field observations during visits of the locality.

Specific Objectives

- To learn about the current adaptation technologies in agriculture and water sectors in project locations.
- To identity CCAT and confer how these technologies can benefit vulnerable women, poor and disadvantaged people at the local level.
- To know about the challenges and potential opportunities on using innovative adaptation technologies for women and disadvantaged sub-populations.

Methods and Steps

To understand the technology needs and identifying gaps, in adaptation technologies for water and agriculture sectors, the research team applied the following techniques and steps:

- Secondary data collection from relevant reports and journals and analysis to develop an inventory on technologies in selected district.
- Consultation with experts, researchers and relevant others.
- Organized consultation workshop at the district and Upazila levels.

- Focus group discussions in union level with community people about agriculture, water, sanitation and hygiene along with livelihood and women empowerment.
- Key informant interview of Chairman/women Member of Union Parishad, Sub Assistant Agriculture Officer, NGO representative, representative from Private Sector, Local School Teachers, experienced people and others.

Defining Adaptation Technology in Agriculture, Water and Women-Friendly Technology

The development of simple tools from wood or shards of rock and the discovery of fire, which provided a way to cook food and create heat and light, was technological developments that allowed people to accomplish tasks more easily and quickly. The definition of technology is science or knowledge put into practical use to solve problems or invent useful tools. An example of technology was during the Stone Age when the first knife or shovel was made from a piece *of stone or obsidian.*[5]

Adaptation to climate change involves identification of and implementation of diverse and wide range of technologies, which can comprise hard-tech such as seawall, improved dykes and irrigation management in the growing salinity and frequent floods as well as software such as change in seeds and cropping patterns, information, skill and knowledge for adaptation. Adaptation is evolving with blending of modern and local knowledge, skill and technological innovation. We have to take into consideration of the needs and priorities of the vulnerable groups as well as the perspectives of technology groups, climate scientists, policymakers, local government and NGOs who are involved with community and actors for development and transfer of adaptation knowledge and technologies (Christiansen et al., 2011). The process of CCAT transfer has to consider the experiential knowledge, institutional arranges (orgware) and the sources of technologies.

The Cancun Agreement on Climate Change in 2010 has adopted the Climate Adaptation Framework and the Technology Mechanism, which provided the opportunities and insight to step up and scale up adaptation technologies. In 2015, the CoP21 and the Paris Climate Agreement (PCA) have provided the way and steps for urgently needed climate technology development and rapid technology transfer for both adaptation and mitigation. The PCA (United Nations [UN], 2015) further strengthened the technology mechanism for research, technology development, demonstration and capacity building for technology transfer in the developing countries. The focuses of the long-term vision of adaptation technology transfer are as follows:

- Identification and development of climate-resilient technologies
- Innovation by blending of local knowledge and indigenous knowledge with external technological knowledge

- Capacity building and enabling environment
- Collaboration, engagement of actors and supports to the vulnerable communities.

The research follows the above-mentioned steps in identification and demonstration of CCAT in agriculture and water in the selected climate-affected zones.

Climate-Resilient and Adaptation Technology

Adaptation requires more resilient infrastructure, more climate-resilient technologies and new agricultural practices to counter the increased climate risks. The climate-resilient technologies mean a new practice that can counter the increased climate risk. Improving access to energy in underserved areas and using low-emission technologies can address the development needs of vulnerable populations while promoting a transition to green, low-emission and climate-resilient development. Adaptation is a process by which strategies to moderate, cope with and take advantage of the consequences of climatic events are enhanced, developed and implemented (UNFCCC, 2006). On the other side, application of technology in order to reduce the vulnerability or enhance the resilience of a nature or human system to the impact of climate change.

Climate change negatively affects human and ecological systems, increases vulnerability and causes a need for adaptation (IPCC, 2007, 2014). The promotion of adaptation technology has been internationally recognized as an important strategy. In order to reduce the vulnerability or enhance the resilience of a community poor, vulnerable women to the impact of climate change, the application of technology is a must requirement.[6]

Adaptation technologies have been divided into three: hardware, software and orgware.

Hardware: Capital goods and equipment like short-duration variety crops are hardware technology.

Software: Capacity, knowledge and know-how to operate are software technology.

The third dimension of adaptation technology is orgware, which is related to the ownership and institutional arrangement of technology development and dissemination. It is important to point out that adaptation technologies do not necessarily have to be new in the sense that they are newly developed. In fact, CCATs are improved, adopted and demonstrated in the changing climate and social contexts. There is a need for blending of local knowledge with external and modern knowledge because none of them are sufficient when they are stand alone. Hence, effective early warning system (EWS) for flood and cyclone can use both local knowledge of weather forecast and information processed through the modern equipment of meteorological office (Alina and Kamruzzaman, 2017).

Many technologies that are being used as adaptation technologies such as rainwater harvesting system (RWHS) and building houses on stilts do in fact date back and have been used since generations to deal with extreme weather condition. The examples for hardware, software and orgware are given below.[7]

Sector	*Hardware*	*Software*	*Orgware*
Agriculture	Selection of crop or crop variety	Farming practices research on new variety	Local institute
Water	Rainwater harvesting system, pond sand filter, low-lift pump, deep tube well	Increase water use efficiency and recycling	Water use associations

A combination of hardware, software and orgware is needed for development and successful application of new and effective adaptation technologies. Stand-alone technology such as an infrastructure or an equipment is not sufficient for effective adaptation. In many cases, the hardwares are not in the central place to address the vulnerability of the groups and communities, rather access to locally available technologies and knowledge, capacity and skill for use of the new CCAT and an enabling institution are the key elements in advancing adaptation in the communities (ibid.). Further, the adaptation capacity largely depends on social and human capitals such as education, awareness, skill, motivation for adaptation, income, access to resources, technologies and services as well as governance (accountability and responsiveness of the local government and duty bearers and sectoral actors). In this regard, Klein (2011) stresses that the effectiveness of the use of adaptation technologies is related to economic, social, institutional and legal process and contexts, where CCAT are demonstrated and scaled up.

The Technology Need Assessment (TNA) project was undertaken by the UNFCCC with the parties in 2010, which assisted the developing countries with methodological, technical and financial supports for identification and assessment of the efficacy of both mitigation and adaptation technologies. Further, the Technology Executive Committee (TEC) of the UNFCCC and the CTCN provide guidance for development and transfer of CCAT in the developing countries. In 2014, the UNEP has conducted a knowledge gap analysis in relation to adaptation technologies in key sectors like agriculture, water and health (UNEP, 2014). The report has identified potential knowledge gaps and challenges in the following three levels:

- Incomplete knowledge production and dissemination for adaptation technologies.
- Inadequate linkages and partnership between and among the institutions and stakeholders.
- Limited diffusion and transfer of knowledge to policy and decision-makers and the users of CCAT for transfer and update (UNEP, 2014; Eisenack et al., 2014).

Adaptive Capacity and Resilience to Climate Change

Adaptive capacity refers to the potential of individuals, communities and societies to be actively involved in the processes of change, in order to minimize negative impacts and maximize any benefits from changes in the climate. This potential is undermined in situations of poverty and inequality, and points to the wider socio-economic factors that influence how households and communities manage risk on a daily basis and relate to the successes and failures of development and poverty reduction.

Resilience means many things to many people, and there is no universally applied definition within the international development and humanitarian sector (Nasreen, 2020). Notions of resilience tend to be similar to adaptive capacity, but often focusing more on shocks and hazards (such as disaster events) than on longer term changes (such as changes to average seasonal conditions over decades). The UNISDR definition of resilience is

> the ability of a system, community or society exposed to hazards to resist, absorb, accommodate to and recover from the effects of a hazard in a timely and efficient manner, including through the preservation and restoration of its essential basic structures and functions.

Climate Change Contexts of Satkhira

Climate Change Impacts in Coastal District

Satkhira is a district in southwestern Bangladesh and is part of Khulna Division. It lies along the border with West Bengal, India. It is on the bank of the Arpangachhia River. The largest city and headquarter of this district is Satkhira. It has a tropical weather with monsoon rainfall from June to August in every year. Satkhira district had a population of about 2 million. Males constitute 49.49% of the population and females 50.51%. Muslims formed 81.86% of the population, Hindus 17.70%, Christians 0.31% and others 0.12%. Satkhira district had a literacy rate of 52.07% for the population 7 years and above. Climate change presents one of the greatest development challenges of our time, and it is now clear that the eradication of Covid-19 or *corona virus* or poverty or inequality cannot be achieved without also addressing the causes and consequences of climate change. Bangladesh is often considered to be the country most vulnerable to climate change. In addition, Bangladesh has high levels of existing poverty and inequality (43.3% of the population lives below the international poverty line of US$1.25 per day), and low per capita GHG emissions at just one ton per person (compared with nearly 20 tons per person in the USA). Therefore, for Bangladesh adapting to climate change is an urgent priority over mitigation action.

FIGURE 11.1 The cyclone "Amphan" badly affected the habitats, water sources and agriculture in Satkira in late May 2020

Most Vulnerable Groups to Climate Change in Satkhira

The vulnerable groups are those who depend on climate-sensitive resources and ecosystems for their families' livelihoods, such as agriculture, fisheries and forests (Figure 11.1). The marginalized and women are more prone to climate disasters since they have limited assets and very political voice to enable them to respond to the impacts of climate change (lower adaptive capacity). The women, socially excluded and disadvantaged groups, are most vulnerable in the district. The vulnerable groups also include wage earners, shrimp collectors, workers in the shrimp farms, forest-dependent communities and people living on the embankments.

Adaptation Technologies in Agriculture in Coastal Region

Bangladesh is one of the most vulnerable countries to climate change because of geographic exposure, low income and greater reliance on climate-sensitive sectors, particularly agriculture. It experiences different types of natural disasters almost every year because of the global warming as well as climate change impacts; these are riverine floods, flash flood, cyclones and storm surges, salinity intrusion, extreme temperature and drought. The climate contexts and social conditions are different in the two ecosystems. There are few common, but many local and context-specific adaptation measures. The government research institutes like Bangladesh Agriculture Research Institute (BARI), BRRI and Department of Agricultural Extension (DAE) as well as NGOs have developed and demonstrated various locally appropriate adaptation measures in crop, fisheries and livestock subsectors of agriculture. The key adaptation responses are changes in crop varieties and cropping patterns (such as salt and draught-tolerant varieties, early crops and flood-resilient rice varieties); floating bed crop farming, raised bed vegetable and fruits growing, ditch and dyke cultivation, drip irrigation, *Kanda* farming in Haor, alternative drying and wetting (ADW), mulching, fresh water storage in the field for small irrigation in the coastal zone, green and compost manuring, agro-forestry etc. (Rahman and Haque, 2013).

The Ministry of Environment and Forest/MoEF (currently named as Ministry of Environment, Forest and Climate Change/MoEFCC) prepared TNA report and an action plan for climate change adaptation in 2012. Linking with the BCCSAP, the report identified several CCAT in agriculture and water sector. The TEC of UNFCCC, prepared the briefing-4 on Adaptation in Agriculture for Bangladesh, has identified a number of adaptation technologies in agriculture, which included seasonal weather forecasting and improving EWSs; water storage and management for irrigation; resilient crop variety development and demonstration with farmers in the climate-affected zones. The report also analyzed the suitability of CCAT in agriculture, barriers and enabling conditions for upscaling adaptation (Alina and Kamruzzaman, 2017). The examples of CCAT in agriculture in the TNA report of the MoEF (2012) are the following:

- Development of salinity-tolerant rice and crop varieties
- Development of drought- and flood-tolerant crop varieties
- Early- and short-duration crop varieties
- Efficient and better water management
- Training on improved farming on irrigation, soil management and land-use planning in the contexts of salinity, drought and floods
- Crop diversification: floating bed and rooftop gardening
- EWS to protect standing crops in the fields.

Adaptation Technology in Agriculture in Satkhira

Bangladesh is largely an agriculture-dependent country and majority of rural people still make up a large part of their livelihoods from agriculture and related activities. Agriculture sector as a whole is again very sensitive to climate change-related stressors. Raising temperature, erratic rainfall, cyclone, increased salinity, tidal surge, flash flood and severe cold with dense fogs are all manifestations of climate change and climate variability that have already started impacting on agriculture and other sectors in Bangladesh. To adapt with the changing situation, a good number of adaptation technologies are being practiced in the agriculture sector over the last few years.

The coastal Satkhira district is adjacent to the Bay of Bengal and the mangrove forest of the Sundarbans. Being in the nearby Bay of Bengal, this district is highly vulnerable to salinity intrusion into agricultural land, cyclone, storm surge and tidal surge, the impact of sea-level rise, drainage congestion, flooding and waterlogging. Every year about 500 hectares of land are converted to saline land. In 2000, the level of salinity was about 23.93 dS/m and in 2009 about 28.64 dS/m in the Shyamnagar

TABLE 11.1 Adaptation options and climate-resilient technologies in agriculture

Sl no.	*Climate-resilient sector*	*Option*
1	Salt-tolerant T. Aman	BR-22 and BR-23, Bina Shail
2	Salt-tolerant BRRI dhan	33, 56, 57 and 62, BRRI dhan-40, 41, 47, 53, 54, 55, 61
3	Salt-tolerant Bina dhan	7 and 16, Bina dhan 8 and 10
4	Salt-tolerant Aus	BRRI dhan 65
5	Salt-tolerant wheat	Bijoy, BARI Gom-25, BAU 1059
6	Salt-tolerant potato	BARI A10, 22, CIP-clone 88, 163
7	Salt-tolerant sweet potato	BARI Mishti Alo-8, 9
8	Salt-tolerant pulses	BARI-Mug-2, 3, 4, 5, 6, BM-01, BM-08, BARI Falon-1, BARI Sola-9
9	Short-duration oil seeds	BARI Sharisha-14, 15, BARI China badam-9, BINA China badam-1, BINA China badam-2, BARI soyabean-6
10	Salinity-resistant jute variety	HC-2, HC-93, CVL-1
11	Saline-tolerant sugarcane variety	ISWARDI-40
12	Salt-tolerant alternative farming	Mele (reed) cultivation, salt-tolerant grass farming
13	Floating vegetable gardening	Ladyfinger, tomato, chili, bitter guard, potato, garlic
14	Dyke cropping	Tomato, sweet gourd, watermelon, cucumber, chili, turmeric, zinger
15	Storing of excess rainwater in ponds	Re-excavation of pond
16	Floating garden (Baira)	Ladyfinger, tomato, chili, bitter gourd, potato, garlic, sweet gourd, cucumber, chili, turmeric, zinger
17	Covering land in dry season to reduce evaporation	Less irrigation
18	Application of ashes and biochar	Less fertilizer
19	Pitcher irrigation	Reduce soil salinity
20	Change in crop production strategy (raised beds, mulches)	Reduce soil salinity
21	Fish culture	Net fishing, case fishing
22	Soft-shell mud crab farming	Case crab farming
23	Resilient livestock rearing	Semi-scavenger housing for goat, semi-scavenger housing for duck, semi-scavenger housing for hen, semi-scavenger houses for sheep
24	Other resilient practices	Land-use changing practice, integrated farming, crab patenting, shifting plantation timing

Source: Climate-resilient agriculture practice in different agro-ecological zones.

TABLE 11.2 Salinity-tolerance levels of rice crops

Sl No.	*Varity*	*Season*	*Salinity (in dS/m)*	
			Seedling stage	*Maturity*
1	BRRI dhan-47	Boro	12–14	8
2	BRRI dhan-40	T. Aman	8	8
3	BRRI dhan-41	T. Aman	8	8
4	Other varieties	Have no reported salt-tolerance level		

Source: Climate Change Cell (2009).

upazila.[10] Agriculture and aquaculture are the principal livelihood options in coastal Bangladesh. But it is most vulnerable to the increasing frequency and intensity of extreme events such as cyclone, storm surges, floods, hailstorms, erratic and heavy rainfall and salinity intrusion (GOB, 2009; Mainuddin et al. 2011). Climate change and climate vulnerability are affecting the land-use patterns, crop systems productivity and optimum agriculture output (GOB, 2009).

Many forms of adaptation practices such as hard and soft adaptation are being practiced and implemented (Asian Development Bank [ADB], 2011). Many agricultural adaptation options are being practiced in coastal Bangladesh to adapt to climate-induced disaster like salinity, flood, waterlogging, drought and so on. To ensure climate-resilient agricultural technology development, many structural and non-structural soft and hard interventions are being developed and demonstrated by BARRI and BARI. Some of these innovations are devised by local communities through their indigenous knowledge and some are planned interventions promoted by different government and non-government organizations. Ruane et al. (2013) mention that floating bed crops/vegetables in the south central and southern areas plant bed raising and dyke cropping at the shrimp *gher* (water bodies which are artificially generated through raising dyke around the lowland) is an old but effective practice used nowadays. Table 11.1 presents the adaptation options and technologies in coastal region.

Table 11.2 presents the salt-tolerant rice and crop varieties in the region. The varieties and technologies were mainly developed and demonstrated by the government departments. NGOs are also working with communities to transfer the technologies.

Brief Descriptions of the Promising Technologies

The sections below described the technologies in details: What is the technology and know-how; how do they reduce risk of climate change and whether they are available in the locality? Rahman and Islam (2013) have presented potential adaptation options and technologies in agriculture. They have suggested a number of adaptation technologies for both coastal region and Haor basin. The work, included in a book entitled "Climate Change Adaptation Actions in Bangladesh" (Shaw), has briefly described adaptation practices and technologies in agriculture, aquaculture, water and DRR. They have touched upon the potentials for further innovation and extension of adaptation technologies for reaching millions of poor, women and vulnerable groups living in the climate-affected zones. There are rice and crop varieties that are resilient to salinity, waterlogging and drought condition. For example, late variety of rain-fed Amon paddy: BR22 and BR23 and IRRI Dhan 46 are best suited in the coastal region. Further, the BINA and BRRI have developed Boro rice (winter-season paddy) that can be grown in moderately salinity-affected soil with profitable yield. Farmers also grow maize, oil seeds, beans and vegetables on the raised beds and embankments of fish ponds in the coastal region (Abedin et al., 2019).

The climate contexts and social conditions are different in the two ecosystems. There are few common, but many local and context-specific adaptation measures. The government research institutes like BARI, BRRI and DAE as well as NGOs have developed and demonstrated various locally appropriate adaptation measures in crop, fisheries and livestock sub-sectors of agriculture. The broad adaptation strategies and responses are as follows:

- Changes in crop varieties and cropping patterns (such as salt- and draught-tolerant varieties, early crops and flood-resilient rice varieties)
- Floating bed crop farming, raised bed vegetable and fruits growing, ditch and dyke cultivation, drip irrigation; and *Kanda* farming in Haor
- ADW, mulching, freshwater storage in the field for small irrigation in the coastal zone, green and compost manuring, agro-forestry etc.

The TNA of the GoB (2021) has identified a number of adaptation technologies in coastal agricultures. These included: development of salinity-tolerant rice and crop varieties; development of floods- and waterlogging-tolerant rice and crop varieties; early variety and short-duration crop varieties so that these could be harvested before cyclone and flood seasons; crop diversification trough floating bed crop cultivation, rooftop and hanging garden crop cultivation; training on improved crop farming, soil management, water and irrigation management and land-use planning etc. The following sections briefly describe the promising adaptation options and CCAT in agriculture.

Saline-Tolerant Crop Varieties

BARI, BIRRI and other research institutions have developed a huge number of saline-tolerant crop varieties to grow crops in the saline land (Figure 11.2). They introduced a number of climate-resilient rice variety technology such as T Aman, BR-22 and BR-23, BRRI dhan 33, 56, 57 and 62, Bina dhan-7, 16, BRRI 47, 61, 67, Bina dhan-8 and 10, T Aman-BRRI dhan 40, 41, 53, 54, salt-tolerant wheat like Bijoy, BARI Gom-25, salt-tolerant potato, BARI Alu-22, CIP clone 88, 163, salt-tolerant sweet potato, BARI Misti Alo-8, 9, salt- and heat-tolerant pulse: BARI

FIGURE 11.2 Saline-tolerant paddy cultivation

Mug 2, 3, 4, 5, 6, BM-01, BM-08, BARI Falon-1, BARI Sola-9, short-duration and salt-tolerant oil seeds, BARI Sharisha-14, 15, BARI China badam-9, BINA China badam-1, BINA China badam-2, BARI Soya bean-6, BARI Til 2, 3, 4, salt-tolerant jute varieties HC-2, Hc 95, CVL-1, salt-tolerant sugarcane varieties ISWARDI-40, Hysum-33, sunflower, jute and sugarcane varieties etc. (Rahman, 2011; Sutradhar et al., 2015). To cope with climate-induced disaster like salinity intrusion, tidal surge and waterlogging, multidimensional resilient option is being implemented.

Majority of the field-level production works are performed by male members with available technologies whereas female members are responsible in every stage of rice farming namely liked seedling, nursing, rice storing, seed preservation, straw drying, weeding, thinning, cleaning, bowling of grain, threshing, drying, husking etc. where technology is less available. Women-rich workers have been able to increase their income through their involvement in rice farming, which have a positive impact on their family life. They can also be involved in decision-making process with their husbands regarding family as well as farming. Through this journey women became more confident and received honor from society also.

Saline-Tolerant Maize Cultivation

Maize is moderately sensitive to salt stress; therefore, soil salinity is a serious threat to its production worldwide. But in coastal area of Bangladesh, farmers have been able to grow maize on saline land in Khulna (a district near to Satkhira). Soil Resource Development Institute initially used 25 bighas of land to cultivate maize in the district. It has been quite a while that a number of studies have been conducted in order to ensure productive use of saline land in the coastal areas of Khulna. As part of this initiative – instructed by the Salinity Management and Research Center (SMRC) – Botiaghata, Khulna, farmers transplanted maize seedlings from full-grown corn crops into saline lands, before water levels rose; they used a "dribbling"

FIGURE 11.3 Saline-tolerant maize cultivation

technique to plant the collected seedlings when saline land remained soft and muddy. Farmers, using the two techniques – transplanting and dribbling – will be able to harvest at least seven-to-eight tons of maize per hectare. Maize crops also can be grown in courtyard or homestead garden in small scale where women can easily take care of the crops field and manage to provide necessary inputs (fertilizer, irrigation, pest control and harvesting) by their own hand (Figure 11.3). During the harvesting period, they also collect the maize and process the crops manually or using women-friendly small husking machine for collection of corn. The straw of maize can be used as cooking fuel, which is a very important element for a rural woman for cooking food. Women can easily control over the resources and preserve the seeds for the next-year cultivation.

Hanging Garden and Floating Bed Vegetable Cultivation

There are some villages in the coastal area that are waterlogged in the majority period of the year. Floating cultivation has been a local practice in some of these villages. It involves planting crops such as compost and other aquatic plants on soilless plastic dram on water and is called a floating garden. This platform provides a base to raise seedlings, vegetables and crops/flowers on it. The plants grown on floating get nutrition and food from organic compost. This provides employment, income, food and nutrition for the farming families and local communities; furthermore, it helps the coastal waterlogged population adapt to the changing environment. Women can easily manage and care the floating garden. They do not need fertilizer and pesticide for cultivating vegetables using the method, and this method is helping them grow vegetables at a lower cost. As the vegetables are being produced in a fully organic way, the demand for these vegetables is higher in the markets. They can use their production for their family consumption and also can sell in the market. Women can earn by using this technology.

FIGURE 11.4 Floating gardens (Baira) of vegetable

FIGURE 11.5 Floating gardens (Baira) of vegetable

Floating Gardens

Local people from the southwest coastal area of Bangladesh have developed floating rafts with a bamboo base, upon which water hyacinth is piled and then covered by the aquatic plants of coconut husk to form a seedbed ready for planting. These floating gardens (Baira), which are cultivated in the rainy season and float above monsoon floods, are now becoming popular in many other areas in the south of Bangladesh (Figures 11.4 and 11.5). This agriculture adaptation practice is mainly seen in sweet waterlogged and flood-prone coastal districts like Pirojpur, Goplaganj, Faridpur, Jalakathi, Barisal zilla but this technology is not applicable in saline-prone area of Bangladesh.

Dyke Cropping

Dyke cropping is a very popular agriculture practice which has combined fish and vegetable cultivation in Shyamnagar and Kaliganj upazila of Satkhira district. Due

FIGURE 11.6 Dyke cropping

to salinity intrusion in coastal area, especially in Satkhira region, most of the land occupied shrimp farm (gher) round the year. The waterlogged *ghers* create opportunity for practicing fish–vegetable dyke cropping in large area. This is a general agriculture adaptation technology practice of seasonal and year-round cultivation of vegetables on the bank of the field. Different types of vegetables like ladyfinger, cucumber, pumpkin, water gourd, bitter gourd, chili and different spinaches are easily grown on it. Women member of a family can easily manage and grow vegetables on it. They can use the vegetables for their family consumption and also sell them to the local market for their income-generating activities (IGAs). This is a popular, women-friendly adaptation technology in coastal area of Bangladesh (Figure 11.6) (Rashid et al., 2017).

Irrigation and Water Management in the Coastal Areas and Storing Excess Rainwater in Ponds/Tanks

Excavation of ponds to store rainwater is a good way to mitigate irrigation water scarcity in the coastal area. Pond water could be the best source of irrigation in dry season. So, during the monsoon, the excess rainwater can be reserved in ponds and can be utilized for irrigation practices. Part of the excess pond water can be used in dry periods for growing vegetables (Figure 11.7). Women household members can collect water from pond and grow vegetables in their homestead for their family consumption. So, pond re-excavation is a good adaption technology in coastal area of Bangladesh.

Covering Land in Dry Season to Reduce Evaporation

Covering land in dry season to reduce evaporation is a technology by which farmers can reduce salinity. Covering land with plastic sheet can be a way of preventing soil evaporation. This adaptation technology is recently used in coastal area to prevent soil salinity (Figure 11.8). Agronomic practices that increase shading of the soil surface, and physical structures that concentrate rainwater, encouraging percolation to deeper layers, also reduce evaporation. Women member of a household can use this technology in their homestead for vegetable cultivation. This is a cheap and easily

FIGURE 11.7 Storing excess rainwater in ponds/tanks

FIGURE 11.8 Cover land to reduce evaporation in dry season

manageable technology and the materials (plastic sheet) used for it are very cheap and available in the local market. It is also a climate-resilient technology in the context of saline-prone area of Bangladesh.

Application of Ashes and Biochar

This is an adaptation technology to bring down the salinity from soil by leaching salts from land. Farmers dug a small pit in the corners of their land which was covered with ashes. The objective of their strategy was to snob salt from water during soil flushing in the rain or excess irrigation. This is scientifically recommended adaptation strategy that can be an application of biochar to amend salt-affected lands. This technology can be used in homestead vegetable garden. The source of ashes mainly comes from cooking stove and generally women of households dump them in a place to use it in agriculture land and also homestead vegetable garden. This technology can be easily handled by women and it is also low cost and climate-resilient (Figure 11.9).

Pitcher Irrigation

Pitcher irrigation uses unglazed baked earthen pots or pitchers buried in the soil to water the garden area. Pitcher irrigation is a popular adaptation practice in coastal

FIGURE 11.9 Application of ashes

FIGURE 11.10 Pitcher irrigation

area because irrigation water is provided from an earthen pitcher having several small holes on its bottom. The earthen pitcher is placed on a raised bed filled with fresh irrigation water (Figure 11.10).

Then jute fibers are entered into the pores. Then, the pitchers are filled with fresh irrigation water to reduce soil salinity, increase the efficiency of irrigation water use, increase land cover as well as soil productivity. Vegetable seeds are planted around the pot. Enough water soaks through the pot into the root zone to supply and maintain plant growth. Excellent results have been achieved with both pumpkins and watermelons with this method.

The technology is popular in case of vegetable cultivation in moderately salt-affected area in Bangladesh. In this circumstances, SMRC of the Soil Resource Development Institute, located in Batiaghata, Khulna, has developed some technologies which can reduce soil salinity, increase the efficiency of irrigation water use, increase land vegetative cover as well as soil productivity. Now the pitcher irrigation technology is popular in Khulna, Satkhira and Bagherhat districts of Bangladesh. Farmers adopt the technology, because thousand hectares of land remain fallow in coastal areas in the dry season. During this period, there are two major problems: soil salinity, water salinity and scarcity of suitable irrigation water. The pitcher irrigation technology is playing an important role to reduce soil salinity as well as increasing the efficiency of irrigation and increasing crop yield.[8] Although this is an excellent method to conservatively use water, it must be noted that care of the irrigation pitchers is very important.

Md. A. Abedin and R. Shaw

Raised Beds and Mulches for Drought Management

BARI has found that using raised beds, mulches and drip irrigation, it is possible to grow high-value horticultural crops like tomato, chili, watermelon and cucumber in saline soils of 4.5–11.0 dS/m (Islam et al., 2019). This is very easy for the women member of the household to plant and nurture this technology. Development of new technology such as crop varieties tolerant to heat and medium-to-high soil salinity is time demanding (Figure 11.11). More emphasis should be given to variety development and to environment-friendly pest and disease controls. This is a woman-friendly and low-cost technology. Because women can easily cultivate vegetables using this technology on their homestead. Using this technology, they can meet domestic demand of vegetable and earn by selling in the market. It could be a step of women empowerment. Women also use suspended pots to produce trees in their homestead area. They can plant small trees, plant, herbs and shrubs in the pots. The greater advantage of using suspended pots is that these pots are easily transferable to a high place when needed (Nasreen, 2018/footnote 7).

Integrated Pest Management Methods

Integrated pest management (IPM) is a sustainable approach to managing pests by combining biological, cultural, physical and chemical tools in a way that minimizes economic, health and environmental risk (National Coalition on Integrated Pest management, 1994). People of all over Bangladesh have indigenous knowledge on pest management. But now farmers are using chemical pesticide in a huge amount. Chemical is controlling pest but reduces the productivity of land. Overuse

FIGURE 11.11 Raised beds and mulches for drought management

of chemical pesticides has a bad impact on soil, environment and human health. According to the Food and Agricultural Organization (FAO) IPM is,

> A pest management system that, in the context of the associated environment and the population dynamics of the pest species, utilizes all suitable techniques and methods in as compatible a manner as possible and maintains the pest populations at levels below those causing economic injury.

IPM has evolved from an exclusively technical approach to a more holistic view of the agricultural production system that links to the long-term sustainability of agricultural production with economic, environmental and social issues, including public health. IPM practices where different from of conventional practices are judicious and need based used chemical pesticides, and integrated all plant protection approaches in an umbrella (Rahman et al., 2009).

The plant protection wing of DAE, the largest public agro-based organization in Bangladesh, directly deals IPM activities. There are some other semi-government organizations, such as Bangladesh Agricultural Development corporation (BADC) and the Bangladesh Academy for Rural Development (BARD) that is engaged in transfer of IPM technologies to the farmers. In order to extend IPM technologies, there is a need to more research and develop some approaches which assist the farmers to get information about IPM, and DAE developed some dissemination techniques like Extension Agent Visit, Farmers Field School, IPM Club and Field Days.

IPM is a potential tool for crop and vegetable farming, but it needs a huge amount of financial investment. If women farmers could have the knowledge of it, they can apply their knowledge to cropping and farming also. Women can easily manage their homestead garden through this tool.

Fish Cage Farming in Coastal Area

A new form of fish farming, known as cage culture, is expanding in flowing water of rivers and canals in various parts of the country especially in Satkhira region of Bangladesh. It is raising hopes for an increased production of fish. Fishes, mostly tilapia, are now farmed in nearly 6,000 cages in rivers where such farming did not exist even a decade ago. The Ministry of Fisheries and Livestock is working to frame a policy on cage culture in inland water, which remains untapped despite the huge potential to augment production of fish. The fish cage culture is suitable in winter season, because, during winter, fishes grow well in rivers because of higher temperature compared to that of in ponds (Figure 11.12). In climate hazard context, winter is less disastrous than summer season, so, there is a huge scope to expand cage fish farming in coastal areas. The cage culture could be a practical solution to boost fish production as land is scarce to make ponds for aquaculture. This technology can easily manage by women, because in the fisherman community men moved to deep sea to catch fish. In that case women takes care of cages and supply

FIGURE 11.12 Fish cage farming in coastal area

FIGURE 11.13 Soft-shell mud crab farming

food time to time, like small snail, rice, wheat and so on, for the fish inside the cages. They can also catch the fish and sell in the market. This technology is low cost and climate-resilient.

Soft-Shell Mud Crab Farming

Soft-shell crab production is very new in Bangladesh and being done on a limited scale in Satkhira regions (Didar-Ul Islam et al., 2015). Mud crabs are also known as Green Crabs or Mangrove Crabs (Shafi and Quddus, 1982). These crabs are omnivorous or scavengers, feeding on dead animal and plant matter. This is also a new adaptation technology in the region (Figure 11.13). There has been an opportunity to involve more women in crab farming activities. Women participate in many activities such as collect crab feeds (snails, low-value fish) and mince and apply to ponds, prepare lime and fertilizers and sometimes apply them, harvest, wash and tic crabs and sell to local buyers. Women will purchase rejected crabs (damaged, underweight with developed gonad) from depots of secondary markets and sell to domestic markets. Some women purchase different sorts of trash fish, other low-value fish from markets. Collectors, fish farmers, *gher* owners then chop them in small pieces as well as sell to the fatteners as crab feed. This sector appeals to achieve the sustainable and substantial improvements of women life as well as to address the gender disparities in access to and control over resources and decision-making. Involvement of women in crab aquaculture allows men to work outside, thereby reducing families' vulnerability. Most important, crab fishery is offering alternative livelihoods to help vulnerable littoral communities be resilient to changing climate condition as an adaptation technology in coastal area of Bangladesh. Unlike other fish and shellfish commodities, there is not a big demand for crab in the domestic market for local consumption because of religious restrictions. But its international demand is high and production from mud crab fisheries has increased gradually mainly related to more coastal people becoming interested in crab farming. There is a paradigm shift in the way that coastal people think about their livelihoods as they shift from shrimp farming to crab farming in response to changing climate. Crab farming serves as a hedge against poverty and provides access to market, food, health services and education. Government initiatives are required to facilitate training, credits, transport and marketing support for crab farming in coastal zone.

Semi-Scavenger Housing for Goat

Semi-scavenger goat rearing is a popular adaptation technology in coastal area of Bangladesh. Goat is known as "poor man's" cow. It contributes to the livelihoods of millions of rural poor. Small and marginal farmers as well as landless, particularly women, are increasingly relying on goat keeping for the socioeconomic upliftment. Goat plays an essential role in connection with food security and rural development as part of an integrated farming system, which is tailored to local needs. Goats not only provided food, but also enabled poor landless people, especially women, to enter the cash economy. Goat rearing can offer more opportunities to the developmental agencies for their further intervention in microcredit, extension and marketing support especially to rural women, landless and small farmers. Goat is an ideal animal species for farming in the rainfed regions where crop production is uncertain, and rearing large ruminants is hampering because of acute feed and fodder scarcity. Goat rearing has distinct economic and managerial advantages over other livestock species because of its less initial investment requirement, low-input requirement, higher prolificacy, early sexual maturity and easy in marketing. Goats can efficiently survive on available shrubs and trees in unfavorable environment. It is possible to boost up the goat production in the country as well as poverty alleviation, women empowerment and employment generation in the coastal area of Bangladesh.

Adaptation Technologies in Water and Sanitation in Coastal Region

Physical impacts of climate change including temperature rise, changes in precipitation, erratic rainfall, drought, increased salinity and climatic extremes like floods, cyclones and tidal surges have profound impacts on sources of water, quality, availability and required access to water for drinking, bathing and domestic uses in Bangladesh. Frequent flooding and erosion, cyclones and salinity have threatened the sources of water and facilities of water supply and sanitation in coastal region. The affected water, sanitation and hygiene (WASH) systems and services have huge consequence on human health, nutrition, medical expenditure and livelihoods of common people, particularly on the vast majority poor in South Asia. Climate change has huge bearing on the commitment made for SDG-6. Access to safe drinking water and sanitation is the most important adaptation measure for long-term resilience and emergency response to get back the back lives on track after a climate extreme (Water Aid, 2019). The climate-induced multiple risks and growing vulnerability in drinking water, sanitation and health have created greater need for wider and effective adaptation in this sector.

The TEC Brief-4 on Adaptation technologies for Bangladesh has prioritized RWHS, borehole and tube well, desalinization, pond sand filters (PSFs) and conservation of water for drinking and domestic uses. The TNA report of GoB (2021) has identified the following CCAT in water and sanitation: rehabilitation of existing embankments and dykes (which protect the live, properties and water sources); management of tidal river systems; managed aquifer recharge (MAR); and artificial aquifer well.

Adaptation Technologies in Water and Sanitation in Satkhira

Bangladesh is considered as one of the most vulnerable countries to climate change impacts. People, particularly the poor, women and marginal communities suffer the most from the climate change. The loss and damage study conducted by BCAS in 2014 revealed that the health and working potentials of women were badly affected by cyclones Sidr and Aila in 2007 and 2009, respectively. Besides these, the sources of water, homestead agriculture including home gardening, employment and income of women and child care are affected by climate change severely. The poor women are the innocent victims of climate change impacts and their unequal social condition. This is deepening the poverty situation.

Groundwater: Generally, it is bacteria free, and it is lifted by using tube well (deep or shallow). However, groundwater could be contaminated with chemical or mineral, arsenic in particular. Process of treating chemical-contaminated water is complex, as well, expensive. It may not be possible to do that at household level. Also, floods may inundate the tube wells and cause difficulties in getting safe water. By raising tube well platforms, communities could minimize this problem.

Surface water: It is available from rivers, ponds and the local water bodies. Generally, surface water is bacteria contaminated. For consumption or washing, it should be treated to make it safe; and it could be done through boiling, filtering or using chemical. At community level, surface water can be filtered through PSF. It is a simple technology and could be managed by the communities. Also, to disinfect water at household level, people could fill in transparent PET bottle and keep that under direct sunlight for six hours. However, neither the PSF nor the PET bottle could remove salt or chemical contamination.

Rain water: To meet the need for safe water, it could be collected and stored during the monsoon. In areas where the annual rainfall is 1,600 mm or more, people could collect enough water to meet their annual needs. Rainwater is safe if used immediately. Stored rainwater often gets bacteria contaminated if not taken proper measures; it should be used after filtration or decontamination.

Understandably, use of safe water depends on its availability. Difficulties in accessing safe water often lead to consumption of unsafe water; and it increases risks of infection. It is important to take measures to prevent contamination of water. Sanitation is primarily concerned with preventing water contamination and disease spread. Crisis of fresh water due to climate change is already visible in many parts of Bangladesh. Fresh water has become scarcer in Bangladesh's drought-prone northwest and in southwest coastal areas, where about 2.5 million poor residents do regularly suffer from shortage of water for drinking and irrigation (World Bank, 2019). Satkhira is a coastal district of Bangladesh where salinity is a big problem. Several adaptation technologies in the water sector can be found in the coastal areas in the country. There are some practiced adaptation technologies in water sector in Satkhira district over the last few years. Those are as follows:

- Desalination
- Rainwater harvesting
- PSF
- Artificial aquifer tube well
- Deep tube well
- MAR
- Reverse osmosis (RO)
- Pipeline water supply (PWS)
- Biosand filter (BSF)

The following sections describe the major WATSAN technologies.

Desalination: Desalination is construction and functioning of community-based water supply system to make safer (non-saline) water available for a community by excavating or re-excavating a pond. Desalinizers attached with piped water supply system are required large capital investments. It requires community land to be established, a permanent structure for its installation, electricity supply and cost for such services, and linkages with technical back-stoppage services, which are not women-friendly yet. The replacement of membranes needs to be handled by technicians, while the frequent washing and maintenance works can be delegated to a salaried supervisor with minimal training, the latter being provided by the technical people. When piped supply system is attached with a desalinizer, the initial cost is further escalated, requiring large-scale subsidies toward offering the safe water supply. In general, the safe water supply system requires both technical support and large-scale financing for replication in areas where non-saline water is still not available. This method is women-friendly, but it needs NGOs or other agencies support for installation and maintenance.

Rainwater Harvesting

Relatively cheap materials can be used for construction of RWHS containers. It has low maintenance cost and requirements. Collected rainwater can be consumed without treatment, if a clean collecting surface is used. Store rainwater in various containers like small pots, drums, jug and pitchers. RWHS provides a supply of safe water close to homes, schools or communities, encourages increased consumption and reduces the time women and children need for collecting water. Some problems are identified to reserve rainwater. Supplies can be contaminated by bird/animal droppings on catchment surfaces and guttering structures unless they are cleaned/flushed before use. However, added filtration system removes such bacterial contamination. Both for household and community levels, RWHSs are very useful as a source for seasonal or round the year fresh water supply and if the groundwater is stressed or quality hazardous. But heavy rain is needed for using this technology (Figure 11.14).

Pond Sand Filter

PSF has been used in the coastal area of Bangladesh since a while. The technology is built around artificially constructed ponds, locally known as "sweet water ponds," which are refilled by rainwater during the monsoon season. The rainwater collected in these ponds is pumped by hand into a storage tank through a filter chamber. The filter chamber is constructed in two parts, the first of which is a pre-filter packed with coconut fibers. This pre-filter reduces the turbidity of the raw water as the raw water flows into the filter chamber. The outflow from the pre-filter flows into the main body of the filter chamber through two overflow pipes. The main filter chamber consists of a layered, sand filter bed, through which the water trickles and

FIGURE 11.14 Rainwater harvesting

in which impurities, including bacteria, are removed in a manner similar to slow sand filtration. The quality of the raw water is further protected through the reservation of the ponds feeding into the filtration system solely for portable water used (UNEP). This technology is women-friendly. A cofinance model where villagers and the implementing NGO shared the cost for the installment has proven to be a good approach that guaranteed local ownership of the PSF while at the same time sharing the financial burden so that the community did not have to cover for it entirely.

Artificial Aquifer Tube Well

Artificial aquifer tube well is an innovation to treat contaminated surface water for domestic use on the saline-prone coast and arsenic- and iron-contaminated parts of Bangladesh. The recharging structure resembles the natural underground aquifer, which is the source of groundwater (Figure 11.15).

A sweet water pond has been selected that contains water throughout the year. A ring well as "recharge zone" has installed within five feet on the bank of that pond. The recharge zone needs to be maintained at least ten feet dipper than the bed of selected pond to continue good recharge. The recharge zone is filled by ¾ inch stone chip up to the first six feet. Forty-feet-diameter ring slab has been installed one after another and sealed by cement. The chamber within the ring slab has been filled by coarse sand having one foot vacuum space at top. The outside of the slab has been sealed by coarse sand. There has a connecting pipe to collect pond water to the recharge zone. The pipe has a filter at the opening end toward the chamber and a simple net at the opening side toward the pond. The pond water siphons to recharged zone through infiltrate system into the artificial horizons sand and stone. The zone of saturation occurs because water infiltrating the soil reaches an impermeable layer of concrete structure as "aquifer" so that it is not able to penetrate any further into the earth. Since the water infiltrates through a layer of soil/sand, it is

FIGURE 11.15 Artificial aquifer tube well

FIGURE 11.16 Deep tube well

significantly free from suspended impurities including microorganisms usually present in surface water. As sweet surface water is the main source of water, it is free from arsenic, iron and saline. Then the water is collected with a hand pump having cement- and glass-made head to avoid rust and friction erosion. The tube well has installed adjacent to the recharge zone having the pumping pipe inside the brick cheap of the recharge zone.

Deep Tube Well

Deep tube well operates in suction mode to withdraw groundwater from the deep aquifer at the potentiometric surface of groundwater. Generally, as the water from the deep tube well is bacteriologically safe, the acceptability of this technology is higher than the shallow hand tube well, which has a risk of bacterial contamination from the nearby pit latrines. Deep tube well is costly for individual household. It can be provided for community water supply and the cost can be minimized by using multiple heads (Figure 11.16).

Managed Aquifer Recharge

MAR consists of adding water, possibly recycled water, to an aquifer under controlled conditions for withdrawal at a later stage. In this technology, water is collected from ponds and roofs and, after passing through a sand filter, is then injected into the shallow saline aquifer through a ring of infiltration wells, creating a lens of fresh water. After the turbidity of the infiltrated water has improved to an acceptable level, water can be abstracted using a standard hand pump yielding water of improved quality (reduced levels of turbidity, coliforms, iron and arsenic). This system remains operational during rainy season only, when rainwater is abundant. The system uses pond water, which is stored rainwater and has been used for drinking and cooking by the families living around the pond. The pond water goes

FIGURE 11.17 Managed aquifer recharge

through a pretreatment process in a chamber containing filter materials, and then is allowed to pass through the recharge wells. The users collect water from this system through tube wells for drinking and cooking (Figure 11.17). Installation cost of this technology is not affordable for a marginalized family.

Pipeline Water Supply

The PWS is normally resilient to disaster shocks and can continue to supply water, if the pond water source is not polluted by salt water intrusion. Piped water systems either use a groundwater source or a pond water nearby. PWS that uses pond water will reduce reliability of pond water for other sources because during the dry period the demand of water could exceed the capacity of the pond. The piped water system is required to be operated by a trained pump operator and the supply lines need to be continuously checked for leakage and uninterrupted water supply.

The system is operated by either electricity or solar energy. A strong institutional arrangement is required for sustainable operation of PWS, which was found absent for most of the existing PWS. The operational and maintenance cost is required to be recovered from the users through tariffs. Piped water supply is a desired option in the coastal area due to its ability to supply water at the doorstep of the households requiring zero or minimum collection effort especially by women. The PWS water is typically arsenic free and bacteriologically safe for drinking, if not polluted through piped system.

Reverse Osmosis

RO is a desalination process and can produce safe drinking water from both surface (i.e., river) water and groundwater as input water irrespective of salinity and other contaminants. This makes this technology option resilient to increasing levels of salinity in the coastal areas of Bangladesh. In this process water is made to pass from the more concentrated solution to a less concentrated one, which is the principle of the process of osmosis. The force necessary to accomplish this is the application of pressure greater than the osmotic pressure of the saline solution. When saline solution is in contact with semipermeable membrane, which is placed under pressure being in excess of its osmotic pressure, water from the solution will flow through the membrane. Water flow will continue till the pressure created by the osmotic head equals the osmotic pressure of the salt solution.

RO reduces the concentration of dissolved solids, including a variety of ions and metals and very fine suspended particles such as asbestos that may be found in water. RO also removes certain organic contaminants, some detergents and specific pesticides. Although RO membranes can remove virtually all microorganisms, it is currently recommended that only microbiologically safe (i.e., coliform negative) water be fed into RO systems.[9]

Acetate membranes have proved most successful to be used for this purpose. The salt content in the water produced can be controlled by reducing the pressure or increasing the number of filtrations. RO technology is very good at purifying saline water but is heavily associated with higher operation and maintenance costs and high capital investment. The cost for energy and chemicals with frequent replacement of parts creates challenges for poor communities and introduces the potential for social discrimination when using this technology in vulnerable communities.

Biosand Filter

A BSF is an adaptation of the traditional slow sand filter, which has been used for community drinking water treatment for 200 years. The BSF is smaller (about 1 m tall, 0.3 m wide on each side) and adapted so that it does not flow continuously, making it suitable for use in people's homes. The filter container can be made of concrete or plastic. It is filled with layers of specially selected and prepared sand and gravel. The sand removes pathogens and suspended solids from contaminated drinking water. A biological community of bacteria and other microorganisms grows in the top 2 cm of sand. This is called the biolayer. The microorganisms in the biolayer eat many of the pathogens in the water, improving the water treatment.

Any kind of water can be used in the BSF – well water, borehole water, pond or river water, tap-stand water or rainwater and the freshly filtered water can collect from the BSF outlet pipe. This makes it very convenient for people because they can use whichever water source is closest to home, make it safe to drink. The water must not have been chlorinated though, or the chlorine will kill the biolayer. The water should also not contain any dangerous chemicals, because the BSF cannot remove most chemicals from water.

Contaminated water is poured into the top of the BSF at least once per day (but not continuously). The water poured into the top of the filter slowly drips through the holes in the diffuser and flows down through the sand and gravel. Treated water flows out of the outlet tube. No power is required – the filter works by gravity. It should take about 1 hour to get 12–18 L of filtered drinking water.

The BSF could reduce a significant amount of microbial contamination in the effluent; it does not meet the microbiological standards by WHO of total coliform and fecal coliform (less than 1CFU per 100 mL). Post-chlorination of the filtrate analysis result meets up the WHO recommendations of having the residual-free chlorine level over 24 hours to protect stored water from recontamination (Kabir et al., 2016). The filter reduces the risks of contamination between the water source and the home, by providing drinking water through treatment at the household level. Overall, the BSF is, and will continue to be a simple, effective and low-cost household level water treatment device for drinking water supply in the hazard-prone hard-to-reach coastal areas in Bangladesh. It is a women-friendly adaptation technology.

Sanitation Technologies in Satkhira

Cyclone or flood seriously damage water supply and sanitation system. Cyclones always come with storm surges. Along with other infrastructure, it washes away latrines and tube wells. Floods inundate latrines and tube wells in the low-lying areas. Latrines collapse; tube wells become contaminated and they become nonfunctional. Tube wells and latrine on higher ground may not go under water; nevertheless, inundated surroundings and approach roads seriously impede people to use these facilities. Also, cyclones or floods force people to leave their houses. Displaced households take shelter on embankments or roadside. More often they do not find enough space or sufficient resources to set latrines or tube wells there. In the project areas there are some adaptable sanitation practices that are appropriate and effective as resilient. In coastal Satkhira, water and sanitation infrastructure damage due to cyclone, tidal surge and inundation of low lands and offshore islands. Many poor households use hanging latrine. Many households use ring slab or sanitary latrine. But when tidal surge comes, all latrines are overflowed by water. There are some innovative technologies used in Satkhira for climate change adaptation. These are discussed below:

Step Toilet

Step toilet is a form of sanitary toilet initiated by *Uttaran* (a local NGO), constructed in a way to prevent flood water from entering the pit. Moreover, the nonporous lining of the toilet helps to prevent any form of leakage, which may contaminate the nearby water body. The cost of building a step toilet can range from 10,000 to 15,000 and is also claimed to be gender sensitive. One of the advantages of step toilets is that they can be repaired by the local people if damaged.

Ring Slab Latrine

Multiple rings are attached to the toilet to increase the base height of the toilet. Ring slab latrine is a cost-effective and women-friendly adaptive technology. Ring slab latrine is widely used as hygienic latrine in Satkhira. It includes digging hole in the ground and putting concrete rings to make the pit, then, placing a slab with integrated ceramic or plastic pan. Depth of the pit is determined considering groundwater table. This technology is hygienic and affordable. This latrine has the characteristics of confinement of feces away from the environment; sealing of the passage between the squat hole and the pit to effectively block the pathways for flies and other insect vectors thereby breaking the cycle of disease transmission; and venting out of foul gases generated in the pit through a properly positioned vent pipe to keep the latrine odor free and encourage continual use of the hygienic latrine.

Floating Composting Toilets

Here the feces of excreta go into the same composting chamber with cleansing material. To adjust the moisture content as well as the C/N ratio, sawdust is used. Here organic household waste can also be added. Composting toilets are used to minimize water consumption and to recycle nutrients available in the feces and urine. The chambers built as vaults are situated above the ground.

Eco-San or UDDT Toilets

These are individual toilets built above the ground using bricks or hollow rocks. This toilet has two separate chambers: one for urine and other for the feces. The floor of the chambers is also sealed with concrete to prevent soil or water coming into contact with the feces. ECO-San toilets help to conserve water and electricity. Setting up ECO-San toilets is extremely affordable as it ranges from 10,000 to 12,000 BDT.

Offset Seepage Pit

The main purpose of this toilet is to increase the user time of the storage chamber. To satisfy water and air distribution, such a bed has four 10-cm-length (4 in) perforated pipes. Durable materials are used for the seepage and storage chambers making sure that they do not collapse during floods or cyclone. Daily operation is minimal. The system operates itself. The dimension of the storage chamber is sufficient for emptying once per year: (annual accumulation of 800 L/year sludge with a family of five). The soak pit will not fill, provided the percolation rate of the soil is more than 15 mm/h with a production of 26 L per household per day. If the percolation rate is less than a soak pit system, it is not feasible. The system requires simple and safe operation routines. Once the storage pit is filled, it needs to be emptied with a desludging device. The sludge will have to be transported and processed at a sludge disposal site.

Women-Friendly Adaptation Technologies in Agriculture, Water and Livelihoods

From the literature review, it has been revealed that women in rural Bangladesh usually play a significant and vital role in all components of agriculture. The nature and extent of women's involvement in agriculture varies greatly from one region to another. However, majority of the subsistence agricultural activities are done by women in addition to their other routine domestic works. These make women doubly or triply burdened, which is related to the long-existing sociocultural gender division of labor between men and women. Selecting adaptation technology related to effects of climate change, therefore, should consider all aspects of gender division of labor. Study reveals that sustainable food production system depends on implementation of resilient agricultural practices, which increase productivity and production, that help maintain ecosystems, that strengthen capacity for adaptation to climate change (Nasreen, 2016). In agro-based economy, rural women perform numerous labor-intensive works such as weeding, hoeing, grass cutting, picking, cotton stick collection, separation of seeds from fiber, keeping of livestock and its other associated activities like milking, milk processing, preparation of ghee and so on.

Researcher pointed out that women's livelihood often termed as resilient due to their home-based year-round activities (Nasreen, 2015). It is evident in the past research that women in Haor region (Sunamganj) adapt with changing climate through practicing traditional and contemporary livelihood activities. Turmeric cultivation was mentioned as profitable followed by vermin compost and vegetable production. Women are involved in almost all the profitable IGAs, especially in vegetable production, vermin compost and poultry rearing. Women in the flood-prone char areas (Gaibandha) are engaged in different IGAs such as homestead agriculture (vegetable cultivation), vermin compost culture and a collective cultivation by women in a group, which they name "fish-chicken-vegetable" production. They are also involved in sheep rearing, milking cow rearing, fish capturing, pigeon rearing and cultivation of paddy, wheat, corn, mustard, irri and Amon rice. In the coastal belt (Dacope, Khulna) shrimp cultivation and agriculture were the occupation for majority of the population since 1990s. However, after the cyclones of Sidr (2007) and Aila (2009), they had to adapt with different kinds of alternative livelihoods as the former two sectors were devastated by disasters, especially due to salinity intrusion, which has been intensified after these two disasters. Women are now involved in homestead agriculture, crab cultivation, poultry raising, tailoring, sheep rearing, petty business, producing salinity-tolerant rice, rainwater harvesting, vermin compost and so on.

Details of activities taken up by women in agriculture and its allied activities are as follows.

Agriculture

Mainly rural women are engaged in agricultural activities in three different ways depending on the socioeconomic status of their family and regional factors. They work as follows:

- Paid laborers.
- Cultivator doing labor on their own land.
- Managers of certain aspects of agricultural production by way of labor supervision and the participation in postharvest operations.

The types of agricultural activities taken up by women include the following:

- Sowing
- Nursery management
- Transplanting
- Weeding
- Irrigation
- Fertilizer application
- Plant protection
- Watering
- Harvesting, winnowing, storing etc.
- Fish cultivation in fellow pond and small-scale water body; cage cultivation
- Bamboo- and cane-made fishing instrument

Livestock

Livestock is the primary livelihood activity used to meet household food needs as well as supplement farm incomes. Rural women earn extra income from the sale of milk and animals. Mostly women are engaged in cattle management activities such as the following:

- Cleaning of animal and sheds
- Watering of cattle
- Milking the animals
- Fodder collection
- Preparing dung cakes or jute stick with cow dung
- Collection farmyard manure

Except grazing, all other livestock management activities are predominantly performed by women. Men, however, share the responsibility of taking care of sick animals. It is evident that the women are playing a dominant role in the livestock production and management activities.

Poultry

Poultry farming is one of the major sources of rural economy. The rate of women participation in poultry farming at household level is central in poultry industry. Duck rearing is much more popular in Haor area and women are mainly involved to take care of duckling/baby duck at the early stage of a duck farm.

Women's Access to Information

Access to information called right to information and freedom of information. If rural marginalized women have advance information of disaster, they can protect their households and crops. This software technology can save people's life and livelihoods. Information can be disseminated through different media, suppose SMS in mobile phone, EWS, miking etc. If the rural women have the access to information, they can take shelter before the disaster and can survive their cattle. They can take measures regarding the disaster management. Women's access to information allows the general public to seek and receive information held by governments (and often by private entities) that perform public duties or receive public funds. It is an important right that serves both governments and their citizens. For governments, an access to information regime helps increase citizen confidence as decision-making becomes more transparent, assists public administration to become more efficient and effective as recordkeeping systems are organized and procedures are established, allows scarce resources to be properly applied and utilized, and can serve to increase foreign investment. Access to information also enables citizens to engage more meaningfully in public life, understand policies and help determine public priorities and use the information to ensure the exercise of other human rights, including the rights to clean water, a safe environment, education and health services. It is a tool that provides the power to ensure that social services reach the most disadvantaged and marginalized people, support true social accountability and promote political and economic empowerment and the protection of individual rights.

EXPERT'S OPINION: KHANDAKER MAINUDDIN, SENIOR FELLOW OF BCAS

To know about adaptive water technology of Satkhira district, Khandaker Mainuddin said, people of coastal and wetland areas in Bangladesh are very accustomed to live with various constraints. He mentions that, in wetland area people mainly depend on deep and shallow tube wells for drinking purpose. They also use river water for domestic purposes. Re-excavation of river and canal could help to improve storage capacity of the surface water. This water could be used in dry season for irrigation and also for domestic purpose. He said, local supply system through pipeline could be a woman-friendly water technology. Rainwater harvesting and elevated tube well for the flooded area are also climate-resilient water technology. He also mentions about defecation practices, which have been categorized as open defecation, hanging latrines, offset pit latrine and some sanitary latrine. For coastal area, pipeline water supply, RWHS, PSF and desalinization technology could be adaptive water technology.

Conclusion and Recommendations

Adaptation is evolving with blending of modern and local knowledge, skill and technological innovation. Effective adaptation must take into consideration of the needs and priorities of the vulnerable groups as well as the perspectives of technology groups, climate scientists, policymakers, local government and NGOs who are involved with community and actors for development and transfer of adaptation knowledge and technologies. The CCAT has three dimensions: hardware, software and orgware. Successful development and demonstration of adaptation technologies must combine these three. For the poor and women, capacity building, skills development for use of CCAT and access to technologies and institutional services from government and the NGOs are the key enabling factors for enhancing gender-responsive adaptation and DRR that may contribute to resilience and gender equality. Responses to climate change include mitigation and adaptation. Mitigation is the reduction of causes of climate change, and adaptation is the reduction of harmful effects of climate change. Both mitigation and adaptation are required to meet the challenges posed by climate change. Both strategies also promote sustainable development. This review has focused on the adaptation technologies in agricultural and water sectors. Expert views and findings from stakeholder consultations have been reflected in the report. Table 11.3 shows the most promising CCAT of agriculture, water and livelihoods.

The current technology matrix and arrangements differ across the regions, sectors and communities. The successful application of the CCAT largely depends on social conditions (level of poverty, awareness, skill and capacity of use of the required technologies and motivation of the poor and women), availability and affordability of the poor and women. The government, NGOs, private sector and market systems and the actors should support the poor, women and socially disadvantaged communities for use of appropriate technologies in agriculture, water, sanitation, health and livelihoods, which are climate-resilient, low cost (considering the affordability of the poor) and available in the locality and women friendly. We believe that CCAT must reduce climate risk, build resilience in the social, human systems and ecosystems as well as lessen burden of the poor women in the long run and thus contribute to gender equality. Hence, the key challenges of pro-poor and gender-responsive technology generation, dissemination and transfer are as follows:

TABLE 11.3 Technology matrix based on literature review and consultations

Regions/CCAT by sector	*Satkhira*
Agriculture	Salinity-tolerant paddy and vegetable growing: BRRI dhan-40, 41, 47 and for T. Aman – BR-22 and BR-23, Bina Shail Potato – BARI A10 and 22, maize and sunflower; crab fattening led by poor women
Water and Sanitation	Rain water harvesting and pond sand filter (PSF) for drinking water; piped water supply Step toilet – a form of sanitary toilet initiated by *Uttaran*, constructed in a way to prevent flood water from entering the pit
Livelihoods	Integrated fish farming and vegetable growing; small trade and business led by women; goat and cattle rearing

- Identification and development of climate-resilient and women-friendly technologies.
- Innovation by blending of local knowledge and indigenous knowledge with external technological knowledge.
- Capacity building and enabling environment.
- Collaboration, engagement of actors and supports to the vulnerable communities.

Further a combination of hardware, software and orgware is needed for development and successful application of new and effective adaptation technologies. The stand-alone technology such as an infrastructure or an equipment is not sufficient for effective adaptation in water and agriculture. The project team feels that the hardwares are not in the central place to address the vulnerability of the groups and communities, rather access to locally available technologies and knowledge, capacity and skill for use of the new CCAT and an enabling institution are the key elements in advancing adaptation in the communities.

It is also viewed that the adaptation capacity largely depends on social and human capitals such as education, awareness, skill, motivation for adaptation, income, access to resources, technologies and services as well as governance, accountability and responsiveness of the local government and duty bearers and sectoral actors. The findings of present research stress on the effectiveness of the use of adaptation technologies in coastal region that are pro-poor, climate-resilient and women-friendly. These technologies are to be developed and demonstrated for both climate risk reduction and economic empowerment. The key focuses of the present research were to take the challenges of the long-term and gender-responsive adaptation technology development and transfer through the following:

- Identification and development of climate-resilient technologies.
- Innovation by blending of local knowledge (indigenous knowledge) with external technological knowledge.

- Providing right information about CCAT, EWS and services to women and poor.
- Capacity building of the poor, women and enabling environment.
- Collaboration and engagement of actors for greater supports to the poor women and most vulnerable communities.

Notes

1 www.preventionweb.net/go/4032
2 www.unb.com.bd/14052020
3 Read more about Satkhira at Annex A.
4 Abedin, M, Awual, Shaw, Rajib. Climate Change Actions in Bangladesh, 2010.
5 www.yourdictionary.com/technology
6 www.adaptation-undp.org/climate-resilient-technologies
7 http://unfccc.int/files/adaptation/methodology_for/vulnerability_and_adaptation
8 Soil Resource Development Institute, Principal scientific officer, Soil Resource Development Institute, Regional office, Khulna, Bangladesh.
9 Drinking Water Assessment Report: Climate-resilient drinking water solutions for Khulna and Satkhira, Prepared for Government of Bangladesh and United Nations Development Program.

References

Abedin, M., Collins, A. E., Habiba, U., & Shaw, R. (2019). Climate change, water scarcity, and health adaptation in southwestern coastal Bangladesh. *International Journal of Disaster Risk Science*, *10*(1), 28–42.

Abedin, M., & Shaw, R. (2013). Agriculture adaptation in coastal zone of Bangladesh. In *Climate change adaptation actions in Bangladesh* (pp. 207–225). Springer, Tokyo.

Alina, S., & Kamruzzaman, M. (2017). Adaptation Technologies in Bangladesh, Report-1. ICCCAD and CCDB, Dhaka.

Asian Development Bank (ADB). (2011). *Adapting to climate change: strengthening the climate resilience of water sector infrastructure in Khulna, Bangladesh*. #Asian Development Bank.

Christiansen, L., Olhoff, A., & Trærup, S. (2011). *Technologies for Adaptation: Perspectives and Practical Experiences*. UNEP Risø Centre, Roskilde, Denmark.

Climate Change Cell. (2009). Characterizing long-term changes of Bangladesh climate in context of agriculture and irrigation. *Department of Environment, Dhaka*.

Didar-Ul Islam, S. M., Bhuiyan, M. A., & Ramanathan, A. L. (2015). Climate change impacts and vulnerability assessment in coastal region of Bangladesh: a case study on Shyamnagar Upazila of Satkhira District. *Journal of Climate Change*, *1*(1–2), 37–45.

Eisenack, K., Moser, S. C., Hoffmann, E., Klein, R. J. T., Oberlack, C., Pechan, A., Rotter, M., & Termeer, C. J. A. M. (2014). Explaining and overcoming barriers to climate change adaptation. *Nature Climate Change*, *4*(10), 867–872. https://doi.org/10.1038/nclimate2350

GOB. (2009). *Bangladesh Climate Change Strategy and Action Plan (BCCSAP) 2009*. Ministry of Environment and Forest• National Designated Authority to GCF. Available at: http://nda.erd.gov.bd/en/c/publication/bangladesh-climate-change-strategy-action-plan-bccsap-2009.

GOB. (2021). *Draft Report (Revised-1) On Training Needs and Capacity Assessment (TN&CA) of Bangladesh Planning Commission (BPC) Strengthening the Capacity of Development Planning Superstructure for Achieving the Developed Country Status*. MoEF http://plancomm.portal.gov.bd/sites/default/files/files/plancomm.portal.gov.bd/notices/0081f067_795f_413c_8fcf_3f1ae97942be/2021-08-25-10-18-96406cc8c60abdc7b5e5e6f38a720d05.pdf

IPCC. (2007). *Climate Change – Impacts, Adaptation and Vulnerability*. WMO and UNEP.

IPCC. (2014). *Climate Change 2014–Impacts, Adaptation and Vulnerability*. Working Group II contribution to the Fifth Assessment Report of the Intergovernmental Panel on Climate Change, Cambridge University Press.

Islam, A. R. M., Shen, S., Yang, S., Hu, Z., & Chu, R. (2019). Assessing recent impacts of climate change on design water requirement of Boro rice season in Bangladesh. *Theoretical and Applied Climatology, 138*(1), 97–113.

Kabir, A. E., Chakraborty, T. K., & Ghosh, G. C. (2016). Bio-Sand Filter (BSF): A simple water treatment device for safe drinking water supply and to promote health in Hazard prone hard-to-reach coastal areas of Bangladesh. *American Journal of Environmental Protection, 5*(5), 107–112.

Klein, R. J. (2011). Adaptation to climate change. In *Climate* (pp. 157–168). Springer, Dordrecht.

MoEF. (2009). *Bangladesh Climate Change Strategy and Action Plan (BCCSAP)*. GoB, Dhaka.

Mainuddin, K., Rahman, A., Islam, N., & Quasem, S. (2011). Planning and costing agriculture's adaptation to climate change in the salinity-prone cropping system of Bangladesh.

Nasreen, M. (2015). *Effectiveness of Resilient Livelihood Framework*. Christian Aid.

Nasreen, M. (2016). *Bangladesh National Conservation Strategy: Gender Issues*. MoEF, GoB-IUCN.

Nasreen, M. (2018). *Integration of Science, Technology and Indigenous Knowledge into Community-based Disaster Risk Management in Bangladesh*. Research Report, International Center for Collaborative Research on DisasterRisk Reduction (ICCR-DRR), Beijing Normal University, China, February.

Nasreen, M. (2020). 'Gender and WASH in Emergency: What happens when a Super Cyclone hits at the Height of a Pandemic?', 7 September, 2020. REACH, Oxford University https://reachwater.org.uk/gender-and-wash-in-emergency-what-happens-when-a-super-cyclone-hits-at-the-height-of-a-pandemic/

National Coalition on Integrated Pest Management (1994*). Toward a goal of 75% cropland under IPM by 2000*. Austin, Texas

Rahman, M. W. (2011). Policies and performances of agricultural/rural credit in Bangladesh: What is the influence on agricultural production? *African Journal of Agricultural Research, 6*(31), 6440–6452.

Rahman, M. A., Chikushi, J., Yoshida, S., & Karim, A. J. M. S. (2009). Growth and yield components of wheat genotypes exposed to high temperature stress under control environment. *Bangladesh Journal of Agricultural Research, 34*(3), 360–372.

Rahman, M. S. & Haque, Z. (2013). Adoption of selected wheat production technologies in two northern districts of Bangladesh. *International Journal of Agricultural Research, Innovation and Technology, 3*(1), 5–11.

Rahman, M. & Islam, A. (2013). Adaptation technologies in practices and future potentials in Bangladesh. In: R. Shaw (ed.), *Climate Change Adaption Actions in Bangladesh*. Springer, Tokyo.

Rashid, M. H., Shirazy, B. J., Ibrahim, M., & Shahidullah, S. M. (2017). Cropping systems and their diversity in Khulna Region. *Bangladesh Rice Journal, 21*(2), 203–215.

Ruane, A. C., Major, D. C., Winston, H. Y., Alam, M., Hussain, S. G., Khan, A. S., ... & Rosenzweig, C. (2013). Multi-factor impact analysis of agricultural production in Bangladesh with climate change. *Global Environmental Change, 23*(1), 338–350.

Saha, D., Hossain, M., Shafayet, S., Mondal, M. S., & Rahman, R. (2016). Agricultural adaptation practices in coastal Bangladesh: Response to climate change impacts. *Journal of Modern Science and Technology*, 4(1), 63–74.

Shaw, R., Mallick, F., & Islam, A. (Eds.). (2013). *Climate Change Adaptation Actions in Bangladesh*. New York: Springer.

Sutradhar, L. C., Bala, S. K., Islam, A. K. M. S., Hasan, M. A., Paul, S., Rhaman, M. M., ... & Billah, M. (2015). A review of good adaptation practices on climate change in Bangladesh. In 5th International Conference on Water & Flood Management, Dhaka (pp. 6–8).

UNEP. (2014). Technology Needs Assessment (TNA) & Knowledge Gap Analysis report, Technology Executive Committee (TEC).

UNFCCC. (2006). *Technologies for Adaptation to Climate Change*. Bonn, Germany.

United Nations (UN). (2015). Paris Agreement, Online English Version.

Water Aid. (2019). *Climate change adaptation and resilience and water, sanitation and hygiene: links between SDG 13 and SDG 6*.

World Bank. (2019). Bangladesh Climate-Smart Agriculture Investment Plan: Investment Opportunities in the Agriculture Sector's Transition to a Climate Resilient Growth Path.

PART IV

Disaster Risk, Human Capital and Preparedness

12

ROLE OF COMMUNITY CLINIC IN REDUCING CLIMATE CHANGE AND DISASTER-INDUCED HEALTH RISKS

An Explorative Study in Coastal Bangladesh

Mohammed Moniruzzaman Khan, Abu-Hena Mostofa Kamal and Nurunnahar Mazumder

Introduction

Currently, the world is experiencing systemic disruptions to the climate and rapid climate change and hence there is a risk to health, well-being, and social functioning (Abedin et al., 2019; Kabir et al., 2016). Climate change is a complex phenomenon and its impact on human health is obvious (Hawlader, 2021). The United Nations Environment Programme reports that 25% of the world's total curable diseases directly occur because of poor environmental quality (Clark, 2007; Manisalidis et al., 2020). Furthermore, it leads to a modified distribution of pathogens and related vectors and malnutrition in people (Ghazali & Guericolas, 2018). In 2018, the World Health Organization (WHO) estimated that climate change annually causes >60,000 deaths, with predictions of >250,000 annual deaths between 2030 and 2050 (WHO, 2018). The literature shows that climate migrants are most often exposed to zoonoses because of their poor living conditions; they have an increased risk of pathologies such as tuberculosis, human immunodeficiency virus, viral hepatitis B and C, viral respiratory infections, vector-borne diseases such as malaria and leishmaniasis, food- and water-borne diseases such as cholera, and skin diseases (Crowley, 2017; Frumkin, 2019; Kabir et al., 2014). Furthermore, climate change and natural disasters have certain short-term and long-term health impacts, including geographic redistribution of communicable diseases, the upsurge of cardiovascular disorders, increase of hypertension-related reproductive health problems for women and adolescents, respiratory illness, and a host of other health hazards (Ahmed, Rehman, & Shahab, 2017; Costello et al., 2009; Crowley, 2017; Ghazali & Guericolas, 2018; Kamal et al., 2018; Luber & Prudent, 2009; Rahman, 2008; WHO, 2018).

DOI: 10.4324/9781003253495-16

Climate change is extensively considered as an important contributing factor for natural disasters for the geographic redistribution of health hazards such as infectious diseases and chronic diseases (Crowley, 2017; Ghazali & Guericolas, 2018). Climate change is one of the important reasons behind the increase of the typical temperature of the troposphere and oceans with impact on the frequency and intensity of different types of extreme weather conditions and natural disasters (Luber & Prudent, 2009). Climate change has a direct impact on the quality of air, water, and food and maybe the gravest threat to the healthcare system during crises (Costello et al., 2009; Crowley, 2017; Ebi et al., 1998; Ghazali & Guericolas, 2018; Luber & Prudent, 2009).

Because of its embedded position in communities, close-to-community (CTC) service providers have been identified as an important cadre to improve universal health coverage and deal with inequalities in the provision of healthcare service (Steege et al., 2018). These service providers are often the first point of contact to the health system and perform promotional, preventive, and curative health services for community members. They are either based in the community or any basic healthcare facility (Haines et al., 2007; Steege et al., 2018). Thus, they are considered to be an important group in health service providers as they have direct contact with communities. The CTC service providers can expand access to services and contribute by improving health outcomes (Hanifi et al., 2020; Khoorshid et al., 2020). They are considered to be 'natural researchers' because of their embedded position in community realities and their ability to relay the practical scenario to outsiders such as policymakers (Ahmed, Rehman, & Shahab, 2017; Pérez & Martinez, 2008;Steege et al., 2018).

The Constitution of Bangladesh is committed to ensure basic health services to its citizens (Visagie et al., 2015). As a signatory state of the Alma-Ata Declaration, the Government of Bangladesh established community clinics (CCs) (one CC for a population of ~6,000). The important aim of establishing a CC was to extend healthcare services to the doorsteps of the people in the community (Khoorshid et al., 2020). CCs were established throughout the country to provide healthcare services for hard-to-reach, remote and isolated communities.

Bangladesh is extremely vulnerable to several climate change–induced disasters, including cyclones, river bank erosion, storm surges, and waterlogging (Banu, 2020). Moreover, because of its geophysical and socio-political characteristics, the coastal areas of Bangladesh are unlike compared to other parts of Bangladesh; embedded characteristics often limit its inhabitants to access endowed resources and mitigate climatic risks and vulnerabilities (Shamsuddoha & Chowdhury, 2007). Climate change is increasing the risk of hydro-metrological disasters in the coastal regions of Bangladesh, which leads to considerable amount of environmental and economic losses such as loss of lives (Shamsuddoha & Chowdhury, 2007). The literature shows that geographically isolated locations, difficulties of communication because of lack of transport, and poor mobile networks make it more difficult to deliver healthcare services in the coastal regions of Bangladesh (Ahmed, 2019; Rawal et al., 2019). The literature shows that it becomes difficult for healthcare providers to deliver services to people during natural disasters and other emergencies (Kamal et al., 2018). Moreover, in the literature, the relationship between natural disasters and health lacks clarity (Kabir et al., 2016; Khoorshid et al., 2020). Furthermore, to date, there is limited information about the socio-economic and cultural context of the epidemiology of climate change, as well as disaster risks in coastal regions. To date, there is no information about the role of CC in mitigating climate-induced health burdens, although, as the first line of care providers, they can play an important role to reduce the disease burden and death toll caused by natural disasters and climate change. Many studies on climate change, natural disasters, and public health have been distinctively performed. However, no comprehensive study has been conducted to examine the role of CCs in reducing climate- and disaster-related health burdens in the coastal regions of Bangladesh. The nexus among climate change, natural disaster, and public health has not been sufficiently addressed in the existing literature (Kabir et al., 2014). Therefore, this study aimed to identify the common diseases in the coastal area of Bangladesh. Moreover, this paper aimed to explore the role of CCs in mitigating health-related vulnerabilities in coastal region. Finally, this paper aimed to examine socio-cultural narratives and attitudes of coastal communities toward the services of CCs. Thus, the results of this article would generate knowledge for NGOs, researchers, the Government of Bangladesh, and countries with similar settings in terms of the role of close-to-door healthcare facilities in reducing climate-induced health hazards.

Methods and Data Sources

Study Site and Study Population

This study deployed a qualitative research method to obtain an in-depth understanding of the perception, attitude, and behavior of healthcare-seeking behavior of coastal region people. The qualitative method examined the role of CCs in reducing climatic health hazards in the study area from the healthcare providers' point of view. Study has been conducted in four coastal unions, namely, Pankhali and Tiladanga Union of Dacope Sub-District from Khulna District, and Gabura and Nurnagar of Shamnagor from Sathkhira District. The important reasons behind purposely selecting study areas are attributed to their locations and close proximity to the coastal area. Secondly, these areas are identified as the worst victim to climate change and natural disasters (Kabir et al., 2016; Parvin et al., 2015). Moreover, one CC from each union has been purposely selected. A total of 13 key informant interviews (KIIs) were conducted with the healthcare providers, including Union Health Inspector (UHI), Community Health Care Provider (CHCP), Health Assistant (HA) and Family Welfare Assistant (FWA), to know about the type of services provided in CCs, key challenges, strengths of CCs in dealing with diseases caused by climate change, and ways to improve services of CCs to treat emergency patients during a disaster. Furthermore, 20 exit interviews were conducted with the healthcare-seeking people who lived close to the CCs (five patients from each CC).

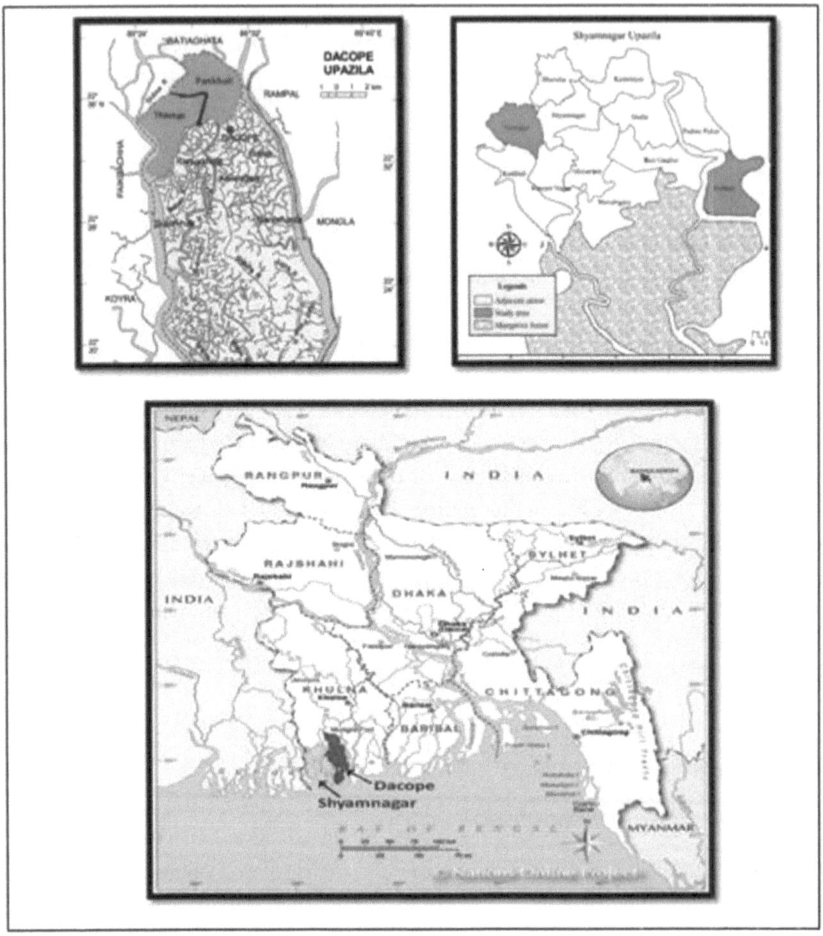

FIGURE 12.1 Map of the study area

The patients were asked about the level of satisfaction with the services received, frequency of visiting CCs, and perception about climate change, natural disasters, and their impact on public health. In addition to interviews, two years of patients register data has been collected (from January 1, 2017 to December 31, 2018) to obtain information about the prevalence of diseases in the study area, seasonality of diseases, and age and gender distribution of patients. Data obtained from the patients' register revealed the most common diseases in the study area and socio-demographic features of patients who visited CCs (Figure 12.1).

Data Collection

The guidelines/checklist for KII and exit interviews were pre-tested with non-participant members of coastal areas, including community healthcare providers and healthcare-seeking community people. Based on pre-testing, these guidelines were revised. The field team collected data using revised guidelines between September 1, 2019 and September 25, 2019.

Data Analysis

The KIIs and exit interviews were audio-recorded with the consent of respondents. Moreover, field notes were obtained during interviews and field notes were reviewed and the jottings of the data expanded at the end of each day of data collection. For any new information in the interview that was not asked in the guideline, the field team incorporated new questions in the guideline for additional exploration. Then, the audio-recorded interviews were transcribed verbatim; the transcriptions and expanded notes were then typed in Microsoft Excel. Then, the typed data were reviewed by the principal investigator and a coding system was developed to identify research themes and concepts. Comparison and triangulation in themes were performed using multiple data collection techniques. In this research, the five stages of the framework approach, that is, familiarization, identifying a thematic framework, indexing, charting, mapping and interpretation, were used to analyze and interpret the results of this study.

Ethical Consideration

The protocol and ethical issues of this study were reviewed and approved by the Research Cell and Review Board of Jagannath University, Dhaka, Bangladesh. Moreover, informed oral consent was obtained from the participants before interviewing them.

Results and Analysis

Demographic Information and Distribution of Diseases

A review of patients' registers shows that whenever a patient visits a CC for treatment, they keep a record of patients' details, including age, gender, description of diseases, and date of visit. The record of the CC in the register shows that a total of 42,905 patients visited four CCs in two years where most of the patients (67.1%) were female; this number is two to three times higher compared to that of their male counterparts. However, >50% of patients (52.5%) were aged between 25 and 49 years and 33.5% of total patients who visited the study CCs between January 1, 2017 and December 31, 2018 were aged 50 years and older. The diseases for service seekers who visited CCs were abdominal pain (14%), physical weaknesses (13.40%), gastric issues (8.72%), and respiratory problems (7.50%) (Table 12.1).

This study shows seasonality and number of patients who visited the CCs. Most patients (39.5%) visited the CCs between November and February. However, the lowest number of patients (25.5%) visited the CCs between March and June

TABLE 12.1 Distribution of disease

Name of the disease	*Sexual identity of the patients*	
	Male (%)	*Female (%)*
Allergy	32.0	68.0
Diarrhea	35.7	64.3
Gastric	32.7	67.3
Dysentery	35.0	65.0
Conjunctivitis	39.4	60.6
Fever	35.2	64.8
Itching/scabies (*daad/chulkani/khos pachra*)	36.7	63.3
Neurological problems (e.g., knee pain, back pain, joint pain and so on)	35.9	64.1
Respiratory problem (e.g., asthma, breathing difficulties and so on)	34.5	65.5
Weakness	27.9	72.1
Urinary tract infection	40.0	60.0
Leucorrhea	0.0	100
Tooth pain	34.5	65.4
Ear pain	27.0	73.0
Anemia	24.6	75.4
Abdominal pain	34.2	65.8
Lactating mother's complexity	0.0	100
Menstrual pain	0.0	100
Tonsil	36.0	64.0
Amnesia	43.4	56.6
Hypertension	0.0	100
Worms	39.0	61.0
Total	**32.9**	**67.1**
Disease information missing	**23.61**	

Source: Community clinic patient register January 1, 2017–December 31, 2018.

(Table 12.2). Furthermore, this study explored how seasonality is related to the increase in the number of patients at CCs in coastal sub-districts.

The results show that the average age of healthcare providers and seekers of this study were 35 and 45 years, respectively; most healthcare providers (8/13) were female. The educational qualification of all caregivers was higher secondary and above, while the healthcare seekers (13/20) were mostly illiterate or housewives (8/20). The results of this study show that the median of the monthly family income of care seekers was BDT 4,000; however, the average distance between the residence of healthcare seekers and CC was 1.97 km (Table 12.3).

Services Available at Community Clinics

This study reports that a union is divided into different clusters and in each cluster there is at least one CC to provide healthcare to the community people. The

TABLE 12.2 Seasonality of disease

Season	*N*	*%*
March–June	10,910	25.5
July–October	15,111	35.0
November–February	16,883	39.5
Total	**42,905**	**100**

Source: Community clinic patient register January 1, 2017–December 31, 2018.

TABLE 12.3 Demographic information of the healthcare provider and healthcare seeker

Demographic information	*Healthcare provider*	*Healthcare seeker*
Age (median) in year	35.00	45.00
Sex		
Male	5	7
Female	8	13
Education		
Illiterate	0	13
Primary	0	2
Secondary	0	3
HSC and above	13	2
Occupation		
Healthcare providers	Healthcare providers (CHCP=4, HA=2, FWA=4, AHI=3)	
Farmer		6
Housewife		8
Day labor		6
Monthly family income of the care seekers (in BDT)		
Mean (range: BDT 2,500– 30,000)		4000
Average distance between CC and residence of the service seekers (in km) (range: km 0.5–5)		1.97

catchment area of every CC is divided into three areas for efficient administrative and better service delivery. Participants of this study reported that CC is locally known as 'clinic', 'free hospital', and 'hospital'. Though there was the provision of one CC for every 6,000 people, healthcare providers stated that they served two to three times higher number of populations in their catchment area.

CCs in Bangladesh initially started with the Extended Programme on Immunization, primary healthcare services and treatment for tuberculosis, diphtheria, whooping cough treatment, and family planning services. The healthcare providers report that primary healthcare services at CCs include treatment for fever, cough, cold, allergy, itching/scabies, headache, stomachache, gastric issues,

FIGURE 12.2 List of medicines available at community clinics

diarrhea, dysentery, seasonal flues, asthma, weakness, and injury and/or wounds. One of the healthcare providers reported that, compared to the initial stage, certain additional healthcare services, such as treatment for pneumonia, hepatitis-B, measles, and rubella, are now available at CCs. All healthcare providers reported that advice in terms of family planning methods for both males and females, supply of contraceptives, primary health care services, first aid, health checkup for pregnant mothers and newborn babies, and worm tablet distribution program is available at CCs. The participants of KII reported that they were responsible to identify poor, helpless, marginal, and landless pregnant women and issued health cards such that they could obtain birth delivery services for free from the Upazila Health Complex (UHC). Healthcare service providers reported that they provided healthcare services for children and adolescents, which were not available at the beginning. While participants reported that CCs provide health education for mothers, they suggested pregnant women and mothers of newborn babies should have nutritious food. All KII participants reported that, in addition to providing treatment for the abovementioned diseases, their responsibilities include referring critical patients to the UHC for improved healthcare. One of the participants described the referral system as follows:

> In case of any serious patients, we are referral to the UHC with a reference slip. The description of the patient is written on a reference slip. Patients having the referral slip avail necessary medicines for free. Furthermore, for

> pregnant women or any other patient with complications, we accompany them to the health complex.

Moreover, healthcare providers reported that they work along with NGOs to raise awareness among community members about disaster preparedness, including tips about ideal health and hygiene practices. The majority of participants reported that they provided water purification tablets and bleaching powder as a disinfectant to spray around households. One of the healthcare providers reported that for any disaster forecast, they used to go to CCs and get ready for an emergency healthcare response. Certain participants reported that there was a team in their locality to deliver emergency healthcare services during a disaster. The members of the team, as per the description of the participants, include an MBBS doctor from the UHC, one CHCP, and one HA from the local CCs. One of the healthcare providers reported the following:

> [We constitute] the emergency medical team, including a MBBS doctor, one CHCP, and one HA to provide healthcare services during a disaster. The team remains on standby till a certain disaster ends. The team leader is the MBBS doctor and we work together during a disaster. We announce through a mike to gather in a place and receive health care service or we visit the affected places, such as cyclone centers, where people shelter. If we identify any critical patient, we refer the patient to the UHC.

All participants of this study reported that there is no such specialized health education program at CC for disaster-prone coastal people, although they trained teenage girls about maintaining hygiene during menstruation. One of the participants reported the following:

> We provide health education for local people. There is another health education program conducted by NGO along with CC where we try to make people aware and provide health tips … we include teenage girls aged between 11 and 19 in health education and make them aware of menstrual hygiene. We suggest them to remain neat and clean.

Prevalence of Diseases

The majority of healthcare providers reported that, on an average, 50–60 patients visit the CCs every day for seeking healthcare. They reported that the number increases during post-disaster periods such as in the post-cyclone and waterlogging conditions. All KII participants reported that fever, cough, cold, allergy, diarrhea, dysentery, jaundice and other water-borne diseases, vector-borne diseases, seasonal flues, food poisoning, and malnutrition are prevalent in the study area. A majority of participants reported that the prevalence of non-communicable diseases such as diabetes, high blood pressure, asthma, cardiovascular disease, stroke, and leukorrhea among women was high in the study areas and incidence of high blood pressure, stroke, and diabetic patients is alarmingly high in the study area. Most respondents reported that the rapid intrusion of salinity in drinking water might have an impact on the increasing incidence of allergy and/or other skin diseases, stroke, and high blood pressure. All participants reported that dysentery was the most prevalent disease in the coastal area. Certain participants reported that pregnant women in the coastal area suffered considerably because of an increased salinity for sleep disturbance and high blood pressure. All healthcare providers reported that by taking bath in dirty water, the females in the study area mostly suffered from uterus-related infections. The healthcare providers report that because of low-calorie intake most women in the study area were anemic and their blood pressure remains low compared to the standard level. Note that, during the summer season, the prevalence of diarrhea was higher; the prevalence of fever is higher during summer and winter in the study area, added by participants. Furthermore, dysentery, gastric issues, joint pain, weakness, and arthritis are common diseases all around the year as per the healthcare providers. One of the healthcare providers reported that out of the total patients, ~30% are diagnosed with dysentery and scabies. Accordingly, the reason behind the prevalence of these two diseases is the increase in salinity in coastal regions.

Table 12.1 reports that the prevalence of diseases among coastal women (67.1%) was more than double compared to their male counterpart (32.9%). All healthcare providers reported that most female patients suffered from physical weakness, anemia, menstrual pain, urinary tract infection, allergic problems, skin diseases, and leukorrhea.

A majority of participants reported that, among the coastal population, salinity intrusion in water and soil was causing the increase in high blood pressure, sleep disturbance, and diabetes. As per their description, climate change contributes to increasing the prevalence of malnutrition, weakness, cold, cough, fever, worms, child pneumonia, dysentery, and other diarrheal diseases. One of the healthcare providers reported that, out of 50 patients, she received at least 30 every day who were suffering from dysentery. Most participants reported that, during and after disasters, the disturbance of snakes and other poisonous insects massively increase. The bites of such poisonous insects and animals lead to the death of many local people. Furthermore, during cyclones and because of waterlogging, the death of children by drowning is a common phenomenon in the study area. One of the participants reported the following:

> during cyclone Aila many of the snake-bitten patients died by not having proper treatment in hospital [CC].

Certain healthcare providers reported that loss of assets and properties because of riverbank erosion and cyclones make people suffer from mental health problems, including sleep disturbance, anxiety, and hypertension.

Challenges for Healthcare Providers

All healthcare providers reported that one of the major challenges for them is old and hazardous CC buildings. The participants felt unsafe to sit under the roof because it might collapse at any time (Figure 12.3). One of the participants reported that patients found it risky to come within the CC. Certain patients stood outside the building while seeking healthcare. Certain participants reported that, during the rainy season, water percolated from the roof and damaged medicines, documents, and other water-sensitive instruments.

Participants report that they failed to provide treatment by not having an adequate and favorable working atmosphere. One of the care providers reported that a part of the ceiling of the CC building fell over the head of a patient and he was severely injured. Then, the local people became violent and healthcare providers were bound to keep the CC closed for a couple of weeks. All healthcare providers reported that the shortage of medicine was the major challenge for them. For these providers, the supply of medicine was insufficient compared to the demand in the study area. One of the participants reported the following:

> we open a carton of medicine in every 3 to 4 months. We have a shortage of medicine compared to the demand of patients. Medicines should be supplied based on the requirement of a particular geographical area. For instance, we require more medicines for treating dysentery and allergenic patients in the coastal area whereas, the supply of these medicines has been reduced here.

They reported that, in the coastal region, a supply of water purification tablets should be available. For healthcare providers, the adequate supply of water purification tablet can reduce water-borne diseases in coastal areas. Certain participants

FIGURE 12.3 Vulnerable infrastructure of CC

reported that the quantity of vitamin tablets has been increased, whereas the quantity of medicines for dysentery and scabies has been reduced. For the healthcare providers, the reduction of medicines for scabies and dysentery treatment put them in trouble as the prevalence of these diseases is higher in the coastal region. Certain participants reported that local people used to bawl by failure of providing their desired medicines. One of the healthcare providers reported the following:

> We are given medicine stocks for three months; however, if we want to maintain 100% medicine dose for each patient, we will not be able to serve 15 days with the given supplies. We know it has a negative health impact on patients who might be drug-resistant but we are helpless.

Certain participants reported that the local people face difficulties to go to a UHC because they have to cross rivers to get treated there. One of the participants reported that the distance of UHC from their locality was ~20 km. Sometimes, inadequate transport facilities and lack of service providers led to maternal and infant mortality, especially during a disaster. Most healthcare providers reported that it becomes extremely difficult for them to move from one place to another during the monsoon as their catchment areas are located at remote places. A female healthcare provider at CC explained her experiences of providing services in the study area:

> it was an extremely remote area; the pregnant mother was facing a complication in birth-delivery. I accompanied the mother along with her family members at night. On our way to the health complex, we had unusual delay to cross the river. Consequently, at first the baby, and then the mother died.

The majority of healthcare providers reported that one of their major challenges for the field-level workers, such as HA and FWA, is workload and an inadequate number of healthcare providers. She said the following:

> I can provide service to five hundred to six hundred couples –supply them contraceptives and necessary medicines. However, because of the lack of human resources, I have to serve 1400 couples in the large catchment area with 1,400 couples. Moreover, because of the shortage of FWA, I performed additional responsibility at another CC where I have another 1400 couples to provide service. Often, it becomes impossible to cover the entire catchment area for me.

Most participants reported that during and after a disaster, a high number of diarrhea and dysentery patients put them in trouble. They could not provide medicine to patients because of limited supplies. Moreover, one of the healthcare providers reported that because of insufficient supply of medicine, they fail to make people satisfied with their service. He added the following:

> People come from distant places to get service but we cannot help them as we do not have adequate medicines … local people ask us why we are sitting here as we do not have adequate medicines?

Certain healthcare providers reported the lack of sufficient space at CCs for providing emergency treatment during a disaster. All participants, except one, reported that they did not receive any special training about treating patients suffering from diseases caused by climate change and/or providing emergency services during a disaster. One of the participants reported that the healthcare providers at CCs require specific training on 'how to provide adequate services to people in the coastal and disaster-prone areas'. One participant has also received a day-long training on emergency disaster management approximately five years ago. However, she has mentioned that it was not adequate to provide quality services during an emergency.

Perception of Care Seekers about CC Services

The study reported that healthcare seekers visit a CC for common diseases; however, for any complication of patients, they preferred visiting a sub-district or district-level hospitals. It has been reported by some people that they receive effective health-related counseling from CCs. In response to the question of how they receive, one of the care seekers reported that the CC treatment helped them to solve health problems such as gastric issues, diarrhea, weakness, dysentery, and allergy. Most patients reported that local people visit CCs four to five times in every month for many types of health-related problems. All participants reported that they visit CC for any health problems and/or if they require any health tips. In an exit interview, a patient noticed that a woman walked for ~90 m to reach the CC to take advice about the first trimester of her pregnancy. For her, at a private clinic, the costs of healthcare services were beyond her affordability and thus she visited a CC to seek advice.

The results of exit interviews demonstrate that a majority of healthcare seekers came to seek treatment for dysentery, gastric issues, scabies, fever, and cold/cough. A number of patients came to seek treatment for physical weakness, high blood pressure, diabetics, breathing difficulties, headache, allergy, gastric issues, worm infestation, frequent urination, chest pain, and maternity-related services, including trimester-based checkups. A great majority of care seekers who participated in this study reported that before having CC in their locality, the people of coastal community had experience in difficulties, including covering long distances for any sort of medical service and expensive treatment. They reported that nowadays community people can receive healthcare services at their doorsteps. The medicines provided by CCs helped them to come round as described by care seekers. One patient reported the following:

> in the past, we had nothing to do except suffer from diseases; we do not have enough money to go to UHC or private clinics. However, because of

> the establishment of CC, any sort of health problem, people can come and get treatment nowadays. For example, the normal cold/cough of children could turn in pneumonia in the past for not getting access to the healthcare facility.

A care seeker reported that the healthcare providers of CC are extremely caring regarding patient's problems. A great majority of care seekers suggested that having an adequate supply of medicine would benefit community people more. Accordingly, a care seeker describes the following:

> I came here to seek treatment for scabies and headache, but received a Napa tablet for headache only; ointment for scabies is currently not available at this CC. Sometimes, health care providers provide a Napa tablet (fever reliever) for all sorts of diseases by not having proper medicine.

In addition to good practices there are certain allegations, including nepotism and favoritism, as reported by a healthcare provider that quality medicines are not distributed among poor people; healthcare providers distribute those medicines among their relatives and elite people in the community. Certain care seekers reported the necessity of specialist doctors and equipment for pathological tests at CCs.

Role of CCs in Emergencies

All healthcare seekers reported that they preferred CCs for most common diseases because it was close to their residence. CCs were located in the walking distance of the community from their residence and did not require transportation. They preferred CCs because counseling and medicine were free. The majority of patients reported that it does not require standing in a queue and waiting a long time for receiving treatment at a CC. For them, it requires at least two days if they want to get treatment from district and even sub-district hospitals; moreover, in a CC, a few minutes are required to get treatment. One of the patients reported the following:

> in case of any physical problem people come to CC at first and the primary reason behind this is poor socio-economic condition of coastal people. Treatment here [at CC] does not require any money that mostly motivates local people.

One of the patients reported that healthcare providers were not responsible for the inadequate supply of medicine. For him, government supply is not adequate as per the demand of the people in the community. A participant reported that people of their community have a positive impression about CC; this is why people come from distant places to receive healthcare services. A healthcare seeker reported that, in the past, health support for the people in the community was ignored and they faced negligence in getting treatment. However, the health education program performed by CC has made people aware of the necessity of having healthcare.

As per majority of healthcare seekers, the prevalence of anemia and malnutrition are high in the coastal region because of poverty. The participants reported that the prevalence of fever and cold, cough etc. increased among people because of climate change and extreme weather. The important reasons behind the most common diseases in the coastal region are the lack of pure drinking water and rapid intrusion of salinity in water. One of the participants reported the presence of iron in tubewell water is another major reason behind the prevalence of dysentery in coastal areas.

All care-seeking people reported that CCs are their first choice for seeking healthcare during a disaster. They reported that most of the time they face difficulties reaching CCs because of the lack of transport and/or waterlogging caused by the disaster. One of the patients reported that oral saline distributed by a CC during a disaster helps local people control diarrheal diseases in the study area. One participant reported that a CC holds a health education program at their locality once in a month where the trainers make local people aware of health and hygiene during a disaster. Certain participants reported that the healthcare providers of CCs attempt their level best to provide treatment during a disaster if people reach a CC. Furthermore, along with MBBS doctors, the local healthcare providers provided healthcare services at cyclone centers at the study areas during and after a disaster as reported by care seekers. All women healthcare seekers reported that health education on sexual and reproductive health for women and adolescents helps them become aware of gender-sensitive health and hygiene practices during a disaster.

Recommendations to Improve Quality Health Care System

Most care providers and health support seekers recommended for pathological instruments, such as equipments for ultrasonography, X-ray, urine test, and blood test, to be available at a CC. For them, the abovementioned equipment would help people receive better treatment at their doorsteps. All healthcare providers reported that having one boat for each CC at the coastal area would help them visit vulnerable households of the community and provide uninterrupted services during a disaster and other emergencies. Moreover, it would help them to transfer critical and emergency patients to sub-district-level hospitals for better treatment. A great majority of care seekers and providers reported that having at least one MBBS doctor, particularly a gynecologist, would help people in the community to receive better treatment. Some care providers suggested the facility of few emergency beds at a CC would help to hospitalize patients during a disaster and other emergencies. One of the healthcare seekers suggested having a specialized female doctor at a CC for female patients.

Discussion

The establishment of ~14,000 CCs for every 6,000 people across the country has made revolutionary changes in Bangladesh and brought healthcare to the community doorstep, particularly for the hard-to-reach communities (Ahmed, Rehman, & Shahab, 2017; Khoorshid et al., 2020; Pérez & Martinez, 2008). The results show that the most common diseases in the study area are fever, cold/cough, allergy, diarrhea, dysentery, jaundice, and other water-borne diseases, vector-borne diseases, seasonal flues, food poisoning, and malnutrition. This study demonstrates that non-communicable diseases such as diabetes, high blood pressure, asthma, cardiovascular disease, stroke, and leukorrhea among women are high in the study areas. The results suggest that the incidence of high blood pressure, stroke, and diabetes is alarmingly increasing in the study area. Furthermore, the most common diseases during and after natural disasters in the study area are dysentery, diarrhea, scabies, allergic problems, gastric issues, physical injury, and wound. The results of this study reveal that people can access CCs for general health-related problems, including family planning, maternal health, and emergency healthcare during and after natural disasters and similar situations.

The research results show that healthcare providers at CCs have a supply of 29 types of medicines to distribute among care seekers. Both healthcare providers and healthcare seekers reported that CCs have been playing an important role to promote primary healthcare services in the study area. Accordingly, the health education program for adolescents is changing the behavior of the community people. However, this study reports that an inadequate supply of medicine is making it hard for care providers to deliver adequate services to people in the community. Furthermore, healthcare providers recommended providing medicines based on the requirement of coastal areas. They reported that medicines, in terms of both quantity and types, should be supplied based on assessing the requirements of coastal people. The medicines supplied to CCs are based on a general understanding of basic health-related problems of the Bangladeshi people. However, the coastal people suffer more from certain particular types of diseases. Moreover, an inadequate supply of medicine increases the possibility of drug resistance, which can be a serious threat for people in coastal areas (Jamrozik & Selgelid, 2020). The supply of medicine based on the analysis of the patient register and other geographical characteristics might help resolve this limitation.

The results of the study show that most patients who visited CCs were female. Furthermore, most female patients who visited CCs were suffering from gastric issues, allergy, hypertension, amnesia, diarrheal diseases, fever, respiratory tract infection, neurological problems, and joint pain. Certain female patients were diagnosed with urinary tract infection, leukorrhea, and irregular menstruation. The results of the prevalence of diseases in the study area coincide with those of several other recent studies (Hutton et al., 2018; Kabir et al., 2016; Kabir & Khan, 2017). Moreover, this study reported that gender identity (females' health status is poorer compared to their male counterpart), lack of monthly family income, and educational status (participants with less or no formal education suffer more with several diseases) are linked with health-related vulnerabilities of the population in coastal regions. The results of this study are similar to those conducted in disaster-prone and coastal areas of different countries (Azad et al., 2013; Fatema et al., 2019; Ginige et al., 2009). For instance, studies conducted in Pakistan, Indonesia, and China reported the prevalence of diarrheal diseases and skin diseases in coastal regions (Baqir et al., 2012; WHO, 2006). The possible reasons behind the health-related vulnerabilities of people in coastal regions in this study might be their inadequate awareness about healthcare and lack of access to healthcare facilities by not having financial ability to bear the treatment expenses (Joarder et al., 2019). Furthermore, the intrusion of salinity in water, air, and soil can be another important reason behind the prevalence of skin diseases and diarrheal diseases in the study area (Chowdhury et al., 2020).

This study reports that people in the community nowadays receive healthcare service at CCs under one roof and within half-an-hour walking distance from their homes even in remote areas (Hanifi et al., 2020; Khoorshid et al., 2020). Healthcare providers reported that they receive 40 patients on an average each day. Furthermore, healthcare providers counsel local people about sexual and reproductive health, health and hygiene, as well as the negative consequences of child marriage. The results reveal that CCs are not only providing health, family, and nutrition services to the rural people but also have become a 'center for health information and consultation'. The location of clinics in the community and free of cost treatment and medicine provided by healthcare providers build the reliability and trust of coastal people in CCs (Khoorshid et al., 2020; WHO, 2019). Moreover, people might feel comfortable with service providers who are known to them (Khoorshid et al., 2020). Khoorshid et al. (2020) report that CCs have good infrastructure that contradicts the results of this study as participants, both healthcare providers and seekers, mentioned about poor infrastructures of clinics. In fact, this study reported that certain healthcare providers were providing service from the local union council and from their private home. In response to the inquiry of poor infrastructure, this study shows though healthcare providers prefer having additional rooms and beds to hospitalize emergency patients, shabby infrastructure of CCs because of old buildings, salinity moisture weather, frequent disaster in coastal regions damage the buildings quicker than non-coastal areas, obstructing the normal function of CCs. Among other challenges, the healthcare providers report risky infrastructure and shortage of healthcare providers. In this study, it was reported that a majority of the CC building was built around two decades ago. Moreover, the water and air salinity of coastal regions have caused severe damage to these buildings. Therefore, the repair and/or reconstruction of buildings has become a dire requirement for CCs in the coastal communities of Bangladesh. Moreover, providing service to the inhabitants of the catchment area with limited manpower is another important challenge for service providers. Although it is reported that there is one CC for 6,000 people (Khoorshid et al., 2020), the results show that the catchment areas comprise two to three times more people. Therefore, quality healthcare services at CCs require additional human resources as recommended by the results of this study.

Emergency healthcare service during a disaster has regularly been severely interrupted because of the crumpled transport and communication (Hossain, 2020; Kamal et al., 2018; Uddin & Mazur, 2015; Ukai, 2009). The results of this study show that neither healthcare seekers nor providers can reach each other to provide healthcare services. Sometimes, transport and communication barriers cost even life of vulnerable people during a disaster and emergencies that have been reported by the participants of the study. In particular, because of the lack of proper roads and transport in the coastal area, Bangladeshi people depend on river-based communication and transport such as *trawlers* and boats to move from one place to another to have healthcare services, which is time-consuming, considering the emergency health conditions of patients. Therefore, the lack of transport and communication system interrupts the emergency healthcare services in coastal areas. Multi-sectoral emergency communication and facilitation of public–private partnership to provide healthcare services and transport facilities for patients and caregivers might reduce the problem as reported by the results of several studies (Bennett et al., 2018; Kuruvilla, 2018; Waluszewski & Snehota, 2019).

Social research dealing with the dynamics of societal living is confronted with many obstacles. For this study, a longitudinal study was first conducted, which could have been more helpful to gather in-depth information about the important challenges to reduce the disease burden in study areas. However, because of budgetary constraints, this study could not be conducted over an extended period. Second, because this is not a macro study, generalization is not possible; the holistic features of the standard of healthcare services provided by community healthcare providers in Bangladesh might not be reflected in this study because the incidence and prevalence of disease and nature of disaster in other parts of Bangladesh are not similar to coastal islands. Despite these limitations, this study, one of very few, in Bangladesh attempted to gather information about the role of CCs in reducing health burdens in the coastal regions of Bangladesh. The results of this study will help the government and its health partners take initiatives to overcome the limitations and provide adequate support to healthcare providers to play an important role at the community level, particularly during disasters and other humanitarian crises.

Conclusion

CCs have made revolutionary changes in delivering healthcare services for the rural people of Bangladesh; the same is happening in coastal disaster-prone areas. Moreover, CCs have the potentiality to serve the people of the coastal regions, although the lack of resources and inadequate supply chain is making the service delivery difficult for healthcare providers. Moreover, traditional epidemiological surveillance systems are not adapted well to the coastal regions to identify vulnerabilities posed by climate change and natural disasters in the field of public health. Additional efforts are required to improve the identification and assessment of risks because it pertains to the different levels of management in the coastal islands of Bangladesh such as environmental changes, population health impact, or health system status.

References

Abedin, A., Collins, A. E., Habiba, U., &Shaw, R. (2019). Climate change, water scarcity, and health adaptation in southwestern coastal Bangladesh. *International Journal of Disaster Risk Science, 10*(1), 28–42. https://doi.org/10.1007/s13753-018-0211-8

Ahmed, J., Rehman, S. U., &Shahab, M. (2017). Community midwives' acceptability in their communities: A qualitative study from two provinces of Pakistan. *Midwifery, 47*, 53–59. https://pubmed.ncbi.nlm.nih.gov/28242494/

Ahmed, J. U. (2019). Friendship hospitals: Healthcare for the riverine people of Bangladesh. *Journal of Developing Societies, 35*(1), 175–194. https://doi.org/10.1177/0169796X19826763

Azad, A. K., Hossain, K. M., &Nasreen, M. (2013). Flood-induced vulnerabilities and problems encountered by women in northern Bangladesh. *International Journal of Disaster Risk Science, 4*(4), 190–199. https://doi.org/10.1007/s13753-013-0020-z

Banu, N. (2020). Climate change and coastal disasters of Bangladesh. *Development in coastal zones and disaster management* (pp. 215–224). Palgrave Macmillan. https://doi.org/10.1007/978-981-15-4294-7_14

Baqir, M., Sobani, Z. A., Bhamani, A., Bham, N. S., Abid, S., Farook, J., &Beg, M. A. (2012). Infectious diseases in the aftermath of monsoon flooding in Pakistan. *Asian Pacific Journal of Tropical Biomedicine, 2*(1), 76–79. https://doi.org/10.1016/S2221-1691(11)60194-9

Bennett, S., Glandon, D., &Rasanathan, K. (2018). Governing multisectoral action for health in low-income and middle- income countries: Unpacking the problem and rising to the challenge. *BMJ Global Health, 3*(Suppl 4), e000880. https://doi.org/10.1136/bmjgh-2018-000880

Chowdhury, A., Hasan, K., Hasan, R., &Bintay, T. (2020). Climate change impacts and adaptations on health of internally displaced people (IDP): An exploratory study on coastal areas of Bangladesh. *Heliyon, 6*(9), e05018. https://doi.org/10.1016/j.heliyon.2020.e05018

Clark, G. (2007). Evolution of the global sustainable consumption and production policy and the United Nations Environment Programme's (UNEP) supporting activities—ScienceDirect. *Journal of Cleaner Production, 15*(6), 492–498. www.sciencedirect.com/science/article/abs/pii/S0959652606001739

Costello, A., Abbas, M., Allen, A., Ball, S., Bell, S., Bellamy, R., ... & Patterson. (2009). Managing the health effects of climate change Related papers. *The Lancet, 373*(9676), 1693–1733.

Crowley, R. A. (2017). Annals of Internal Medicine Climate Change and Health: A position paper of the American College of Physicians. *Annals of Internal Medicine, 164*(9), 608–610. https://doi.org/10.7326/M15-2766

Ebi, K. L., Exuzides, K. A., Lau, E., Kelsh, M., &Barnston, A. (1998). Weather changes associated with hospitalizations for cardiovascular diseases and stroke in California, 1983–1998. *International Journal of Biometeorology, 49*(1), 48–58. https://doi.org/10.1007/s00484-004-0207-5

Fatema, S. R., Islam, S., East, L., &Usher, K. (2019). Women' s health-related vulnerabilities in natural disasters: A systematic review protocol. *BMJ Open, 9*(12), 1–5. https://doi.org/10.1136/bmjopen-2019-032079

Frumkin, H. (2019). Managed retreat as a strategy for climate change adaptation in small communities: Public health implications. *Climatic Change, 153*, 1–14. https://link.springer.com/content/pdf/10.1007/s10584-019-02382-0.pdf

Ghazali, D. A., & Guericolas, M. (2018). Climate change impacts on disaster and emergency medicine focusing on mitigation of disruptive effects: An international perspective. *International Journal of Environmental Research and Public Health, 15*(7), 1–13. https://doi.org/10.3390/ijerph15071379

Ginige, K., Amaratunga, D., & Haigh, R. (2009). Mainstreaming gender in disaster reduction: Why and how? *Disaster Prevention and Management: An International Journal, 18*(1), 23–34. https://doi.org/10.1108/09653560910938510

Haines, A., Sanders, D., Lehmann, U., Rowe, A. K., Lawn, J. E., Jan, S., Damian G.W., &Bhutta, Z. (2007). Achieving child survival goals: Potential contribution of community health workers. *The Lancet, 369*(9579), 2121–2131. https://doi.org/10.1016/S0140-6736(07)60325-0

Hanifi, S. M. A., Hossain, A., Chowdhury, A. H., Hoque, S., &Selim, M. A. (2020). Do community scorecards improve utilisation of health services in community clinics: experience from a rural area of Bangladesh. *International Journal for Equity in Health, 19*(1), 1–11. https://doi.org/https://doi.org/10.1186/s12939-020-01266-5

Hawlader, M. D. H. (2021). Climate change and its impact on health in Bangladesh. In M. J. & M. N. Islam (Eds.), *Climate change in Bangladesh* (1st ed., pp. 85–92). Cham: Springer. https://link.springer.com/content/pdf/10.1007%2F978-3-030-75825-7.pdf

Hossain, B. (2020). Role of organizations in preparedness and emergency response to flood disaster in Bangladesh. *Geoenvironmental Disasters*, 7(33). https://doi.org/https://doi.org/10.1186/s40677-020-00167-7

Hutton, C.W., Adger, W. N., Hanson, S. E., &Salehin, M. (2018). *Ecosystem services for well-being in deltas: Integrated assessment for policy analysis*. Springer Nature. https://link.springer.com/book/10.1007/978-3-319-71093-8

Jamrozik, E., &Selgelid, M. J. (2020). Ethics and drug resistance: Collective responsibility for global public health. In Michael J. Selgelid (ed.), *Ethics and drug resistance: Collective responsibility for global public health* (pp. 3–18). Cham: Springer.

Joarder, T., Chaudhury, T. Z., &Mannan, I. (2019). Universal health coverage in Bangladesh: Activities, challenges, and suggestions. *Advances in Public Health, 2019*, 1–3. https://doi.org/https://doi.org/10.1155/2019/4954095

Kabir, R., Khan, H.T., Ball, E., &Caldwell, K. (2014). Climate change and public health situations in the coastal areas of Bangladesh. *International Journal of Social Science Studies, 2*(3), 109. https://doi.org/10.11114/ijsss.v2i3.426

Kabir, I., Rahman, B., Smith, W., Fatima, M. A., &Milton, A. H. (2016). Climate change and health in Bangladesh: A baseline cross-sectional survey. *Global Health Action, 9*(1), 26609. https://doi.org/10.3402/gha.v9.29609

Kabir, R., Khan, H.T.A., Ball, E., &Caldwell, K. (2016). Climate change impact: The experience of the coastal areas of Bangladesh affected by cyclones Sidr and Aila. *Journal of Environmental and Public Health, 2016*. https://doi.org/10.1155/2016/9654753

Kabir, R., &Khan, H.T.A. (2017). Study on the health status of coastal people in Bangladesh after cyclone Sidr and Aila. *European Scientific Journal, 13*(15), 10–21. https://doi.org/10.19044/esj.2017.v13n15p10

Kamal,. A.H. M., Umama, U., Roman, S., &Khan, M. M. (2018). *Impact of flood on women' s sexual and reproductive health: An empirical evidence from northern Bangladesh*. Global J Med Res, *18*. https://www.kuet.ac.bd/webportal/ppmv2/uploads/15370815296-Impact-of-Flood-on-Womens.pdf

Khoorshid, B., Ali, L., Akhtar, S., Islam, Z., Rumana, K., &Hossain, S. (2020). Community clinics in Bangladesh: A unique example of public-private partnership. *Heliyon, 6*(5), e03950. https://doi.org/10.1016/j.heliyon.2020.e03950

Kuruvilla, S. (2018). Business not as usual: How multisectoral collaboration can promote transformative change for health and sustainable development. *BMJ*, 363. https://doi.org/10.1136/bmj.k4771

Luber, G., &Prudent, N. (2009). Climate change and human health. *Transactions of the American Clinical and Climatological Association, 120*, 113–117. www.ncbi.nlm.nih.gov/pmc/articles/PMC2744549/

Manisalidis, I., Stavropoulou, E., &Stavropoulos, A. (2020). Environmental and health impacts of air pollution: A review. *Froniers in Public Health, 8*(February), 1–13. https://doi.org/10.3389/fpubh.2020.00014

Pérez, L. M., &Martinez, J. (2008). Community health workers: Social justice and policy advocates for community health and well-being. *American Journal of Public Health, 98*(1), 11–14. https://doi.org/10.2105/AJPH.2006.100842

Rabbani, M.G., Rahman, A. A., Shoef, I. J., & Khan, Z. M. (2015). Climate change and food security in vulnerable coastal zones of Bangladesh. In U. Habiba, M.A. Abedin, A.W.R. Hassan, & R. Shaw (eds.), *Food security and risk reduction in Bangladesh* (pp. 173–185). Springer, Tokyo. www.springer.com/series/11575%0Ahttp://link.springer.com/10.1007/978-4-431-55411-0

Rahman, A. (2008). Climate change and its impact on health in Bangladesh. *Regional Health Forum, 12*(1), 16–62. www.indiaenvironmentportal.org.in/files/Regional_Health_Forum_Volume_12_No_1_RHF-vol12.pdf#page=24

Rawal, L. B., Kanda, K., Biswas, T., Tanim, I., Poudel, P., Renzaho, A. M. N., & Abdullah, A. S. (2019). Non-communicable disease (NCD) corners in public sector health facilities in Bangladesh: A qualitative study assessing challenges and opportunities for improving NCD services at the primary healthcare level. *BMJ Open, 9*(10), e029562. https://doi.org/10.1136/bmjopen-2019-029562

Shamsuddoha, M., & Chowdhury, R. K. (2007). Climate change impact and disaster vulnerabilities in the coastal areas of Bangladesh. *COAST Trust, Dhaka*, 40–48. www.coastbd.org

Steege, R., Taegtmeyer, M., Mccollum, R., Hawkins, K., Ormel, H., Kok, M., Rashid, S., Otiso, L., Sidat, M., Chikaphupha, K., Gemechu, D., Ahmed, R., Tolhurst, R., Gomez, W., &Theobald, S. (2018). How do gender relations affect the working lives of close to community health service providers? Empirical research, a review and conceptual framework. *Social Science & Medicine, 209*, 1–13. https://doi.org/10.1016/j.socscimed.2018.05.002

Uddin, J., &Mazur, R. E. (2015). Socioeconomic factors differentiating health-care utilization of cyclone survivors in rural Bangladesh: A case study of cyclone Sidr. *Health Policy and Planning, 30*, 782–790. https://doi.org/10.1093/heapol/czu057

Ukai, T. (2009). The Great Hanshin-Awaji Earthquake and the problems with emergency medical care. *Renal Failure, 19*(5), 633–645. https://doi.org/10.3109/08860229709109029

Visagie, S., Schneider, M., Scheffler, E., &Schneider, M. (2015). The impact of health service variables on healthcare access in a low resourced urban setting in the WesternCape, South Africa. *African Journal of Primary Health Care and Family Medicine*, 7(1), 1–11. https://doi.org/10.4102/phcfm.v7i1.820

Waluszewski, A., &Snehota, I. (2019). The public-private partnership (PPP) disaster of a new hospital – expected political and existing business interaction patterns. *Journal of Business & Industrial Marketing, 5*, 1119–1130. https://doi.org/10.1108/JBIM-12-2018-0377

Wei Ni, Guoyo Ding, Yifei Li, H. L., & Baofa Jiang. (2014). Impacts of floods on dysentery in Xinxiang city, China, during 2004–2010: A time-series Poisson analysis. *Global Health Action*, 7(1), 9716. https://doi.org/10.3402/gha.v7.23904

WHO. (2006). *Communicable diseases following natural disasters risk assessment and priority interventions*. www.who.int/diseasecontrol_emergencies/guidelines/CD_Disasters_26_06.pdf

WHO. (2018). *Climate change and health*. www.who.int/news-room/fact-sheets/detail/climate-change-and-health

WHO. (2019). *Independent evaluation of community based health services in Bangladesh*. www.who.int/docs/default-source/searo/bangladesh/pdf-reports/2019-20/cbhs---report-2019.pdf?sfvrsn=fdecaade_2

13

ROLE OF HUMAN CAPITAL ON DISASTER PREPAREDNESS IN THE SOUTHWESTERN COASTAL BANGLADESH

Mamunur Rahman and Md. Nasif Ahsan

Introduction

Bangladesh is a low-lying deltaic country in South Asia. Geographically, Bangladesh is a slice of the part of the Bengal basin and is one of the largest deltas in the world shaped mainly by the Ganges-Brahmaputra-Meghan (GBM) River system, which mainly comprises low and flat lands (Nahar and Khan, 2014; Mallick et al., 2005). It becomes more prone to natural hazard–triggered disasters due to many hydro-geographic and geological features like geographical location, flat deltaic topography consists of the low elevation of the sea, tropical climate variability, unsteady tropical atmospheric condition, and regional background (Habiba et al., 2013; Paul and Routray, 2011). Currently, Bangladesh is rated as one of the most disaster-prone countries globally, and roughly 97 percent of her total land area and all populations are severely exposed to several natural hazards like tropical cyclones, floods, droughts, landslides, and riverbank erosion (Alam, 2016). The IPCC's assessment in 2006on climatic disaster risks throughout the world stated that Bangladesh ranked second among the ten extremely vulnerable countries for disasters (Habiba et al., 2013). Disasters cause destructive socio-economic, environmental, and physical impacts like the death of many people every year, risks of poverty, economic hardship, population displacement, unemployment, disempowerment, food crisis, damages of property, structural displacement and destructions, damage to the ecosystems, family disintegration, and many other short- and long-term social problems in Bangladesh (Nahar and Khan, 2014; Khan, 2008; Elahi, 1991).

In Bangladesh, coastal people are exposed to the dreadful impacts of disasters. They made relentless efforts to cope with disasters' impacts by adopting preparedness actions against natural hazard–triggered disasters through their indigenous knowledge and practices. Again, in response to these frequent disasters, short,

DOI: 10.4324/9781003253495-17

intermediate-, and long-term prevention and mitigation actions are framed in Bangladesh through the initiatives of government and NGOs (Paul and Routray, 2011; Khan, 2008; Alam, 2016). Several studies claim a significant positive impact of education (including formal, informal, and non-formal) on disaster risk reduction through adopting preparedness actions against disasters (Shaw et al., 2011; Habiba et al., 2013; Khan, 2008).

One of the examples representing the power of knowledge and education in the context of disaster risk management is the story of a 10-year-old British schoolgirl, Tilly Smith, who warned the tourists to flee to safe areas before the Indian Ocean tsunami hit the coast and saved over 100 tourists in Thailand on 26 December 2004. Tilly Smith could recognize the signs of an approaching tsunami after learning about the phenomenon in her geography lessons at school just weeks before visiting Thailand (Shaw et al., 2011; UNISDR, 2006b). Another example, during the same tsunami, which represents the power of indigenous knowledge acquired from the previous experience of disasters, only seven lives were lost out of a total population of about 83,000 on the Simeulue Island located off the coast of Sumatra that was only 100 km away from the epicenter of the massive earthquake that turned into tsunami. This community had previously experienced a tsunami in the year 1907. The knowledge of tsunami had been transferred from generation to generation through cultural practices such as songs and poems. These experiences of facing disasters in the past helped them prepare themselves against the tsunami and mitigate the negative impacts (Shaw et al., 2011; UNISDR, 2006a). There is hardly any study in Bangladesh that explicitly addressed the role of human capital in terms of knowledge, experience, and education on disaster risk reduction. Considering the facts, we intend to explore the impact of human capital on disaster preparedness in coastal Bangladesh.

Theoretical Considerations

In recent years, disasters in terms of intensity and frequency of cyclones, floods, and droughts have increased worldwide (IPCC, 2001; Khan and Rahman, 2007). An additional increase in global temperature might also increase tropical cyclones with substantial compensations (Emanuel, 2005). Unfortunately, coastal Bangladesh is well known for severe exposure to natural hazard–triggered disasters in recent years (Nahar and Khan, 2014; Paul, 2009). An assessment on climatic risks worldwide published by IPCC (2006) stated Bangladesh as the second among the ten extremely vulnerable countries for disaster. Previous records show that Bangladesh is almost annually affected by disasters, and occasionally faces several disasters in the same year in recent times (Habiba et al., 2013; Khan, 2008).

Disasters have severe vicious impacts on Bangladesh's socio-economic factors, structural and non-structural issues (Nahar and Khan, 2014; Paul and Routray, 2011; Paul and Dutt, 2010; Khan, 2008). In his study, Khan (2008) claimed that disasters adversely affect Bangladesh's economy and impede development. Disasters result in the death of many lives every year, increased poverty, economic loss, food crisis, unemployment, disempowerment, population displacement, family disintegration, damage to the ecosystem, structural destruction, and displacement, and many other short- and long-term social problems in Bangladesh (Elahi, 1991; Khan, 2008; Nahar and Khan, 2014). Several studies suggest that several catastrophic cyclones hit Bangladesh in 1822, 1876, 1961, 1965, 1970, 1991, 2007, and 2009 (Blaikie et al., 1994; Dube et al., 1997; GOB, 2008; Paul and Dutt, 2010; Paul and Routray, 2011; Nahar and Khan, 2014; Alam, 2016). Several studies and reports also claimed that about 500,000;138,000;3406; and 191 people were killed, respectively, during the cyclones of 1970, 1991, 2007, and 2009 (Ali, 1980; Haider et al., 1991; GOB, 2008; Paul, 2009). Again, some studies revealed that economic loss estimated as US$86.4 million, US$2 billion, US$437.6 million, and US$269.28 million during the cyclones of 1970, 1991, 2007, and 2009, respectively (Blaikie et al., 1994; Dube et al., 1997; GOB, 2008; Paul and Dutt, 2010; Paul and Routray, 2011; Nahar and Khan, 2014; Alam, 2016).

Different short-, medium-, and long-term plans were adopted as preventive and mitigating actions to deal with these frequent disasters in Bangladesh. Prevention and mitigation measures may reduce the vulnerability of hazards and disasters in formulating those action plans while frequencies of disasters remain unavoidable (Khan, 2008). Disaster preparedness comprises different levels of preparation of a counter-disaster plan, forecasting of a disaster, warning about a disaster, preservation, conservation of resources needed during and after a disaster, and training of the related personnel in Bangladesh (Rahman, 2001). It emphasizes pre-disaster management-related issues, including planning and preparedness actions, organizational training, planning, information management, public relations, public awareness building, and disaster education in recent times in Bangladesh (Habiba et al., 2013).

To prepare a community against disasters, the essential elements are knowledge, awareness, resources, and an efficient framework. These elements need a long period to develop through a focused goal and a predetermined strategy. From both government and non-government sectors, disaster preparedness education and awareness-building programs have been emphasized more recently in Bangladesh (Khan, 2008; Mallick et al., 2005; Alam, 2016; Nahar and Khan, 2014; Habiba et al., 2013). The government and non-government institutions have introduced several formal and non-formal education programs in Bangladesh to foster the disaster management processes (Khan, 2008; Habiba et al., 2013). The Hyogo Framework for Action (HFA), 2005–2015, priority for action-3 gives importance to knowledge and education, highlighting formal and non-formal education to create awareness toward disasters and is labeled awareness-raising as an essential mechanism for disaster risk reduction (Habiba et al., 2013). The priority for action-1 (understanding disaster risk) of the Sendai Framework for disaster risk reduction 2015–2030 as the successor instrument to the HFA also gives the importance to understanding disaster risk in all its dimension of vulnerability, capacity, exposure of people and their assets, disaster characteristics, and the environment. Moreover, such knowledge can be implemented in the context of risk management, prevention, mitigation, preparedness, and response (UNDRR, 2021). The National Curriculum and Textbook Board (NCTB) of Bangladesh had disseminated disaster education through course curriculum in 1994/1995. Furthermore, Disaster Management Bureau (DMB) and Comprehensive Disaster Management Program (CDMP) had reviewed the books of NCTB focusing on the need for preparedness measures at family and community levels against disasters, emphasizing the need for disaster risk reduction in 2006 (Habiba et al., 2013). Formal education about the disaster has been introduced in different levels operated by both government and non-government institutions like adult literacy program, disaster education at primary level, secondary level, and tertiary level and the disaster education courses consist of different disaster preparedness issues like pre-disaster preparedness and vulnerability reduction, post-disaster response, relief, and rehabilitation and disaster mitigation and long-term development (SEMP, 2003; Khan, 2008). The leading NGOs working in disaster management and education are Bangladesh Rural Advancement Committee (BRAC), Proshika, Gono Shahajjo Shongstha (GSS), Dhaka Ahsania Mission, and Disaster Management Forum (DMF) in Bangladesh (Khan, 2008). Non-formal education of disaster initiated by the government, mainly public awareness and training programs conducted by the DMB to promote an informed, alert, self-reliant, and sustainable community that can actively collaborate all disaster management activities. Information about disaster preparedness, including disaster risks and basic community needs, availability of assistance programs, seasonal preparedness reminders, and post-disaster tips disseminate to the vulnerable people by the government and NGOs (Khan, 2008;Biswas and Reza, 2000). The information is delivered and disseminated to the people through study circles, workshops and seminars, public talks and presentations, leaflets, posters, photographs, radio and TV spots, features and dramas, films and videos, and open-air dramas, namely path dramas and folk songs (Khan, 2008). Non-formal education about disaster preparedness is the more widespread and operative way of making people aware of the disasters and providing them the knowledge to deal with the adversities of the catastrophes (Khan, 2008; Habiba et al., 2013; Paul and Routray, 2011; Shaw et al., 2011).

Disaster education, disaster risk education, or disaster prevention education are different terminologies that mainly mean disaster risk reduction education or preparedness education. Disaster education refers to education that can facilitate and contribute to creating a culture of prevention and preparing for a disaster for the people and communities (Habiba et al., 2013; Bernal, 2019). In their research, Shaw et al. (2011) suggested that education plays a dynamic role in reducing disasters and achieving human security in achieving sustainable development. A study conducted by Vaughan et al. (2003) in Costa Rica uncovered that if the children are conducted through proper disaster-related education, then parents and other adults in their families could also benefit from making sense of disasters and disaster preparedness. Petal and Izadkhah (2008) recommended that the success of disaster risk reduction education influences the goal of developing disaster-resilient communities and further enlightened that integration of formal and non-formal/informal education can ensure the sustainability of learned knowledge from generation to generation. Shaw and Kobayashi (2001) stated that it is recognized that school plays a significant role in raising awareness among students, teachers, and parents in the context of formal education. Petal and Izadkhah (2008) again suggested that formal curriculum integration can be framed in elective courses or modules about disasters into the existing courses that can foster disaster risk reduction education. Ronan et al. (2001) claimed in their research based on the information of school children in New Zealand that students who received disaster education were more aware of the disaster, and they have effectively disseminated their knowledge to their family and community to prepare themselves against disaster, which helped them to prepare against disaster significantly. Again, to facilitate the relationship between risk perception and risk reduction behavior, community and social psychological factors must be included in disaster education, and disaster education should be incorporated with community development initiatives to increase resilience and to decrease the risks of disasters (Paton and Johnston, 2001; Paton, 2005). So, it can be stated that education plays a pivotal role in promoting disaster risk reduction knowledge and awareness, enhancing the sustainable development of a disaster-prone country (Shaw et al., 2011).

Shaw et al. (2011) claimed that experiences of facing disasters in the past positively affect disaster risk management. Studies conducted by Lidstone (1996) and Lidstone and Nielsen (1999) claimed that real-life learning, reflective learning, and situated learning could enhance disaster risk management in a disaster-prone community as well as a disaster-prone area or a country. Studies conducted by Bubeck et al. (2013); Lawrence et al. (2014); Lindell and Hwang (2008); Sattler et al. (2000); Horney et al. (2008); and McGee and Russell (2003) claimed that people who have experiences of facing disasters like floods, earthquakes, hurricanes, and wildfires are more likely to take preparedness actions for future disasters. Again, having an experience of being affected by and surviving a disaster may increase the awareness about possible damages of the disasters, evacuation decisions, preparedness measures, knowledge for recovering the aftermath of disasters, and knowledge on how to cope with the adversities of disasters (Sattler et al., 2000). A study on two villages in Barguna district and another two villages in Patuakhali district of Bangladesh conducted by Paul and Routray (2011) revealed relentless efforts or strategies or preparedness measures of the people of the study area to cope with the severe effects of natural hazards like cyclones, storm surges, and floods based on their experiences of previously faced disasters, age-old indigenous knowledge, and practices. The study by Paul and Routray (2011) demonstrated that the preparedness measures against disasters implemented by the people of that areas were like unique design and construction methods of houses, building *Machan* and *Pataton*, temple-shaped house and support four sides of the house with guy ropes tied to bamboo poles or trees to protect their houses. People used to dig *Safety-hole* for preserving and storing food items and valuable assets to retrieve those after disasters. They also put valuable materials in a net or jute bag, kept it in a shallow pond, or tied it with strong trees with cloths or ropes to retrieve them after the disasters.

Human capital is a less tangible asset of a human being. Human capital denotes the inherent dynamic competencies of human beings. Human capital is formed by persons' changes that result in skills and capabilities that make them able to act in new approaches. These skills and capabilities can be augmented through investment in education, on-the-job training, health, etc. Human capital is viewed as an asset that generates a flow of services, most often measured as earnings. However, more comprehensive output methods are also applied to bring out comprehensive results (Coleman, 1988; Eide and Showalter, 2010). In this study, we tried to determine the impact of human capital on disaster preparedness. Here, we considered the educational status of male and female household heads, the experience of disasters, number of disasters faced, and number of participated or got discussion session or training on disaster preparedness as the human capital of the sampled households. Besides, we also incorporated some other human capital and socio-economic-related aspects like age, monthly income, household size, number of the male member, number of the female member, number of children (0–14 years), and number of older (>65 years) person in the family, duration of current residence, loss from disasters, and expenses in adopting preparedness measures to examine the impact of these issues on disaster preparedness of the sampled households.

Study Location and Methods

Study Location

Satkhira district in southwestern coastal Bangladesh was selected for conducting this study. The total area of the district is 3817.29 square kilometers. It is bounded by Jessore district on the north, the Bay of Bengal on the south, Khulna district on the east, the state of West Bengal, India on the west. There are seven Upazilas and 79 unions in Satkhira district (BBS, 2017). For this study, we selected Assasuni Upazila among the seven Upazilas of Satkhira district as the said Upazila is a significantly disaster-affected area. From Assasuni Upazila, we randomly selected Sreeula union, and from this union, six villages, namely – Mariala, Hazrakhali, Bokchar, Gazipur, Sreeula, and Mohishkur were finally chosen. Figure 13.1 shows the locations of the sampled villages considered in this study.

Data Collection

This study is mainly based on primary data. Primary data had been collected through a structured questionnaire, which was designed to collect the primary data for constructing and achieving the objectives of this study. The necessary qualitative and quantitative data and information had been collected according to the structured questionnaire through face-to-face survey with the respondents about livelihood, assets and quality of life, information about their status of disaster preparedness, disaster education-related status, and some other opinions of the respondents about disaster-related issues like ideas regarding the possible occurrence of major climatic

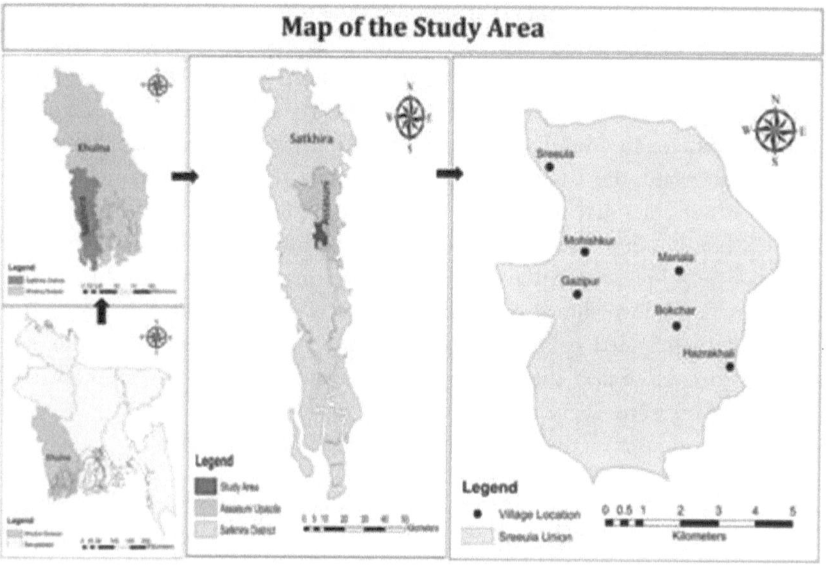

FIGURE 13.1 Map of the study location
Source: HDX (2020).

extreme events in their community, their opinions about disaster preparedness measures which government and NGOs can implement in their locality based on their experiences and perceptions, etc. from the purposively selected 20 respondents from each of the six villages mentioned previously, that is, total 120 respondents. This study was conducted during the month of January–June in 2019.

Analytical Approach

Descriptive statistics was used to analyze the socio-economic and demographic characteristics of the respondents like age, education, household size, housing condition, housing ownership status, etc. Again, other information related to the disasters and disaster preparedness like the location of the respondent's house, duration of current residence, number of disasters faced, financial loss in disasters faced, expenses in taking preparedness measures, etc. involved with the sampled households were analyzed using descriptive statistics.

We also used two econometric models named logit and Poisson regression for the two dependent variables named 'adopting preparedness measures' (taken as binary) and 'number of taken preparedness measures', respectively. A logit regression model was used to investigate the impact of education and experience on disaster preparedness. In the logit regression model, the variable termed 'adopting preparedness measures' was considered the first dependent variable. Here, the dependent variable is a binary outcome variable coded as one (1) – if the respondent has taken any preparedness measure against disasters and zero (0) – otherwise. Explanatory variables such as education that was divided into two parts as the educational status of the male household head and educational status of the female household head, while the experience was used at first baseline model to explore the impacts of education and experience on adopting preparedness measures against disasters. Afterward, the baseline model is extended through incorporating other explanatory variables such as age, monthly income, household size, number of male members, number of female members, number of children (0–14 years),number of older (>65 years) persons in the household, duration of current residence, number of disasters faced, loss from disasters, expenses in adopting preparedness measures, number of participated or got discussion session or training about disaster preparedness to explore also the impact of those explanatory variables on adopting preparedness measures against disasters.

The standard logit regression model is framed as follows:

$$L = In\left[\frac{p_i}{1 - P_i}\right] = \beta_0 + \beta_i X_i + \mu_i \tag{13.1}$$

Here, p_i = Probability of happening i^{th} event whose value always lies between 0 and 1, β_0 = Intercept term, β_i = Coefficient of ith number of variables, X_i = i^{th} number of explanatory variables, i = 1…n, and μ_i = Error terms.

The second dependent variable, 'number of taken preparedness measures,' is count data that took a finite number of values. So, here, Poisson regression is used to analyze such count data. Estimating the degree of relationship of the number of preparedness measures taken with education is divided into two parts: the educational status of the male/female household head and disaster experience. A baseline Poisson regression model was applied first, followed by its extension by incorporating other explanatory variables such as age, monthly income, household size, number of male members, number of female members, number of children (0–14 years), number of older (>65 years) persons in the family, duration of current residence, number of disasters faced, loss from disasters, expenses in adopting preparedness measures, and number of participated or got discussion session or training about disaster preparedness to explore also the impact of those explanatory variables on the number of taken preparedness measures against disasters.

Following Rao and Markos (2015), the standard function of a Poisson regression model is as follows,

$$Prob\left(V_i = v \mid X_i\right) = \frac{exp^{-\lambda_i} \lambda_i^{v_i}}{v!}; \qquad V = 0, 1, 2, 3\ldots\ldots \tag{13.2}$$

where, $Prob\left(V_i = v_i\right)$ denotes the probability of the number of preparedness measures taken that takes non-negative integer values, and its' mean values and variance are the same: E (V) = λ and variance (V) = λ. Moreover, it is notable that the parameter of this exponential has to be $\lambda_i > 0$. In this study, the assumption of

Poisson regression was not satisfied. That was why, to address the problem of over-dispersion, the alternative approach negative binomial model was undertaken as an estimator of the number of preparedness measures taken by the respondents.

So, the form of negative binomial model is:

$$Prob\left(V_i = v_i | \lambda, \propto\right) = \frac{r\left(v_i + \propto\right)}{v!r\left(\propto\right)}\left(\frac{\propto}{\propto + \lambda_i}\right)\propto\left(\frac{\lambda_i}{\propto + \lambda_i}\right)^{v_i} \tag{13.3}$$

Where $\propto$ denotes dispersion parameter, which measures over-dispersion. Here, the model's mean remains the same as the Poisson model; E (V) = λ but variance (V) = $\lambda / \propto$.

Table 13.1 presents the variables used in this study with their definitions, measuring units, and adapted sources.

TABLE 13.1 Description of the variables

Variables	*Explanation of the variables*	*Unit of measurement*	*Adapted sources*
Gender	Male-1, female-2, third gender-3	-	Hoffmann and Muttarak (2017)
Age	Age of the respondent	Years	Hoffmann and Muttarak (2017)
Occupation	Business-1, service holder-2, shrimp cultivator-3, crop cultivator-4, fisherman-5, day laborer-6	-	Paul and Routray (2011)
Education	1.Educational status of the male household head 2.Educational status of the female household head	Years of schooling	Hoffmann and Muttarak (2017)
Monthly income	Monthly income of the respondent	BDT	Hoffmann and Muttarak (2017)
Household size	Number of a family member	Number	-
Number of male members	Number of male members in respondent's family	Number	-
Number of female members	Number of female members in respondent's family	Number	-
Number of children	Number of children in the family (0–14 years)	Number	-
Number of older persons	Number of older persons in the family (>65 years)	Number	-
Housing ownership status	Owned-0, rent-1, shared-2	-	Paul and Routray (2011); Khan, 2008

(*continued*)

TABLE 13.1 Cont.

Variables	*Explanation of the variables*	*Unit of measurement*	*Adapted sources*
Housing condition	Kacha-0, semi-pacca-1, pacca-2	-	-
Location of the house of the household		Kilometer and meter	Hoffmann and Muttarak (2017)
Duration of current residence	Duration of residing in current residence	Years	Hoffmann and Muttarak (2017)
Experience of disaster	Experience of disasters faced in the past (yes = 1, otherwise = 0)	Dummy	Hoffmann and Muttarak (2017)
Number of disasters faced	Number of disasters faced in last ten years	Number	-
Financial loss or damages in disasters faced	All kinds of financial loss or damages in disasters faced by respondent	BDT	Nahar and Khan (2014); Alam (2016)
Adopting preparedness measures	Yes = 1, otherwise = 0	Dummy	Hoffmann and Muttarak (2017); Khan (2008)
Number of taken preparedness measures	Number of taken preparedness measures against disasters	Number	Hoffmann and Muttarak (2017)
Expenses in preparedness measures	Amount of expenses incurred to take preparedness measures till now	BDT	Shaw et al. (2011); Habiba et al. (2013); Alam (2016)
Number of discussion session or training got or participated	Number of discussion sessions or training got or participated on disaster-related issues	Number	-

Source: Author's compilation from literature review and field survey.

Major Findings

Socio-economic Characteristics of the Respondents

The average household size of the sampled household was 7.19; the average number of male members was 2.15; the average number of female members was about 2; the average number of children was about 2; and the average number of older persons was 1.4. We found that the average age of the people is 39.03 years. It is found from the analysis that 18.33 percent, 39.17 percent, 27.5 percent, and 15 percent of respondents have ages between 25 and 35 years, 36 and 45 years, 46 and 55 years, and 56 and 65 years, respectively. In the case of occupation, 29.17 percent of the sample household was found as shrimp cultivators, which was the highest among different occupations. The other categories of occupation of the sampled household were business, service holder, crop cultivator, fisherman, day laborer, etc. We found that the average years of schooling of male and female household heads are eight and seven years, respectively. Again, it is found that 42.5 percent, 47 percent, 4 percent, and 18 percent of the male household heads have the education till primary level, secondary level, higher secondary level, and graduation level, respectively. We have also found that 47.5 percent, 40 percent, 10 percent, and 2.5 percent of the female household heads have the education till primary, secondary, higher secondary, and graduation levels, respectively. Analysis revealed that 19.17 percent, 15 percent, 40.83 percent, 13.33 percent, 10.83 percent, and 0.83 percent of sampled households have monthly income between the ranges of BDT 5,000 and 10,000 (≈US$58.23–116.46), BDT 10,001 and 15,000 (≈US$116.48–174.70), BDT 15,001 and 20,000 (≈US$174.71–232.93), BDT 20,001 and 25,000 (≈US$232.94–291.16), BDT 25,001 and 30,000 (≈US$291.17–349.39), and BDT 30,001 and 35,000 (≈US$349.40–407.62), respectively. The average monthly income of the sampled households was found to be BDT 18,338(≈US$ 216.36). The minimum and maximum monthly incomes of the respondents are BDT 8,000(≈US$ 94.39) and BDT 35,000(≈US$ 412.95), respectively. Approximately 73.33 percent of sampled households owned houses, 15.83 percent shared houses, and 10.83 percent rented houses. We also found that 26.67 percent have *kacha*, 24.17 percent have *semi-pacca*, and 59 percent have *pacca* houses of the sampled household.

Disaster-Related Issues for the Households

Considering the exposure, we figured out that 24.17 percent, 35.83 percent, 10 percent, 11.67 percent, 8.33 percent, and 10 percent of sampled respondent's houses have located near the coast (<2 km), near a river (<200 m), reasonably close to the coast (in between 2 and 5 km), reasonably close to the river (in between 200 m and 1.5 km), reasonably far from the coast (>5 km), and reasonably far from the river (>1.5 km), respectively. We considered the duration of current residence of the respondents as a factor of adopting preparedness measures and the number of preparedness measures taken by the households against disaster. The average duration of the current residence of the respondents is just above 16 (±9.2) years. The majority (92.50 percent) of sampled households suffered and experienced different disasters in the last three years. Cyclones, riverbank erosion, and coastal erosion were the main natural hazards that triggered disasters in the last three years. As a coastal area, our study areas were battered by cyclone *Amphan* in 2020, cyclone *Fani* in 2019, cyclone *Titli* in 2018, cyclone *Mora* in 2017, and cyclone *Roanu* in 2016. Another common and destructive natural hazard was riverbank erosion, which eventually turned into disasters suffered by the coastal people twice or thrice a year, resulting in floods and substantial damages to coastal people and their properties. The average yearly damages or losses caused by disasters was estimated as BDT 50,681 (≈US$595.12) incurred by the sampled respondents who suffered disasters. About 56 percent of sampled households adopted three to five preparedness measures against disasters. Among the adopted measures – food storage, structural upgrades, preparing an emergency kit, family emergency plan, and shifting of residence were notable. Structural upgradation was the preparedness measure adopted by a large number of people. Our results suggest that the average amount of expenses incurred to adopt those preparedness measures against is just over BDT 1,47,000 (≈US$ 1734.41).

Our results suggest that about 60 percent of people had ideas about climate change and disaster risk–related issues. They got information about climate change and extreme climatic events from government sources, NGOs, TV programs, radio programs, books, newspapers, training, friends, and neighbors. Most of the sampled respondents claimed a high possibility of placing major extreme events in their community in the following five years. We explored that about 60 percent of people knew about disaster education. Their opinions about disaster education refer to knowledge about disaster warning messages, pre-, during, and post-disaster preparedness, warning flags, concepts about risks, hazards maps, communication with emergency operation centers, making awareness to their family and neighbors, disseminating information of emergencies to their community, and work together to adapt and mitigate the disaster shocks and damages in their community. They gained this disaster education–related knowledge and ideas from government sources, NGOs, TV programs, radio programs, books, newspapers, training, friends, and neighbors. We attempted to explore the types of preparedness measures against disasters taken by the respondents after gathering or acknowledging the knowledge about disaster and disaster preparedness that they did not take before. Our result implies that most people had taken structural upgradation as a measure against disasters after acknowledging the facts of disaster. Having gained knowledge on disaster and preparedness, they executed other preparedness measures like food storage, preparing an emergency kit, family emergency plan, and evacuating to safe havens. None took or planned to make disaster insurance a preparedness measure against disasters.

About 70 percent of the respondents reported that they participated in or got discussion sessions or training arranged by the government, NGOs, community clubs, friends, and neighbors about extreme events and disaster preparedness. They claimed that the issues like natural hazards, disaster risk, and disaster preparedness had been discussed or presented in those discussion sessions or training. Respondents also claimed that discussion sessions or training they participated in helped them and their families to substantially take and execute disaster preparedness measures. We found a notable issue that experiences of encountering disasters influenced and facilitated people and their community to take preparedness measures against disasters.

We explored several preparedness measures taken and executed by government institutions, NGOs, and local community-based institutions. These measures are building cyclone shelters, afforestation, blocking of vulnerable parts of the embankment, CPP volunteers, Union Disaster Management Committee (UDMC), dragging signal flags, announcements of emergencies pre-disaster time by local government bodies, training and workshops on disaster preparedness by government institutions and NGOs, short path dramas, and making risk and hazard maps.

Human Capital and Disaster Preparedness: Empirical Analysis

A logit Regression Analysis

Alogit regression model was used to investigate the impact of education and experience on disaster preparedness (Equation 13.1 in Analytical Approach sub-section). In this regression model, the variable termed 'adopting preparedness measures' was considered the dependent variable, a binary variable coded as one (1) – if the respondent had taken any preparedness measure against disasters and zero (0) – otherwise. Several explanatory variables were used (see in Table 13.2). Among these explanatory variables, education was divided into two parts: the educational status of the male and female household head and disaster experience are used at first baseline model to explore the impacts of education and experience on disaster preparedness. Result suggests that with the increase in the schooling of household heads (male and female), the probability of adopting preparedness measures would also increase.

From the analysis, we also found that experience of disasters would more likely escalate the likelihood of adopting preparedness measures against disasters. So, there seemed to be a positive impact of education and experience of disasters on adopting preparedness measures against disasters. Educational status of the people influenced and helped positively to adopt preparedness measures against disasters. The probability of taking preparedness measures against disasters will increase with household heads' education levels (both male and female). So, it can be stated that the more the schooling years of household heads (both male and female), the more the

TABLE 13.2 Impact of education and experience on adopting preparedness measures

Explanatory variables	*Marginal effects*
Educational status of the male household head (years of schooling)	0.5273186***(0.0070189)
Educational status of the female household head (years of schooling)	0.020876***(0.0066272)
Experience of disaster (dummy: yes = 1, otherwise= 0)	4.9263387***(0.0322501)
Constant	1.02911***(0.10425)
Number of observations	120
Prob > chi^2	0.0000
Pseudo R^2	0.7437
Dependent variable: Adopting preparedness measures (dummy: yes = 1, otherwise= 0)	

Note: Standard errors in parentheses.

*** $p<0.01$, ** $p<0.05$, * $p<0.1$

tendency of taking preparedness of those households. We found that the male and female household heads from the sampled households who have the education level of graduation, higher secondary, and secondary level have the higher probability of taking preparedness measures against disasters. People having past experiences of disasters also tended to take preparedness measures against disasters more than their counterparts.

We attempted to determine the impact of several socio-economic and disaster-related aspects on adopting disaster preparedness using the logit regression model in Table 13.3. The results suggest that age, monthly income, financial loss, or damages in disaster faced by people and the likelihood of adopting preparedness measures are more likely related with each other, which means one unit increase of age (one year), monthly income (BDT1), and financial loss or damages in disaster faced (BDT1) will increase the probability of adopting preparedness measures and vice versa and the results are statistically significant. More elaborately, the findings show that the sampled households with a higher monthly income had a higher probability of taking preparedness measures. Again, those with higher financial loss or damages in disaster faced had the higher possibility of taking preparedness measures against disastrous events and shocks. Expenses in preparedness measures are less likely related to the probability of adopting preparedness measures, which indicate that one unit (BDT1) increase in expenses in preparedness measures would decrease the likelihood of adopting preparedness measures. Therefore, expenses in

TABLE 13.3 Impact of several socio-economic and disaster-related aspects on adopting preparedness measures

Explanatory variables	*Marginal effects*
Age (years)	0.03676 (0.0059293)
Monthly income (in BDT)	0.000124**(4.6406)
Household size (number)	0.4174154 (0 .0307495)
Male members in house (number)	0.4204907 (0.0554732)
Female members in house (number)	0 .302849 (0.0501046)
Number of children (number)	1.757178 (0.0560341)
Older persons in house (number)	0.447(1.969)
Duration of current residence (in years)	0.170108**(0.0041704)
Number of disasters faced (number)	0 .3478***(0.0047164)
Financial loss or damages in disasters suffered(BDT)	5.9407***(3.3007)
Expenses in preparedness measures (BDT)	−2.1506**(3.5207)
Training participation(number)	0.1042589 (0.0052768)
Constant	3.00921***(0.00435)
Number of observations	120
Prob >chi^2	0.0000
Pseudo R^2	0.8437
Dependent variable: Adopting preparedness measures (dummy: yes = 1, otherwise= 0)	

Note: Standard errors in parentheses.

*** p<0.01, ** p<0.05, * p<0.1

taking preparedness measures are negatively related to taking preparedness measures against disasters faced by the households.

Again, our results suggest that household size, number of male members, number of female members, number of children, and number of older persons in the family seemed to be more likely to affect the likelihood of adopting preparedness measures against disasters though these were statistically significant. However, a notable outcome is explored from the examination that duration of the current residence, the number of disasters faced, and the number of participated or got discussion sessions or training on disaster-related issues by the people have a positive impact on the probability of adopting preparedness measures against disasters. The sampled households living in their current residence for a higher number of years (for a long time) had a higher possibility of taking preparedness measures. Again, the sampled households who faced a higher number of disasters had larger household sizes; higher male, female, children, and older persons in their family had a higher probability of taking preparedness measures.

Poisson Regression Analysis

At this stage, we applied a Poisson regression model (Equation 13.2). The dependent variable, number of taken preparedness measures, is count data that takes a finite number of values. However, as the assumption of Poisson regression was not satisfied, a negative binomial model (Equation 13.3) was applied to overcome the assumption-related challenges as mentioned above. Table 13.4 presents the results from this analysis.

TABLE 13.4 Impact of education and experience on number of taken preparedness measures

Explanatory variables	*Poisson regression*	*Negative binomial regression*
	Marginal effects	*Marginal effects*
Educational status of the male household head (years of schooling)	0.0391761(0.05272)	0.0391761(0.05272)
Educational status of the female household head (years of schooling)	0.0526457*(0.073)	0.0526457(0.073)
Experience of disaster (dummy: yes = 1, otherwise= 0)	5.064***(1.34907)	4.034**(1.34907)
Constant	1.21772**(.5524691)	1.21772**(.5524691)
Number of observations	120	120
Prob >chi^2	0.0000	0.0000
Pseudo R^2	0.748	0.689
Dependent variable: Number of taken preparedness measures		

Note: Standard errors in parentheses.

*** p<0.01, ** p<0.05, * p<0.1

TABLE 13.5 Impact of several socio-economic and disaster-related issues on number of taken preparedness measures

Explanatory variables	*Poisson regression*	*Negative binomial regression*
	Marginal effects	*Marginal effects*
Age (years)	0.0215102(0.0303)	0.0215102 (0.0303099)
Monthly income (BDT)	0.000184**(0.00004)	0.00034(0.0000304)
Household size (number)	0.0447318(0.2897)	0.0447318(0.289708)
Number of male members (number)	0.0632827(0.56407)	0.0632827(0.5640672)
Number of female members (number)	0.0942103(0.41378)	0.0942103(0.4137886)
Number of children (number)	0.083243(0.406854)	0.083243(0.4068536)
Number of older persons (number)	0.302849(0.05011)	0.302849(0.0501046)
Duration of current residence (years)	0.03315*(0.026969)	0.03315 (0.026969)
Number of disasters faced (number)	0.1522571*(0.0869)	0.132561(0.075842)
Number of discussion session or training got or participated (number)	0.01717**(0.05641)	0.01318(0.0464058)
Financial loss or damages in disasters faced (BDT)	1.825***(2.3006)	1.306*(1.2226)
Expenses in preparedness measures (BDT)	−2.3006*(2.8006)	−2.3006(2.8006)
Constant	0.3668049**(.5258)	0.3668049**(.5258)
Number of observations	120	120
Prob > chi^2	0.0000	0.0000
Pseudo R^2	0.708	0.679
Dependent variable: Number of taken preparedness measures		

Note: Standard errors in parentheses.

*** p<0.01, ** p<0.05, p<0.1

The educational status of the household heads (both male and female) positively affected the number of preparedness measures, which implies that the number of years of schooling of male and female household heads would increase the degree (number) of preparedness measures. Result also suggests that a higher number of disaster experiences positively affected the number of preparedness measures adopted.

Table 13.5 also presents results from another Poisson regression model incorporating other explanatory variables to explore the impact on the number of preparedness measures taken against disaster. This model suggests that age, monthly income, financial loss, or damages in disaster faced by people positively affected the number of adopted preparedness measures against disasters.

Our result suggests that expenses in preparedness measures negatively affected the number of preparedness measures taken, implying that a one-unit increase in expenses would decrease the number of preparedness measures adopted. Again, we found that household size, number of male members, number of female members, number of children, and number of older persons in the respondents' household positively affected the magnitude of preparedness measures though not statistically significant. While the duration of the residence of disaster-affected respondents, the number of interactions for disaster-related training/workshop positively affected the number of preparedness measures adopted against disasters.

Conclusion

Bangladesh is one of the most disaster-prone countries in the world. Natural hazards like tropical cyclones, storm surges, floods, drought, riverbank erosion, etc. frequently hit Bangladesh. In this context, in recent years, disaster preparedness is a burning issue in Bangladesh to adapt and mitigate the shocks and damages done by disasters. This study has attempted to reveal the impact of human capital in terms of education, the experience of disasters, and different socio-economic parameters on disaster preparedness in coastal Bangladesh.

We have found a positive impact of education and experience of disasters on the likelihood of adopting preparedness measures against disasters. People with previous experiences of disasters have exhibited a higher degree of preparedness against disasters. Notably, the educational status of the people is prompting them to take preparedness measures against disasters. There is also a substantial positive impact of age, monthly income, household size, number of male members, number of female members, number of children and number of older persons in the family, duration of current residence, number of disasters faced, financial loss or damages in disasters faced, and the number of participated or got discussion session or training on disaster-related issues like climate change, the impact of climate change, climate change–induced hazards and disasters, and disaster preparedness on the likelihood of adopting preparedness measures and on the number of adopted preparedness measures against disasters. In this case, our results suggest that people have used the sources like government institutions, NGOs, TV programs, radio programs, books, newspapers, training, friends, and neighbors to know and learn about disaster-related issues in their areas. On the other hand, we have found that expenses in preparedness measures negatively impact the likelihood of adopting preparedness. As preparedness measures, people have adopted food storage, structural upgrades, preparing emergency kits, family emergency plans, and shifting residence against disasters. None of the sampled households has taken disaster insurance as preparedness measures against disasters.

We recommend, based on our empirical findings, which can be implanted by the government and NGOs working in this context, to increase the number of discussion sessions or training on disaster-related issues like climate change, the impact of climate change, climate change–induced hazards, and disasters and disaster preparedness as there exists positive impact of number of participated or got discussion session or training on disaster-related issues on taking preparedness measures and on number of taken preparedness measures. We also recommend increasing the financial support or aid to the affected people in this study areas as well as people exposed to natural disasters in southwestern Bangladesh to lessen their loss or damage in disasters faced and to lessen their cost of taking preparedness measures as there is a negative impact of these two factors on taking preparedness measures and the number of adopted preparedness measures against disasters. Finally, we recommend that government and NGOs can implement disaster insurance programs as preparedness measures against disasters besides other programs in the coastal areas in Bangladesh.

References

Ahsan, M. N., Takeuchi, K., Vink, K., and Warner, J. (2016). Factors affecting the evacuation decisions of coastal households during cyclone Aila in Bangladesh. *Environmental Hazards, 15*(1), 16–42. doi:10.1080/17477891.2015.1114912

Alam, E. (2016). Earthquake and tsunami knowledge, risk perception and preparedness in the SE Bangladesh. *Journal of Geography and Natural Disasters, 5*(154).doi:10.4172/2167-0587.1000154

Ali, A. (1980). *'Storm surges in the Bay of Bengal and their numerical modeling'*, School Accountability Report Card (SARC) Report No. 125/80, Atomic Energy Commission, Dhaka, Bangladesh.

BBS (2017). 'District Statistics 2017', Bangladesh Bureau of Statistics (BBS), Ministry of Planning, Government of the People's Republic of Bangladesh, Dhaka.

Bernal, L.A. (2019). 'Disaster education', Training and Prevention Center for Emergency and Environmental Management.www.preventionweb.net/files/3442_DisasterEducation.pdf

Biswas, D.K. and Reza, A. (2000). *'Durjoge Amader Koronio (in Bangla)'*, Proshika ManobikUnnayanKendro, Dhaka, Bangladesh.

Blaikie. P., Cannon, T., Davis, I., and Wisner, B. (1994). *At risk: Natural hazards, people's* vulnerability, and disasters. London: Routledge.

Bubeck, P., Botzen, W. J. W., Kreibich, H., and Aerts, J. C. J. H. (2013). Detailed insights into the influence of flood-coping appraisals on mitigation behavior. *Global Environmental Change*, 23(5), 1327–1338.

Coleman, J. S. (1988). Social capital in the creation of human capital. *American Journal of Sociology, 94*, 95–120.

Dube, S.K., Rao, A.D., Sinha, P.C., Murthy, T.S., and Bahulayan, N. (1997). Storm surge in the Bay of Bengal and Arabian Sea: The problem and its prediction. *Mausam, 48*, 283–304.

Eide, E. R. and Showalter, M. H. (2010). *Human capital*. Elsevier, pp. 282–287.

Elahi, K.M. (1991). Impacts of riverbank erosion and flood in Bangladesh: An introduction. *Riverbank erosion impact study*. Dhaka: Jahangirnagar University.

Emanuel, K. (2005). Increasing destructiveness of tropical cyclones over the past 30 years. *Nature, 436*(4), 686–688.

GOB. (2008). *'Cyclone Sidr in Bangladesh: Damage, loss, and needs assessment for disaster recovery and reconstruction'*, a report prepared by the Government of Bangladesh assisted by the International Development Community with financial support from the European Commission. https://reliefweb.int/report/bangladesh/cyclone-sidr-bangladesh-damage-loss-and-needs-assessment-disaster-recovery-and

Habiba, U., Abedin, M. A., and Shaw, R. (2013). Disaster education in Bangladesh: Opportunities and challenges. *Disaster Risk Reduction Approaches in Bangladesh*, 307–330.

Haider, R., Rahman, A., and Huq, S. (1991). *'Cyclone'91: An environmental and perceptual study'*, Bangladesh Centre for Advanced Studies (BCAS), Dhaka, Bangladesh.

HDX. (2020). Bangladesh - Subnational Administrative Boundaries. Retrieved from https://data.humdata.org/dataset/cod-ab-bgd (accessed on 22 February 2022).

Hoffmann, R. and Muttarak, R. (2017). Learn from the past, prepare for the future: Impacts of education and experience on disaster preparedness in the Philippines and Thailand. *World Development, 96*, 32–51.

Horney, J., Snider, C., Malone, S., Gammons, L., and Ramsey, S. (2008). Factors associated with hurricane preparedness: Results of a pre-hurricane assessment. *Journal of Disaster Research, 3*(2), pp. 143–149.

IPCC. (2001). '*Climate change 2001: The scientific basis, contribution of working group-I to the IPCC Third Assessment Report*', Intergovernmental Panel on Climate Change (IPCC), Geneva, Switzerland.

Islam, A. (1998). *'Alternative strategies for flood hazard mitigation programme in Bangladesh: Floodplain management approach'*, The University Press Limited, Dhaka, Bangladesh.

Khan, M.R. and Rahman, E.M. (2007). Partnership approach to disaster management in Bangladesh: A critical policy assessment. *Natural Hazards, 41*(2), 359–378.

Khan, M. Shah Alam. (2008). Disaster preparedness for sustainable development in Bangladesh. *Disaster Prevention and Management, 17*(5), 662–671.

Lawrence, J., Quade, D., and Becker, J. (2014). Integrating the effects of flood experience on risk perception with responses to changing climate risk. *Natural Hazards, 74*(3), 1773–1794.

Lidstone, J. (1996). 'Disaster education: Where we are and where we should be', in Lidstone, J. (Ed.), *International perspectives on teaching about hazards and disasters*. Philadelphia: Channel View Publications, p. 3.

Lidstone, J. and Nielsen, S. (1999). Public education and disaster management: Is there any guiding theory? *Australian Journal of Emergency Management, 13*(3), 14–19.

Lindell, M. K. and Hwang, S. N. (2008). Households perceived personal risk and responses in a multi hazard environment. *Risk Analysis, 28*(2), 539–556.

Mahmood, M.N., Dhakal, S.P., and Keast, R. (2014). The state of multi-purpose cyclone shelters in Bangladesh. *Facilities, 32*(9/10), 522–532.

Mallick, D.L., Rahman, A., Alam, M., Juel, A.S.M., Ahmad, A.N., and Alam, S.S. (2005). Floods in Bangladesh: A shift from disaster management towards disaster preparedness. *Institute of Development Studies, 36*(4), 53–70.

McGee, T. K. and Russell, S. (2003). "It's just a natural way of life…" an investigation of wildfire preparedness in rural Australia. *Global Environmental Change Part B: Environmental Hazards, 5*(1–2), 1–12.

Mohibbullah, M., Gain, A. K., and Ahsan, M. N. (2021). Examining local institutional networks for sustainable disaster management: Empirical evidence from the South-West coastal areas in Bangladesh. *Environmental Science & Policy, 124*, 433–440. doi: https://doi.org/10.1016/j.envsci.2021.07.016

Nahar, N. and Khan, M. M. H. (2014). Disasters: Socioeconomic impacts in Bangladesh. *Banglavision, 13*(1), 58–67.

Paton, D. (2005). 'Community resilience: Integrating hazard management and community engagement', in Proceedings of the International Conference on Engaging Communities, Queensland Government/UNESCO, Brisbane. Available at: www.engagingcommunities2005.org/abstracts/Paton-Douglas-final.pdf (accessed on April 2, 2019).

Paton, D. and Johnston, D. (2001). Disaster and communities: Vulnerability, resilience and preparedness. *Disaster Prevention and Management, 10*(4), 270–277.

Paul, B.K. (2009). Why relatively fewer people died? The case of Bangladesh's cyclone Sidr. *Natural Hazards, 50*(2), 289–304.

Paul, B. K. and Dutt, S. (2010). Hazard warnings and responses to evacuation orders: The case of Bangladesh's cyclone Sidr. *Geographical Review, 100*(3), 336–355.

Paul, S. K. and Routray, J. K. (2011). Household response to cyclone and induced surge in coastal Bangladesh: Coping strategies and explanatory variables. *Natural Hazards, 57*(2), 477–499.

Petal, M. and Izadkhah, Y.O. (2008). 'Concept note: Formal and informal education for disaster risk reduction', in Proceedings of the International Conference on School Safety, Islamabad, Pakistan, 1416. Available at: http://citeseerx.ist.psu.edu/viewdoc/download?doi=10.1.1.371.284&rep=rep1&type=pdf (accessed on February 7, 2019).

Rahman, M.S. (2001), 'Disaster management and public awareness in Bangladesh', in Nizamuddin, K. (Ed.), *Disaster in Bangladesh, selected readings*. Dhaka: Disaster Research Training and Management Centre, University of Dhaka.

Ronan, K.R., Johnson, D.M., Daly M., and Fairley R. (2001). School children's risk perceptions and preparedness: A hazards education survey. *Australasian Journal of Disaster and Trauma Studies*. https://ci.nii.ac.jp/naid/10021902088/

Rao, D.P.N. and Markos, W. (2015). Demand determinants and use value of Gudumale recreational area, in Hawassa City, Ethiopia. *International Journal of Business Quantitative Economics and Applied management Research, 1*(11).

Sattler, D. N., Kaiser, C. F., and Hittner, J. B. (2000). Disaster preparedness: Relationships among prior experience, personal characteristics, and distress. *Journal of Applied Social Psychology, 30*(7), 1396–1420.

SEMP. (2003). *'DurjogeBecheThaka (in Bangla) campaign for popular education'*, Sustainable Environmental Management Program (SEMP), Ministry of Environment and Forest, Government of Bangladesh, Dhaka, Bangladesh.

Shaw, R. and Kobayashi, M. (2001). The role of schools in creating earthquake-safer environment. Organisation for Economic Co-operation, and Development (OECD) Workshop, Thessaloniki, 2001.

Shaw, R., Takeuchi, Y., Ru Gwee, Q., and Shiwaku, K. (2011). Disaster education: An introduction. *Disaster Education*, 7, 1–22.

UNDRR. (2021). What is the Sendai Framework for disaster risk reduction [online]. Available at: www.undrr.org/implementing-sendai-framework/what-sendai-framework (accessed on August 25, 2021).

UNISDR. (2006a). *'Newsletter ISDR Inform-Latin American and the Caribbean'*, Report No.13, United Nations International Strategy for Disaster Reduction (UNISDR), Geneva, Switzerland.

UNISDR. (2006b). *'Disaster risk reduction begins at School'*, World Disaster Reduction Campaign, United Nations International Strategy for Disaster Reduction (UNISDR), Geneva, Switzerland. Available at: www.unisdr.org/we/inform/publications/2105

Vaughan, C., Gack, J., Solorazano, H., and Ray, R. (2003). The effect of environmental education on schoolchildren, their parents, and community members: A study of intergenerational and intercommunity learning. *Journal of Environmental Education, 34*(3), 12.

14

NATURE-BASED SOLUTIONS IN COASTAL DISASTER MANAGEMENT IN BANGLADESH

Rezaur Rahman

Introduction

The coastal zone of Bangladesh consists of almost one-fifth of its land mass and is part of the largest delta in the world, the Bengal Delta. The 710-km-long coast consists mostly of the lower part of the Ganges-Brahmaputra-Meghna (GBM) basin (Figure 14.1). The land is eroding as well as accreting in the south in a very dynamic environment. Its location atop the Bay of Bengal causes it to regularly face a number of sea-borne hazards such as tidal flood, salinity, cyclones, storm surge, etc.

The coastal area of 27,738 sq km holds 22.4 million people with a population density of 807 persons/sqkm (GED, 2018). Most of the population in the coast are engaged in highly vulnerable occupations such as small-scale farming, artisanal fishing, and wage laboring (CEGIS, 2004). Lives and livelihoods are at extreme risk to various hazards especially to cyclonic storm surges. The world's deadliest cyclone occurred in the coast of the country in 1970 killing nearly half-a-million people.

Since 1960s, various protection measures both structural as well as non-structural, have been taken such as polders, shelters, afforestation, cyclone preparedness program, etc. Due to environmental impact of structural measures, non-structural options gradually started to receive more attention and nature-based solutions (NbSs) have gained traction in recent times. However, the NbSs are going through a host of problems while trying to establish themselves. This chapter describes the background of two major NbSs adopted in the coast: the barriers they have been facing and the way forward to amass full benefits of NbSs.

DOI: 10.4324/9781003253495-18

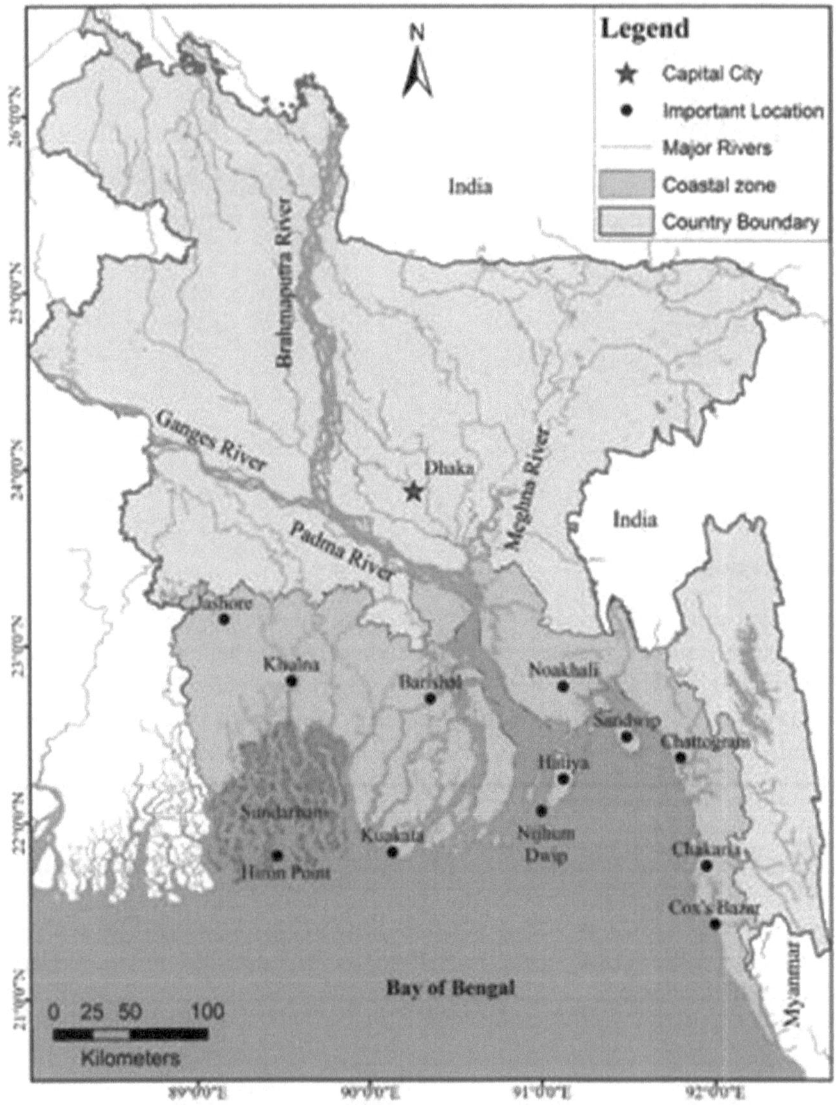

FIGURE 14.1 Coastal zone of Bangladesh showing locations referred in this chapter

Tidal Flood

Occurrence

The coastal region is extremely low and flat. Most of the land lie within 1 m of sea level. The entire coast is crisscrossed with numerous rivers, which carry tide and sediment inland from the sea. Tides are diurnal with normal period of about 12 hours. So, there are two high tides and two low tides every day. During dry season, when flow from upstream is less, tidal water pushes far inland and inundates surrounding land during high tide. The mean tidal range at Hiron Point near the Sundarbans in west coast is around 3 m, increases up to 6 m near Sandwip at central coast and then decreases to 3.6m further east at Cox's Bazar (Haque and Nicholls, 2018).

Protections

Saline water from the sea traverses the entire coast along with the high tide. Without embankments, tidal water will inundate the lower lands at regular intervals. The salt content of the tidal water is often high relative to crop tolerance causing damage to standing crops. To sustain agriculture, famers used to build low-height earthen embankments that allowed them to cultivate one crop a year. These embankments stayed in place for eight months during the drier period and would be washed away by high flow of the rivers during monsoon needing reconstruction next year. Because of their temporary nature, their impact on river-floodplain environment was minimal.

The low earthen embankments were constructed and maintained under the aegis of the Zamindars (landlords) who were vested with the responsibility of collecting revenue from the cultivators until the end of the British rule in 1947. However, following the departure of the British from the sub-continent, the Zamindari system was abolished and consequently the conventional system of embankment management came to an end (Rahman et al., 2021).

These low embankments deteriorated for lack of maintenance in absence of Zamindars and the need for protection in the coastal area was recognized by the then government (Leedshill, 1968). Accordingly, Coastal Embankment Project (CEP) was initiated in the 1960s, under which 139 polders were eventually built covering 1.28 million hectares of coastal land with 5,700km of embankments (GED, 2018). As can be seen from Figure 14.2, polders have been built throughout the coast with a higher preponderance in the southwestern part.

Poldering is a Dutch concept in which areas surrounded by rivers are protected by embankments on all sides. Sluices are provided at appropriate locations to drain excess rainwater from within poldered area especially during monsoon. Sluices are unidirectional hydraulic structures and do not allow any tidal flow inside from the rivers. This is the way the polders cutoff the tidal floodplain from the rivers impeding the natural sedimentation on the floodplain.

When there was no obstruction, rivers brought sediments during high tide and left them behind on surrounding floodplain during low tide. This is the natural process of building of Bengal Delta, the largest delta on the earth. Once disrupted, the sedimentation started occurring on the riverbeds. The siltation of rivers caused drainage congestion in vast areas since 1980s mainly in southwest region.

Auerbach et al. (2015) report that areas in southwest Bangladesh, enclosed by embankments in the 1960s, have lost 1.0–1.5m of elevation. They attribute this elevation loss to interruption of sedimentation inside the embankments, combined

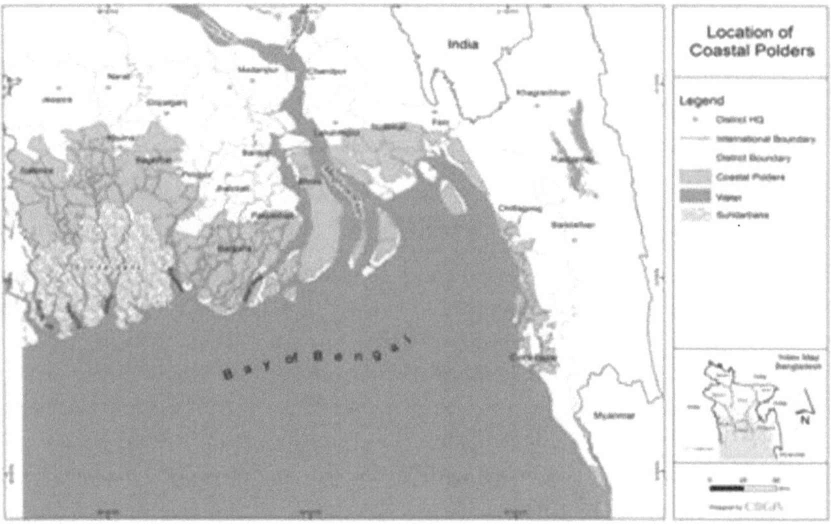

FIGURE 14.2 Spread of polders in the coastal zone

with accelerated compaction, removal of forest biomass, and a regionally increased tidal range. The feasibility study of CEP by Leedshill (1968) while investigating the impact of the projects on river sedimentation did not foresee large-scale sedimentation problem of the coastal rivers, which has become apparent by now.

Deadly Cyclones

> The destructive elements of a cyclone are the violent winds and the water heaped up under the center. In these alluvial coastal areas, the land is only just above the sea level at ordinary high tides, so that an additional 8 or 9ft of water is concentrated in the same manner as a tidal bore and rises even higher. When the cyclone has passed on, the heaped-up water retreats back again to the sea, and swimming men and cattle are carried out to sea with it. Villages in the path of the storm are flattened by the violent winds combined with the water. Thousands of men and cattle are drowned, crops and food supplies are destroyed, worst of all, wells and water reservoirs are filled with salt water (Hatch-Barnwell, 2011).

Occurrence

The Bay of Bengal is one of the most favored places on earth for generation of cyclone. About 10% of the global total of tropical cyclones forming in different regions of the tropics occur in the Bay of Bengal. Cyclones and storm surges are therefore common in the coast of Bangladesh causing large-scale casualties every now and then. Historically, eight out of ten deadliest tropical cyclones in the world have originated over the Bay Bengal. In fact, out of the 36 most deadly tropical cyclones in recorded history, 26 have been over the Bay Bengal. The Bay of Bengal occupying just 0.6% of the global ocean area is estimated to be responsible for four out of five cyclone-related deaths in the world (Pinto, 2021).

BUET & BIDS (1993) provides an account of past cyclones and related damage in the coast. Ain-E-Akbari written by courtier Abul Fazal during the reign of the Mughal Emperor Akbar contains the description of havocs in Chittagong coast caused by cyclonic storm surges. It mentions that in 1584, about 200,000 people were killed in Barishal by cyclonic storm surge. In 1790, severe cyclonic disaster occurred in Noakhali. More than 70,000 people of Barishal and 95% population of Hatiya were killed in 1822 by another cyclone.

The 1970 cyclone, the most fatal storm in the world history, killed an estimated 300,000 to 500,000 people when it made landfall in Bangladesh near Barishal, bringing a devastating storm surge of up to 9 m to the coast. The 1991 cyclone made a landfall in Chattogram coast with a storm surge up to 7.5 m and killed 140,000 people. The wind speed in both cyclones reached nearly 225 km/h (ibid). There are number of reasons for such cyclone proneness of the coast. Important among those are funnel-shaped coast, shallow bathymetry of the Bay of Bengal, and interaction of surge with tides. The triangular shape of the Bay of Bengal causes a funneling effect resulting in massive storm surges. The shallow bay allows to build the surge further. And if the landfall coincides with the high tide, it easily adds up another 1–2 m of surge height. So, while cyclone damages occur due to gusty winds and torrential rain, the most devastating damage on life and properties are caused by storm surges.

The frequency of cyclonic storms forming over Bay of Bengal is also very high. The two significant factors behind cyclone formation – sea surface temperatures and humidity – are often quite favorable over the Bay of Bengal during both pre-and post-monsoon. The sluggish winds around the Bay keep the surface temperatures relatively high at about 28–31°C, while constant rainfall keeps the humidity abundant (Pinto, 2021).Due to global warming, sea surface temperature (SST) in the Bay of Bengal is rising. Rahman and Uddin (2022)show that near the southwestern coast (around Hiron Point), the SST has a highly significant ($P < 0.001$) increasing decadal trend of over 1.3°C in the monsoon (April, May) and about 0.7°C in the post-monsoon period (October, November). Near the southeastern coast, SST has significant rising trends (0.8°C per 10 years in the monsoon and about 0.5°C per 10 years in the post-monsoon period). Since formation of cyclone is directly dependent on SSTs, therefore such rise in SST is expected to give rise to more cyclones in the Bay. Most likely as a precursor, a series of cyclones have made landfall in the coast in recent time such as Roanu in 2016, Mora in 2017, Fani and Bulbul in 2019, Amphan in 2020, and Yash in 2021.

Protection

Bangladesh is a champion in cyclone preparedness by undertaking series of measures over the years. This has resulted in dramatic fall in casualties since 1991 cyclone. For example, while cyclone Sidr in 2007is climatologically comparable with the 1970 and 1991 events, the consequence in terms of human casualties have been much less. Around 5,000 people died in cyclone Sidr, which is attributed to the enhanced disaster management practices imparted institutionally over last few decades in Bangladesh (Mallick and Rahman, 2013). Among various measures in practice, the following measures provide direct protection to the coastal population.

Shelter

After the 1970's cyclone, Bangladesh started building cyclone shelters in the coast to provide safe haven to the evacuees. After the 1991 cyclone, the need for more and better shelters was recognized further. Accordingly, a Multi-purpose Cyclone Shelter Programme (BUET & BIDS, 1993) was initiated. Under this program, the concept of multi-purpose use of shelter, especially, as school during normal time, was instituted. Currently, there are more than 4,000 shelters in the coast (GED, 2020). Although there is still need for more shelters, large-scale evacuations during cyclones made possible by the presence of the current shelters have been responsible in reducing death figures during cyclones.

Polders

Building embankments strong and high enough for protection against storm surge is not cost-effective except in economically important areas. Nearly 6,000 km of embankments under different polders, however, offer some degree of protection to the coastal inhabitants. These embankments are not designed to prevent overtopping or to resist the destructive wave surges that accompany cyclonic storms (Leedshill, 1968). However, if the embankments are overtopped or even breached, the obstruction of the embankments provides a reduction in the height and intensity of the storm wave within the project area and a delay in time of the flooding, which offer some protection to polder inhabitants and their property.

Forests

Sundarbans mangrove forest provides great protection against cyclone by lowering wind speed as well as reducing the surge impacts. Recognizing this lifesaving role of forests in the coast, Bangladesh has created a green belt of mangrove forest over 200,000 hectares of land along the coast since 1960s (BFD, 2016). Dasgupta et al. (2019) show that mangroves have the capacity to reduce the rise in seawater levels between 4 and 16.5 cm and water inflow speed to a range between 29% and 92% for Sidr-like cyclone contributing significant savings in rehabilitation and maintenance costs by protecting embankments from breaching, toe-erosion, and other damage. Mita (2019) found that in general, presence of afforestation on the foreshore of the land can significantly reduce surge propagation extent and surge depth on the land. Density and width of forest play a pivotal role in determining the effectiveness of mangrove afforestation in attenuating surge propagation and reducing thrust force on embankment in the hinterland.

Despite various protection measures, the coastal region remains highly vulnerable to cyclone due to very high population density (GED, 2018), subsistence living (CEGIS, 2004), and increasing sea surface temperature due to climate change (Rahman and Uddin, 2022). Although warning capability has improved and shelter capacity has increased over the years, still these prove inadequate during time of crisis. There are also social and cultural barriers like access, safety, and security of women, toward full utilization of safe haven capacity that increases the vulnerability of the people at risk.

Nature-Based Solution

The definition of NbS by IUCN – the principal proponent of this types of solution, is "actions to protect, sustainably manage, and restore natural or modified ecosystems that address societal challenges effectively and adaptively, simultaneously providing human well-being and biodiversity benefits". NbSs are globally gaining popularity and are increasingly seen as a strategy to address the most pressing societal challenges such as climate change and food security. At least 66% of Paris Agreement signatories included NbS in some form to help achieve their climate change mitigation and/or adaptation goals (Sinha and Bimson, 2021).

In the coast, two measures namely tidal river management (TRM) and afforestation can be considered as NbS by assessing them against eight principles of NbS (Gain et al., 2022). The evolution of these measures, their status, and the required improvements to firmly establish them as NbS are discussed below.

Tidal River Management

Background

The polders in the coast initially were considered a success as it raised crop production within the polders by protecting the crops from saline inundation. However, siltation of the rivers caused severe drainage congestion in many places especially in Jashore and Khulna districts. Large tracts of land became unfit for cultivation due to prolonged waterlogging. The implementing agency, Bangladesh Water Development Board (BWDB) undertook a project named Khulna-Jashore Drainage Rehabilitation Project (KJDRP) during the period of 1994–2004 with the objective to relieve the project area of 100,000 hectares from waterlogging. The options they considered initially were first dredging all the silted-up rivers then building a big regulator downstream to stop tidal flow to the project area. The idea was that if the tidal flow is stopped, then the source of sediment is cut off and as a result, rivers inside the project area will be free from any chance of sedimentation.

However, the local people were not convinced about this regulator option. Their apprehension was that it was just shifting the problem downstream. They feared that there will be sedimentation in front of the regulator eventually blocking the upstream flow from project area. As a result, the drainage congestion will reappear soon. The local population called for a sustainable solution to drainage congestion and suggested TRM. TRM is an indigenous solution based on local knowledge of living with tidal environment. In KJDRP area (Figure 14.3) several *beels*

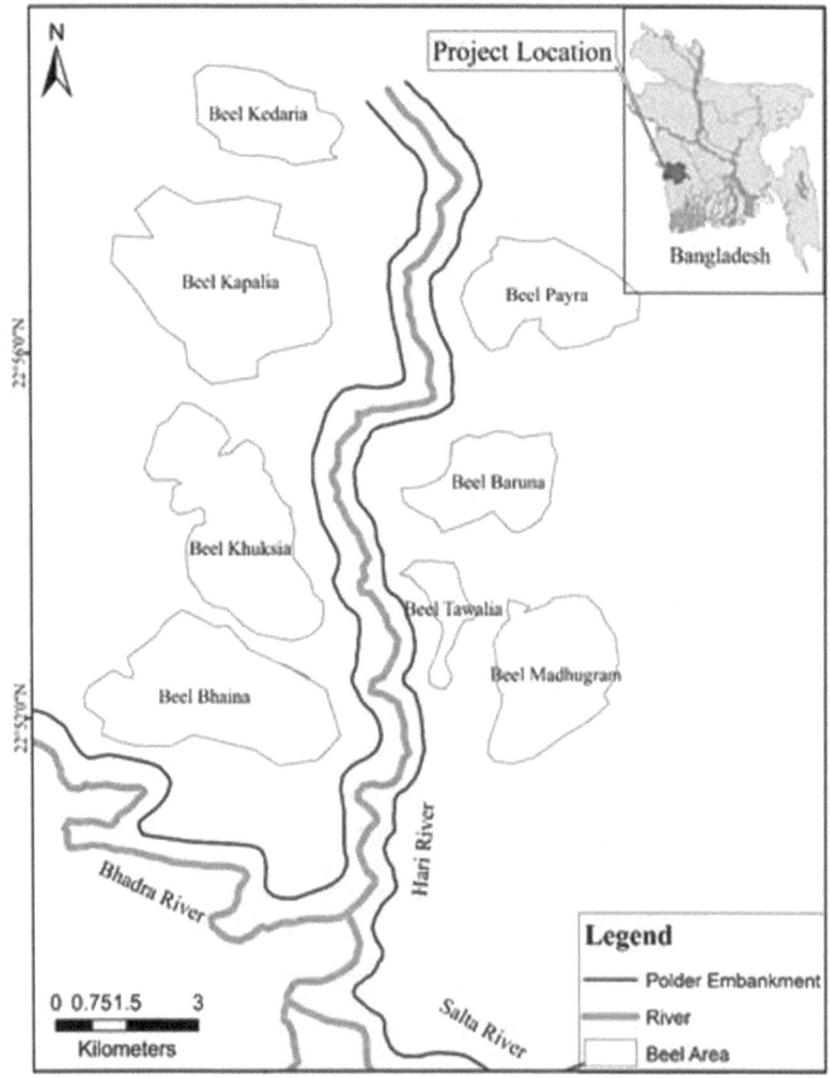

FIGURE 14.3 Potential beels for TRM in KJDRP area

were preliminarily selected for TRM to relieve the drainage congestion in the Hari River system. *Beel* is the local term for natural depression in the landscape that holds some water at the deepest pocket round the year.

TRM is temporary de-poldering in tidal river. A cut is made at a suitable place on the embankment and tidal water is allowed to enter the polder via a natural or man-made canal and flow to a *beel* that then starts acting as tidal basin. Sediments are deposited on these *beels* rather than in the rivers and these lowlands are gradually raised mimicking the delta building process. Once these lowlands are sufficiently raised and become fit for perennial agriculture, then the cut on the embankment is closed and another *beel* is brought under tidal operation – a process named rotation, which is carried out for 4–5 years in each *beel*.

When a *beel* is being is used for TRM, the farmers (crop or fish) of that *beel* cannot use their land for any productive purposes. This causes some hardship for them. But the local people agreed to this temporary suffering for permanent solution to the waterlogging problem. By rotating TRM among the *beels*, it was assumed that the sufferings will be shared among the project beneficiaries.

Problems during Rotations

While BWDB was developing structural measures such as a bigger regulator downstream to restrict tidal flow to the project area, people got impatient and despite legal threats from BWDB like filing cases, forcefully opened Beel Bhaina to natural tidal flow in 1997. The effect was dramatic. Sediment deposition on the Beel Bhaina tidal basin rapidly raised its land to make it fit for cultivation within 2 to 3 years. The entire project area became free from drainage congestion as rivers reclaimed its conveyance capacity.

The advocates of TRM became more confident about this local solution and started opposing any further structural intervention more forcefully. BWDB eventually relented and adopted TRM under KJDRP. They prepared Beel Kedaria for TRM operation and opened in 2002. The TRM operation successfully ran till 2005 as per plan keeping the project area free from waterlogging. However, to the disappointment of farmers of Beel Kedaria area, unlike Beel Bhaina, the land at Beel Kedaria was not raised much. As Die (2013) points out this might be caused by the fact that Beel Kedaria is positioned approximately 19 km upstream from the confluence of the Hari and the upper Bhadra river, which makes tidal influence in Beel Kedaria less than Beel Bhaina.

By that time, next *beel* in rotation Beel Khukshia was not ready and people of Beel Khukshia were not agreeable to lend their *beel*. As a result, TRM could not be started immediately leading to silting up of Hari River almost immediately. Waterlogging returned to the project area with a vengeance. Seeing this effect, which also affected them greatly, Beel Khukshia inhabitants reluctantly agreed and BWDB could start next round of TRM in Beel Khukshia near the end of 2006. The famers, however, demanded compensation for TRM in their land. But the compensation failed to materialize, and famers got frustrated. Operation in Beel Khukshia eventually stopped in 2013 after overrunning.

Next rotation was due in Beel Kapalia. Fish famers of Beel Kapalia, learning from experience with Beel Khukshia demanded compensation in advance. But due to various administrative problems, which could not be overcome, this could not be arranged and people turned against TRM. A violent protest ensued and twelve government and BWDB cars were burnt during demonstration in June 12. Die (2013) found that although overwhelming majority was in favor in temporary inundation of *beels* to prevent waterlogging, an ineffective compensation mechanism and the absence of accountable water management institutions were found to be the root causes of this conflict.

Till now, TRM remains suspended in KJDRP area. The project area is suffering from waterlogging after every heavy rainfall. BWDB has tried to convince famers of Beel Kapalia to hand over their land for TRM but with no avail. So, to keep the project area free from waterlogging, BWDB is dredging the rivers continuously and even engaged pumping at few places. These are expensive options and not sustainable in the long run. Such piecemeal approach instead of reviving TRM is being criticised (e.g., see editorial of highly influential Bengali daily – Prothom Alo, 2021).

Improvements

Although TRM is an indigenous solution, it is technically very sound. Adnan et al. (2020) identified 106 *beels* suitable for TRM in southwest region. Their modeling of potential sediment deposition predicted that the consequent increase in land elevation could be up to 1.4m in five years, which would alleviate land subsidence and reduce flood susceptibility. The study concludes that while during the implementation of TRM agriculture must cease for a few years, a systematic program of TRM could result in a long-term increase in agricultural production by reducing flood susceptibility of agricultural lands in delta regions.

The coastal region is also extremely vulnerable to sea-level rise (SLR). TRM can also be an effective tool to counteract the effects of SLR (Cornwall, 2018). Islam et al. (2021) conclude that owing to the abundant availability of sediment, application of controlled flooding like TRM provides an opportunity to counteract the impact of SLR by means of land raising, particularly along the tidal river reaches in the GBM delta.

TRM is a locally developed solution, and it was expected that it will be locally managed. The selection of *beels* for next round of rotation was vested on local people to reduce social conflicts. BWDB was responsible for preparation of the *beels*. There were failures on both sides. Water Management Organizations, which were apolitical and formed during the project phase, became marginalized by political forces (Dewan et al., 2015) or dysfunctional after the end of the KJDRP project (Rahman, 2008). BWDB also failed to prepare *beels* for tidal operation in due time due to administrative procedures that are simply not suitable for urgency that is required for such time-bound works (ibid). As a result, TRM could not be properly implemented, and waterlogging remained as a recurrent problem.

Beels are ecologically important sites. They support fish and migratory birds. So, siltation of all *beels* is not a desirable option although farmers are happy with raising their lands. Shampa and Pramanik (2012) suggest that siltation should be allowed in a certain amount like 1m or 1.5m, which also ensures the required environmental flow in the *beel*. Sediment inside the *beel* could be dredged yearly and by this way one *beel* can act as tidal basin for many years. It saves the biodiversity of other *beels* in the study area and at the same time keeps the river drainable and navigable for many years.

Afforestation

Background

According to BUET & BIDS (1993), in the late 1960s, satellite images revealed that almost 5,000 sqkm of newly formed lands were rising in the coast. To stabilize and consolidate these extremely fragile and unstable accretions, government decided to take a massive afforestation program with the following major objectives.

- Accelerate the process of siltation and stabilization of the soil.
- Create forest shelter belts to protect life and property inland from tidal bores.
- Create an urgently needed resource to add to the national wealth.
- Create job opportunities for the rural communities.
- Create an environment for wildlife, fishes, and other estuarine and marine fauna.

The forest department started the afforestation program in 1966 in modest scale. Later it was accelerated under different government programs with assistance from donor agencies like World Bank and Asian Development Bank. So far, more than 200,000 hectares of plantations have been established. However, at present, only 61,574 hectare of mature plantations have been found to be in existence, the rest have either been eroded by the sea, returned to Land Department, or destroyed under biotic pressure (BFD, 2016).

Problems at successions: the afforestation program has faced varieties of constraints and problems during its course. Most importantly, the government requirement that the land be returned by Forest Department to Land Department 20 years after plantation. The land is then cleared of forests and used for making new settlement and economic activities such as agriculture, fisheries, and salt farming. It is also true that after some time, land is raised so much that it becomes unsuitable for mangrove forest. Since people know that land will eventually be cleared of forests, they start encroaching much before 20 years to establish land rights and violent conflicts often results between different sections of people. Local musclemen first establish dominance and because of difficulty of access to remote lands, a rule of lawlessness prevails. They are often behind inciting poor people to encroach forests for resource abstraction. Such encroachment has been the primary cause of mangrove loss after they have been established.

Saenger and Siddiqi (1993)noted four types of land stability problems in a very dynamic coast that have caused loss of mangroves from existing plantations in the coast. They are burial of mangrove seedlings when and where sediment accretion rates are particularly high; smothering of seedlings by sand in areas where wave action reworks large volumes of sand shore-wards; the winnowing of fine sediment (clay and silt) from the plantation site during prolonged stormy periods, leaving a mobile sandy lag deposit; and the erosion of plantation margins through bank slumping by migrating tidal and river channels.

CASE STUDY –NIJHUMDWIP (ISLAM ET AL., 2021)

NijhumDwip National Park (NDNP) has a total area of 163.5 sq km located in the southern part of Bangladesh. The land is mostly made up of intertidal mudflats and sandy flat lands. The land cover of this national park has been changed dramatically over the years, particularly the forest cover, because of human interaction, such as illegal harvesting and conversion of forest land to agricultural land.

More than 42% of the total forest cover area (1,781 hectares) has been deforested over the decades between 1990 and 2011. The study also reported that nearly 1,235 hectares and over 856 hectares of forest area disappeared from 4217.5 hectares in 1990 to 2982.5 hectares and 2126.16 hectares in 2001 and 2011, respectively. In contrast, an increase of 310 hectares of forest area has been recorded in the last decade, from 2011 to 2020.

This research revealed that the forest cover changed drastically from 1990 to 2020, particularly in the period of 1990 to 2001, nearly one-third of the forest cover area was deforested. Despite an increase of 310.32 hectares in forest cover from 2011 to 2020, deforestation has continued until the last decade, resulting in large areas of forest cover loss. The most significant positive change was that the total land area nearly doubled from 6487.83 hectares (39.70%) in 1990 to 11745.81 hectares (71.90%) in 2020, indicating a higher possibility of new accreted land in the future if we can ensure better management and protection of this mangrove forest.

As newly accreted lands become more stabilized and ecological conditions change (soil type, frequency, duration of inundation, etc.), these become increasingly unsuitable for the initial pioneer species such as Keora (*Sonneratiaapetala*). Such conditions typically start to occur after 4–5 years of initial plantation. Mature Keora trees, which make up most coastal plantations, are particularly susceptible to pests such as stem borer, which usually come from neighboring agricultural land. This creates large gaps in monoculture plantations where pest infestations can spread quickly. These gaps can no longer be filled through new Keora saplings (i.e., through natural regeneration of Keora), as the ecological conditions that were originally suitable for Keora have changed. However, in the absence of a nearby source of alternative mangrove species suited to these changed ecological conditions, the gaps persist, become larger and/or are invaded by weed species. Losses of between 50% and 80% of planted trees due to pest infestations resulting in big gaps in greenbelt structures have been reported. Susceptibility to pests and diseases will be further aggravated by climate change. Thus, after a few years, mangrove plantations that were established with a handful of pioneer species require increasing levels of management to promote "artificial succession" if they are to avoid such losses of trees (UNDP, 2015).

Shrimp culture started in the 1980s when the afforestation program was gaining momentum. There was large-scale destruction of both natural and man-made mangrove forests. Most famous clearance was the Chakaria Sundarban in the Chittagong coast – the only remaining natural mangrove forest in that region. Clearing of that forest is blamed for very high loss of human lives during 1991 cyclone (BUET & BIDS, 1993).

Community participation in forest management is now being encouraged for maintenance of the plantations afterward and discourage encroachment into the forest and abstraction of forest resources. In order to foster local participation, Forest Department is engaging local villagers as laborers and training them in nursery and plantation techniques. Alternative income-generating activities such as livestock rearing, vegetable cultivation, etc. are being promoted to reduce pressure on forests. Community-level activities are also being promoted, which include establishing cooperatives to improve marketing of local products, ecotourism, and making energy-efficient cook stoves (Chow et al., 2019).

Improvement

One of the major goals of the recently formulated Forestry Master Plan 2017–2036 (BFD, 2016) is creation of a strong coastal shelterbelt of climate-resilient plantations on newly accreted islands and other unused public lands. Approximately 100,000 hectares of coastal plantations of climate-resilient mangrove and other species on unused public lands and rising islands shall be carried out. Land may be acquired for coastal plantations if suitable public land is not available. Institutional areas, roadsides, embankments, dikes, and all available land shall be planted to complete the greenbelt. Main features of the coastal afforestation and the greenbelt initiative shall be as follows:

- Continue to establish coastal plantations with climate-resilient mangrove and non-mangrove species with good carbon sequestering abilities and create a multilayered multispecies tree formation, which also allows accrual of some benefits to the neighboring communities while continuing to perform their protective role.

- Create a permanent "greenbelt" with mangrove and non-mangrove species on newly accreted land as well as acquired land where accreted land is not available.
- Ensure protection of these plantations with support and cooperation from the local communities and, through the creation of plantation protection units, whose sole function will be to ensure protection of the coastal plantations including greenbelts.
- Review the current rules governing the coastal plantations and make appropriate changes to make them more effective. In addition, as the shelterbelt is a permanent protective feature, current policy on 20-year vesting of forest with the Forest Department will have to be revised to ensure that there is a provision in the policy for a permanent retention of land under the greenbelt by the Forest Department.

Afforestation is recognized as a very important adaptation as well as mitigation measure in the climate change context. Bangladesh is one of the most vulnerable countries to increased threat of SLR and cyclonic storms due to climate change. Coastal greenbelts can be an armor against such looming threats. Afforestation also has great mitigation potential. Mangroves play a significant role in sequestering of carbon and reducing greenhouse gases (Islam and Rahman, 2015). The Forest Investment Plan (FIP) formulated in 2017 promotes sustainable forests and livelihoods for carbon sequestration, resilience to climate change, and biodiversity conservation and improved livelihoods.

Conclusions

In the coast, two measures that are in practice can be considered as NbS. One is TRM which is an indigenous solution to waterlogging resulting from massive polderization in the coast. Another one is creation of a centrally planned greenbelt to protect the coastal population from cyclonic storm surges. These NbSs are in practice for considerable period all the while facing a series of problems. For example, in case of TRM, every rotation of beels and in case of afforestation, every succession of plantation gave rise to new problem.

The problems are technical, social, and institutional in nature. This is inevitable in a fragile delta and overpopulated land-scarce country. NbSs facing host of problems is not unusual (e.g., see Sarabi, 2020). Despite the problems faced by TRM and afforestation, they have gained traction in the coast. To make them properly perform their intended function, a proactive management approach needs to be adopted. They need to be continuously monitored and actions are to be taken to rectify any ensuing problem as soon as they appear. In other words, adaptive management is required in order to make these NbSs fully functional.

When TRM or coastal afforestation was launched, they were not known as NbS. Now that the benefits of NbSs are well known, therefore proper steps should be taken to meet the various criteria of NbS for each measure so that full benefits from these measures as NbS can be harnessed. For example, instead of seeing coastal afforestation as a mere extension of plantation, it needs to be reimagined as a harbor of biodiversity.

Due to climate change, the coast of Bangladesh is at great risk to SLR and cyclonic storm surges. Various studies have shown great potential of TRM and afforestation in mitigating the risk of SLR and cyclonic storm surges, respectively. This is one more reason that makes it so imperative to nurture them as NbSs.

References

Adnan, M.S.G., Talchabhadel, R., Nakagawa, H., & Hall, J.W. (2020). The potential of Tidal River Management for flood alleviation in Southwestern Bangladesh. *Science of the Total Environment*, 731, 138747.

Auerbach, L.W., Goodbred Jr, S.L., et al. (2015). Flood risk of natural and embanked landscapes on the Ganges–Brahmaputra tidal delta plain. *Nature Climate Change*, 5: 153–157.

BFD (2016). Bangladesh Forestry Master Plan 2017–2036; (Draft Final). Bangladesh Forest Department, December.

BUET & BIDS (1993). Multipurpose Cyclone Shelter Programme. Final Report, part 4, Volume IV, Planning and Implementation Issues. Planning Commission, Government of Bangladesh.

CEGIS (2004). Vulnerability Analysis of Major Livelihood Groups in the Coastal Zone of Bangladesh. Final Report submitted to Food and Agriculture Organization (FAO). Center for Environmental and Geographic Information Services, Dhaka, April.

Chow, J., Khanom, T., Hossain, R., & Khadim, J. (2019). *Forest management for climate change adaptation in Bangladesh*. In: S. Huq et al. (eds.), *Confronting climate change in Bangladesh: Policy strategies for adaptation and resilience*. Springer. 39–50.

Cornwall, W. (2018). Bending to the water's will. *Science*, 359(6379): 980–985. DOI:10.1126/science.359.6379.980

Dasgupta, S., Islam, M. S., Huq, M., Huque Khan, Z., & Hasib, M.R. (2019). Quantifying the protective capacity of mangroves from storm surges in coastal Bangladesh. *PLoS ONE*, 14(3): e0214079.

Dewan, C., Mukherji, A., & Buisson, M.C. (2015). Evolution of water management in coastal Bangladesh: from temporary earthen embankments to depoliticized community-managed polders. *Water International*, 40(3): 401–416.

Die L.D. (2013). *Tidal River Management: Temporary depoldering to mitigate drainage congestion in the southwest delta of Bangladesh*. Minor thesis Water Resources Management submitted in partial fulfillment of the degree of Master of Science in International Land and Water Management at Wageningen University, the Netherlands.

Gain, A. et al. (2022). Overcoming challenges for implementing nature-based solutions in deltaic environments: insights from the Ganges-Brahmaputra delta in Bangladesh. *Environmental Research Letters*, 17: 064052.

GED (2018). Bangladesh Delta Plan 2100. General Economics Division (GED), Bangladesh Planning Commission, Government of the People's Republic of Bangladesh.

GED (2020). 8th Five Year Plan. General Economics Division (GED), Bangladesh Planning Commission, Government of the People's Republic of Bangladesh.

Hatch-Barnwell, S. (2011). *The last guardian: Memoirs of Hatch-Barnwell, ICS of Bengal*. The University Press Limited, Dhaka, Bangladesh.

Haque, A. & Nicholls, R.J. (2018). Floods and the Ganges-Brahmaputra-Meghna Delta. In: R.J. Nicholls et al. (eds.), *Ecosystem services for well being in deltas*. Palgrave Macmillan.

Islam, M.R., Khan, M.N.I., Khan, M.Z., & Roy, B. (2021). A three decade assessment of forest cover changes in Nijhumdwip national park using remote sensing and GIS. *Environmental Challenges*, 4:100162.

Islam, M.F., Middelkoop, H., Schot, P.P., Dekker, S.C., & Griffioen, J. (2021). Spatial and seasonal variability of sediment accumulation potential through controlled flooding of the beels located in the polders of the Ganges-Brahmaputra-Meghna delta of Southwest Bangladesh. *Hydrological Processes*, 35:e14119.

Islam, S.A. & Rahman, M.M. (2015). Coastal afforestation in Bangladesh to combat climate change induced hazards. *Journal of Science, Technology & Environment Informatics*, 2(1):13–25.

Leedshill-De Leuw Engineers (1968) Coastal Embankment Project. Engineering and Economic Evaluation: Volume 1. Dacca.

Mallick, F. & Rahman, A. (2013). Cyclone and tornado risk and reduction approaches in Bangladesh. In: R. Shaw et al. (eds.), *Disaster risk reduction approaches in Bangladesh*. Springer.

Mita, K.S. (2019). *Role of mangrove plantation in improving embankment safety and supporting local livelihoods*. M.Sc. thesis in Water Resources Development. Institute of Water and Flood Management, Bangladesh University of Engineering and Technology, Dhaka.

Pinto, D.N. (2021). A brief history of deadly Bay of Bengal cyclones that have caused 80% of world's cyclone-related deaths, 26 May, TWC India.

ProthomAlo (2013). Water logging at Bhabadaha– draining out water is not effective, build tidal basin. *Editorial*, June 13.

Rahman, A. & Uddin, M.N. (2022). Challenges and opportunities for saline agriculture in coastal Bangladesh. In: K. Negacz et al. (eds.), *Future of sustainable agriculture in saline environments*. Boca Raton, Florida: CRC Press.

Rahman, M.A., Dawes, L., Donehue, P., Rahman, M.R. (2021). Cross-temporal analysis of disaster vulnerability of the southwest coastal communities in Bangladesh. *Regional Environmental Change*, 21:67.

Rahman, R. (2008). *Investigation into replicability of good practices in flood management*. Research Report Institute of Water and Flood Management, Bangladesh University of Engineering and Technology, Dhaka.

Saenger, P. & Siddiqi, N.A. (1993). Land from the sea: The mangrove afforestation program of Bangladesh. *Ocean and Coastal Management*, 20(1):23–39.

Sarabi, S. (2020). Uptake and implementation of nature-based solutions: An analysis of barriers using interpretive structural modeling. *Journal of Environmental Management*, 270: 110749.

Shampa & Pramanik, M.I.M. (2012). Tidal River Management (TRM) for selected coastal area of Bangladesh to mitigate drainage congestion. *International Journal of Scientific & Technology Research*, 1(5).

Sinha, V.R. & Bimson, K. (eds.) (2021). *Nature-based solutions in the Ganges Brahmaputra Meghna (GBM) river basin: Case studies and lessons learned*. Thailand: IUCN.

UNDP (2015). *Integrating community based adaptation into afforestation and reforestation programmes in Bangladesh*. Project Document.

15

BLUE REVOLUTION IN COASTAL BANGLADESH

A Call for an Inclusive Policy and Sustainable Governance

Choyon Kumar Saha and Md. Masud-All-Kamal

Introduction

Shrimp aquaculture has gained significant traction in coastal Bangladesh since late 1970s due to increased demand of shrimp in international markets. The growth of shrimp farming in coastal zones has directly been supported by the government's policies that offer economic incentives, technology, and infrastructural advancement. The government of Bangladesh first acknowledged shrimp aquaculture as an export-oriented sector in the Second Five Year Plan (1980–1985), which explicitly prescribed the expansion of shrimp production across coastal zones (Haque, 1994). This official endorsement was driven by the government's efforts to foster economic growth, employment generation, rural vitalization, food security, agrarian development, and rural poverty alleviation (Ravallion, 1990). As a result, the area of shrimp farms increased from about 115,000 hectares to 215,305 hectares between 1986 and 2015 (Farmer, 1989; The World Bank, 1985; Belton, 2016). The coastal regions have appeared as a shrimp production hub, containing roughly 40,000 shrimp farms (Ahsan and Brandt, 2015). Such a rapid expansion was also explicitly driven by the involvement of transnational development actors and financial institutions, who financed to upgrade the production processes, and build shrimp estates, technological infrastructures, and hatcheries through the implementation of aquaculture development projects.

The large private entrepreneurs in Bangladesh started to engage in commercial shrimp production because they foresaw the potential of earning high profits from this export-oriented industry. Governments' market-oriented policies (i.e., shrimp policies in 1992, 1998, 2013, and 2014) have expanded the scope for the rise and engagement of these actors in advancing and spreading shrimp

DOI: 10.4324/9781003253495-19

aquaculture. Additionally, the increase in shrimp prices in international market had driven the upswing of petty businessmen and vested interest groups in this lucrative sector. Bangladesh is one of the largest shrimp-producing nations, which shares about 2.5 percent of the global shrimp trade and earns around USD550 million in 2016–2017 FY. It provides employment and livelihood opportunities to 1.2 million people, supports approximately 4.8 million dependents (EJF, 2014), and increases coastal households' income up to 36 percent (Ray et al., 2021). Within the broader landscape of 'blue economy' (Wenhai et al., 2019), the rapid growth of shrimp farming and its significant contributions to socioeconomic development and human well-being is described as the 'blue revolution'(Deb, 1998; Garlock et al., 2020).

While a burgeoning body of scholarship to date has paid attention to the adverse impacts of shrimp farming, there remains a lack of analysis about the driving forces that have spurred the diffusion and intensification of shrimp farming throughout coastal Bangladesh. In particular, an analysis is absent about how government policies have buttressed the export-oriented shrimp aquaculture. Given this context, this chapter examines sociopolitical and economic drivers of the blue revolution and how different policy initiatives have intensified the blue revolution. Additionally, by synthesizing the existing literature on adverse outcomes of the blue revolution, this paper assesses the extent to which the so-called blue revolution undermines the vulnerability of disaster-inflicted coastal communities. Considering the evolving effects of blue revolution, this research underscores the necessity of restructuring current policies to make shrimp aquaculture socially inclusive and equitable, as well as environmentally sustainable that are the underlying characteristics of the blue economy (Smith-Godfrey, 2016; Bennett et al., 2019). An inclusive policy framework can ensure the effective governance of coast-based shrimp aquaculture, and thereby creating conditions for a socially just and sustainable blue economy (Cisneros-Montemayor et al., 2021).

We conducted a systematic literature review to gather and analyze data. Various electronic databases (e.g., Web of Science and Science Direct) have been utilized to collect scholarly and gray literatures. To examine the early development of shrimp farming, we gathered information from archival sources since no research has been conducted to date to explain the drivers of blue revolution and the relation of government policies to the spread of shrimp aquaculture. Data related to shrimp trade were gathered from the Bangladesh Frozen Foods Exporters Association (BFFEA), Export Promotion Bureau of Bangladesh (EPBB), and Fisheries Resources Survey System (FRSS). In the following section, we describe shrimp policies that have inspired driving forces in diffusing shrimp aquaculture. We then broadly examine various sociopolitical and economic forces that shape and reinforce the expansion of coastal shrimp aquaculture, one of the core frontiers of blue economy. We then present effects of blue revolution that undermine vulnerable conditions of coastal people and the environment. Finally, we call for creating an inclusive policy framework to govern the shrimp industry more equitably and sustainably.

Overview of Shrimp Aquaculture Policies in Bangladesh

The government formulated 'Shrimp Mohal Management Policy' in 1992 that supported the conversion of *khas* land (state-owned degraded land) into industrial shrimp farms (Pokrant and Reeves, 2005). As demand for shrimp and its prices continued to grow, the government revised shrimp policies after three years for the modernization of production processes, market expansion, and technological advancement. In 1998, the government introduced a new comprehensive National Fisheries Policy, which recognized shrimp as an exportable and profitable commodity that encouraged entrepreneurs and private sectors to be engaged in shrimp production and market expansion (MoFL, 1998). It widely inspired market actors through providing various incentives such as infrastructural development, shrimp insurance, laboratory, and technological facilities (Table 15.1). The revised policy of 1998 drove the rapid rise of business actors in this lucrative sector by the end of 1990s.

TABLE 15.1 Shrimp policies and growth-oriented features

Shrimp policies	*Key features*
Shrimp Mohal Management Policy 1992	• Leasing government land to local shrimp farmers • Providing loan with low interest • Exemption of business tax
National Fisheries Policy 1998	• Recognizing shrimp as an exportable commodity • Infrastructure development in shrimp farming areas • Encouraging private sectors for building commercial hatchery • Strengthening markets, modernization of laboratory and technological facilities • Collaborating with technologically developed nations
The Shrimp Plot Lease, Renewal, Management and Development Policy 2013	• Introduced health regulations for exportable products • Promising to boost shrimp-based foreign currency • Leasing of 7021.76 acres of government land to private producers and entrepreneurs • Technological development for shrimp farmers and entrepreneurs
National Shrimp Policy 2014	• Introduced modern technologies • Providing economic incentives to producers who face unexpected loss from natural hazards • Encouraging the investment of private sectors, cooperatives, and financial institutions • Expansion of international markets • Emphasizing the standard of exportable products • Encouraging financial institutions to finance with flexible conditions • Introducing shrimp insurance • Encouraging local actors to engage in farming

Source: Authors' compilation.

To expand shrimp farming in southeast coastal areas, the government created a special policy in 2013 known as Shrimp Plot Lease, Renewal, Management and Development Policy. This policy offered the leasing of a thousand acres of government land to private entrepreneurs that previously leased to local farmers (direct producers) (MoFL, 2013). This specialized policy expanded the opportunity to form an overarching National Shrimp Policy in 2014 that welcomed cooperatives, credit-providing institutions, commercial banks, insurance companies, and private sectors to be engaged in the expansion of shrimp industry (MoFL, 2014).

Drivers of Blue Revolution

Markets and Foreign Currency

The government policies adopted over the last three decades have directly promoted shrimp production and expanded opportunities for creating local and international markets through offering diverse incentives (e.g., tax exemption) to producers and investors. The process eventually leads to the earning of foreign currency from exporting shrimp and its related products to international markets. The country exports about 97 percent of its farmed shrimp in the United Kingdom (UK), Canada, Japan, United States of America (USA), China, France, Russian Federation, Saudi Arabia, Australia, and the European Union (EU) (SNI, 2014; MoFL, 2018). The UK is the most significant international market as it imports 70 percent of the total shrimp from Bangladesh (Molla, 2018). Bangladesh currently exports frozen shrimp products to more than 50 international markets and the annual growth rate of exporting shrimp is about 4.67 percent. Shrimp earns a large volume of revenue, which makes this sector one of the country's largest earners of foreign currency. The government considered Bangladesh is in a comparative advantage position for shrimp aquaculture because of copious shrimp fries, suitable climate, and cheap laborers.

It is estimated (Table 15.2) that foreign currency from exporting shrimp products has steadily been increased since the end of the 1990s. The sector faced a considerable decline in 2001–2002, 2002–2003, 2007–2008, 2008–2009, 2009–2010, 2012–2013, 2014–2015, 2015–2016, and 2016–2017. It continued to fall to USD348 million in 2018–2019 (Hossan, 2021). Despite significant declines in foreign earning, the opportunity of profits and benefits has encouraged financial institutions and market actors to continue to expand shrimp farming. The demand for homegrown shrimp has increased rapidly in domestic markets, despite local consumers view it as an expensive product. One estimation shows that shrimp price in local markets ranges between TK 350 and 375 per kilogram (Parvez, 2015). The increase in purchasing power of local consumers has created a strong impetus among firms, producers, and interest groups to expand shrimp farming areas and its production across the coast. The expansion of markets in home and abroad, however, has appeared as one of the most powerful economic forces that inspires business actors and financial institutions to invest in shrimp production for profits, incentives, and benefits.

TABLE 15.2 Foreign currency earned from shrimp exports

Year	*Export value (million USD)*
1999–2000	322.43
2000–2001	349.75
2001–2002	252.18
2002–2003	297.04
2003–2004	362.87
2004–2005	365.82
2005–2006	403.58
2006–2007	456.98
2007–2008	368.92
2008–2009	353.48
2009–2010	371.66
2010–2011	459.64
2011–2012	471.67
2012–2013	454.93
2013–2014	550.16
2014–2015	509.72
2015–2016	472.37
2016–2017	446.04

Source: Data compiled from BFFEA (2008), FRSS (2015), and EPBB (2017).

Development Actors, Financial Institutions, and Government Agencies

In the 1980s, Bangladesh embraced structural adjustment programs (SAPs) promoted by transnational financial institutions such as the World Bank and the International Monetary Fund (IMF) who have long been influenced the country's development strategies and pressured for adopting neoliberal policies (Bhattacharya et al., 1999; Sobhan, 2007). The 'derivative policy prescriptions' (Adnan, 2013, p. 105) of international financial institutions emerged as pressure for the government to adopt neoliberal policies such as trade liberalization and privatization of state-owned lands for promoting export-oriented shrimp aquaculture (Guimaraes, 2002; Bhattacharya et al., 2005). SAPs, which were designed for economic reforms that aimed to foster war-torn economy, generate employment, attenuate rural poverty, and promote agrarian development (see Ravallion, 1990), were supported by the World Bank's credits that emphasized on promoting export-oriented industries such as shrimp. Despite the negative effects of SAPs on people's lives and 'geographical composition' in sub-Saharan Africa (Riddell, 1992), they were implemented in Bangladesh.

In 1981, however, the World Bank pushed the government to consider shrimp farming as a potential source for fish. The presence of international development actors in promoting shrimp aquaculture became visible since the early 1980s. For example, the World Bank and Asian Development Bank (ADB) took two different projects namely *Shrimp Culture Project* and *Fisheries Culture Development Project*, respectively. Under these projects, the Ministry of Agriculture and Forest (MoAF) handed over 5,000 and 2,000 acres of state-owned land in 1978 and 1982, respectively, to the Department of Fisheries (DoF) that aimed at expanding private sectors–led shrimp farming at Rampur Mouza of Chakaria Upazila in coastal Cox's Bazar (MoFL, 2013). The total amount of shrimp estate at Rampur was 7021.76 acres that was allocated to private shrimp cultivators and large entrepreneurs.

In 1986, the *Shrimp Culture Project*, which spanned for five years and amounted to USD36.7 million, was launched by a consortium formed by the IDA credit (USD22 million), United Nations Development Program (UNDP) grant (USD4.5 million), the Government of Bangladesh's (GoB) contribution (USD4.5 million), credit from local commercial banks (e.g., the Bangladesh Bank, Bangladesh Krishi (agriculture) Bank, Sonali Bank and Agrani Bank)(USD3.8 million), and private investors (USD1.9 million) (The World Bank,1994). These local and transnational financial institutions, development agencies, government entities (e.g., the DoF, the Ministry of Fisheries and Livestock, the Ministry of Finance, the Ministry of Irrigation, Water Development and Flood Control, and the Bangladesh Water Development Board [BWDB]) and private investors were directly involved in financing and implementing this project on nearly 7,000 hectares of land and preparing additional 2,500 hectares of areas for shrimp aquaculture in coastal areas of Khulna and Cox's Bazar districts.

The 1986 consortium project motivated other development organizations (e.g., the UK-based Overseas Development Administration [ODA] and the Department for International Development [DFID]) to finance and develop new projects such as the Third Fisheries Project in 1991 and the Northwest Fisheries Extension Project in 1987 (Guimaraes, 2002; Islam and Haque, 2010). The United States and the United States Agency for International Development (USAID) have also been financing to develop shrimp aquaculture technologies, infrastructures, and research programs through a nongovernmental organization (i.e., BRAC) based in Bangladesh. Likewise, transnational development actors (i.e., the EU and the Norwegian Agency for Development Cooperation) and the government have also jointly financed in the BEST-Better Fisheries Quality (BEST-BFQ) project from 2010 to 2014 to meet the safety and quality requirements for exporting shrimps in the EU markets (BEST-BFQ, 2014). This project was implemented by the DoF and the United Nations Industrial Development Organization (UNIDO).

It is also evident that international research organizations are involved in promoting shrimp production. For instance, the World Fish has implemented, in partnership with the DoF and the Bangladesh Fisheries Research Institute (BFRI), a five-year USAID-funded project amounting to USD25 million from 2011 to 2016 (Pijl, 2014). In addition, the UNDP and the Food and Agriculture Organization (FAO) have been supporting the DoF since 1989 to promote shrimp culture through training and improving the quality of fish farms. In this way, transnational financial institutions, development agencies, intergovernmental organizations, local banks, research organizations, government agencies, and private investors have been historically financed and promoted shrimp farming, which arguably generates adequate revenue and economic incentives for many of these actors and institutions.

Business Actors

The shrimp industry has created ample economic benefits for owners, farmers, communities, processors, seafood associations (e.g., Bangladesh Frozen Foods Exporters Association), intermediaries, retailers, sellers, buyers, and exporters. It is estimated that 250,000–300,000 farmers are currently producing about 170,000m tons of shrimp annually in the southwest coastal Bangladesh (Kruijssen et al., 2012). The economic benefit is estimated by Shawon et al. (2018), which indicates that small-scale farmers' income from per acre of land is approximately TK228,000 (USD2655). A recent study shows that shrimp farming in southern coastal communities helped to significantly increase household incomes up to 36 percent and coastal residents view shrimp production is more lucrative and profitable than agriculture, particularly for higher production (Ray et al., 2021). Small-scale producers are more benefitted from higher production than others, which in turn generates higher profits, and net returns are larger than total expenditure (Islam et al., 2005). Producers also earn greater income in terms of low-input costs. The voice of an organic shrimp farmer manifests economic benefits:

> Before switching to organic we faced problems every month to care for the family, which has a total of 20 members...the price I got for the shrimp also varied a lot; sometimes it was very high, sometimes very low. In the organic system I regularly receive cash in hand and there is no need to worry. Switching to organic production I could increase my production, which combined with the lower input cost, allows a higher overall income.
>
> *(Hensler, 2013, p. 15)*

Local communities also widely engage in shrimp production that provides economic incentives and better opportunities for community development. Ray et al. (2021) illustrate the rise of household incomes from shrimp farming among coastal communities. The benefits of community people are reflected in the following remark of the president of one local community:

> The main advantages of organic shrimp in my view can be seen in the whole community; before the people were not solvent and engaged in illegal activities in order to have enough to eat. There were a lot of thieves. Now they have enough money, and they can buy food and clothing.
>
> *(Hensler, 2013, p. 15)*

Huq (2017) notes that 'the owners of enclosures have made millions and so have the owners of the shrimp-processing factories in Khulna (often also enclosure owners themselves) by exporting shrimp to Europe, US and Japan'. Some owners have bought property in London, New York, and Geneva through accumulating profits from shrimp production and export. Gain (2010) identifies that shrimp aquaculture in Chakaria Sundarban provides significant benefits to non-local individuals, organizations, and companies. It generates huge profits for absentee farm owners, business communities, and seafood exporters (Akber et al., 2017). Likewise, operators of shrimp hatcheries, who partly constitute shrimp supply chain, also earn higher profits by ensuring better management. It is estimated that net annual return of a hatchery operator in Cox's Bazar and Jessore is TK 39.6 (USD0.5) million and TK 36.7 (USD0.46) million, respectively (Islam et al., 2011). Besides, intermediaries, who are usually profit-seekers, play very significant role in the distribution of shrimp seeds from hatcheries to farmers. For instance, hatcheries sell 80 percent of shrimp seeds to *aratdars* or *beparies* (a group of intermediaries), who are the commission agents assisting seed traders and earning monies from them by charging a fixed fee of TK 25–30 (USD0.35) (Islam et al., 2011). In sum, shrimp production and business offer profits and economic incentives to actors involved in shrimp supply chain that eventually encourage the spread of shrimp farming in coastal regions.

Power Politics and Local Elites

Leaders of major political parties (e.g., Bangladesh Awami League [BAL], Bangladesh Nationalist Party [BNP], and Jatiya Party) exert their power to amass wealth and control local resources for their own benefits and material interests (Khan, 2003; Wohab and Akhter, 2004; Rahman, 2010). For instance, Mahbubul Alam Hanif, one of the powerful leaders and the Joint General Secretary of BAL, has established a shrimp farm, Quest Aquaculture and Fish Processing Limited, on 35 acres of land at Sutarkhali union of Dacope upazila in Khulna district (Mahmood, 2017). The farm is jointly funded by two state-sponsored financial institutions: the Bangladesh Bank and the Investment Corporation of Bangladesh. Although the submission of environmental impact assessment (EIA) report to and receiving an environmental clearance certificate (ECC) from the Department of Environment (DoE) are mandatory for building shrimp enclosure over five acres of land, this farm was established without any such approval(Mahmood, 2017). The farm is located at ecologically critical area (ECA) of the Sundarbans (the world's largest mangrove forest) and thus this farm has breached the Environment Conservation Act of 1999 that proclaims no commercial activities will be allowed within 10-km area of the Sundarbans. The farm owner's strong local political ally at Dacope is the vice-president of Bangladesh Jatiya Sramik League (a national worker's league that is politically and ideologically tied to the BAL) and the caretaker of that farm, who thinks that an ECC is not required to establish a shrimp farm (Mahmood, 2017). According to the 1989 Land Reform Board Act, though state-owned land should be allocated to landless poor people, however, politically powerful shrimp farmers and leaders of political parties have been capturing government land for their own sake (Deb, 1998). It is also reported that miscreants backed by political leaders and allies have captured 16 shrimp farms of a local cooperative in Cox's Bazar (EJF, 2003).

Local influential elites, middlemen, affluent interest groups, and large entrepreneurs exert their social, political, and economic power to convert the vast amount of public land into shrimp farms (The Daily Star, 2009a, Akber et al., 2017). Local elites often restrain the development by controlling land, capturing the flow of external resources, grabbing rural institutions, and using the power structure in pursuit of their own vested interests (Lewis and Hossain, 2008). Besides, affluent interest groups exert their economic and social influences on the rural-based shrimp supply chain (Pokrant, 2014). These local elites and affluent interest groups create a patron-client system in rural Bangladesh (Blair, 2005). A group of smallholders in southwestern regions fails to successfully mobilize themselves during anti-shrimp movements particularly against large-scale farming and absentee farm owners, who are backed by politically empowered local elites (Afroz et al., 2017). Likewise, intermediaries' role in shrimp farming is usually profit-oriented as they comprise marketing channel and get economic benefits from selling shrimp seeds to producers and farmers (Islam et al., 2011). Thus, local elites and powerful political leaders have played a crucial role in the spread of shrimp farming.

Government Officials and Bureaucracy

In Bangladesh, the complicity of political leaders, unscrupulous bureaucrats, government officials, law enforcement agents, and large landowners is increasingly evident in capturing and grabbing *char* (state-owned degraded land) lands (Feldman and Geisler, 2012) where they often build shrimp farms (Adnan, 2013). Instead of resolving conflicts and violence, officials of local civil administration in Noakhali (e.g., the Deputy Commissioner and Superintendent of Police) not only served the interest of rich and powerful political groups but also they became one of the influential beneficiaries by allotting shrimp farming land for themselves (Adnan, 2013). Although land grabbing for shrimp production is increasingly reported by national media, few actions have been taken by law-enforcing agencies or local civil administration against the 'manipulation of land records by powerful interest groups' (Adnan, 2013, p. 106). Moreover, outsiders, who are rooted in urban centers, invest in rural and coast-based lucrative shrimp aquaculture because 'they are very powerful individuals who have direct links with government bureaucracy and political parties' (Alauddin and Hamid, 1999, p.58). Relatives of bureaucrats such as powerful shrimp farmers grab state-owned land that is usually entitled for landless people (Deb, 1998). Large shrimp farm owners easily get lease approval of public land and water bodies from local government officials (e.g., Upazila Nirbhahi Officer [Chief Executive Officer of a sub-district]) and executives of the BWDB, who permit producers to open sluice gates for saltwater required for shrimp production (Afrozet al., 2017). The problems of land grabbing and dispossession that local communities encounter for shrimp farming remain unheard and have received little attention by government officials; rather farm owners use law-enforcing agencies and officials for their own sake. In addition, the lack of implementation of existing laws related to land grabbing and dispossession encourages powerful politicians, local elites, and beneficiaries to the expansion of shrimp aquaculture along the coasts.

Effects of Blue Revolution on the Coastal Society and Environment

Socioeconomic Effects

Though shrimp aquaculture has been contributed to economic growth, agricultural diversification, and coastal livelihoods' development (Rahman and Hossain, 2009), it has resulted in land grabbing and dispossession of poor peasants by vested interest groups. In southeastern coast of Bangladesh (e.g., Noakhali), grabbing of local *char* land to construct shrimp farms was led by political leaders (e.g. the Member of Parliament and Minister), bosses of local political parties, and chairmen and members of *Union Parishad* (smallest administrative unit) (Adnan, 2013). Under the Shrimp Mohal Management Policy, the Deputy Commissioner of Noakhali declared a total of 11955.59 acres of state-owned land for shrimp production. Of this land, nearly 2,039 acres were inhabited by landless people. It encouraged local politicians, producers, large entrepreneurs, businessmen, and politically empowered land grabbers to forcefully evict around 40,000 poor landless people from their own land by means of intimidation, torture, abuse, threats, and forged allegation of criminal cases (BLAST, 2004).

Similarly, a large volume of *khas* land in southwestern coastal regions has been unlawfully grabbed by powerful political leaders and local influential people (Azad et al., 2009). Powerful grabbers occupy *khas* land by forcible eviction, intimidation, and corruption (e.g., producing forged documents to confirm ownership) (Feldman and Geisler, 2012). Affluent groups and urban dwellers backed by local administrators and political leaders forcibly grab agricultural land to convert into shrimp farms that result in marginalization, dispossession, and social exclusion (EJF, 2014). Small and medium coastal peasants have been lost control over their own land as they have been forced to lease out agricultural land to powerful shrimp producers (Afroz et al., 2017). This has transformed agrarian land relations through which powerful vested interest groups gain more control over the land.

Land grabbing generates displacement, eviction, social unrest, and violent conflicts among local communities. Swapan and Gavin (2011) identify about 150 incidents of shrimp-related land conflicts in Bedkashi union of southwest Khulna district. Conflicts often stem from nonpayment of lease money and setting shrimp farms within locality against the willingness of communities (Swapan and Gavin, 2011), which caused 150 deaths in the last decade (EJF, 2003). It is increasingly reported that robbery, murder, kidnapping, claiming ransom, and killing are common in coastal shrimp farming areas (The Daily Star, 2008, 2011, 2012, 2013).

For instance, 'police recovered the beheaded body of a shrimp trader at a shrimp enclosure at Shitarampur village under Dumuria upazila of Khulna district' (The Daily Star, 2009c) and 'robbers contacted the family members over phone and asked them to pay TK 2 lakh per head for their release within the next three days, otherwise they would be killed, and their bodies would be dumped in the river' (The Daily Star, 2013). Shrimp culture has created competition for accessing to and controlling over resources generating tensions in local communities, vested interest groups, agribusiness investors, and absentee landowners (Hossain et al., 2013; Afroz et al., 2017; Akber et al., 2017). Resistance and movements of local farmers have increased against powerful owners of large shrimp enclosures at Chandpai union of Mongla upazila in coastal Bagerhat, who have illegally released saline water into agricultural land using sluice gates (The Daily Star, 2009b). Movements against land grabbing for shrimp culture are organized by local people, left-wing organizations, civil society groups, nongovernmental organizations, and local political parties driven by vested interests (Guhathakurta, 2008; Pokrant, 2014; Paprocki, 2021). Shrimp culture is increasingly plagued by injustice, discriminatory attitudes, violence against women, and the denial of workers' rights. It manifests in an earlier study that argues 'It [shrimp farming] is a brutal process by which the protein is extracted from the poor people [...] this extraction is to feed the bloated stomachs of the rich. This is certainly a violation of the right to life' (EJF, 2003, p. 31). In many cases, the shrimp industry breaches International Labour Organization's (ILO) labor laws related to eight-hour work per day and involves forced labor and unpaid overtime as well as failures to provide healthcare facilities, maternal leave, safety measures, and sanitary facilities (Datta et al., 2010).

Though the country has ratified the ILO's Worst Forms of Child Labour Convention, child labor still manifests in shrimp fry collections and children's work with poor wages (EJF, 2014). Female workers frequently subject to verbal abuse and work over specified hours set by the ILO, which is reflected in the voice of a woman–'we work for 12 hours but often get paid for 10 hours' (Islam, 2008, p. 228). They do not have specific working hours, minimum wages, and appointment letters. Instead, they regularly encounter sexual abuse, indecent remarks, physical assaults and harassment by owners, supervisors, guards, and male workers (Alam et al., 2012; EJF, 2014). Around 30 and 150 women are being kidnapped and raped, respectively, around shrimp farming areas in coastal Bagerhat (EJF, 2003). For instance, a teenage girl named Rafiza was raped by an employee at Khulna and encountered threat of retaliation by rapist and shrimp farmer instead of getting justice (SSNC, 2011). Victims cannot raise voice against sexual violence as their perpetrators are politically and financially powerful, and local law-enforcing agencies favor powerful convicts. Females who constitute almost half workforce in the shrimp industry experience threat of losing job, nonpayment for sick leave, long working hours, delayed payment, misconduct, and wage discrimination (a female worker gets TK50 whereas a male gets TK70 per day) (The Daily Star, 2007). Unequal wage payment generates profits for actors and beneficiaries in shrimp supply chain.

While the spread of shrimp farming has been boosted revenue and foreign currency for supply chain actors, adverse socioeconomic effects have blurred the sector's vital contribution to diversifying coastal livelihoods and rural agrarian economy. These spillover socioeconomic outcomes have created double burden for coastal communities who are already impacted by and vulnerable to natural calamities. Social effects like dispossession and displacement will potentially exacerbate the evolving 'social vulnerability' of coastal communities resulted from natural hazards over the last two decades (Saha, 2015; Masud-All-Kamal and Hassan, 2018; Rabby et al., 2019).

Environmental Effects

Industrial shrimp aquaculture has altered local land-use patterns through the forcible transformation of agricultural land and vegetation coverage into shrimp farms (Swapan and Gavin, 2011). Large entrepreneurs, business communities, and affluent producers exert their political and economic power on rice farmers to take lease of agricultural land for shrimp production (Pokrant, 2014). A study reveals that 274 hectares (79 percent) of its good arable land in Damarpota village of southwest Satkhira have been transformed into shrimp farms between 1985 and 2003 (Ali, 2006). Many farmers have converted agricultural land into shrimp enclosure due to its potential to generate more profit than that of rice farming (Islam et al., 2003; Ray et al., 2021). As shrimp culture requires saltwater, it affects the production of rice, fruits, and seasonal vegetables because most farms have been built around agricultural land and locality by financially powerful producers (Sohel and Ullah, 2012). These powerful producers, who are engaged in local politics, hire goons (locally known as saline water terrorists who are paid by hirers) to break coastal dams due to forcible release of saltwater into shrimp enclosures that weakens dams' natural stability (The Daily Star, 2009b). It is reported that brackish water is one of the leading causes for increasing about 26 percent soil salinity in southwest coastal regions (Parvez, 2011). In addition, the implementation of shrimp projects had been led to the clearance of Sundarbans mangrove forest. In a project completion report, ADB reveals that:

> [A]bout 800 ha of mangrove forest was (or is being) cleared to culture brackish water shrimp. A large portion of approximately 100,000 ha of land now being utilised for shrimp culture in Bangladesh was originally mangrove forest...The clearing of 800 ha of mangrove in the Chokoria Sundarban under the Project has clearly reduced due to shrimp/fish breeding and nursery grounds in the area.
>
> *(Gain, 2010)*

The Sundarbans plays a defensive role to buffer cyclones and storm surges that protect life and livelihoods of millions of coastal people. For instance, it protected local communities of southwestern Bangladesh from super cyclone Sidr in 2007 (Akber

FIGURE 15.1 The intrusion of saline water into shrimp farms through a canal at Munshiganj in Satkhira

et al., 2018). The destruction of mangrove forest for shrimp farming will exacerbate flooding in coastal Bangladesh (Nasreen, 2009; Deb and Ferreira, 2017). This in turn will undermine resilience of coastal communities.

In short, the expansion of shrimp farming has led to deforestation, agricultural land conversion (Figure15.1), and salinity intrusion, which will underpin the existing vulnerability of coastal people. In other words, commercial shrimp farming has far-reaching negative effects on coastal communities' adaptive capacity to confront climate variability and extremes (Masud-All-Kamal and Nursey-Bray, 2021).

Conclusion

This chapter demonstrates that shrimp aquaculture in coastal Bangladesh has gained a significant traction due to numerous policy support and incentives (e.g., tax exemption, leasing state-owned land, and credit with low interest). Given the contribution of shrimp aquaculture in earning foreign currency and promoting economic growth, government policies recognize it as a valuable export-oriented commodity. Policy incentives have created a space for diverse actors to be engaged in shrimp supply chain. The possibility of high profit has tempted urban-based entrepreneurs and powerful local elites to invest in shrimp farming. They have been maintained a strong lobby with bureaucrats and political leaders in power, which allows them to secure a 'social license' to operate shrimp farming, thereby evading environmental protection policies and regulations. Officials of local civil administration and law-enforcing agencies have served the interest of vested groups. This interactive dynamics dramatically diffuses shrimp farming across coastal regions, which contributes to livelihoods' diversification, income generation, and government revenue. Given these positive roles, the rapid diffusion has termed as blue revolution in the broader landscape of blue economy that highlights environmentally sustainable and socially equitable growth of marine-based economic activities (e.g., shrimp aquaculture), contributing to the development and human well-being (Bennett et al., 2019).

While the policy frameworks have attracted multiple actors to be involved in shrimp aquaculture, they have failed to develop a robust regulatory mechanism for sustainable governance of shrimp aquaculture in coastal areas. Our analysis reflects that the current policy frameworks are exclusive, inequitable, and unjust, and so are their implementation. The policy frameworks generate revenue and profit for shrimp producers and investors, but they remain mute in addressing injustice, exploitation, discrimination, dispossession, eviction, human rights violation, land transformation, forest destruction, and environmental degradation. Instead, local civil administrations have appeared as important patrons of shrimp farming. They favor politically and financially powerful people, which in turn deepens local communities' vulnerability to risks associated with socioecological change. The exiting policy frameworks, in other words, undermine the adaptive capacity of coastal communities to withstand future climatic and 'anthropogenic' stresses that residents of climate vulnerable regions (e.g., low-lying island-chars in northern Bangladesh) face (Saha, 2017). While shrimp aquaculture appears to be a frontier of blue economy (Smith-Godfrey, 2016), exiting policy frameworks and their implementation have created a blurred understanding whether it can be designated as blue economy that is socially just, equitable, and sustainable.

Given the negative socioeconomic and environmental impacts of shrimp aquaculture on disaster-stressed coastal communities, we call for restructuring the existing shrimp policies either to check the unbridled growth or incorporate necessary guidelines to address adverse consequences. It entails the formulation of an inclusive and equitable shrimp policy that will be designed with principles, not just promoting investors and beneficiaries alone, simultaneously integrating effects and concerns of local communities and their coastal environment. An inclusive policy framework and its proper implementation can address the existing controversies and challenges around this sector, as well as ensure the environmentally sustainable and socially just governance of shrimp aquaculture. The construction of this framework lies beyond the scope of this chapter. Further research could highlight on building a policy framework that will underpin equity and sustainability in shrimp aquaculture. Such policies can promote the country's coast-based shrimp aquaculture toward being a true frontier of blue economy.

References

Adnan, S. (2013). Land grabs and primitive accumulation in deltaic Bangladesh: Interactions between neoliberal globalization, state interventions, power relations and peasant resistance. *The Journal of Peasant Studies*, 40(1): 87–128.

Afroz, S., Cramb, R., Grunbuhel, C. (2017). Exclusion and counter-exclusion: The struggle over shrimp farming in a coastal village in Bangladesh. *Development and Change*, 48(4): 692–720.

Ahsan, D., Brandt, U.S. (2015). Climate change and coastal aquaculture farmers' risk perceptions: Experiences from Bangladesh and Denmark. *Journal of Environmental Planning and Management*, 58(9): 1649–1665.

Akber, M.A., Islam, M.A., Ahmed, M., Rahman, M.M., Rahman, M.R. (2017). Changes of shrimp farming in southwest coastal Bangladesh. *Aquaculture International*, 25(5): 1883–1899.

Akber, M.A., Patwary, M.M., Islam, M.A., Rahman, M.R. (2018). Storm protection service of the Sundarbans mangrove forest, Bangladesh. *Natural Hazards*, 94(1): 405–418.

Alam, T., Shahi, M., Shifuzzaman, L. (2012). Economic prospects as well as human rights violation at shrimp farming: A study based in south west coastal region of Bangladesh. *IOSR Journal of Humanities and Social Science*, 1(1): 45–49.

Alauddin, M., Hamid, M.A. (1999). Shrimp culture in Bangladesh with emphasis on social and economic aspects. In: P.T. Smith (ed), *Towards sustainable shrimp culture in Thailand and the region*. ACIAR Proceedings No 90, Canberra: Australian Centre for International Agricultural Research, 53–62.

Ali, A.M.S. (2006). Rice to shrimp: Land use/land cover changes and soil degradation in Southwestern Bangladesh. *Land Use Policy*, 23(4): 421–435.

Azad, A.K., Jensen, K.R., Lin, C.K. (2009). Coastal aquaculture development in Bangladesh: Unsustainable and sustainable experiences. *Environmental Management*, 44(4): 800–809.

Belton, B. (2016). Shrimp, prawn and the political economy of social wellbeing in rural Bangladesh. *Journal of Rural Studies*, 45: 230–242.

Bennett, N.J., Cisneros-Montemayor, A.M., Blythe, J., Silver, J.J., Singh, G., Andrews, N., Calò, A., … Sumaila, U. R. (2019). Towards a sustainable and equitable blue economy. *Nature Sustainability*, 2(11): 991–993.

BEST-BFQ (BEST-Better Fisheries Quality) (2014). 'Simplification of shrimp supply chain launched at paranpur, satkhira. Better fisheries quality (BEST-BFQ), Bangladesh'. Available at: https://bfqbangladesh.wordpress.com/2014/06/20/simplification-of-shrimp-supply-chain-launched-at-paranpur-satkhira/

BFFEA (Bangladesh Frozen Foods Exporters Association) (2008). Shrimp and fish news: News letter of Bangladesh Frozen Foods Exporters Association. April-June, Dhaka, Bangladesh, 2–41.

Bhattacharya, D., Rahman, M., Khatun, F. (1999). Environmental consequences of structural adjustment: Towards sustainable shrimp culture in Bangladesh. Centre for Policy Dialogue Occasional Paper Series 2.

Bhattacharya, D., Rahman, M., Khatun. F. (2005). Environmental impact of structural adjustment policies: The case of export-oriented shrimp culture in Bangladesh. In: D. Bhattacharya, M. Rahman, F. Khatun, A. Raihan, and U.K. Deb (eds), *Bangladesh in the global trade regime: Labour, environment, agriculture, export and trade negotiations*. Dhaka, Bangladesh: Pathak Shamabesh and the Centre for Policy Dialogue, 44–85.

Blair, H., 2005. Civil society and propoor initiatives in rural Bangladesh: Finding a workable strategy. *World Development*, 33(6): 921–936.

BLAST (Bangladesh Legal Aid and Services Trust) (2004). 'Noakhali char eviction case or shrimp farming case'. Available at: www.blast.org.bd/issues/shelter/241-5194of2004

Cisneros-Montemayor, A.M., Moreno-Báez, M., Reygondeau, G., Cheung, W.W., Crosman, K.M., González-Espinosa, P.C., Lam, V.W., Oyinlola, M.A., Singh, G.G., Swartz, W., Zheng, C.W. (2021). Enabling conditions for an equitable and sustainable blue economy. *Nature*, 591(7850): 396–401.

Datta, D.K., Roy, K., Hassan, N. (2010). Shrimp culture: Trend, consequences, and sustainability in the south-western coastal region of Bangladesh. In: A.L. Ramanathan, P. Bhattacharya, T. Dittmar, M.B.K. Prasad, and B.R. Neupane (eds). *Management and sustainable development of coastal zone environments*. Dordrecht: Springer.

Deb, A.K. (1998). Fake blue revolution: Environmental and socio-economic impacts of shrimp culture in the coastal areas of Bangladesh. *Ocean & Coastal Management*, 41(1): 63–88.

Deb, M., Ferreira, C. M. (2017). Potential impacts of the Sunderban mangrove degradation on future coastal flooding in Bangladesh. *Journal of Hydro-Environment Research*, 17:30–46.

EJF (Environmental Justice Foundation) (2003). *Smash & grab: Conflict, corruption and human rights abuses in the shrimp farming industry*. London: Environmental Justice Foundation.

EJF (Environmental Justice Foundation) (2014). *Impossibly cheap: Abuse and injustice in Bangladesh's shrimp industry.* London: Environmental Justice Foundation.

EPBB (Export Promotion Bureau of Bangladesh) (2017). *Export statistics of frozen shrimp and other fishes*. Government of Bangladesh, Dhaka.

Farmer, A.S.D. (1989). Proposed research programme for the brackish water station to be established at Paikgacha. Final report on World Bank second agricultural research project. FAO/UNDP TA Project BGD/83/010.

Feldman, S., & Geisler, C. (2012). Land expropriation and displacement in Bangladesh. *The Journal of Peasant Studies*, 39(3–4): 971–993.

FRSS (Fisheries Resources Survey System) (2015). Fisheries statistical report of Bangladesh. Department of Fisheries, Bangladesh. Available at: https://drive.google.com/file/d/0B6XjGVIRbmt0dGtyWDM5WUpHa1U/view

Gain, P. (2010). Monoculture destroys coast and forests. Available at: http://archive.thedailystar.net/suppliments/2010/02/ds19/segment2/forests.html

Garlock, T., Asche, F., Anderson, J., Bjørndal, T., Kumar, G., Lorenzen, K., Ropicki, A., Smith, M.D., Tveterås, R. (2020). A global blue revolution: Aquaculture growth across regions, species, and countries. *Reviews in Fisheries Science & Aquaculture*, 28(1): 107–116.

Guhathakurta, M. (2008). Globalization, class and gender relations: The shrimp industry in southwestern Bangladesh. *Development*, 51(2): 212–219.

Guimaraes, J.P. de C. (2002). *A landscape of contracts: The organisation and impact of shrimp culture in southwest Bangladesh*. Thesis (PhD), University of Utrecht.

Haque, S.M. (1994). Annual report of Bangladesh Frozen Foods Exporters Associations (BFFEA), BFFEA Special Bulletin 5.

Hensler, L. (2013). A sustainable future for shrimp production in Bangladesh? An ethical perspective on the conventional and organic supply chain of shrimp aquaculture in Bangladesh. Sustaining Ethical Aquaculture Trade (SEAT).

Hossain, M.S., Uddin, M.J., Fakhruddin, A.N.M. (2013). Impacts of shrimp farming on the coastal environment of Bangladesh and approach for management. *Review of Environmental Science and Biotechnology*, 12(3): 313–332.

Hossan, S. (2021). Shrimp farming in Bangladesh: Why shrimp sector is lagging behind? Business Inspection. Available at: https://businessinspection.com.bd/why-the-shrimp-sector-of-bangladesh-is-lagging-behind/

Huq, S. (2017).The many faces of adaptation. *The Daily Star*. 26 July. Available at: www.thedailystar.net/opinion/politics-climate-change/the-many-faces-adaptation-1438603

Islam, M.R., Haque, M.R. (2010). Impacts of northwest fisheries extension project (NFEP) on pond fish farming in improving livelihood approach. *Journal of Bangladesh Agriculture University*, 8(2): 305–311.

Islam, M.S. (2008). From pond to plate: Towards a twin-driven commodity chain in Bangladesh shrimp aquaculture. *Food Policy*, 33(3): 209–223.

Islam, M.S., Milstein, A., Wahab, M.A., Kamal, A.H.M., Dewan, S. (2005). Production and economic return of shrimp aquaculture in coastal ponds of different sizes and with different management regimes. *Aquaculture International*, 13: 489–500.

Islam, M.S., Rahman, M.S., Haque, M.M., Sharmin, S. (2011). Economic study on production and marketing of shrimp and prawn seed in Bangladesh. *Journal of Bangladesh Agricultural University*, 9(2): 247–256.

Islam, M.S., Wahab, M.A., Miah, A.A., Kamal, A.H.M.M. (2003). Impacts of shrimp farming on the socioeconomic and environmental conditions in the coastal regions of Bangladesh. *Pakistan Journal of Biological Science*, 6(24): 2058–2067.

Khan, M.M. (2003). State of governance in Bangladesh. *The Round Table: The Commonwealth Journal of International Affairs*, 92(370): 391–405.

Kruijssen, F., Kelling, I., MeeChee, H., Jespersen, K., Ponte, S. (2012). *Value chains of selected aquatic products from four Asian countries.* A review of literature and secondary data.

Lewis, D., Hossain, A. (2008). Beyond 'the net'? Institutions, elites and the changing power structure in rural Bangladesh. In: D. Gellner and K. Hachhethu (eds), *Local democracy in South Asia: Microprocesses of democratization in Nepal and its neighbours. Governance, conflict, and civic action.* Vol. 1, New Delhi: Sage.

Mahmood, I. (2017). AL Hanif sets up shrimp farm near Sundarbans. *Prothom Alo*, 11 July. Available at: http://gads.prothom-alo.com/bangladesh/news/153429/Hanif-sets-up-shrimp-farm-near-Sundarbans

Masud-All-Kamal, M., Hassan, S. M. (2018). The link between social capital and disaster recovery: Evidence from coastal communities in Bangladesh. *Natural Hazards*, 93(3):1547–1564.

Masud-All-Kamal, M., Nursey-Bray, M. (2021). Socially just community-based climate change adaptation? Insights from Bangladesh. *Local Environment*, 26(9): 1092–1108.

MoFL (Ministry of Fisheries and Livestock) (1998). National fisheries policy. The Government of the People's Republic of Bangladesh, Dhaka.

MoFL (Ministry of Fisheries and Livestock) (2013). Cox's Bazar zilar Chakariaupazilar Rampur mouzar motsho odhidoptoradin chingri plot ijara, ijaranobayan, bebosthapona o unnayannitimala 2013 (in Bengali). The Government of the People's Republic of Bangladesh, Dhaka.

MoFL (Ministry of Fisheries and Livestock) (2014). National shrimp policy 2014. Government of the People's Republic of Bangladesh, Dhaka.

MoFL (Ministry of Fisheries and Livestock) (2018). Annual Report 2018. Department of Fisheries, Bangladesh, Dhaka.

Molla, M.M. (2018). Bangladesh struggles with adulterated shrimp. the third pole.net: Understanding Asia's water crisis. Available at: www.thethirdpole.net/en/2018/02/08/bangladesh-struggles-with-adulterated-shrimp/

Nasreen, M. (2009). Disasters and disaster research: A case study of floods in Bangladesh. *Understanding the Global Environment*, 265–287.

Paprocki, K. (2021). *Threatening dystopias: The global politics of climate change adaptation in Bangladesh.* New York: Cornell University Press.

Parvez, S. (2011). Salinity chokes croplands. *The Daily Star*, 3 April. Available at: www.thedailystar.net/news-detail-180223

Parvez, S. (2015). Shrimp farmers turn to local market to recover costs. *The Daily Star*, 2 June. Available at: www.thedailystar.net/business/shrimp-farmers-turn-local-market-recover-costs-91066

Pijl, W. van der. (2014). *An update of shrimp and prawn supply chain initiatives in Bangladesh: Recommendations for inclusive shrimp supply chain development for the STDF project.* LEI: Wageningen UR (University & Research Centre), 14–28.

Pokrant, B. (2014). Brackish water shrimp farming and the growth of aquatic monocultures in coastal Bangladesh. In: J. Christensen and M. Tull (eds), *Historical perspectives of fisheries exploitation in the indo-pacific.* Dordrecht: Springer, 107–132.

Pokrant, B., Reeves, P. (2005). *From fish and forest: The changing nature of coastal development policy and its impact on coastal resources and communities in southeast Bangladesh*. Proceedings of the Center for Maritime Research Conference, People and the Sea III, July 7–9, Amsterdam, Netherland.

Rabby, Y.W., Hossain, M.B., Hasan, M.U. (2019). Social vulnerability in the coastal region of Bangladesh: An investigation of social vulnerability index and scalar change effects. *International Journal of Disaster Risk Reduction*, 41: 101329.

Rahman, M.M., Hossain, M.M. (2009). Production and export of shrimp of Bangladesh: Problems and prospects. *Progressive Agriculture*, 20(1&2): 163–171.

Rahman, M.S. (2010). *Institutionalization of democracy in the political parties in Bangladesh: Does culture matter?* Thesis (Master), North South University, Bangladesh.

Ravallion, M. (1990). The challenging arithmetic of poverty in Bangladesh. *The Bangladesh Development Studies*, 18(3): 35–53.

Ray, S., Mondal, P., Paul, A.K., Iqbal, S., Atique, U., Islam, M.S., Mahboob, S., … Begum, S. (2021). Role of shrimp farming in socio-economic elevation and professional satisfaction in coastal communities of Southern Bangladesh. *Aquaculture Reports*, 20: 100708.

Riddell, J.B. (1992). Things fall apart again: Structural adjustment programmes in sub-Saharan Africa. *The journal of modern African studies*, 30(1): 53–68.

Saha, C.K. (2015). Dynamics of disaster-induced risk in southwestern coastal Bangladesh: An analysis on tropical Cyclone Aila 2009. *Natural Hazards*, 75(1): 727–754.

Saha, C.K. (2017). Dynamics of climatic and anthropogenic stressors in risking island-char livelihoods: A case of northwestern Bangladesh. *Asian Geographer*, 34(2): 107–129.

Shawon, N.A., Prodhan, M., Haque, M., Khan, M.A., Mitra, S. (2018). Financial profitability of small-scale shrimp farming in a coastal area of Bangladesh. *Journal of Bangladesh Agricultural University*, 16(1): 104–110.

Smith-Godfrey, S. (2016). Defining the blue economy. *Maritime Affairs: Journal of the National Maritime Foundation of India*, 12(1): 58–64.

SNI (Shrimp News International) (2014). Bangladesh: Organic shrimp farming. Available at: www.shrimpnews.com/FreeReportsFolder/NewsReportsFolder/BangladeshOrganicShrimp.html

Sobhan, R. (2007). *The political economy of malgovernance in Bangladesh: Collected works of Rehman Sobhan*. Dhaka: Centre for Policy Dialogue.

Sohel, S.I., Ullah, M.H. (2012). Ecohydrology: A framework for overcoming the environmental impacts of shrimp aquaculture on the coastal zone of Bangladesh. *Ocean & Coastal Management*, 63: 67–68.

SSNC (Swedish Society for Nature Conservation) (2011). Murky waters: The environmental and social impacts of shrimp farming in Bangladesh and Ecuador. Available at: http://nijerakori.org/wp-content/uploads/2016/03/murky_waters-The-environmental-and-social-impacts.pdf

Swapan, M.S.H., Gavin, H. (2011). A desert in the delta: Participatory assessment of changing livelihoods induced by commercial shrimp farming in southwest Bangladesh. *Ocean & Coastal Management*, 54(1): 45–54.

The Daily Star (2007). Reduce gender wage gap in shrimp cultivation field. *The Daily Star*, 30 November. Available at: www.thedailystar.net/news-detail-13653

The Daily Star (2008). Shrimp trader stabbed to death. *The Daily Star*, 25 October. Available at: www.thedailystar.net/news-detail-60260

The Daily Star (2009a). Jalmahal policy won't help grassroots fishermen. *The Daily Star*, 25 July. Available at: www.thedailystar.net/news-detail-98638

The Daily Star (2009b). Farmers resist saline water based shrimp farming. *The Daily Star*, 11 May. Available at: www.thedailystar.net/news-detail-87724

The Daily Star (2009c). Body of shrimp trader found. *The Daily Star*, 30 November. Available at: www.thedailystar.net/news-detail-115900

The Daily Star (2011). Shrimp enclosure owner killed. *The Daily Star*, 8 June. Available at: www.thedailystar.net/news-detail-189078

The Daily Star (2012). Shrimp trader murdered. *The Daily Star*, 29 June. Available at: www.thedailystar.net/news-detail-240192

The Daily Star (2013). Forest robbers kidnap 4 shrimp enclosure owners for ransom. *The Daily Star*, 31 July. Available at: www.thedailystar.net/news/forest-robbers-kidnap-4-shrimp-enclosure-owners-for-ransom

The World Bank (1985). Bangladesh shrimp culture project. Report no 5561. Washington: The World Bank.

The World Bank (1994). Project completion report on Bangladesh shrimp culture project. Report no 13679. Washington: The World Bank.

Wenhai, L., Cusack, C., Baker, M., Tao, W., Mingbao, C., Paige, K., Xiaofan, Z., … Yue, Y. (2019). Successful blue economy examples with an emphasis on international perspectives. *Frontiers in Marine Science*, 6: 261.

Wohab, M.A., Akhter, S. (2004). Local level politics in Bangladesh: Organization and process. *BRAC University Journal*, 1(1): 23–32.

PART V

Behavioural Change and Community Participation

16

COMMUNITY PRIORITIES IN DISASTER RISK REDUCTION INTERVENTIONS

A Critical Perspective from Bangladesh

Tuhin Roy, Mahbuba Nasreen, Md. Kamrul Hasan, and M.M. Abdullah Al Mamun Sony

Introduction

Over the past couple of decades, disaster risk reduction (DRR) has evolved as a key disaster management strategy. It is now well recognized that DRR can be applied to build disaster-resilient communities. The United Nations International Strategy for Disaster Reduction (UNISDR, 2009) defined DRR as "the concept and practice of reducing disaster risk through reduced exposure to hazards, lessened vulnerability of people and property, wise management of land and the environment and improved preparedness for adverse events". Rather than eradicating disasters and/or disaster-related risks, the DRR approach encompasses a wide spectrum of policies, strategies, and practices applied to reduce or prevent vulnerabilities and risks resulting from disasters (Matsuoka, 2013).

The late 1980s marked a shift toward a DRR approach following an increasing understanding that developmental gains had been lost due to disasters (Shaw et al., 2013, p. 8). This paradigm shift was coincided with the United Nations declaring the 1990s as the International Decade for Natural Disaster Reduction (IDNDR) (Shaw et al., 2013, p. 9). This marked a watershed moment in thinking about, and formulating policies and strategies in relation to, disaster management. Later, the Sendai Framework for Disaster Risk Reduction (SFDRR, 2015–2030) was adopted at the Third United Nations World Conference on Disaster Risk Reduction, held from 14 to 18 March 2015 in Sendai, Japan. This framework succeeded the Hyogo Framework for Action (HFA), and it marked a shift away from disaster management to pre-disaster risk reduction. The HFA emphasized DRR as a priority for action (Shaw et al., 2013, p. 14). It called for reducing risk and preventing disaster through multi-stakeholder engagement. The initiative was supported by the United Nations Office for Disaster Risk Reduction (UNISDR, 2015). The Sendai

DOI: 10.4324/9781003253495-21

framework stressed the importance of community-focused DRR strategies. Clause 7 of the framework document stresses multi-sectoral engagement and collaboration:

> There has to be a broader and a more people-centered preventive approach to disaster risk. Disaster risk reduction practices need to be multi-hazard and multi-sectoral, inclusive and accessible in order to be efficient and effective.
>
> *UNISDR, 2015, p. 10*

In tandem with the Sendai framework, the United Nations Sustainable Development Goals (SDGs), which reflect current development philosophy, underscored the need for reducing disaster-related risks. The SDGs recognize that disasters undermine efforts to achieve sustainable development (UNISDR, 2015). The United Nations Conference on Sustainable Development, held in 2012 and entitled "The future we want", called for DRR and for building resilience to disasters with a renewed sense of urgency in the context of sustainable development and poverty reduction. Furthermore, the UN stressed that DRR principles must be incorporated at multiple levels, thus reaffirming and endorsing the goals of sustainable development articulated at the Rio Declaration on Environment and Development (UNISDR, 2015, p. 11).

Importantly, DRR aims at addressing disaster-related vulnerabilities and building disaster-resilient communities (Islam et al., 2020). Resilience is technically defined as "capacity to recover", "degree of preparedness", and the ability to deal with adverse changes and shocks (Nasreen, 2019). Resilience refers to the ability to bounce back from disasters. It is the capacity of individuals, groups, communities, societies, or systems to recover from disastrous events and the ability to return to normal functioning (Shaw et al., 2013, p. 6). To materialize this aim, it is crucially important to allow for community participation, consultation, and coordination among all stakeholders involved in the design, implementation, and evaluation of DRR intervention programs.

Against the background outlined above, this chapter critically examines DRR as a disaster reduction strategy, with reference to Bangladesh. To do this, it tracks the evolution of the concept of DRR and then outlines some of the key components of DRR including community ownership, participation, coordination, and gender. The authors gathered and synthesized evidence-based research from the communities living in the selected southwest coastal areas of Bangladesh. Special attention was directed to whether or not such communities get their voices heard in national DRR interventions. Next, it draws on field research experiences in southwestern Bangladesh and on the relevant literatures to illustrate how DRR has been implemented in Bangladesh. Finally, the chapter ends with some concluding remarks with respect to making DRR more effective in Bangladesh.

The Evolution of DRR

As indicated, DRR has developed as a mainstream strategy that focuses on reducing risks and addressing disaster-related risks, by replacing previous focus on conventional hazard management (Benge & Neef, 2020). The declaration of the IDNDR in 1989 by the United Nations stressed on the need to reduce loss of life and properties, especially in developing countries through managing disasters (Lechat, 1990). The Declaration also highlighted the importance of creating better knowledge about the cause of disasters, establishing early warning systems and developing disaster-related information-sharing mechanisms (Lechat, 1990).

During the IDNDR era, critics argued that there was relatively little engagement of policymakers at the national level and that most of the researchers and leading members were from the More Economic Developed Countries (MEDC) that promoted high-tech solutions but were not informed well about the ground realities of the most at-risk countries (Katayama, 1993). Katayama (1993) mentioned that the mandate of IDNDR was too broad to the national institutions that may trigger procrastination and inaction. For this reason, disaster management efforts were not fully successful during the IDNDR.

Successively, a paradigm shift in the field of disaster research and disaster policies to manage disasters occurred in the past couple of decades and the concept of "risk reduction" was introduced (Nasreen, 2012/2019). Critics of "environmental determinism" and the "accident" thesis (Hewitt, 1983) urged to reconstruct disaster knowledge by considering disaster as an unavoidable event. Subsequently, several novel concepts such as "resilience" gained currency. For instance, the concept of building resilience through reducing vulnerabilities was proposed (Gitz & Meybeck, 2012; Orimoloye et al., 2021). It was found that the failure of development initiatives in terms of social, economic, political, psychological, and environmental variables relied largely on the national and international political economy, and such failures had increased vulnerabilities (Benge & Neef, 2020). Therefore, disasters were conceptualized in the first quarter of the twenty-first century as a failure of governmental efforts. At this point, global policymakers felt the importance of introducing DRR interventions.

Currently, DRR is a widely used terminology in the field of disaster management. Several scholars have defined this term in different ways. Nevertheless, their essence is mostly the same. Among them Twigg's (2015) definition has been considered for this chapter because it covers specific points:

> the broader sense to cover policy, strategic, institutional, and operational issues …, whilst the term Disaster Risk Management (DRM) is used more specifically to refer to aspects of operational practice. But a key point about both terms is that they describe a very broad-based approach to the causes of disasters and dealing with their consequences.
>
> *Twigg, 2015, pp. 6–9*

Therefore, DRR has evolved as a policy measure to increase the success of development projects, which aims at risk management. It is a systematic approach to identifying, assessing, and reducing risks associated with hazards and human activities (Twigg, 2015, p. 9). The basic principles of DRR were articulated in the HFA (2005–2015) and subsequently, in the SFDRR (2015–2030).

The HFA was considered the first international formal outcome-based framework of DRR within which efforts were made to overcome the limitations of IDNDR by setting some strategic goals. This blueprint first considered the DRR as an integral part of the existing policies and infrastructures that are aligned with the SDGs. To succeed in this initiative, some priorities for actions were formulated in the Hyogo Conference. Although the HFA had given priority to strengthening national and local institutions, the ultimate success remained limited due to a lack of the integration of different sectors such as agriculture, infrastructure, health, education, energy, tourism, and so on (Briceño, 2015). Certainly, the lack of opportunities of local governments to collaborate with international and national policymakers was another limitation that made it difficult to implement successful DRR interventions (McGee et al., 2016).

Subsequently, the SFDRR has evolved as a successor of the HFA and as a holistic model for the sustainable reduction of disaster risks through building resilience and reducing vulnerabilities wherein a wide range of multi-sectoral cooperation is emphasized. The principle of empowering local authorities and local communities to reduce disaster risk occupies a unique position in the SFDRR, compared to the previous frameworks (Kelman & Glantz, 2015). This framework has emphasized on disaster risk management through identifying, assessing, and reducing new risks, as well as the old ones which can be natural, human-made, biological, and so forth. It also stressed on integrating various stakeholders by formulating policies and strengthening infrastructures not only at the national level, but also at the local level (Wahlström, 2015). Thus, DRR intervention has become a major disaster-related approach among development partners, policymakers, researchers, and others (Nasreen, 2012/2019).

DRR and Community's Prioritization

In line with the spirit of the SFDRR, the popularity of the bottom-up approaches has increased over the past few years among practitioners, as well as within the scholarly communities, because the core value underpinning this approach emphasizes community priorities (Asif et al., 2013). From this approach, the community is viewed as the ultimate beneficiary of development interventions, and it is also argued that each community has its own knowledge and strategies to address its problems (Paton & Johnston, 2001). DRR from a community perspective not only means encouraging the establishment of community-based organizations, but also promoting local knowledge and expertise, identification of local needs, building local capacity, strengthening local interventions, and so on (Shaw, 2012). Several studies such as Pandey and Okazaki (2005), Victoria (2003), and Cuthbertson et al. (2019) have shown the effectiveness of this approach, which promoted a sense of ownership in development programs among community members.

In Bangladesh, research has demonstrated the importance of integrating local wisdom in mitigating disaster risks. To give an example, Islam (2018)showed that in the flood-prone low-laying areas of Bangladesh, the use of local wisdom-based community practices helped to reduce flood and river erosion-related hazards. Furthermore, the applications of low-cost local resource-based technologies were largely effective in reducing these disaster-related risks and in building more resilient communities in the Ganges floodplains area of Bangladesh (Islam, 2018). Such evidence hints at the effectiveness of community priorities in the existing DRR programs.

Nonetheless, some scholars (i.e., Baker & Grant Ludwig, 2018; Dynes, 1994; Hardoy et al., 2011) have not supported the wide involvement of community organizations in DRR interventions without good governance.They oppose giving communities a full control over identifying, assessing, and reducing risks. They reasoned that an unwillingness to engage in collaborative work among informal actors and formal actors was a barrier in making DRR successful at the community level (Hardoy et al., 2011; Twigg & Mosel, 2017). Moreover, research from the disaster-prone southwestern Bangladesh reveals that the community people have not been given the opportunity to have their voices heard. In some studies (i.e., Islam et al., 2020; Mojid, 2020; Roy et al., 2020), it was also evident that until recently a lack of community priorities resulted in limited success of DRR intervention programs. For instance, in Bangladesh, scant attention was paid to gender-specific local needs (Nasreen, 1995, 2008) in floodplains and in the coastal DRR policies and for this reason, earlier flood and cyclone shelter projects did not produce their intended results (Shahin et al., 2020).

DRR in Bangladesh

Bangladesh is regarded as a high-risk country susceptible to a wide array of natural hazards including floods, cyclones, tornadoes, tidal surges, tsunamis, and earthquakes. Currently, it is grappling with the Covid-19 pandemic, which the sociologist Raewyn Connell (2020) has recently described as a "social disaster". Despite this, the country has made commendable progress over the past several decades in managing disasters. Concurrently, in regard to policy, it moved from a "reactive humanitarian relief" approach after a disaster to a "proactive risk management" before the disaster (Nasreen, 2012/2019; Walker, 2013, p. viii). Consequently, deaths and devastation from disastrous events have significantly decreased in recent years. While almost half a million people died from the 1970 Bhola Cyclone, the rate of mortality went down to several hundred in the cases of Cyclone Aila in 2009 and Sidr in 2007 (Walker, 2013, p. ix).

To combat disasters and to make the community resilient, there is a relatively well-developed institutional setup in Bangladesh. All DRR functions governed by the past Ministry of Relief and Rehabilitation (MoR&R). This institutional structure was formed with the Standing Orders on Disaster (SOD) in 1997 with responsibilities vested on multiple committees, ministries, and organizations (MoEF, 2009; Shaw et al., 2013). Along with this institutional structure, DRR activities are also coordinated in Bangladesh by the National Disaster Management Council (NDMC), National Disaster Management Advisory Committee (NDMAC) in coordinated efforts with the Inter-ministerial Disaster Management Coordination Committee (IMDMCC).

The Bangladesh National Plan for Disaster Management (NPDM) envisioned a future in which to achieve a paradigm shift in disaster management from conventional responses and recovery to a more informed risk reduction approach (Government of Bangladesh 2005, as cited in Shaw et al., 2013). It outlined strategic goals, institutional arrangements, and conceptual framework that inform the disaster and emergency management of the Government. The Government implements its DRR intervention programs through the Ministry of Disaster Management and Relief (Islam et al., 2020, p. 4). It is interesting to note that while this Ministry implements its DRR intervention programs through the Department of Disaster Management, the Ministry of Environment and Forest leads the country's climate change adaptation (CCA) programs by involving the governmental bodies under its jurisdiction such as the Department of Environment and the Department of Forest in Bangladesh (Islam et al., 2020).

Concurrently, other governmental bodies at the lower levels of the public administration such as the district or sub-district level offices and international non-government organizations (NGOs) play a vital role in the implementation of DRR and climate change–related activities in the country. Research on the integration of DRR and CAA interventions by the Government highlighted that "all the influential decision makers in these organizations have competing interests in controlling decisions regarding the integration of DRR and CCA" (Islam et al., 2020, p. 5), which has resulted in a lack of coordination and synergy between competing governmental bodies and ministries.

As regards legal instruments, Bangladesh also has a Disaster Management Act (2012) that followed the National Disaster Management Policy (2008). At the mainstreaming strategies of these legal instruments, the community has placed at the very last stage where a furnished package of the capacity building comes through maintaining a hierarchical channel. The policy planning and the advocacy of awareness-raising have used to take place at the very beginning of this channel which consisted of the national- and district-level strategies. Besides this, to address each of the individual hazards, Bangladesh has some interconnected agencies to manage disasters (MoDMR, MoEF, 2012). According to Mall et al. (2019), the climate change–related vulnerabilities in Bangladesh are generally composed of different hydro meteorological and socioeconomic factors. Therefore, the country has multi-sectoral committees and cells such as the National Environmental Council, the National Steering Committee on Climate Change, climate change cells in all ministries, and the Climate Change Secretariat under the Ministry of Environment and Forests (MoEF, 2012).

Along with policies, though Bangladesh has a well-structured institution for disaster risk responses, it has always remained questionable as to what extent such structures function effectively. As evidence of the failure of such structural initiatives was observed in different disasters such as during the super cyclone Aila (2009), when massive losses to physical and environmental assets occurred. According to an estimation, 190 casualties, 24,300 homeless families, and 77,000 acres of destroyed farmland were recorded immediately after Aila (Saha, 2015). Apart from migration, loss of livelihood pattern and insecurity, and change in social structure were also observed in the southwest coastal region of Bangladesh (Islam & Hasan, 2016; Roy & Sony, 2019). However, after these intervention programs were introduced, several studies such as those by Roy and Sony (2019) reported intergroup conflicts in the southwest coastal areas of Bangladesh. This, in turn, created vulnerabilities in terms of weak social integrity and value of social responsibility. It can be argued that if communities were prioritized in DRR interventions, such social anomalies would not have arisen in the southwest coastal areas of Bangladesh.

Consultation with Local People

To identify local needs and to involve local people in DRR, consultation with local people is viewed as an effective technique. Schwartz and Deruyttere (1996) wrote, "the community, planners and lending agency staff enter a dialogue in which the community's ideas and priorities help shape projects. The final design of the project reflects that community responses received during consultative dialogues" (p. 8). Through the consultative process, local community's gain their authority and power over the development process and thus some of the barriers in policy formulation, implementation and evaluation can be removed (Schwartz &Deruyttere, 1996). A clear emphasis on consultation with local people for successful DRR interventions was also placed in clause 27(f) of the SFDRR:

> it is important … to assign, as appropriate, clear roles and tasks to community representatives within disaster risk management institutions and processes and decision-making through relevant legal frameworks, and undertake comprehensive public and community consultations during the development of such laws and regulations to support their implementation.
>
> *UNISDR, 2015, p. 17*

Turning to the question of community consultation in Bangladeshi DRR interventions strategies, research shows that some other parts of Bangladesh like flood-prone areas there are consultation with local people for flood risk reduction program formulation (Islam, 2018). However, there was much little evidence about community consultation occurring in southwest coastal DRR intervention programs. Roy (2019) found that without much coordination with local people, the household disaster risk identification activities (Table 16.1) were designed and implemented in the southwest coastal areas of Bangladesh through the DRR policy and the SOD prioritized community consultation.

Table 16.1 presents southwest coastal villagers' responses about household disaster risk identification activities by local GOs and NGOs (Roy, 2019). They lived in three separate villages. According to Table 16.1, in 2007 and 2009, *Jaliakhali* villagers had no plan for identifying hazard risks. A similar pattern was also observed

TABLE 16.1 Information regarding household disaster risk identification activities

HH disaster risk identification activities [Codes: 0= No, 1=Just planning, 2= Planning and initiation, 3=Yes, 4 = Done, 5=Well done everything]	*HH level performance*								
	2007			*2009*			*2015*		
	Jaliakhali	*Rayenda*	*Moddhom Khalisabunia*	*Jaliakhali,*	*Rayenda*	*Moddhom Khalisabunia*	*Jaliakhali*	*Rayenda*	*Moddhom Khalisabunia*
1. Identifying various types of hazards that may affect household	0 (100%)	0 (64.3%)	0 (92.5%)	0 (71.7%)	2 (55.8%)	2 (63.3%)	2 (38.3%)	3 (76.7%)	2 (43.3%)
2. Identifying probable hazard risks for respondent family	0 (100%)	0 (69.2%)	0 (90.8%)	0 (71.7%)	3 (47.5%)	2(43.3%)	2(40.0%)	3 (75.8%)	2 (36.7%)
3. Plan to assess household (HH) vulnerability issues	0 (100%)	0 (81.7%)	0 (95.0%)	0 (70.0%)	3 (36.7%)	2 (57.5%)	2 (41.7%)	3 (60.8%)	2 (45.0%)
4. Plan to identify the vulnerable places beside HH	0 (95.8%)	0 (82.5%)	0 (95.8%)	0 (69.2%)	3 (41.7%)	2 (59.2%)	2 (41.7%)	3 (49.2%)	2 (37.5%)
5. Risk that may affect respondents' assets or savings	0 (90.0%)	0 (73.3%)	0 (96.7%)	0 (66.7%)	3 (37.5%)	2 (53.3%)	2 (42.5%)	3 (51.7%)	3 (26.7%)
6. Knowledge about risk ranking to identify disaster risk	0 (95.8%)	0 (66.7%)	0 (90.0%)	0 (58.3%)	3 (35.8%)	2 (56.7%)	2 (45.0%)	3 (54.2%)	3 (31.7%)
7. Institutional connections/ networking to identify risk	0 (93.3%)	0 (81.7%)	0 (95.0%)	0 (69.2%)	3 (44.2%)	2 (60.0%)	2 (31.7%)	3 (65.8%)	3 (37.5%)
8. Documentation/documents to identify risks	0 (96.7%)	0 (89.2%)	0 (98.3%)	0 (73.3%)	3 (36.7%)	2 (40.0%)	2 (40.0%)	3 (63.3%)	3 (34.2%)

9. Getting training on risk identification	0 (92.5%)	0 (85.0%)	0 (90.8%)	0 (65.8%)	3 (41.7%)	2 (40.0%)	2 (45.8%)	3 (54.2%)	3 (42.5%)
10. Risk identification planning in HH level	0 (90.0%)	0(78.3%)	0 (100%)	0 (56.7%)	3 (33.3%)	2 (50.0%)	2 (34.2%)	3 (58.3%)	3 (33.3%)
11. Knowledge about the intensity of natural disaster	0 (89.2%)	0 (67.5%)	0 (89.2%)	0 (58.3%)	3 (35.0%)	2 (65.0%)	3 (37.5%)	3 (67.5%)	3 (38.3%)
12. Idea about the frequency of natural hazards	0 (88.3%)	0 (79.2%)	0 (94.2%)	0 (49.2%)	3 (35.8%)	2 (60.0%)	2 (35.8%)	3 (65.8%)	2 (42.5%)

Data Source: Roy (2019).

in the *Moddhom Khalishabunia* village and *Rayenda* village (Roy, 2019). Nevertheless, Roy (2019) found that in 2015 the majority of the respondents in *Jaliakhali* were planning and initiating various household DRR/mitigation activities (Table 16.1). Roy (2019) further found that none of the respondents had taken part in planning and assessing activities of disaster risk identification activities and subsequently, between 2007 and 2015 their knowledge of risk identification was also found very poor. Roy (2019) wrote:

> Nobody thinks about the locality and the previous job history of the local people; everybody comes with their pre-planned, profitable livelihood option without considering [the needs] … of local people.
>
> *p. 61*

Similarly, Habiba et al. (2013)presented three major community-based programs in southwest coastal areas such as cyclone preparedness program (CPP), floating garden, and rainwater harvest programs without much consultation with community people. Subsequently, Roy (2019) stated that "as a result, at the end of the project, the local people gain little sustainability" (p. 61). Therefore, research has shown that there was limited consultation with local people for DRR interventions. This signals that there were little opportunities for community prioritizations in DRR interventions in the southwest coastal areas of Bangladesh.

Local People's Ownership

Creating ownership among the local community members is another proven DRR intervention technique. Therefore, it was placed at one of the second priorities of action of the SFDRR. The clear indication of local people's ownership has found in SFDRR at clause 27 (a):

> To mainstream and integrate disaster risk reduction within and across all sectors and review and promote the coherence and further development, as appropriate, of national and local frameworks of laws, regulations and public policies, which, by defining roles and responsibilities, guide the public and private sectors in: (i) addressing disaster risk in publicly owned, managed or regulated services and infrastructures; (ii) promoting and providing incentives, as relevant, for actions by persons, households, communities and businesses.
>
> *UNISDR, 2015, p. 17*

Research by Scott and Few (2016) revealed that community ownership played a significant role in capacity building at the local level and thus it reduced dependence on centralized bureaucratic structures. Local people's sense of ownership tends to increase community control over its own resources (Azad et al., 2019). The significance of creating a sense of ownership among the local community people is that it enhances partnership between various stakeholders. External interventions without local consultation and community participation reduced the efficacy of DRR programs (Scott & Few, 2016).

A successful outcome of creating a sense of local ownership is shown by Akter (2020)who found that in the south-central coastal area of Bangladesh, a strong sense of ownership motivated local people to take part in the Coastal Embankment Improvement Project in Bangladesh not only physically, but also financially. It was found that local people contributed between US$35 and US$65 each (Akter, 2020). But it was also seen that the arsenic mitigation program in Bangladesh has failed due to a lack of maintenance, which happened due to a lack of sense of ownership (Shafiquzzaman et al., 2009). Likewise, in southwest coastal areas, due to lack of sense of ownership the local people were more willing to take shelter in neighboring houses and embankments rather than cyclone shelters (Mallick, 2014). This evidence suggests that there is a gap in community prioritization in most of the DRR interventions in Bangladesh, especially in southwest coastal areas.

The Value of Local People's Knowledge

The value of local people's knowledge to understand the disaster risk can be found at the first priority of SFDRR at clause 24 (i):

> To ensure the use of traditional, indigenous, and local knowledge and practices, as appropriate, to complement scientific knowledge in disaster risk assessment and the development and implementation of policies, strategies, plans and programs of specific sectors, with a cross-sectoral approach, which should be tailored to localities and to the context.
>
> *UNISDR, 2015, p. 15*

Local people generally accumulate their knowledge through experience over generations to cope with disastrous events. Over the past few decades, some authors (i.e., Garai, 2017; Sony, 2018; Sultana & Mallick, 2015) called for integrating local people coping mechanisms in DRR, as well as giving value to local knowledge in policy formulation. Plenty of literature also highlight the effectiveness of local coping mechanisms in disaster management. For instance, Garai (2017) has explored how the local coastal people of Bangladesh built a unique housing structure in order to withstand cyclones. Similarly, it was also found that local people have some sort of knowledge through which they can predict any natural events (Garai, 2017). Besides this, the local people have knowledge through which they used to cope with economic vulnerabilities and be prepared for future hazards (Garai, 2017; Sony, 2018; Sultana & Mallick, 2015).

Table 16.2 presents a comprehensive picture of local coping mechanisms in the three phases of disaster management practiced by the southwest coastal people of Bangladesh. However, in some literatures (i.e., Garai, 2017; Mohibbullah et al., 2021; Roy, 2019; Roy & Sony, 2019; Sultana & Mallick, 2015), it has been found that though local knowledge could contribute to mitigating disaster risks, the undervaluing and the predominance of external interventions may not tape into the local knowledge, skills, and capacities that community people can bring to the table. Such evidence depicts that there is a lack of community prioritization in the national DRR intervention programs in Bangladesh.

Table 16.2 presents some local disaster coping mechanisms of the southwest coastal areas of Bangladesh (Sony, 2018). These local disaster coping strategies were found highly effective in the three phases of disaseter. For instance, at the post-disaster phase, relief provided by relatives and local elites (Table 16.2) as an initial support after disaster in the southwest coastal areas of Bangladesh was found effective in different studies such asMohibbullah et al. (2021), Garai (2017), and Sony (2018). Similarly, Garai (2017) has explored that the local people can predict a serious natural event based on their own knowledge. For instance:

> Drizzling lasting for several days … and winds blowing from the east side bring the message of cyclone or tidal surge according to them. If this wind blowing continues harshly, then the water level of the river may increase up to 10 feet, overflow the embankment and enter saline water into the locality. The densely black clouds in the western corner … of the sky also indicate the coming northwester (cyclone) according to the respondents of the locality.
>
> *Garai, 2017, p. 429*

Notwithstanding, few studies have found that local knowledge had been valued in formulating national DRR. Therefore, it is important to consider local knowledge for sustainable DRR interventions in the near future.

Coordination, Accountability, and Transparency of Stakeholders

The effectiveness of any DRR strategy, to a large extent, depends on multi-stakeholder consultations at various levels including the United Nations, regional intergovernmental bodies, national governments, private sector, NGOs, community-based organizations, universities, researchers, and the media, among others (Shaw et al., 2013, p. 3; Seddiky et al., 2020). Multi-level consultation, collaboration, and teamwork are crucially important for the success of any DRR intervention (Shaw et al., 2013, p. 3).

> No single agency or actor can deal with DRR issues alone. Effective DRR relies on the efforts of many different stakeholders, including UN agencies, regional and international organizations, CSOs, private sectors, media and academics. The collaboration and cooperation among all stakeholders are crucial in order to improve the resilience of communities.
>
> *Shaw et al., 2013, p. 18*

TABLE 16.2 Local coping strategies of coastal people

Patterns		*Efficiency and effectiveness*
Pre-natural hazard activities		
Perception and prediction indicators		
• Pitch-black cloud and cold wind flow on northwestern corner is the sign of natural hazard • Most disaster occurs in the months of March to April (especially *Kalboyshakhi*) • Abnormal animal behavior is another sign of natural disaster, i.e., ants' upward movement is sign of tidal surge, and egrets' flying is sign of storm • Drizzling lasting from several days (mostly seven to twelve days) is another sign of disaster		Raise awareness among community and make themselves prepare to combat disaster.
Wave of river		Especially elderly people perceive it. Raise awareness among community and make themselves prepare to combat disaster.
Mass media weather forecast (radio, television, mobile SMS)		Raise awareness among community and make themselves prepare to combat disaster.
Pre- and during natural activity		
Preventive measures		
Community awareness	Person to person Young people to community Union council's disaster management committee Through institutions, i.e., mike, bell (religious institutions)	To make people aware and to minimize the losses and combat disaster.
Binding their tin/thatch house roofs by rope with a tree (specifically coconut tree) and bamboo to protect its flying in the wind		Safe their house from storm.
Untie cattle		They could save themselves.
Taking cattle on embankment		Reduce the death/destruction of livestock from cyclone.
Keeping important document, money, gold, and silver under floors		Reduce the loss and protect wealth from distraction.
Taking shelter on embankment, religious institutions, bricks building, school, cyclone shelter		Save themselves and protect family members' life.
Using boats		Rescue people and save family members.
Climbing on trees		Save own self from tidal surge.
Repair the embankment		To protect crops, resist saline water and surge.

(continued)

TABLE 16.2 Cont.

Patterns	*Efficiency and effectiveness*
During and post-natural hazards activities Coping techniques	
Taking children, women, old-aged person in religious institution, embankment, brick-built house, and union *parishad* building during natural hazards	To save human life from hazards.
Store food and seeds	To resist tidal surge.
• By making *Gola* is a traditional separate food store room made of bamboo, tin on a high platform	To support during a time.
• Putting food under the floor by using foil packet	
Store food in plastic pots and clay pots and put it on a platform	
Store rain water in a big water tank (plastic water tank, clay water tank, concrete water tank)	To drink in crisis period.
Purifying water (used to boil, tablets, alum, sunlight, and modern filter machine)	To protect themselves from water-borne diseases.
Storing straw and some dry food for cattle	Provide cattle temporary means for survival.
Using water from pond	To get cooking water after disaster.
Building huts on embankment	To live temporarily after disaster.
Building brick building	To resist future disaster.
Source of income like fishing, Government's projects, going to forest	To solve emergency economic problem.
Using leaf, grass, turmeric, basil, spider net, lime as medicine	To cure little injury during and after disaster.
Migration	To solve emergency economic problem.
Relief provided by relatives and local elites	To give initial support after disaster.

Data source: Sony (2018).

Similarly, Twigg (2007) also argued that no single group or organization can address every aspect of DRR which sees disasters as complex problems demanding a collective response from different disciplinary perspectives and diverse groups. The level of follow-up actions of the International Strategy for Disaster Reduction (ISDR) system to promote HFA implementation includes global, regional, national, and thematic levels by emphasizing the multi-stakeholder partnership.

A lack of coordination and consultation among various stakeholders may result in weak and ineffective DRR programs because of duplication, confusion, and inefficient use of scarce resources (Islam et al., 2020). DRR programs are more likely to fail, if these do not consider community peoples' priorities, voices, and unless these tap into community peoples' wisdom and knowledge accumulated over time and passed down across generations.

Research has shown that there was a lack of collaboration and coordination between different governmental bodies such as ministries and departments (Islam et al., 2020). In Bangladesh, many of the disasters are closely linked to climate change; numerous challenges lie in the integration of DRR and CCA. Research by Islam et al. (2020) highlighted that the lack of integration of CCA and DRR interventions has led to inefficient use of scarce resources and overlapping programs implemented. Their research further pointed to a lack of coordination and collaboration, poor governance, competing actors and institutions, lack of information and communication, little or no opportunities for knowledge sharing and community involvement, among other issues and problems in DRR. The authors wrote,

> Lack of collaboration … and a lack of coordination are common between DRR and CCA approaches. Coordination and collaboration across sectors and stakeholders are considered necessary for the effective integration of DRR and CCA.
>
> *Islam et al., 2020, p. 6*

Findings from a study of the role of NGOs in disaster such as floods argued that there was a need for a multi-stakeholder's involvement in line with global best practices aimed at reducing disaster risk (Seddiky et al., 2020). The study showed that although NGOs have potential to play a pivotal role in DRR in Bangladesh, their involvement has been limited due to lack of multi-stakeholder consultation and funding's constraints (Seddiky et al., 2020).

Another important component of CPDRRI is the accountability and transparency of, and coordination among, all stakeholders. Most of the time, especially in the international sphere, the accountability and transparency of stakeholders in disaster management are used to refer to disaster risk governance. For community prioritization, the accountability and transparency of the stakeholders to the community members are crucially important to formulate realistic and implementable prevention, preparedness, response, and recovery plans for sustainable DRR interventions (Ikeda & Nagasaka, 2011; Jones et al., 2015). Howsoever, in the context of southwest coastal areas, any kinds of evidence of accountability and transparency of stakeholders including GOs and NGOs empirically in the southwest coastal areas have not found (Roy, 2019). Therefore, the possibility of corruption has been speculated by Kabir et al. (2021) due to the involvement of different non-state actors. Despite, since disaster risk governance is a multi-sector approach and critical to generalize for all the community members, the mass uses of ICT could be a solution to such challenges proposed by Mondal et al. (2021). This component of CPDRRI relatively depends on the above components. Besides, without valuing community members' voice and decision-making as well as participation such as accountability and transparency of the actors are not ensured. This study does not find any kind of significant evidence presenting the successful disaster risk governance in southwest coastal areas of Bangladesh. Therefore, further study is needed to address this literature gap.

Gender Priorities

Hardly any DRR intervention program could be successful without prioritizing gender needs, since women, men, and other genders have different experiences in disasters, and the levels of vulnerabilities vary along the gender lines, alongside other axes of social differences such as class, place, ethnicity, and social status. Some research in Bangladesh has shown how gender influences the experiences of women when faced with the impacts of climate change including floods and abnormal weather conditions. For instance, Ikeda (2009) stated, "As women and men have different experiences in disaster, gender concerns should be fully addressed by the community and integrated into the action they take up to reduce disaster risks" (p. 443). In a line with this view, much of the literature, like Shahin et al. (2020), Ahmed and Kiester (2021), and so on, have emphasized on the need to mainstream gender in DRR interventions.

A pioneering study on gender and disaster conducted by Mahbuba Nasreen (1995) indicates that disasters affect all segment of population but the burden of coping with disasters falls heavily on the shoulders of women. In another study, conducted with intersectional categories of women, examined that gender-based violence increases during and post-disaster situations, which have not studied properly (Nasreen, 2008). She also argues that despite challenges women contributions are major in household resilience (Nasreen, 2012/2019). The negative consequences of lack of gender priorities in different DRR intervention programs in coastal Bangladesh have been presented in different scholarly materials. For instance, due to a lack of consideration of women's needs during cyclones, women have safety and security concerns when they are relocated to cyclone shelters, but these concerns were not adequately addressed (Shahin et al., 2020). Subsequently, most of the female members who have past bitter experiences were not willing to take shelter at most cyclone shelters (Ayeb-Karlsson, 2020; Shahin et al., 2020).

Lack of such priorities also leads to unequal adaptive capacity based on gender differences. For example, Ahmed and Kiester (2021) found that in the southwest coastal areas of Bangladesh, most of the female-headed farmers were relatively vulnerable to male-headed farmers in terms of social mobility and securing alternative livelihood strategies in the times of crisis and emergencies. Nevertheless, the gender-specific livelihood condition could be improved through an effective legal framework and gender-sensitive DRR planning (Hasan et al., 2019). Thus, the above discussions suggest that without proper priorities of gender-specific needs, the community prioritization in DRR intervention may not be effective. Overall, this chapter indicates that in the southwest coastal areas of Bangladesh, community priorities have been addressed poorly in the DRR interventions.

Conclusion

Over the past couple of decades, the community-based disaster risk reduction (CBDRR) has become a well-accepted model of engaging community for the purpose of addressing risks and vulnerabilities arising from multiple natural and human-made hazards. In line with this view, several researchers emphasized the effectiveness of the bottom-up approach of DRR intervention in which community actors through their participation influence DRR interventions. For this reason, it is important to understand to what extent community's voices are heard in DRR programs. From this viewpoint, this chapter has sought to discuss how much community priorities have taken place in DRR intervention programs in Bangladesh. As DRR covers a vast area, it is unrealistic to discuss all aspects of DRR intervention program at the same time. Therefore, this chapter has highlighted some key components of community participation in DRR in the context of southwestern Bangladesh in particular.

To assess how much local needs have been addressed in the national DRR policies and how the mission of the SFDRR framework could be successful through addressing the strengths and weaknesses of these policies, a critical review of literatures was undertaken. In addition, the authors drew on their field experiences of working with local communities affected by natural hazards in Bangladesh. They considered the components of the CPDRRI model such as consultation with local people, local people's ownership, the value of local people's knowledge, accountability and transparency of stakeholders, and gender priorities in DRR.

Accessing the five key components mentioned above, this chapter reveals, except for the creation of local ownership, that there are few evidences to suggest that DRR programs have consulted with local people, valued local knowledge, and assessed gender needs. Furthermore, there was no evidence of any stakeholders' accountability to local people in the southwest coastal areas of Bangladesh. To address these gaps, further empirical study is needed. Through this study it is also recommended that necessary initiatives should be taken by the policymakers to make room for community consultation and participation in the DRR programs to increase community resilience and decrease their vulnerabilities. Furthermore, the existing DRR needs to be multi-sectoral, more engaging, and built around the principles of community, equity, strengths, resilience, and partnership. Finally, since NGOs play a vital part in the Bangladesh society and in its development efforts, they need to be more actively involved in DRR intervention alongside the government.

References

Abedin, M. A., Habiba, U., & Shaw, R. (2014). Community perception and adaptation to safe drinking water scarcity: Salinity, arsenic, and drought risks in coastal Bangladesh. *International Journal of Disaster Risk Science*, *5*(2), 110–124. https://doi.org/10.1007/s13753-014-0021-6

Abedin, M. A., & Shaw, R. (2015). The role of university networks in disaster risk reduction: Perspective from coastal Bangladesh. *International Journal of Disaster Risk Reduction*, *13*, 381–389. https://doi.org/10.1016/j.ijdrr.2015.08.001

Ahmed, S., & Kiester, E. (2021). Do gender differences lead to unequal access to climate adaptation strategies in an agrarian context? Perceptions from coastal Bangladesh. *Local Environment*, *26*(5), 650–665. https://doi.org/10.1080/13549839.2021.1916901

Akter, S. (2020). Social cohesion and willingness to pay for cyclone risk reduction: The case for the Coastal Embankment Improvement Project in Bangladesh. *International Journal of Disaster Risk Reduction, 48*, 101579.https://doi.org/10.1016/j.ijdrr.2020.101579

Ali, I., Hatta, Z. A., & Azman, A. (2014). Transforming the local capacity on natural disaster risk reduction in Bangladeshi communities: A social work perspective, *8*(1), 34–42. https://doi.org/10.1111/aswp.12023

Asif, M., Searcy, C., Zutshi, A., & Fisscher, O. A. (2013). An integrated management systems approach to corporate social responsibility. *Journal of Cleaner Production, 56*, 7–17. https://doi.org/10.1016/j.jclepro.2011.10.034

Ayeb-Karlsson, S. (2020). 'I do not like her going to the shelter': Stories on gendered disaster (im)mobility and wellbeing loss in coastal Bangladesh. *International Journal of Disaster Risk Reduction, 50*, 101904.https://doi.org/10.1016/j.ijdrr.2020.101904

Azad, M. A. K., Uddin, M. S., Zaman, S., & Ashraf, M. A. (2019). Community-based disaster management and its salient features: A policy approach to people-centred risk reduction in Bangladesh. *Asia-Pacific Journal of Rural Development, 29*(2), 135–160. https://doi.org/10.1177/1018529119898036

Baker, N. D., & Grant Ludwig, L. (2018). Disaster preparedness as social control. *Critical Policy Studies, 12*(1), 24–43. https://doi.org/10.1080/19460171.2016.1214075

Benge, L., & Neef, A. (2020). Linking disaster risk reduction to development: The evolution of 'building back better' in International Disaster Management Frameworks. In A. Neef & N. Pauli (Eds.), *Climate-induced disasters in the Asia-Pacific region: Response, recovery, adaptation* (Vol. 22, pp. 11–24). Emerald Publishing Limited. https://doi.org/10.1108/S2040-726220200000022002

Briceño, S. (2015). Looking back and beyond Sendai: 25 years of international policy experience on disaster risk reduction. *International Journal of Disaster Risk Science, 6*(1), 1–7. https://doi.org/10.1007/s13753-015-0040-y

Connell, R. (2020). COVID-19/sociology. *Journal of Sociology, 56*(4), 745–751.https://doi.org/10.1177%2F1440783320943262

Cuthbertson, J., Rodriguez-Llanes, J. M., Robertson, A., &Archer, F. (2019). Current and emerging disaster risks perceptions in Oceania: Key stakeholders recommendations for disaster management and resilience building. *International Journal of Environmental Research and Public Health, 16*(3), 460. https://www.mdpi.com/1660-4601/16/3/460

Dynes, R. (1994). Community emergency planning: False assumptions and inappropriate analogies. *International Journal of Mass Emergencies and Disasters, 12*(2), 141–158.

Garai, J. (2017). Qualitative analysis of coping strategies of cyclone disaster in coastal area of Bangladesh. *Natural Hazards, 85*(1), 425–435. https://doi.org/10.1007/s11069-016-2574-8

Gitz, V., & Meybeck, A. (2012). Risks, vulnerabilities and resilience in a context of climate change. In A. Meybeck, J. Lankoski, S. Redfern, N. Azzu, & V. Gitz (Eds.), *Building resilience for adaptation to climate change in the agriculture sector* (pp. 19–36). Rome: FAO.

Habiba, U., Shaw, R., & Abedin, M. A. (2013). Community-based disaster risk reduction approaches in Bangladesh. In R. Shaw, F. Mallick, & A. Islam (Eds.), *Disaster risk reduction approaches in Bangladesh* (pp. 259–279). Springer. https://doi.org/10.1007/978-4-431-54252-0_12

Hardoy, J., Pandiella, G., & Barrero, L. S. V. (2011). Local disaster risk reduction in Latin American urban areas. *Environment and Urbanization, 23*(2), 401–413. https://doi.org/10.1177/0956247811416435

Hasan, M. R., Nasreen, M., &Chowdhury, M. A. (2019). Gender-inclusive disaster management policy in Bangladesh: A content analysis of national and international regulatory frameworks. *International Journal of Disaster Risk Reduction, 41*, 101324.https://doi.org/10.1016/j.ijdrr.2019.101324

Hewitt, K. (1983). The idea of calamity in a technocratic age. In K. Hewitt (Ed.), *Interpretations of calamity* (Vol. 1, pp. 3–32). Routledge.

Ikeda, K. (2009). How women's concerns are shaped in community-based disaster risk management in Bangladesh. *Contemporary South Asia, 17*(1), 65–78. https://doi.org/10.1080/09584930802624679

Ikeda, S., &Nagasaka, T. J. (2011). An emergent framework of disaster risk governance towards innovating coping capability for reducing disaster risks in local communities. *International Journal of Disaster Risk Science, 2*(2), 1–9.

Islam, M. N. (2018). Community-based responses to flood and river erosion hazards in the active Ganges floodplain of Bangladesh. In R. Shaw, K. Shiwaku, & T. Izumi (Eds.), *Science and technology in disaster risk reduction in Asia* (pp. 301–325). Academic Press. https://doi.org/10.1016/B978-0-12-812711-7.00018-3

Islam, M. R., &Hasan, M. (2016). Climate-induced human displacement: A case study of Cyclone Aila in the south-west coastal region of Bangladesh. *Natural Hazards, 81*(2), 1051–1071. https://doi.org/10.1007/s11069-015-2119-6

Islam, S., Chu, C., &Smart, J. C. (2020). Challenges in integrating disaster risk reduction and climate change adaptation: Exploring the Bangladesh case. *International Journal of Disaster Risk Reduction, 47*, 101540.https://doi.org/10.1016/j.ijdrr.2020.101540

Jones, S., Manyena, B., &Walsh, S. (2015). Disaster risk governance: Evolution and influences. In J. F. Shroder, A. E. Collins, S. Jones, B. Manyena, & J. Jayawickrama (Eds.), *Hazards, risks and disasters in society* (pp. 45–61). Academic Press. https://doi.org/10.1016/B978-0-12-396451-9.00004-4

Kabir, M. R., Khan, S., Chowdhury, S., Jahan, S., Islam, K. A., & Zayed, N. M. (2021). Corruption possibilities in the climate financing sector and role of the civil societies in Bangladesh. *Journal of Southwest Jiaotong University, 56*(2), 55–65. https://doi.org/10.35741/issn.0258-2724.56.2.6

Katayama, T. (1993). International decade for natural disaster reduction: Are we chasing a dream? Proceedings of IECON 1993–*19th Annual Conference of IEEE Industrial Electronics*.

Kelman, I., &Glantz, M. H. (2015). Analyzing the Sendai framework for disaster risk reduction. *International Journal of Disaster Risk Science, 6*(2), 105–116. https://doi.org/10.1007/s13753-015-0056-3

Lechat, M. F. (1990). The international decade for natural disaster reduction: Background and objectives. *Disasters, 1*(14), 1–6.

Mall, R. K., Srivastava, R. K., Banerjee, T., Mishra, O. P., Bhatt, D., & Sonkar, G. (2019). Disaster risk reduction including climate change adaptation over South Asia: Challenges and ways forward. *International Journal of Disaster Risk Science, 10*(1), 14–27. https://doi.org/10.1007/s13753-018-0210-9

Mallick, B. (2014). Cyclone shelters and their locational suitability: An empirical analysis from coastal Bangladesh. *Disasters, 38*(3), 654–671. https://doi.org/10.1111/disa.12062

Matsuoka, Y. (2013). *Enhancing local government's capacities to cope with urban disasters in Makati, Philippines*. PhD Thesis, International Environment and Disaster Management Laboratory, Graduate School of Global Environment Studies, Kyoto University, Kyoto.

McGee, S., Frittman, J., Ahn, S. J., & Murray, S. (2016). Implications of cascading effects for the Hyogo Framework. *International Journal of Disaster Resilience in the Built Environment*, 7(2), 144–157. https://doi.org/10.1108/IJDRBE-03-2015-0012

MoEF, G. (2009). *National Adaptation Programme of Action (NAPA)*. Dhaka: MoEF, Government of the People's Republic of Bangladesh.

MoEF, G. (2012). *Introduction to Ministry of Environment and Forest.* Dhaka: MoEF, Government of the People's Republic of Bangladesh.

Mohibbullah, M., Gain, A. K., & Ahsan, M. N. (2021). Examining local institutional networks for sustainable disaster management: Empirical evidence from the South-West coastal areas in Bangladesh. *Environmental Science & Policy, 124*, 433–440. https://doi.org/10.1016/j.envsci.2021.07.016

Mojid, M. (2020). *Climate change-induced challenges to sustainable development in Bangladesh.* IOP Conference Series: Earth and Environmental Science.

Mondal, T., Pramanik, S., Pramanik, P., Datta, K. N., Paul, P. S., Saha, S., & Nandi, S. (2021). Emergency communication and use of ICT in disaster management. In M. Sakurai & R. Shaw (Eds.), *Emerging technologies for disaster resilience: Practical cases and theories* (pp. 161–197). Singapore: Springer. https://doi.org/10.1007/978-981-16-0360-0_10

Nasreen, M. (1995). *Coping with floods: the experience of rural women in Bangladesh.* Unpublished PhD Dissertation. Massey University, New Zealand.

Nasreen, M. (2008). *Violence against women during floods and post-flood situations in Bangladesh.* ActionAid International and ActionAid Bangladesh.

Nasreen, M. (2012/2019, 2nd ed.). *Women and girls: Vulnerable or resilient?* Dhaka: Institute of Disaster Management and Vulnerability Studies, University of Dhaka, Bangladesh.

Nasreen, M. (2019). *Resilience.* Humanitarian Encyclopedia. https://humanitarianencyclopedia.org/concept/resilience

Orimoloye, I., Belle, J., & Ololade, O. (2021). Exploring the emerging evolution trends of disaster risk reduction research: A global scenario. *International Journal of Environmental Science Technology, 18*(3), 673–690. https://doi.org/10.1007/s13762-020-02847-1

Pandey, B., &Okazaki, K. (2005). Community-based disaster management: Empowering communities to cope with disaster risks. *Regional Development Dialogue, 26*(2), 52–60.

Paton, D., &Johnston, D. (2001). Disasters and communities: Vulnerability, resilience and preparedness. *Disaster Prevention Management, 10*(9), 270–277. https://doi.org/10.1108/EUM0000000005930

Roy, T. (2019). Disaster risk reduction interventions in Bangladesh: A case study of selected villages of southwest region. *Journal of the Asiatic Society of Bangladesh (Hum.), 64*(1), 43–69.

Roy, T., Chandra, D., Sony, M. A. A. M., &Rahman, M. S. (2020). Impact of salinity intrusion on health of coastal people: Reflections from Dacopeupazila of Khulna district, Bangladesh. *Khulna University Studies, 17*(1 & 2), 57–66.

Roy, T., &Sony, M. M. A. A. M. (2019). Social structural changes in post-disaster area from Marxist point of view: Reflections from Dacope, Khulna, Bangladesh. *NIU International Journal of Human Rights, 6*(1), 3–13.

Saha, C. K. (2015). Dynamics of disaster-induced risk in southwestern coastal Bangladesh: An analysis on tropical Cyclone Aila 2009. *Natural Hazards, 75*(1), 727–754. https://doi.org/10.1007/s11069-014-1343-9

Schwartz, N., &Deruyttere, A. (1996). *Community consultation, sustainable development and the Inter-American Development Bank.* Washington, DC: Concept Paper for the Inter-American Development Bank.

Scott, Z., &Few, R. (2016). Strengthening capacities for disaster risk management I: Insights from existing research and practice. *International Journal of Disaster Risk Reduction, 20*, 145–153. https://doi.org/10.1016/j.ijdrr.2016.04.010

Seddiky, M. A., Giggins, H., & Gajendran, T. (2020). International principles of disaster risk reduction informing NGOs strategies for community based DRR mainstreaming: The Bangladesh context. *International Journal of Disaster Risk Reduction, 48*, 101580.https://doi.org/10.1016/j.ijdrr.2020.101580

Shafiquzzaman, M., Azam, M. S., Mishima, I., & Nakajima, J. (2009). Technical and social evaluation of arsenic mitigation in rural Bangladesh. *Journal of Health, Population, and Nutrition, 27*(5), 674. https://dx.doi.org/10.3329%2Fjhpn.v27i5.3779

Shahin, M., Billah, M., Islam, M. M., Parvez, A., &Zaman, A. K. M. M. (2020). Cyclone shelters need sustainable development. *International Journal of Disaster Resilience in the Built Environment, 11*(5), 659–678. https://doi.org/10.1108/IJDRBE-12-2015-0061

Shaw, R. (2012). Chapter 1 overview of community-based disaster risk reduction. In R. Shaw (Ed.), *Community-based disaster risk reduction* (Vol. 10, pp. 3–17). Emerald Group Publishing Limited. https://doi.org/10.1108/S2040-7262(2012)0000010007

Shaw, R., Islam, A., & Mallick, F. (2013). National perspectives of disaster risk reduction in Bangladesh. In R. Shaw, F. Mallick, & A. Islam (Eds.), *Disaster risk reduction approaches in Bangladesh* (pp. 45–62). Springer. https://doi.org/10.1007/978-4-431-54252-0_3

Sony, M. M. A. A. M. (2018). *Indigenous coping mechanism and role of religion in disaster management: An evidence from Dacope* (Publication Number 141642). Khulna: Khulna University.

Sultana, Z., &Mallick, B. (2015). Adaptation strategies after cyclone in southwest coastal Bangladesh: Pro poor policy choices. *American Journal of Rural Development, 3*(2), 24–33.

Twigg, J. (2007). *Characteristics of a disaster-resilient community: A guidance note*. Department for International Development (DFID).

Twigg, J. (2015). *Disaster risk reduction*. Overseas Development Institute.

Twigg, J., & Mosel, I. (2017). Emergent groups and spontaneous volunteers in urban disaster response. *Environment Urbanization, 29*(2), 443–458. https://doi.org/10.1177%2F0956247817721413

UNISDR. (2009). *UNISDR Terminology on Disaster Risk Reduction*. Geneva: United Nations Office on Disaster Risk Reduction, Geneva. https://www.unisdr.org/files/7817_UNISDRTerminologyEnglish.pdf

UNISDR. (2015). *Sendai framework for disaster risk reduction 2015–2030*. Geneva: UNISDR. www.unisdr.org/files/43291_sendaiframeworkfordrren.pdf

Victoria, L. P. (2003). Community-based disaster management in the Philippines: Making a difference in people's lives. *Philippine Sociological Review, 51*, 65–80. www.jstor.org/stable/44243073

Wahlström, M. (2015). New Sendai framework strengthens focus on reducing disaster risk. *International Journal of Disaster Risk Science, 6*(2), 200–201. https://doi.org/10.1007/s13753-015-0057-2

Walker, N. (2013). Message: The challenging road to disaster resilience in Bangladesh.In R. Shar, F. Mallick, &A. Islam (Eds.), *Disaster risk reduction approaches in Bangladesh* (pp. vii–ix). Tokyo: Springer.

Yoon, D. K., Kang, J. E., & Brody, S. D. (2016). A measurement of community disaster resilience in Korea. *Journal of Environmental Planning and Management, 59*(3), 436–460. https://doi.org/10.1080/09640568.2015.1016142

17

DEVELOPING A BEHAVIOUR CHANGE INTERVENTION USING THE BEHAVIOUR CHANGE WHEEL TO INCREASE COMMUNITY PARTICIPATION IN THE LOCAL LEVEL DISASTER MANAGEMENT PROCESS IN BANGLADESH

Sanjoy Kumar Chanda

Introduction

The United Nations Sustainable Development Goals (SDGs) explicitly target risk reduction under Goal 13: Taking urgent action to combat climate change and its impacts (United Nations, 2015a). The Sendai Framework for Disaster Risk Reduction 2015–2030 states in its post-2015 agenda for disaster that while disasters considerably hinder progress towards sustainable development, effective disaster risk management contributes to sustainable development (United Nations, 2015b). Disasters occur first at the local level and affect local people, and consequently, the responsibility for managing such risks requires the linkage of the local level with high importance (Cutter et al., 2012). Disaster management plans are currently being discussed to ensure the participation of local people around the world. However, questions remain as to how much this participation establishes actual planning problems, for example, how many people effectively participate in the decision-making process, how the participation process is conducted, who the decision makers are and how far they use their will in the planning process (Mallick & Vogt, 2011).

The Climate Risk Index 2021 shows that Bangladesh is one of the most disaster-prone countries in the world (Eckstein et al., 2021). Frequent natural calamities have destroyed the socioeconomic structure of the country. Disaster management has been now considered as an event rather than a development process (Islam, 2013). Linkage with ongoing development programmes and community participation in programme planning and implementation will enhance local capacity and readiness (Huq, 2016).

DOI: 10.4324/9781003253495-22

The term community participation refers to the involvement of local people in any project to mitigate their problems or to develop their socioeconomic conditions (Harvey et al., 2002). In addition, the disaster management process includes several programmes, plans and actions that are taken by the government to mitigate disasters. Thus, community members can participate in disaster management goal setting and planning, programme preparation, implementation and evaluation (Harvey et al., 2002). This format is also successful in some European and Asian countries, such as Switzerland, Germany, China, Korea and Japan in responding to natural disasters (Fawcett & Plan International, 2012). Disasters need to be addressed through proper planning and action, along with the active participation of affected people in efforts to improve the sustainable development process as well as reduce property damage and human sufferings (Hossain, 2013). The active participation of communities will be more effective in enhancing local capacity building to recover the disaster-affected coastal people in Bangladesh (Islam & Hasan, 2015). Local disaster management activities with community people boost confidence and capabilities to carry out development activities and disaster preparedness at the local level (Islam, 2018).

The government of Bangladesh has initiated the local level disaster management process through several programmes, such as relief and response to disaster risk reduction (DRR), community-based disaster management (CBDM) and enhancing community resilience to natural disasters (Azad et al., 2019) to reduce the impact of disasters. However, the participation of Bangladeshi communities in the local disaster management process was not significant (Islam, 2018), for example, only 14% were involved with the earlier disaster management (BBC Media Action, 2016) and for this reason, the process was not fully successful. Disaster human vulnerabilities are increasing day by day and new strategies are a necessity in disaster management (Rahman et al., 2017). Thus, targeted interventions now need to be identified to overcome barriers and improve the participation of the community people in the local level disaster management process in Bangladesh.

A Theory Informed Intervention Design

Oliver-smith (2009) and Nasreen (2004) point out that disasters occur frequently in Bangladesh, but the application of theoretical frameworks in disaster research is limited. This study intended to design an intervention for disaster management based on a behaviour change theory. Although several frameworks exist to classify behaviour change interventions, no one is comprehensive and conceptually coherent in designing effective interventions (Michie et al., 2011). It is an emerging field of enquiry to develop behaviour change interventions, which are also known as complex interventions (O'Cathain et al., 2019). However, no appropriate process is reported yet within the literature (Connell et al., 2015; Truelove et al., 2020). The guidelines established by the United Kingdom Medical Research Council (MRC) framework for the development and evaluation of complex interventions (Craig

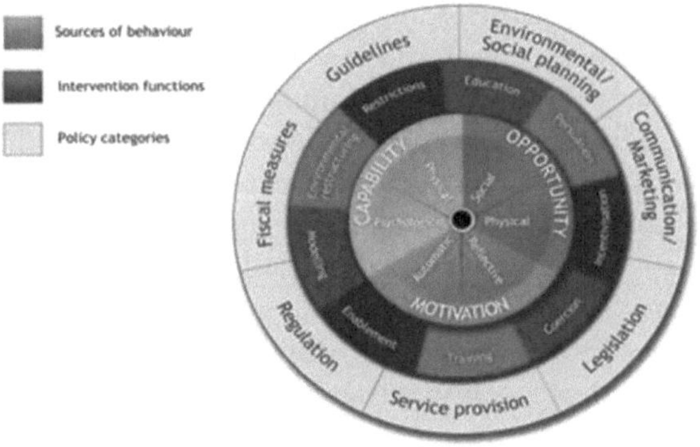

FIGURE 17.1 The Behavioural Change Wheel
Source: Michie et al. (2014).

et al., 2008; Skivington et al., 2021) recognize the application of theory as a fundamental element in the development of interventions. This study has used the theoretically driven Behaviour Change Wheel (BCW) (Figure 17.1) framework developed by Michie et al. (2014) to underpin and develop a behaviour change intervention to improve community participation in the local disaster management process in Bangladesh. The BCW provides a comprehensive and useful framework for the development of interventions and greater control over the replication and evaluation of interventions (Webb et al., 2016). This framework synthesizes 19 behaviour change theories with the Capability, Opportunity and Motivation of Behaviour (COM-B) model sitting at its centre (Webb et al., 2016). Appropriate theories and models can also assist to explain behaviour and show the way of developing more effective ways to influence and change behaviour (Glanz & Bishop, 2010). Thus, the COM-B model has been used for this study because it offers a logical approach to behaviour change and was successfully used to design interventions in order to change the behaviour of the target population (Connell et al., 2015). Where more detail is needed, COM-B can be sub-divided into domains within the Theoretical Domains Framework (TDF). Exploring a given behaviour associated with COM-B and/or TDF components helps to decide which determinants need to be addressed first to achieve behaviour change (Fulton et al., 2016). The BCW framework then guides the selection of intervention functions and policy categories, if required, and specific Behaviour Change Techniques (BCTs) from the Behaviour Change Technique Taxonomy (BCTTv1) (Michie et al., 2013) on which the intervention content is based.

Aims

To explore barriers to community participation in the local level disaster management process in order to identify:

1. The target behaviour for a behaviour change intervention to increase community participation.
2. The development of an intervention to achieve change in the target behaviour.

Methods

Study Setting

Bangladesh is located in South Asia, with an estimated population of 168.2 million (Bangladesh Bureau of Statistics [BBS], 2021). It is one of the most densely populated countries in the world, with a population density of 1140 per square kilometre (BBS, 2021), and about 62% of them lived in rural areas in 2020 (World Bank, 2021). Bangladesh is a deltaic country highly exposed to the risk of disaster (Roy, 2019). The country is geographically located in a floodplain, close to one of the most active seismic zones in the world, exposed to a large coastline with frequent cyclones and storms (Ministry of Disaster Management and Relief, 2020). According to the Global Climate Risk Index 2021, Bangladesh ranks the seventh most affected and vulnerable country in the world due to the impact of long-term climatic disasters (Eckstein et al., 2021). The country faces several major disasters every year including floods, cyclones, droughts, tidal surges, tornadoes, earthquakes, river erosion, fire, infrastructure collapse, arsenic contamination of groundwater, waterlogging, water and soil salinity, cold wave, building collapse, epidemic and various forms of pollution (Rahman et al., 2017). Between1996 and 2015, around 185 disaster events occurred in Bangladesh which incurred an average of 0.732% of GDP loss per annum and a total economic loss of $2.283 billion (Ministry of Disaster Management and Relief, 2020).

Study Design

Figure 17.2 shows the BCW intervention design process that includes three key stages: Stage 1: Understanding the target behaviour; Stage 2: Identifying the intervention functions; and Stage 3: Identifying content and implementation options. These stages can be further sub-divided into eight specific steps, including (i) defining the problem in behavioural terms; (ii) selecting the target behaviour; (iii) specifying the target behaviour; (iv) identifying what needs to change; (v) identifying appropriate intervention functions; (vi) identifying policy categories; (vii) identifying behaviour change techniques; and (viii) determining the mode of intervention delivery.

The steps of the BCW are as follows:

1–3. Steps one to three of the BCW design process guide intervention designers to identify a specific behaviour to change. The intervention behaviour was identified from the review findings and expert opinion. A comprehensive review of the evidence was completed prior to the intervention design. A literature

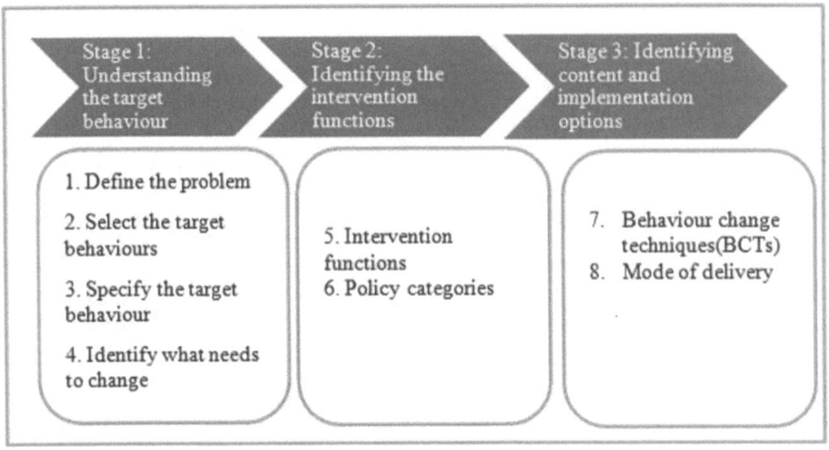

FIGURE 17.2 The BCW intervention design process

Source: Michie et al. (2014).

search was undertaken with Scopus and Web of Science databases as well as Google Scholar and Google to identify relevant papers. A search strategy was developed using free texts and subject headings relating to the areas of barriers to community participation in the local level disaster management process in Bangladesh.

4. The COM-B model was used to identify what needs to change in the behaviour.
5. The relevant intervention functions were selected from the list of nine intervention functions based upon an evaluation of the APEASE criteria (affordability, practicability, effectiveness, acceptability, side effects and safety, and equity) (see below the APEASE criteria in detail).
6. The relevant policy categories to support the delivery of the identified intervention functions were selected based upon intervention functions.
7. The relevant BCTs from the BCTT were selected based on an evaluation against the intervention functions and policy categories that were selected in previous steps.
8. Finally, the mode(s) of delivery based upon an evaluation against the APEASE criteria and confirmation of the intervention content were selected.

Prioritising the Target Behaviour Process and APEASE Criteria

A four-step criteria developed by Michie et al. (2014) was followed to select the target behaviour in step 2 based on a decision-making rule for each criterion. These criteria are as follows:

1. Impact of behaviour change – how much of an impact changing the behaviour will have on the desired outcome.

2. Likelihood of behaviour change – how likely it is that the behaviour can be changed.
3. Spill over score – how likely it is that the behaviour will have a positive or negative impact on other related behaviours.
4. Measurement score – how easy it will be to measure the behaviour.

In addition, criteria were rated as (i) unacceptable, (ii) unpromising but worth considering, (iii) promising and (iv) very promising.

Moreover, the APEASE criteria developed by Michie et al. (2014) were used in order to assess the feasibility of the intervention functions in step 5 and the potential mode of delivery in step 8. The APEASE criteria have been developed to assist researchers in designing and evaluating interventions (Michie et al., 2014). In this study, the feasibility of the intervention functions based on the APEASE criteria was selected as it is the beginning of stage 2, and if the intervention functions are considered with strategic judgments, then the next steps can be designed accordingly. In addition, as the mode of delivery is the final step of the BCW, the APEASE criteria in this step were used in selecting the preferred mode of delivery.

Findings

The findings section is set out in what way each of the eight steps, grouped into three broad stages, informed the subsequent in the intervention design process.

Stage 1: Identify Intervention Options

This stage includes steps 1–4: defining problem, selecting target behaviours, specifying the target behaviour and understanding what needs to change.

Step 1. Define the Problem

The lack of community participation in the local level disaster management process in Bangladesh was identified as a key behaviour problem through the existing literature (Huq, 2016; Islam, 2018; Mahalder et al., 2017; Rahman, 2014) and that interventions were needed to improve community participation in disaster management.

Step 2. Select the Target Behaviours

Considering the decision to increase community participation in addressing the need for a change of overall behaviour, step 2 of the BCW process involved considering all possible behaviours that could be targeted in the interventions to increase community participation. Having defined the behavioural problem, a list of four target behaviours based on consultation with some disaster experts and a

TABLE 17.1 Potential target behaviours by prioritizing criteria

Potential target behaviours relevant to increasing community participation in the local level disaster management process in Bangladesh	*Criteria to prioritise behaviours*			
	Impact of behaviour change	*Likelihood of changing behaviour*	*Spillover score*	*Measurement score*
Raising community awareness and understanding to increase active participation of people in the local disaster management process	Very promising	Promising	Promising	Promising
Developing disaster management plans to ensure much participation of local communities	Promising	Unpromising but worth considering	Promising	Unacceptable
Disseminating the success stories of reducing disaster risks widely among the local people	Promising	Unpromising but worth considering	Promising	Unpromising but worth considering
Increasing fund of the government to launch disaster management programme with communities	Unpromising but worth considering	Promising	Promising	Unacceptable

literature review was identified that could increase the participation of communities in the local disaster management process. The potential target behaviours are listed in the left column of Table 17.1 and behaviours were prioritized based on a four-step criterion developed by Michie et al. (2014).

Following the 'less is ideal' principle when selecting the target behaviour to design effective intervention (Michie et al., 2014), the study selected one target behaviour from the list of behaviours. Based on prioritizing criteria, 'raising community awareness and understanding to increase active participation of people in the local disaster management process' was selected as the specific target behaviour because this behaviour only met all criteria to prioritize behaviours. The behaviour was very promising in terms of the impact because it would have a great impact on improving community participation in disaster management. In addition, the behaviour was also rated as promising in connection with a likelihood of changing behaviour as it was thought that it would be easier to change community participating behaviour. It was also estimated promising in the case of spillover, as community members involved in disaster management would probably persuade other members to do the same. Finally, it was judged to be promising in terms of ease of measurement because it would be easy to monitor behaviour by others.

Step 3. Specify the Target Behaviour

Step 3 described the details of the behaviour, persons involved with the behaviour, activities, frequency, time and with whom it needs to perform for the behaviour to be carried out (Table 17.2). People in the community were targeted for their active participation in the local disaster management process. To achieve the desired goal, people in the community need to increase their capabilities, use opportunities and increase motivation.

Step 4. Identify What Needs to Change

This step involved identifying barriers to the target behaviour, and which of these barriers were considered to change with a view to achieving the desired change in behaviour of the community people in participating the disaster management process. In doing so, the COM-B model and TDF domains (Michie et al., 2014)

TABLE 17.2 Specifying the target behaviour and content

Target behaviour	*Raising community awareness and understanding to increase active participation of people in the local level disaster management process in Bangladesh*
Who needs to perform the behaviour?	Community people
What do they need to do differently to achieve the desired change?	• They need to know the local disaster management process • They need to know how to involve with the process • They need to understand what the local disaster management process offers • They need to use opportunities available to get them involved • They need the motivation to be aware of participation
When do they need to do it?	Immediately before the cyclone season/National Disaster Preparedness Day (NDPD)/preferred time
Where do they need to do it?	Local union council/local cyclone shelters
How often do they need to do it?	Once they meet the Union Disaster Management Committee (UDMC) such as Upazila Nirbahi Officer (UNO) or Union Parishad Chairman (UPC)/researcher
With whom do they need to do it?	UDMC/researcher/Department of Disaster Management (DDM), nongovernmental organizations (NGOs)/international nongovernmental organizations (INGOs)

were used to identify the barriers to the target behaviour and to guide the choice of intervention. This included undertaking a comprehensive review of the existing literature.

A literature search of seven articles (BBC Media Action, 2016; Huq, 2016; Islam, 2018; Kabeer, 2000; Mahalder et al., 2017; Rahman, 2014; Uddin et al., 2021) focusing on factors impeding the target behaviour of 'raising community awareness and understanding to increase active participation of people in the local disaster management process in Bangladesh' revealed ten barriers (see the first column of Table 17.3). The barriers were mostly related to individual(s), organisational and motivational factors, and these were needed to change to raise community awareness and understanding. According to the BCW framework guidelines, data identified from the literature review were analysed by using the COM-B components and TDF domains in order to identify what needs to change to the target behaviour. Initially, the review findings were used with capability, opportunity and motivation components of the COM-B model to analyse behaviour in the context for designing behaviour change intervention. These findings were further mapped onto the six TDF constructs: Knowledge; Environmental context and resources; Social influences; Professional/social role and identity; Beliefs and consequences; and Reinforcement. The TDF domains were involved providing a level of detail of the behaviour in Table 17.3.

Stage 2: Identify Intervention Options

This stage includes two steps: identifying intervention functions and identifying policy categories.

Step 5. Identify Appropriate Intervention Functions

The BCW instructs on which types of intervention functions are likely to initiate behaviour change in each COM-B component and the corresponding TDF domains. Intervention functions, which best met the APEASE criteria, were considered for inclusion in this step. Out of a possible nine intervention functions, Table 17.4 shows how the following five were selected based on the APEASE criteria as most useful for addressing the identified barriers: (i) education; (ii) persuasion; (iii) modelling; (iv) environmental restructuring; and (v) enablement. In addition, Table 17.4 illustrates how the selected intervention functions relate to the corresponding COM-B and TDF components, and how these functions can potentially bring about the desired behaviour change.

Step 6. Identify Policy Categories

Five policy categories were identified in Table 17.3 as potential policy categories that could form the focus of the behaviour change intervention: (i) Communication/marketing; (ii) Guidelines; (iii) Regulation; (iv) Environmental/social planning; and (v) Service provision.

TABLE 17.3 Matrix of links between the summary of barriers, COM-B components, TDF domains intervention functions, policy categories and BCTs for the behaviour change intervention

Lack of knowledge about disaster preparedness decisions: Community members felt that they did not have enough access to information and/or did not know how decisions were made about local disaster preparedness in their communities that prevent them from participating in the local level disaster management process (BBC Media Action, 2016; Islam, 2018).	Capability	Knowledge	Education– providing information to gain knowledge Persuasion– persuading verbally to boost self-confidence	Communication/ marketing guidelines Regulations Service provision	4.1 Instruction on how to perform a behaviour 5.2 Salience of consequences 5.3 Information about social and environmental consequences 9.1 Credible source 15.1 Verbal persuasion about capability	*Mode of delivery*: Face-to-face at both individual and community level *Who will deliver and how*: The UDMC will invite community members to attend a meeting and/ or provide disaster preparedness decisions timely to communities through leaflets/ posters/verbally so that they can be informed about the local disaster management plan.
Lack of awareness and understanding of the local disaster management process: Role, mandate and functioning of the disaster management committee were unknown to local communities (BBC Media Action, 2016; Islam, 2018).			Education– providing information to gain knowledge Persuasion – persuading verbally to boost self-confidence	Communication/ marketing guidelines Regulations Service provision	4.1 Instruction on how to perform a behaviour 5.2 Salience of consequences 5.3 Information about social and environmental consequences 9.1 Credible source 15.1 Verbal persuasion about capability	*Mode of delivery*: Face-to-face at both individual and community level *Who will deliver and how*: The researcher will provide information to the community through leaflets/posters/ verbally so that they can be aware of the local disaster management process.

(*continued*)

TABLE 17.3 Cont.

Lack of knowledge about involvement with disaster management activities: Community members did not have proper knowledge of the existing disaster management process, where the process takes place, when they can go, how to get contact (Islam, 2018).			Education– providing information to gain knowledge Persuasion – persuading verbally to boost self-confidence	Communication/ marketing guidelines Regulations Service provision	4.1 Instruction on how to perform a behaviour 5.2. Salience of consequences 5.3 Information about social and environmental consequences 9.1 Credible source 15.1 Verbal persuasion about capability	*Mode of delivery*: Face-to-face at both individual and community level *Who will deliver and how*: The researcher will provide information to the community through leaflets/posters/verbally so that they can gain knowledge on how to get involved in the local disaster management process.
Lack of resources: Lack of resources needed to launch community level discussion groups and regular committee meetings with community representation (BBC Media Action, 2016; Mahalder et al., 2017).	Opportunity	Environmental context and resources	Persuasion – persuading verbally to boost self-confidence Enablement – increasing capability to be aware	Communication/ marketing guidelines Regulations Service provision	4.1 Instruction on how to perform a behaviour 9.1 Credible source 15.1 Verbal persuasion about capability	*Mode of delivery*: Face-to-face at both individual and community level *Who will deliver and how*: The UDMC will provide information to the community through leaflets/posters/verbally so that they can be aware of discussions on the local disaster management.

Lack of timely warning system: There was a lack of timely or early warning system about the impending disaster in communities (Huq, 2016; Islam, 2018; Rahman, 2014).		Environmental restructuring– changing the social environment	Guidelines Environmental/ social planning Service provision		*Mode of delivery*: Face-to-face at both individual and community level *Who will deliver and how*: The UDMC will take the necessary steps to set up an early disaster warning through miking/ leaflets to inform the community.
Lack of time: Some community members did not have enough time to meet other members of the same community and they became more individualistic (Mahalder et al., 2017; Rahman, 2014).		Enablement – increasing capability to identify barriers and to select strategies	Guidelines Regulations	1.2 Problem solving 2.3 Self-monitoring of behaviour	*Mode of delivery*: Face-to-face at both individual and community level *Who will deliver and how*: The researcher will motivate community members to identify barriers to time management and to choose strategies for overcoming impending barriers so that they can be involved in the local disaster management process.

(*continued*)

TABLE 17.3 Cont.

Poor social networks of the rural people: When a community is hit by floods, families are locally advised to move to safer places for a while and consequently, all the social networks are likely to be affected during this period (Mahalder et al., 2017; Rahman, 2014).	Social influences	Persuasion – persuading verbally to boost self-confidence Enablement – increasing capability to cope with the adverse situations	Communication/ marketing	3.1 Social support (unspecified) 5.3 Information about social and environmental consequences	*Mode of delivery*: Face-to-face at both individual and community level *Who will deliver and how*: The researcher will advise the community to manage the situation with patience and to keep the cell numbers of relatives, friends or buddies so that they can make a call to them in case of emergency.
Cultural barrier: Some women did not want to move to safe places, such as nearby disaster centres or schools during a disaster due to the culture of *purdah* or female seclusion. The practice of *purdah* sometimes inhibits the mobility of some Bangladeshi women (Huq, 2016; Kabeer, 2000; Uddin et al., 2021).		Education – providing information to gain knowledge Persuasion – persuading verbally to boost self-confidence Modelling – sharing folk songs and dramas about the benefits of participation in the disaster management process	Communication/ marketing guidelines Regulations	4.1 Instruction on how to perform a behaviour 5.2. Salience of consequences 5.3 Information about social and environmental consequences 9.1 Credible source 15.1 Verbal persuasion about capability	*Mode of delivery*: Face-to-face at both individual and community level *Who will deliver and how*: The researcher will provide information to the community through leaflets/posters/ verbally/staging folk songs and dramas so that women can be aware to get involved in the local disaster management process.

Lack of motivation to be involved in the community activities: Community members perceived that participating in a joint/community initiative would not benefit them personally (BBC Media Action, 2016).	Motivation	1. Professional/social role and identity 2. Beliefs and consequences	Education – providing information to gain knowledge Persuasion – persuading verbally to boost self-confidence Modelling – sharing folk songs and dramas about the benefits of participation in the disaster management process	Communication/marketing guidelines Regulations	4.1 Instruction on how to perform a behaviour 5.2. Salience of consequences 5.3 Information about social and environmental consequences 9.1 Credible source 15.1 Verbal persuasion about capability	*Mode of delivery*: Face-to-face at both individual and community level *Who will deliver and how*: The researcher will provide information to the community through leaflets/posters/verbally/staging folk songs and dramas so that they can be motivated to know the benefits of participation in the local disaster management process.
Lack of confidence in the current disaster forecasting system: Around 35% of the community members were not able to understand the existing forecasting system (Mahalder et al., 2017)and were accustomed to getting disaster forecasting from nature as well as local beliefs and practices (Rahman, 2014).		Emotion	Education – providing information to gain knowledge Persuasion – persuading verbally to boost self-confidence	Communication/marketing guidelines Regulations Service provision	4.1 Instruction on how to perform a behaviour 5.2. Salience of consequences 5.3 Information about social and environmental consequences 9.1 Credible source 15.1 Verbal persuasion about capability	*Mode of delivery*: Face-to-face at both individual and community level *Who will deliver and how*: The researcher will provide information to the community through leaflets/posters/verbally so that they can understand the modern forecasting system.

TABLE 17.4 Selection of intervention functions based on the APEASE criteria

Candidate intervention function	*Does the intervention function meet the APEASE criteria?*
Education	Yes
Persuasion	Yes
Incentivisation	No – not affordable, practicable or cost-effective
Coercion	No – not practical or acceptable
Training	No – not practical or acceptable
Restriction	No – not practicable as there are no options to restrict in this context
Environmental restructuring	Yes
Modelling	Yes
Enablement	Yes

Stage 3: Identifying Content and Implementation Options

Stage 3 includes the final two steps: BCT identification and selecting the mode of delivery.

Step 7. Identifying Behaviour Change Techniques

The behaviour change techniques (BCTs) associated with the five selected intervention functions were identified from the behaviour change technique taxonomy (BCTTv1), which lists 93 BCTs with descriptions and examples of their application. Following training in the identification and coding of BCTs, and in accordance with BCW guidelines, a list of all possible BCTs was generated. Afterwards, the list was considered which BCTs could most feasibly be applied within the context of intervention development and would be most useful for addressing the identified barriers to the target behaviour. In total, the following nine BCTs were identified as relevant: (i) 1.2 Problem solving; (ii) 2.3 Self-monitoring of behaviour; (iii) 3.1 Social support (unspecified); (iv) 4.1 Instruction on how to perform a behaviour; (v) 5.2. Salience of consequences; (vi) 5.3 Information about social and environmental consequences; (vii) 9.1 Credible source; (viii) 12.1 Restructuring the physical environment; and (ix) 15.1 Verbal persuasion about capability.

Step 8. Selecting the Mode of Delivery

The mode of delivery that met the APEASE criteria was only considered (Table 17.5). Face-to-face communication was selected as an appropriate mode of delivery for this intervention because it was familiar to the community people and met the APEASE criteria. In addition, strategies, such as providing leaflets and/or posters/verbal persuading/staging folk songs and dramas were chosen to raise community awareness and understanding as these were practicable and cost-effective. Table 17.5 shows the application of the APEASE criteria in relation to increasing the

TABLE 17.5 Mode of delivery based on the APEASE criteria

Mode of intervention delivery				*Do the modes of intervention delivery meet the APEASE criteria in the context of raising community awareness and understanding to increase active participation of people in the local level disaster management process in Bangladesh?*
Face-to-face	Individual (strategies: providing leaflets and/or posters/verbal persuading/staging folk songs and dramas)			Yes
	Group (strategies: providing leaflets and/or posters/verbal persuading/staging folk songs and dramas)			Yes
Distance	Population level	Broadcast media	Television	These modes of interventions delivery are less relevant as communities are unlikely to have access to phones, computers or be exposed to other forms of media.
			Radio	
		Outdoor media	Billboard	
			Poster	
		Print media	Newspaper	
			Leaflet	
		Digital media	Internet	
			Mobile phone app	
	Individual level	Phone	Phone helpline	
			Mobile phone text	
		Individually accessed computer programme		

participation of communities in the local level disaster management process in Bangladesh.

Discussion

This paper describes my experience of designing a complex behaviour change intervention that aims at increasing community participation in the local level disaster management process. This paper gives a step-by-step description of how evidence such as barriers identified from the review, theory such as COM-B and a validated assessment tool such as the APEASE criteria were incorporated into the development of the intervention design. The use of the BCW suggests a systematic process for designing a theory-driven intervention. The process has started with a behavioural diagnosis of what behaviour needs to be changed, followed by selecting the potential intervention functions, policy categories and BCTs to diagnose that change.

The implementation problems, where barriers and facilitators are not assessed by using a theoretical framework, are limited to direct pragmatic strategies rather than theoretically informed (Rothman, 2004). The BCW framework shows the appropriate way to design interventions so that these can meet the demands of the beneficiaries (Truelove et al., 2020). In this study, a behavioural diagnosis using the BCW revealed that Bangladeshi communities faced barriers in their capability, opportunity and motivation to be involved people in the local level disaster management process. This led to the selection of five potential intervention functions, five policy categories and nine overarching BCTs to bring about change in the target behaviour. Michie et al. (2014) provided a clear link between the COM-B model placed at the hub of the BCW, intervention functions that form the inner ring and policy categories that form the outer ring. This systematic approach shows why interventions directed by the BCW need to be designed. Connecting all components of the BCW, such as selecting the appropriate intervention functions, policies and the relevant BCTs is necessary to increase the likelihood of behaviour change when designing an intervention. For example, in the study, in order to improve communities' psychological capability to be active, it was deemed necessary to intervene by education, persuasion and modelling (intervention functions) and focus on communication/marketing, guidelines and regulations (policy categories). In addition, specific BCTs were chosen to be integrated to complement the intervention function and policy category were as follows: 4.1 Instruction on how to perform a behaviour; 5.2. Salience of consequences; 5.3 Information about social and environmental consequences; 9.1 Credible source; and 15.1 Verbal persuasions about capability. Interventions to change one's behaviour are less likely to be successful if there is a lack of linking between the various stages of the BCW (Truelove et al., 2020).

Several potential intervention delivery modes are present in the local setting for most behaviour change techniques (Michie et al., 2008), including educational meetings, educational detailing, reminder systems and feedback (Bero et al., 1998). We suggest the face-to-face mode of intervention delivery technique that is considered appropriate in the Bangladeshi context as well as acceptable and feasible among the target group such as the local population. The mode of intervention techniques can be delivered to the Bangladeshi community by either the researcher or the Union Disaster Management Committee. Also, in most cases, face-to-face awareness campaigns and/or providing information through providing leaflets and/or posters/persuading verbally/staging folk songs and dramas are chosen as strategies to change the target behaviour, raising community awareness and understanding to increase participation of people in the local level disaster management process. These culturally competent strategies were found effective in Bangladesh and other developing countries to increase disaster awareness among local communities (Disaster Management Bureau &

Disaster Management & Relief Division, 2010; Disaster Preparedness Training Programme, 2000).

Strengths and Limitations

Developing interventions is a complex task. The strengths of the study include its theoretical and practical basis. This study strengthens the theoretical foundations on which to develop community participation in the local level disaster management process. It brings together two important factors: (i) cultural understanding and rigour (Gould et al., 2014) and (ii) application of the TDF, COM-B and BCW (Truelove et al., 2020) in the local context to increase community participation for disaster management in Bangladesh. In order to design the proposed intervention to be realistic, the researcher needed to have a thorough understanding of BCW. Guidance of the BCW aided the researcher to propose a simplified approach which begins by mapping barriers to the target behaviour identified from the literature review with theoretical domains and ends by identifying the relevant BCTs based on the cultural context for implementation. However, the study includes some limitations too. The process of using the BCW was laborious and time consuming. Also, how successful the BCW can design interventions is sometimes unclear. The proposed process of designing interventions of combining research evidence, mapping links between different components of the BCW and feasibility information is also subject to bias. This is because, it requires more empirical work to test the method in different contexts and to examine the connections between theoretical evaluation and BCTs. In addition, this study included key findings based on only literature review, but it would be more practicable to have findings from empirical data.

Conclusion

This is the first known behaviour change intervention design using the BCW guidelines in the local context of the disaster management process in Bangladesh. A culturally competent approach that includes face-to-face, providing information, verbal persuading and staging performances to increase awareness and understanding of community members in order to increase their participation in the local disaster management process in Bangladesh. The use of the BCW framework has facilitated in rationally informing a targeted intervention based on the barriers identified from the literature review. This study offers a practical guide for future scholars who want to apply a similar methodological approach to designing an intervention in the local setting of the disaster management process.

References

Azad, M. A. K., Uddin, M. S., Zaman, S., &Ashraf, M. A. (2019). Community-based disaster management and its salient features: A policy approach to people-centred risk reduction in Bangladesh. *Asia-Pacific Journal of Rural Development, 29*(2), 135–160. https://doi.org/10.1177/1018529119898036

Bangladesh Bureau of Statistics [BBS]. (2021). *Report on Bangladesh sample vital statistics 2020.* www.bbs.gov.bd/site/page/b588b454-0f88-4679-bf20-90e06dc1d10b/-

BBC Media Action. (2016). *What are the barriers to community involvement in local level disaster preparedness measures in Bangladesh?*http://downloads.bbc.co.uk/mediaaction/pdf/research-summaries/local-disaster-preparedness-august-2016.pdf

Bero, L. A., Grilli, R., Grimshaw, J. M., Harvey, E., Oxman, A. D., & Thomson, M. A. (1998). Closing the gap between research and practice: An overview of systematic reviews of interventions to promote the implementation of research findings The Cochrane Effective Practice and Organization of Care Review Group. *BMJ, 317*(7156), 465–468.

Connell, L. A., McMahon, N. E., Redfern, J., Watkins, C. L., & Eng, J. J. (2015). Development of a behaviour change intervention to increase upper limb exercise in stroke rehabilitation. *Implementation Science, 10*(34), 1–9. https://doi.org/10.1186/s13012-015-0223-3

Craig, P., Dieppe, P., Macintyre, S., Michie, S., Nazareth, I., & Petticrew, M. (2008). Developing and evaluating complex interventions: The new Medical Research Council guidance. *BMJ, 337*, a1655.https://doi.org/10.1136/bmj.a1655

Cutter, S., Osman-Elasha, B., Campbell, J., Cheong, S. M., McCormick, S., Pulwarty, R., … Ziervogel, G. (2012). Managing the risks from climate extremes at the local level. In C. B. Field, V. Barros, T. F. Stocker, D. Qin, D. J. Dokken, K. L. Ebi, M. D. Mastrandrea, K. J. …, & P. M. Midgley (Eds.), *Managing the risks of extreme events and disasters to advance climate change adaptation* (pp. 291–338). Cambridge University Press.

Disaster Management Bureau & Disaster Management & Relief Division. (2010). *National plan for disaster management 2010–2015.* http://nda.erd.gov.bd/en/c/publication/national-plan-for-disaster-management-2010-2015

Disaster Preparedness Training Programme. (2000). *Increasing community disaster awareness*.www.ifrc.org/Global/Inccdp.pdf

Eckstein, D., Kunzel, V., & Schafer, L. (2021). *Global Climate Risk Index 2021: Who suffers most from extreme weather events? Weather-related loss events in 2019 and 2000–2019.* Bonn: Germanwatch e.V. www.germanwatch.org/eng/cri

Fawcett, A. & Plan International. (2012). *Climate extreme: How young people can respond to disaster in a changing world.* www.childreninachangingclimate.org/

Fulton, E. A., Brown, K. E., Kwah, K. L., & Wild, S. (2016). StopApp: Using the Behaviour Change Wheel to develop an App to increase uptake and attendance at NHS stop smoking services. *Healthcare (Basel), 4*(2). https://doi.org/10.3390/healthcare4020031

Glanz, K. & Bishop, D. B. (2010). The role of behavioral science theory in development and implementation of public health interventions. *Annual Review of Public Health, 31*, 399–418. https://doi.org/10.1146/annurev.publhealth.012809.103604

Gould, G. S., Watt, K., Stevenson, L., McEwen, A., Cadet-James, Y., & Clough, A. R. (2014). Developing anti-tobacco messages for Australian Aboriginal and Torres Strait Islander peoples: Evidence from a national cross-sectional survey. *BMC Public Health, 14*(250), 1–13.

Harvey, P. A., Baghri, S., & Reed, R. A. (2002). *Emergency sanition: Assessment and programme design.* Loughborough: WEDC, Loughborough University.

Hossain, M. A. (2013). Community participation in disaster management: Role of social work to enhance participation. *Antrocom Online Journal of Anthropology, 9*(1).

Huq, S. M. S. (2016). Community based disaster management strategy in Bangladesh: Present status, future prospects and challenges. *European Journal of Research in Social Sciences, 4*(2), 22–35.

Islam, K. F. (2013). *Community based disaster management in Bangladesh*. www.researchgate.net/publication/333972153_Community_Based_Disaster_Management_in_Bangladesh

Islam, M. R. & Hasan, M. (2015). Climate-induced human displacement: A case study of Cyclone Aila in the south-west coastal region of Bangladesh. *Natural Hazards, 81*(2), 1051–1071. https://doi.org/10.1007/s11069-015-2119-6

Islam, M. T. (2018). How does local government cope with disaster in Bangladesh? https://blogs.lse.ac.uk/southasia/2018/01/08/how-does-local-government-cope-with-disaster-in-bangladesh/

Kabeer, N. (2000). *The power to choose: Bangladeshi women and labour market decisions in London and Dhaka*. London: Verso.

Mahalder, B. K., Islam, M. S., Hossain, B. M. R., & Nabi, F. D. (2017). A study on disaster management practices for vulnerable rural communities in Bangladesh. *Bangladesh Journal of Scientific Research, 30*(1&2), 55–67.

Mallick, B. & Vogt, J. (2011). Social supremacy and its role in local level disaster mitigation planning in Bangladesh. *Disaster Prevention and Management: An International Journal, 20*(5), 543–556. https://doi.org/10.1108/09653561111178970

Michie, S., Atkins, L., & West, R. (2014). *The behaviour change wheel: A guide to designing interventions*. London: Silverback Publishing.

Michie, S., Johnston, M., Francis, J., Hardeman, W., & Eccles, M. (2008). From theory to intervention: Mapping theoretically derived behavioural determinants to behaviour change techniques. *Applied Psychology, 57*(4), 660–680.

Michie, S., Richardson, M., Johnston, M., Abraham, C., Francis, J., Hardeman, W., … Wood, C. E. (2013). The behavior change technique taxonomy (v1) of 93 hierarchically clustered techniques: Building an international consensus for the reporting of behavior change interventions. *Annals of Behavioral Medicine, 46*(1), 81–95. https://doi.org/10.1007/s12160-013-9486-6

Michie, S., van Stralen, M. M., & West, R. (2011). The behaviour change wheel: A new method for characterising and designing behaviour change interventions. *Implementation Science, 6*, 42.https://doi.org/10.1186/1748-5908-6-42

Ministry of Disaster Management and Relief. (2020). *National plan for disaster management (2021–2025): Action for disaster risk management towards resilient nation*. https://floodresilience.net/resources/item/national-plan-for-disaster-management-2021-2025/

Nasreen, M. (2004). Disaster research: Exploring sociological approach to disaster in Bangladesh. *Bangladesh e-Journal of Sociology, 1*(2), 21–28.

O'Cathain, A., Croot, L., Duncan, E., Rousseau, N., Sworn, K., Turner, K. M., … Hoddinott, P. (2019). Guidance on how to develop complex interventions to improve health and healthcare. *BMJ Open, 9*(8), e029954.https://doi.org/10.1136/bmjopen-2019-029954 %J BMJ Open

Oliver-smith, A. (2009). *Sea level rise and the vulnerability of coastal peoples: Responding to the local challenges of global climate change in the 21st century*. Bonn: UNU-EHS. https://collections.unu.edu/view/UNU:1861

Rahman, M., Rahman, M. S., & Rahman, M. (2017). Disasters in Bangladesh: Mitigation and management. *Barisal University Journal, 4*(1), 139–163.

Rahman, S. U. (2014). *Impacts of flood on the lives and livelihoods of people in Bangladesh: A case study of a village in Manikganj district*. Dhaka: BRAC University. http://dspace.bracu.ac.bd/bitstream/handle/10361/3802/13168004.pdf?sequence=1

Rothman, A. J. (2004). Is there nothing more practical than a good theory?: Why innovations and advances in health behavior change will arise if interventions are used to test and refine theory. *International Journal of Behavioral Nutrition and Physical Activity, 1*(11), 1–7.

Roy, T. (2019). Disaster risk reduction interventions in Bangladesh: A case study of selected villages of southwest region. *Journal of the Asiatic Society of Bangladesh (Hum.), 64*(1), 43–69.

Skivington, K., Matthews, L., Simpson, S. A., Craig, P., Baird, J., Blazeby, J. M., … Moore, L. (2021). A new framework for developing and evaluating complex interventions: Update of Medical Research Council guidance. *BMJ, 374*, n2061.https://doi.org/10.1136/bmj.n2061

Truelove, S., Vanderloo, L. M., Tucker, P., Di Sebastiano, K. M., & Faulkner, G. (2020). The use of the behaviour change wheel in the development of ParticipACTION's physical activity app. *Preventive Medicine Reports, 20*, 101224.https://doi.org/https://doi.org/10.1016/j.pmedr.2020.101224

Uddin, M. S., Haque, C. E., & Khan, M. N. (2021). Good governance and local level policy implementation for disaster-risk-reduction: Actual, perceptual and contested perspectives in coastal communities in Bangladesh. *Disaster Prevention and Management: An International Journal, 30*(2), 94–111. https://doi.org/10.1108/DPM-03-2020-0069

United Nations. (2015a). *Transforming our world: The 2030 agenda for sustainable development*. https://sdgs.un.org/2030agenda

United Nations. (2015b). *Sendai framework for disaster risk reduction 2015–2030*. www.preventionweb.net/files/43291_sendaiframeworkfordrren.pdf

Webb, J., Foster, J., & Poulter, E. (2016). Increasing the frequency of physical activity very brief advice for cancer patients. Development of an intervention using the behaviour change wheel. *Public Health, 133*, 45–56. https://doi.org/https://doi.org/10.1016/j.puhe.2015.12.009

World Bank. (2021). *Rural population (% of total population)—Bangladesh*. https://data.worldbank.org/indicator/EN.URB.LCTY.UR.ZS?locations=BD

18

DEMOGRAPHIC DETERMINANTS OF DISASTER PREPAREDNESS BEHAVIOR AMONG THE INHABITANTS OF SOUTHWEST COASTAL BANGLADESH

Shahana Akter and Md. Mujibor Rahman

Introduction

There are many studies that demonstrate that anthropogenic actions are more important causes of disaster than environmental phenomena (Kelman et al. 2016). In recent years, natural disasters are increasingly seen as a barrier to achieve sustainable development goals (SDGs) (Watson et al. 2007). Human endeavors to overcome disaster and minimize risks as old as human history. Disasters occur due to a combination of hazard and vulnerability. Where hazards refer to input from nature and vulnerability is mostly anthropogenic (Kelman et al. 2016). Recent studies have shown that coastal Bangladesh is highly vulnerable to disaster impacts (Mallick & Vogt 2012) and southwestern coastal Bangladesh is recognized as a hotspot. Evidence suggests wide ranges of variations in disaster risk in coastal Bangladesh across spatial and temporal scale (Murshed et al. 2022).

Vulnerability attributed to disaster can be traced back to many factors and demographic patterns are one of the important ones. In no cases hazards can be prevented but their damages can be mitigated, and risk can be reduced by various means such as improved warning system changes in disaster preparedness behavior (DPB) (Najafi et al. 2015). But the fact is there are very few comprehensive studies to explore the role of demographics on DPB. Some roles of demographic factors in disaster preparedness behavior are still unclear. Disaster preparedness is defined as actions that ensure resources necessary to carry out an effective response are available before any disaster. DPB comprises a series of those actions and their performance rely on many factors. DPBs include preparedness perception, awareness, societal norms, community participation and empowerment, social trust, coping style, available resources and sociodemographic attributes like age, sex, education, occupation, income level, marital status, etc. (Najafi et al. 2017).

DOI: 10.4324/9781003253495-23

This study therefore seeks to identify traditional coping mechanisms to disaster and to express an overview of overall idea about the role of demographic determinants for DPB risk reduction. The aim of this study is to determine the role of human capital for DPB among the people of southwest Bangladesh. Similar kind of study was found by Najafi et al.(2017). The study area was selected because of the severity of natural disasters like tropical cyclones, salt, river erosion, and storm surges, the village of Borunpara in the Gangarampur union in Batiagata Upazilla in the Khulna district has been observed recently. For the previous few decades, the village has located 10 kilometers from the Pashur River's eastern bank. This river has been degraded severely. The village of Borunpara has been separated into two pieces in recent decades due to river erosion, known as Borunpara and Dip Borunpara/Mativanga. It became 6 kilometers after the detour. A total of 462 households are represented by nine municipalities. In this community, there are 200 households. Most of the tribes in this village work in agriculture. Fishermen, boatmen, and shopkeepers are scarce. The area has the scarcity of educational facilities and proper treatment facilities. There is no cyclone protection in this village. Residents of this village use schools near other villages as cyclone shelters. Findings of the study will be able to throw a new light to be helpful to design strategy for disaster management by focusing how demographic parameters affect the population of the study area on DPB. It will also be helpful lfor the relevant authorities with formation and adaption of mitigation measures and introduction of policy interventions. The most significant gap in this study is women's participation in DPB, as women are more involved in their jobs and also have to care for their families and children. As a result, they have a better understanding of when and how to respond to a certain disaster. Because the women in the research region do not properly demonstrate their enthusiasm to participate in this study, and also lack of education, these could be considered as substantial barriers to community disaster preparedness activities.

Materials and Methods

General Information and Selection of Research Area

Bangladesh's southwest coast is especially prone to calamities. In southern Bangladesh, the districts of Khulna, Bagerhat, and Satkhira are regarded as the most disaster-prone. Borunpara is the village of Gangarampur union under Batiaghata Upazilla of Khulna district. Borunpara is a small village in the Gangarampur union, about 22 kilometers from Khulna. No. 9 Borunpara village is the municipality of Gangarampur union. The study areas ofBorunpara and Dip-Borunpara villages are about 3.4 square kilometers and the population is over 300. The study regions are located between the longitudes of 22°37'59" and 22°38'59" north and 89°30'40" and 89°31'23" east. The maps of the study area are given in Figure 18.1.

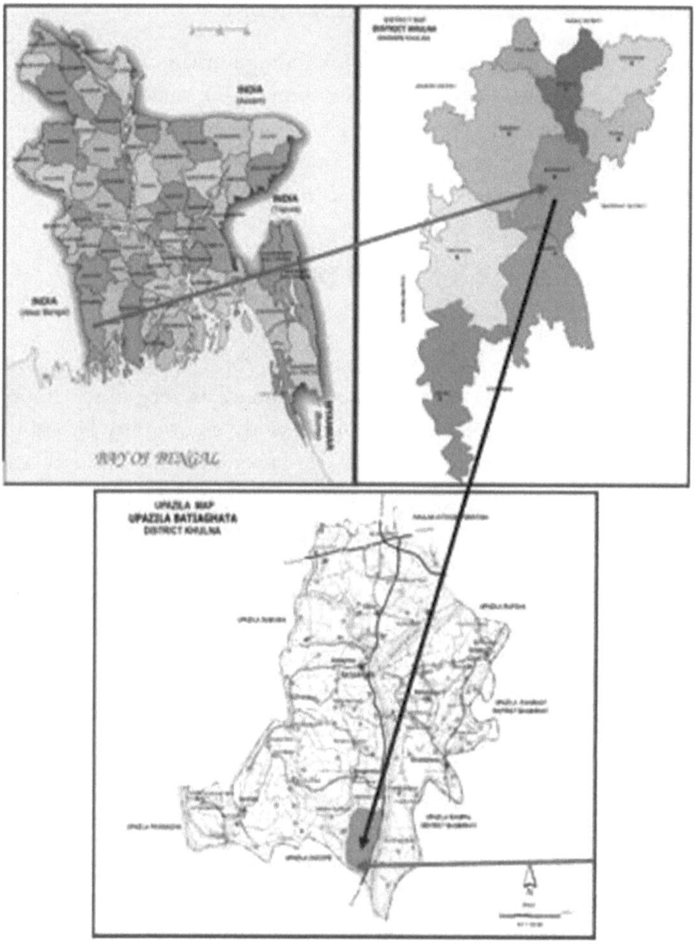

FIGURE 18.1 Map of the study area (Borunpara village of Batiaghata Upazilla under Khulna district)

Source: Gangarampur Union Council.

Identifying Disaster Preparedness among the Inhabitants

To fulfill the objectives of this study it's important to identify the disaster preparedness among the inhabitants of the research area. For identification of the preparedness, several factors are considered for the analysis like incidence of natural hazards, frequency of natural hazards, DPB scores for the study participants, independent sample t-test for differences between the groups showing the relative contribution of each variable to predict DPB score, etc. Similar studies are found in Primer (n.d.)

Knowledge Identification for Risk Assessment

Indigenous knowledge is very important in building human capital for DPB phases. From questionnaire survey, focus group discussion (FGD) information analyzed risk zoning, hazard ranking and risk valuation, and disaster vulnerability score (DVS) for hazard ranking. Similar studies were conducted in *Public Awareness and Public Education for Disaster Risk Reduction: A Guide*(n.d.).

Community Capacity and Natural Disaster Resilience Identification

Several analyses were conducted in this study for the identification for risk assessment. Community capacity for hazard resilience is very much important in building human capital for DPB phases. In this study, community hazard resilience capacity for disaster preparedness was analyzed for community capacity assessment for preparedness. Natural hazard resilience capacity for disaster preparedness is used to determine levels of resilience (Table 18.5). The toolkit minimum to maximum score is 1–5 (Goal toolkit, 2015). The higher the number, the higher the resilience (Table 18.5). This methodology has also applied in *Basics of Capacity Development for Disaster Risk Reduction*(n.d.).

Data Collection: Sampling, Analysis, and Interpretation

The village of Borunpara, under the Gangarampur union in Batiaghata Upazilla, was selected as a good representative of the disaster area. A household was used as a sampling unit. Households were randomly selected from the villagers. A household member over the age of 30 with extensive knowledge and experience in disasters is considered as a sample unit. A total of 190 households from the village of Borunpara were recorded. Correlation matrices of the major sociodemographic characteristics of the village of Borunpara and all other statistical analyses were performed by using SPSS, Minitab, and MS Office (2013) statistical softwares. For the sampling, Kothari equation (1990) was followed to determine the sample size and it was also observed that ten households in Borunpara were lying outside the cluster. The sample size for the rest of the households (180 for Borunpara) was calculated as follows:

$$n = \frac{z^2 pqN}{e^2 (N - 1) + z^2 pq}$$

where,

N = Population size; n = sample size; p = sample proportion; q = 1−p; z = standard variation at a given confidence level; e = acceptable error (the precision) at 95% confidence level

$$z = 1.96; p = 0.25; q = (1 - 0.25) = 0.75; e = 0.05$$

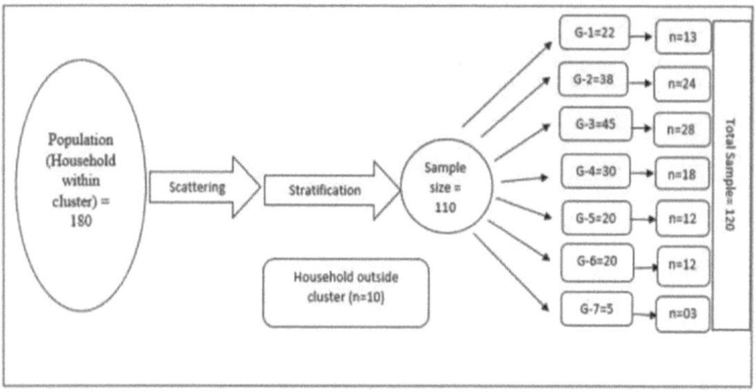

FIGURE 18.2 Stratified educational groups in Borunpara, Gangarampur union, Batiaghata, Khulna

Source: Field survey, 2018.

Here,

N= 180 within cluster for Borunpara village. By putting the values in the equation and calculation, it is found that the estimated sample size is 110 for Borunpara village (Figure 18.2).

The total sample size of the household survey was found to be 120 in Borunpara village. The random sampling method was chosen for this study. This is to avoid bias and ensure that the sample represents the population.

Results and Discussion

Knowledge and Community Practice Governing Human Capital in Disaster Preparedness

In-depth qualitative data provide a more complete contextual study of the trends seen in the survey to better understand the level of human capital within the community. The following are the study's main findings:

Incidence of Natural Hazards

About 45% of respondents could recognize the natural risks like cyclone, huge storms, and salinity; 35% said river erosion, 15% said health diseases, and 5% said earthquake, fire, and drought among the total (120) respondents (Figure 18.3).

Frequency of Natural Hazards

If the main hazard in their community had changed in frequency over the past 20 years,51% of survey respondents said that it had become more often, 25% said

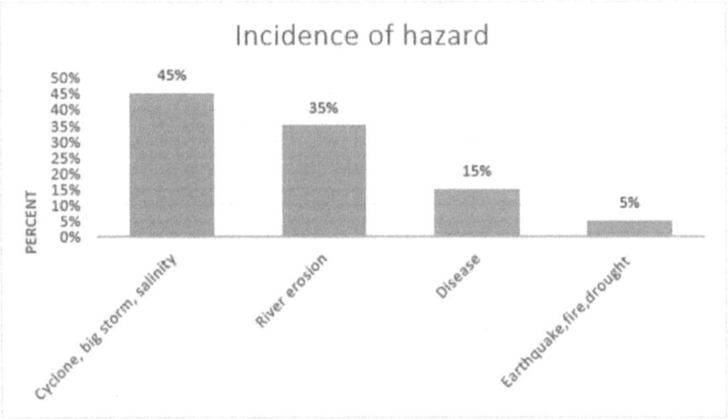

FIGURE 18.3 Proportion of individual respondents (%) identifying different hazard types occurring in Borunpara village

Source: Field survey, 2018.

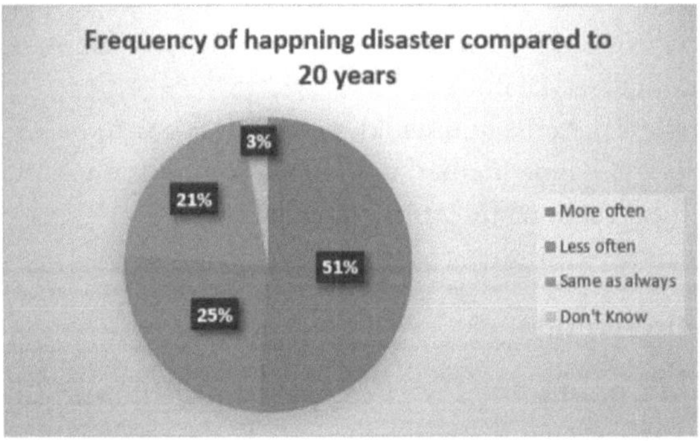

FIGURE 18.4 Proportion (%) of individual respondent reporting increase in hazard frequency over the past 20 years

Source: Field survey, 2018.

it had become less often, 21% said it had remained the same, and 3% said they did not know (Figure 18.4).

Disaster Preparedness Behavior Scores for the Study Participants

The DPB scores of the participants are shown in Table 18.1. The information was structured using a public-ready indexing system. The community of practices was compared to demographic factors such as gender, age, occupation, education, and income level; 90% of those polled had a DPB score of less than or equal to 8. It also reveals that 49.2% of those surveyed had done nothing to prepare for a future disaster.

TABLE 18.1 Disaster preparedness behavior scores for the participant

DPB score	*Frequency*	*Percent*	*Cumulative percent*
0	59	49.2	49.2
1	19	15.8	65.0
2	8	6.7	71.7
3	13	10.8	82.5
4	10	8.3	90.8
5	2	1.7	92.5
6	2	1.7	94.2
7	7	5.8	100.0

Independent Sample t-test for Differences between the Groups

An independent sample t-test was performed to determine whether there was a significant difference between the two independent groups (Table 18.2).

Men were shown to be more active in DPB than women in this study. Significant independent t-test shows that men have high significant value (<0.05) than others in the perspective of age (among 21–45 years).Male respondents have high significant value where it shows that they are more involving in DPB score than female respondents. Similarly in terms of education, the secondary education has too significant values than the P value. So, it had a great impact on the preparedness behavior. Farmers are more engaging in DPB score apart from other professions according to the significant t-test value. The married people are more aware of DPB score as it has a significant t value (0.002). Likewise, the respondents who have income between 5000 and 10,000 and10,000 and 20,000 have the significant value, which shows that they are more prepared for DPB than other income groups. It was also revealed that those with a high level of education, income, and occupation value were more prepared than those with none of these attributes. Currently, employed participants scored much higher than unemployed participants on DPB. DPB score had no link to marital status or other variables.

Community Capacity for Hazard Resilience Livelihoods

Community Capacity Assessment for Preparedness

Capacity assessment is an iterative process of design-application-learning-adjustment, rather than a one-time intervention. It contributes to the creation of a unified framework for a systematic response to capacity development.

The research reveals the community's behavior in terms of preparedness. First, capacity assessment of the community toward assets and needs with four core issues and five factors which shows low community capacity. And the factor between the desired level and the existing level among the community was found out (0.60) (UNDP, 2008) which shows low community practice according to the core issues (Table 18.3). Compared capacity assessment with the desired level and the existing

TABLE 18.2 Independent sample t-test for differences between the groups

Variable	*Groups*	*N (sample size)*	*Mean of DPB score*	*SD*	*t (t-test)*	*t-test (total)*	*Sig.(2-tailed)*	*90% CI (confidence interval)*
Age	21–45 years	115	1.51	2.09	2.62	5.06	.010	.9002–.1259
	46–65 years	5	2.20	1.09	2.44		.070	2.5602–.1602
Sex	Male	103	1.39	1.91	2.11	4.23	.037	.7716–.0245
	Female	17	2.41	2.7	2.12		.050	2.8205–.0030
Education	Illiterate	1	1.0000			6.329		
	Primary	86	1.91	2.21	−.341		.734	.3932–.5560
	Secondary	33	.575	1.22	6.67		.000	−.9897–1.8588
Occupation	Farmer	45	.5556	1.21	−7.965	−6.86	.000	−1.0790– −(1.8099)
	Labor	49	1.73	1.84	−1.006		.319	.2649– −.7955
	Businessman	13	2.84	2.60	1.169		.265	2.422– −.7305
	Service holder	2	1.50	2.12	−.333		.795	18.5593– −195593
	Housewife	11	3.18	3.09	1.267		.235	3.2594– −.8958
Marital status	Single	8	3.75	2.12	2.33	−0.87	.052	3.5235– −.0235
	Married	111	1.39	1.98	−3.20		.002	−.2306– −.9766
	Others (widow, divorced)	1	.000					
Income Level	<5000	1	1.0000			−2.83		
	5000–10,000	84	1.0357	1.63	−5.417		.000	−.6103– −1.3183
	10,000–20,000	31	3.1290	2.43	2.584		.015	2.0212– .2369
	>20,000	4	.0000	.00000				

TABLE 18.3 Capacity assessment of the capacity assets and needs for enabling environment

Point of entry enabling environment	*Capacity assessment*									
Core issues	*Functional capacities*									
	Engage stakeholder		*Assess a situation and define a vision and mandate*		*Formulate policies and strategies*		*Budget, manage, implement*		*Evaluate*	
	Desired level	*Existing level*	*Desired level*	*Existing level*	*Desired level*	*Existing level*	*Desired level*	*Existing level*	*Desired level*	*Existing level*
1.0 Institutional arrangements	4.00	3.00	4.50	3.23	4.00	3.21	4.50	2.50	3.50	2.90
2.0 Leadership										
3.0 Knowledge										
4.0 Accountability										
Overall	4.00	3.00	4.50	3.23	4.00	3.21	4.50	2.50	3.50	2.90
Difference	1.00		1.27		.79		1.00		.60	

Source: UNDP (2008).

TABLE 18.4 Capacity assessment of the capacity assets and needs for organization

Point of entry organizational	*Capacity assessment*									
Core issues	*Functional capacities*									
	Engage stakeholders		*Assess a situation and define a vision and mandate*		*Formulate policies and strategies*		*Budget, manage, and implement*		*Evaluate*	
	Desired level	*Existing level*	*Desired level*	*Existing level*	*Desired level*	*Existing level*	*Desired level*	*Existing level*	*Desired level*	*Existing level*
1.0 Institutional arrangements										
2.0 Leadership										
3.0 Knowledge										
4.0 Accountability	4.00	2.00	3.00	1.00	5.00	3.21	4.00	3.50	5.00	3.50
Overall	4.00	2.00	3.00	1.00	5.00	3.21	4.00	3.50	5.00	2.90
Difference	2.00		2.00		1.79		0.50		1.50	

Source: Goal toolkit (2015).

level which is low that shows less preparedness in the budget, manage, and implement phase during preparedness of disaster(0.50) (Table 18.4).An evaluation will help identify the gap between the level of capacity needed and that currently exists in Northern Ireland's health sector. The results of the evaluation will be used to build consensus on urgent capacity requirements and to outline a path ahead in designing capacity development actions.

Community Hazard Resilience Capacity for Disaster Preparedness

The GOAL Toolkit Disaster Resilience Guidance assesses hazard resilience. GOAL defines resilience as a community's ability to foresee and adapt to hazards, as well as absorb, respond to, and recover from shocks and pressures. This toolkit's measurement of community resilience can be interpreted in two ways. It can provide an estimated percentage of resilience, or it can be used to assess resilience levels (Goal toolkit, 2015).

Table 18.5 demonstrates the community's catastrophe preparedness hazard resilience capacity. The community resilience score is 60%, which indicates moderate resilience. It is done to find out hazard resilience capacity for disaster preparedness between communities and categorized into 0%−100%, and 81−100, which is level 5 and indicates high resilience. And after the result found out to be 60%, which is medium resilience and the act of this is the development and implementation of solutions capacity to act is improved substantial, interventions are more numerous and longterm.

TABLE 18.5 Hazard resilience capacity for disaster preparedness

%	*Level*	*Category*	*Description*
0–20	1	Minimal resilience	Little awareness of the issue(s) or motivation to address them. Actions limited to crisis response.
21–40	2	Low resilience	Awareness of the issue(s) and willingness to address them. Capacity to act (knowledge and skills, human, material and other resources) remains limited. Interventions tend to be one-off, piecemeal, and shortterm.
41–60	3	Medium resilience	Development and implementation of solutions. Capacity to act is improved and substantial. Interventions are more numerous and longterm.
61–80	4	Resilient	Coherence and integration. Interventions are extensive, covering all main aspects of the problem, and they are linked within a coherent long-term strategy.
81–100	5	High resilience	A "culture of safety" exists among all stakeholders, where DRR is embedded in all relevant policy, planning, practice, attitudes, and behavior.
The community resilience score			60%
Category			Medium resilience

Source: GOAL Toolkit (2015).

Indigenous Knowledge for Disaster Risk Assessment

Indigenous knowledge has a critical role in the development of human capital for DPB stages. In-depth qualitative data for a more extensive textual analysis of the trends seen in the survey are needed to understand the degree of human capital within the community. The following are the study's main findings:

Hazard Ranking and Risk Valuation

In this location, a focus group was held with relevant and educated persons to assess the danger zone based on the qualities listed above.

It is a common occurrence with the potential for catastrophic and harmful consequences. It is a major priority for the community. Immediate mitigation and emergency response plans are required. Borunpara is now classified as a high-risk zone, and the village is extremely vulnerable to river erosion. Many events had occurred in that area and according to the questionnaire and ranking out of 10, the total risk valuation was 868 (Goal toolkit, 2015) (Table 18.6).

Disaster Vulnerability Score

DVS is calculated by combining waterlogging, cyclone, flood, river erosion and storm surge, and salinity intrusion score using the equation.

DVS= (0.1 × Waterlogging score + 0.15 × Cyclone score + 0.225 × Flooding score + 0.3× River erosion score + 0.1 × Storm surge + 0.125 × Salinity score)/6

The percentages of areas with various disaster scores are displayed in Table 18.7. The research area's vulnerability score ranges from 0.90 to 1.14. The lower the value, the less vulnerable you are, and the higher the value, the more vulnerable you are. Most of the places (33.70%) have a fire score of 0.25–0.5, indicating that most of the research areas are somewhat low vulnerable to disaster. Areas with a catastrophe score of less than 0.25 are considered low vulnerable, whereas areas with a disaster value of less than 0.75 are considered moderately highly vulnerable. About 15.73% of the area is in the high-risk category for disaster.

TABLE 18.6 Risk zoning of the study area

SL	*Event*	*Hazard ranking*	*Vulnerability indexing*	*Social cost prioritization*	*Risk valuation*
1	River erosion	8	7	6	336
2	Salinity	5	6	6	180
3	Flooding	4	4	5	80
4	Cyclone	6	6	5	180
5	Storm surge	4	5	4	80
6	Waterlogging	2	3	2	12
Total					868

TABLE 18.7 Distribution of areas according to disaster vulnerability score along with FEMA score

Vulnerability score	*FEMA score*	*Vulnerability category*	*Frequency*	*Percentage*
0.91–0.96	0.25	Low vulnerability	20	22.47%
0.97–1.02	0.5	Moderately low	30	33.70%
1.03–1.08	0.75	Moderate vulnerability	25	28.08%
1.09–1.14	1	High vulnerability	14	15.73%
Total				100

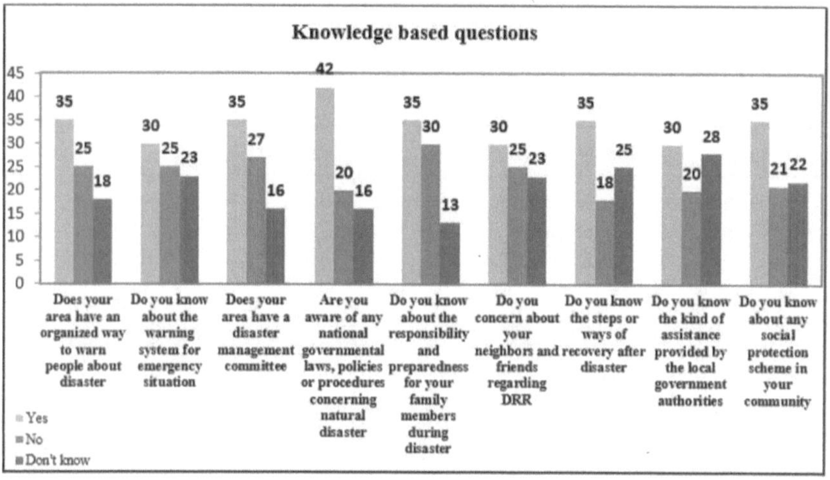

FIGURE 18.5 Frequency distribution of individuals regarding knowledge-based questions on disaster preparedness by multiple response

Source: Field survey, 2018.

Knowledge Scores Concerning to Disaster Preparedness

Every respondent in the villages was asked nine individual knowledge-based questions about disaster preparedness (Figure 18.5) to determine their level of awareness about catastrophe occurrence. Knowledge rating is based on public readiness indexing, which aids in measuring knowledge aspects during an emergency or further scoring, the frequency of respondents in villages is calculated.

For knowledge scoring, frequency is transformed to a percentage. According to the poll questionnaire, 44% said yes, 30% said no, and the remaining 26% said they couldn't answer any of the questions (Figure 18.6).

Demographic Analysis

Community practices were compared to demographic factors such as gender, age, occupation, education, and income level. Males made up 85.8% of the total

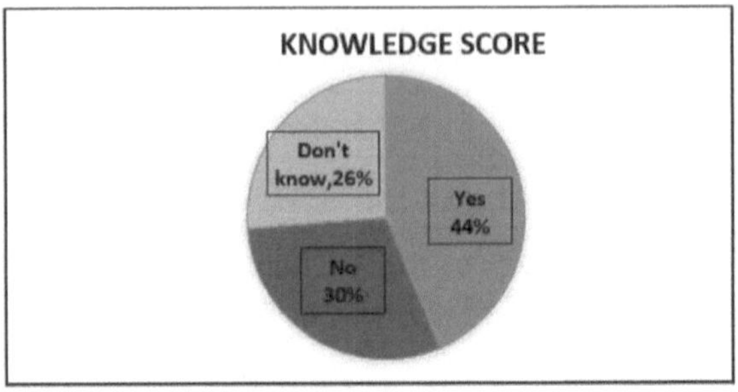

FIGURE 18.6 Proportion (%) of individuals obtaining knowledge score

Source: Field survey, 2018.

TABLE 18.8 Demographic analysis of the respondents

Variable	*Groups*	*Frequency*	*Percent*
Age	21–45 years	115	95.8
	46–65 years	5	4.2
Sex	Male	103	85.8
	Female	17	14.2
Education	No education	7	6
	Primary	81	68
	Secondary	32	26
Occupation	Farmer	45	37.5
	Labor	49	40.8
	Businessman	13	10.8
	Service holder	2	1.7
	Housewife	11	9.2
Marital status	Single	8	7
	Married	107	89
	Others (widow, divorced)	5	4
Income level	<5000	4	4
	5000–10,000	75	62
	10,000-20,000	38	25
	>20,000	7	6

respondents, while females made up 14.2%. Only 68% of those polled had completed primary school. A quarter of the respondents had a medium level of income. Most of the participants were low-income earners (Table 18.8).

Discussion

The knowledge score obtained from this study denotes that the knowledge level gives much more training opportunities to warning systems. The dimensions of knowledge were not adequate about disaster among the respondents of the

Borunpara village. Human capital within a community influences the dimensions of demographic factors for DPB. This helped the community immensely starting from passing indigenous knowledge and skills for risk reduction. Similar studies found that only 10% of the participants had DPB score of 5 or more, which defined as prepared persons (Najafi et al., 2017), whereas our findings show that 90% of the respondents had DPB score less than or equal to 8. It also shows that 49.2% respondents had not done anything to protect themselves from possible future disasters. Moreover, we found that 44% of the respondents have the indigenous knowledge to identify the risk of disaster and finally the 60% community participants have the capacity that represents medium resilience. Effects of demographic determinants on DPB were not properly found in abundance in the study area because of culture, values, and living in isolation. After a period during disaster situation, demographic factors on preparedness behavior become less active and sometimes breakdown due to competition and conflict over access to external recovery support. Thus, the role of human capital decreases at disaster situation among the villagers or neighborhood. Due to this, adequacy of factors is not observed for disaster response, recovery, relief, and preparedness. From the above statistical analysis it is shown that the most effective demographic factors are occupation, income, and education.

Conclusion

People with higher incomes are expected to have more access to qualified properties and live in more disaster-resistant areas, according to previous research, whereas the opposite is true for low-income families. Poorer individuals are also less likely to reduce the consequences of dangers because they lack a sense of personal control over probable outcomes, according to the study. Another factor of DPB evaluated in the study was marital status. Married folks are usually more concerned about their safety and security. People who are married have better vision and productive thinking than those who are not married. The amount of danger that individuals believe a catastrophe and hazards pose may be characterized as risk perception. Education was found to be a factor of DPB in this study. It is thought that formal education might help people prepare for disasters because it improves their cognitive and learning skills as well as their access to information. The findings revealed that employment, marital status, income level, and education are important variables in Bangladesh's southwest coastal area specially Borunpara village under Khulna district. As a result, it is critical that government officials, emergency agencies, community leaders, and educators give greater attention to low-income persons, various professional groups, and education to attain better levels of individual disaster preparation.

References

Kelman, I., Gaillard, C. J., Lewis, J., & Mercer, J. (2016). Learning from the history of disaster vulnerability and resilience research and practice for climate change. *Natural Hazards*, 82(1), 129–143.

Mallick, B., & Vogt, J. (2012). Cyclone, coastal society and migration: Empirical evidence from Bangladesh. *Economics*, 34(3), 217–240. doi: 10.3828/idpr.2012.16

Murshed, S., Griffin, A., Islam., A. M., Wang., H. X., & Paull., D. (2022). Assessing multi-climate-hazard threat in the coastal region of Bangladesh by combining influential environmental and anthropogenic factors. *Progress in Disaster Science*, 16(100261), 18. https://doi.ong/10.1016/j.pdisas.2022.100261.

Najafi, M., Ardalan, A., Akbarisari, A., Noorbala, A. A., & Jabbari, H. (2015). Demographic determinants of disaster preparedness behaviors amongst Tehran inhabitants, Iran. *PLoS Currents*, 7.

Najafi, M., Ardalan, A., Akbarisari, A., Noorbala, A. A., & Elmi, H. (2017). The theory of planned behavior and disaster preparedness. *PLoS Currents*, *9*, 1–16. https://doi.org/10.1371/currents.dis.4da18e0f1479bf6c0a94b29e0dbf4a72

Primer, A. U. (n.d.). *United Nations Development Programme Capacity Development: A UNDP Primer. Public Awareness and Public Education for Disaster Risk Reduction: A Guide.*

Watson, J. T., Gayer, M., & Connolly, M. A. (2007). Epidemics after natural disasters. *Emerging Infectious Diseases*, *13*(1), 1.

Wilson, S. L., & Kershaw, M. A. (2008). Caring for young children after a hurricane: Florida's childcare workers reflect on support and training needs. *Children Youth and Environments*, *18*(1), 237–253.

PART VI

Disaster Risk Governance and Policy Implementation

19

"GAPS ON THE GROUND"

Local-Level Policy Implementation for Disaster Risk Reduction in Coastal Bangladesh

C. Emdad Haque and M. Salim Uddin

Introduction

Disaster management by local institutions incorporates various dimensions of decision-making and implementation, including applying national policies and guidelines to the local community level (Choudhury et al. 2019; Tierney 2012). The primary goal of disaster management is to achieve disaster risk reduction (DRR) through enhancing public awareness to address disaster risk and vulnerability. Attaining this objective requires the necessary political will and allocation of appropriate resources by governing institutions (Fan 2015; UNDP 2010; Ahrens and Rudolph 2006). The role of institutions is critical here due to their central role in policy formulation, legitimizing practices, and establishing norms and rules of conduct. In the local community context, they are critical to accessing resources, engaging stakeholders, ensuring public participation, reducing risk, and managing emergencies and disasters (Yaro et al. 2015; Tierney 2012; UNDP 2008). Reflecting this, the recent UN Sendai Framework for Disaster Risk Reduction, 2015–2030, emphasized the importance of local-level institutions as a means by which local communities can self-organize and build resilience to natural hazard (UNISDR 2015). The rationale for such a locally focused strategy is grounded in the observation that most successes in risk mitigation and disaster management are directly tied to the effectiveness of local-level institutions' governance structures and implementation of national DRR policy (Jones et al. 2013).

Recent global DRR strategies, such as the UNISDR (2004) and Hyogo Framework for Action (2005–2015), are heavily influenced by initiatives undertaken during the 1990s as part of the United Nations International Decade for Natural Disaster Reduction (Tierney 2012; Raju and da Costa 2018). These international initiatives emphasize a shift away from response-oriented approaches to DRR and toward cultivating greater involvement of local-level institutions in

DOI: 10.4324/9781003253495-25

resilience-building and policy formulation and implementation (Rahman 2018; Islam et al. 2019). Similarly, the Sendai Framework for Disaster Risk Reduction (2015–2030) and the UNISDR (2015) place particular emphasis on local-level institutional 'good disaster governance' (Raju and da Costa, 2018; Ashu and van Niekerk, 2019), with the Sendai Framework stating: "it is necessary to continue strengthening good governance in disaster risk reduction strategies at the national, regional and global levels and improving preparedness and national coordination for disaster response, rehabilitation and reconstruction, and to use post-disaster recovery and reconstruction"(p. 10) for enabling society back on its feet and for transforming into positive socioeconomic prosperity.

The Hyogo and Sendai frameworks for DRR highlight local government institutions as the main locus of National Disaster Management Policy implementation, as they are uniquely situated to empower local communities and build local disaster resilience capacity (UNISDR, 2015; Ashu and van Niekerk, 2019; Choudhury et al. 2019). Recognizing the significance of local-level institutions, a considerable number of countries have begun decentralizing their national disaster management structures (Ikeda and Nagasaka 2011), includingVietnam (Adger 2000; Chau et al. 2014),Australia (Bosomworth 2018), Ghana (Yaro et al. 2015), and Chile (Valdivieso et al. 2017).However, several empirical studies have revealed that, in many countries, local institutions often fail to effectively implement national policy at the local level and arethus unable to uphold disaster governance principles (Fan 2015;Villagra and Quintana 2017; Rahman 2018; Raju and da Costa 2018; Choudhury et al. 2019). For example, analyzing the progress of several African countries against the Hyogo Framework, van Niekerk (2015)found that widespread deficiencies in accountability, transparency, efficiency, responsiveness, predictability, and trust have severely impeded implementation of national disaster management initiatives at the local level. However, the reasons for such failures are not well understood (Fan 2015;Villagra and Quintana 2017). Bangladesh, as one of the most vulnerable countries to extreme weather events, such as cyclones and associated tidal surges, is a case of particular interest in the study of local-level disaster policy implementation (IPCC 2012; Islam et al. 2019; Hasan et al. 2020).

In recent decades Bangladesh has experienced many catastrophic cyclones, storm surges, and floods (Rahman et al. 2019; Sattar and Cheung 2019), and the country's vulnerability to natural disasters, especially along the coastal regions (Rahman et al. 2019; Ahmed et al. 2016; Choudhury et al. 2019), is well-known internationally. Bangladesh has experienced 104 cyclones over the past 250 years (68 classified as severe), and the fatalities have ranged from as low as 50 to as high as 500,000 per disaster (Haque and Uddin 2013; Ahmed 2019). In the 1970–2013 period alone, approximately 700,000 people lost their lives due to nature-triggered disasters, with economic losses totaling around US$7.5 billion (Haque and Uddin 2013; Masud-All-Kamal 2013).

There have been significant shifts in the policy and institutional domains, especially since the 1990s, emphasizing the decentralization of emergency and disaster management to local-level institutions (Islam and Walkerden 2017;Vij et al. 2018;

Hasan et al. 2019). A key objective of these new policies is to enhance the capacity of local institutions at both the Union (lowest-level administrative unit) and *Upazila* (sub-district; second lowest administrative unit) levels to respond effectively to environmental disasters. Our understanding of the extent to which policies of this type have achieved their key objectives, especially in low- and middle-income countries (LMIC) like Bangladesh, is still very limited.

Noticeable progress in public policy formulation by the Government of Bangladesh (GoB) led to the development of elaborate frameworks and disaster response mechanisms that resulted in lowering the loss of human lives and property related to extreme weather events (Ahmed 2019; Islam and Walkerden 2017; Haque and Uddin 2013). Analysis of recent GoB policy enactments reveals that the national disaster management focus has been shifting away from a response and post-disaster relief-oriented approach and toward a more comprehensive approach embodying risk and vulnerability reduction, decentralization of risk management provisions and authority, and the enhancement of local community resilience (Vij et al. 2018; Hasan et al. 2019; Ahmed 2019; Haque and Uddin 2013).

A trend of continuous progress can be seen in the GoB's introduction of the National Plan for Disaster Management (NPDM) (2010–2015; 2016–2021), Standing Orders on Disaster (SOD) (1997, last updated 2010), the Disaster Management Act (DMA) (2012), the National Disaster Management Policy (NDMP) (2015), the Comprehensive Disaster Management Programme (CDMP) (2004–2009; 2010–2014), and DRR-associated policy documents (Islam and Walkerden 2017; Ahmed 2019; Hasan et al. 2019). The core principles of the NPDM (2010–2015; 2016–2020) and SOD are as follows: (i) promotion of national ownership of disaster management strategies by ensuring the participation of the government, NGOs, and civil society; (ii) increased focus on local-level benefits for marginalized and disadvantaged groups; (iii) adoption of a comprehensive, all-hazard approach to risk reduction; (iv) involvement of development partners; and (v) adoption of long-term perspectives on risk reduction. Together with the DMA (2012) and NDMP (2015), these aim to ensure the effective participation of local communities (Ahmed 2019; Hasan, Nasreen and Chowdhury 2019). The NDMP and SOD primarily offer guidelines and outline the roles and responsibilities of every organizational level involved in DRR and emergency response management. Under these provisions, a committee operates at each administrative level to manage the response to impending disasters and post-disaster impacts (Uddin et al. 2020; Choudhury et al. 2019).

In addition to such national initiatives, the UNDP-assisted CDMP was launched under the Disaster Management and Relief Division of the Ministry of Food and Disaster Management to institutionalize a comprehensive DRR approach. CDMP Phase I (2004–2009) pioneered several innovative tools, including Community Risk Assessment, Risk Reduction Action Plans, and Local Risk Reduction Funds, which aided in the mainstreaming of the Disaster Management System and in strengthening community institutional supports. Based on the success and lessons learned from Phase I, CDMP Phase II (2010–2014) aimed to further strengthen the national capacity for DRR and improve response and recovery processes (Choudhury et al. 2019; Hasan et al. 2019).

The above-noted efforts have indeed made great strides in implementing and improving national-level policies; most of the shortcomings in disaster management plans are the result of poor implementation at the local level (Rahman 2018; Vij et al. 2018; Ahmed 2019). However, evidence of poor implementation of national policies in LMICs, including Bangladesh, is scant. This gap calls for empirical investigation and critical assessment of the implementation process, and for improving our understanding of the existing barriers to executing national disaster management policies and programs at the local level in Bangladesh.

Considering this backdrop, the present study is an empirical investigation set in the local communities of coastal Bangladesh. We posit that developing effective local disaster management structures and enhancing community resilience to disaster shocks are difficult without effective implementation of national policy at the local level. In order to test this assertion, we set the specific objectives to:

1. examine the function and effectiveness of local-level disaster management institutions in coastal regions of Bangladesh.
2. assess the extent to which national disaster management policies have been implemented by these local institutions (e.g., Union *Parishad* and locally active NGOs); and
3. identify the major barriers against their successful implementation of national policies on DRR.

Conceptual Considerations: Local Institutions and Disaster Management at the Local Community Level

In the context of our study, the nexus of local institutions, disasters, and resilience is the most relevant conceptual domain (Figure 19.1). In this paper, we use the term institutional adaptationto refer to modifications and alterations made within institutions or their components to adjust to change in the wider physical and social environment in their implementation of policy (Adger 2000; Nyamwanza 2018). We find merit in applying Ostrom's (2008) notion of *institutions*, referring primarily to the rules-in-use, regulations, norms, beliefs, and customs along with their associated organizational infrastructure. In the context of the present study, only nature-triggered *disasters*, such as tropical cyclones, storm surges, floods, and droughts and the risk they impose upon humans and their resources are relevant. *Resilience* is defined as the "capacity of a system to absorb disturbances and reorganize while undergoing change so as to still retain essentially the same function, structure, identity, and feedbacks" (Walker and Holling 2004, 4). In the DRR context, such capacity is associated with a community's local institutions for disaster management and their ability to self-organize (Berkes 2007).

The recognition of the critical importance of local-level institutions in formulating and, more so in, implementing national disaster management policies

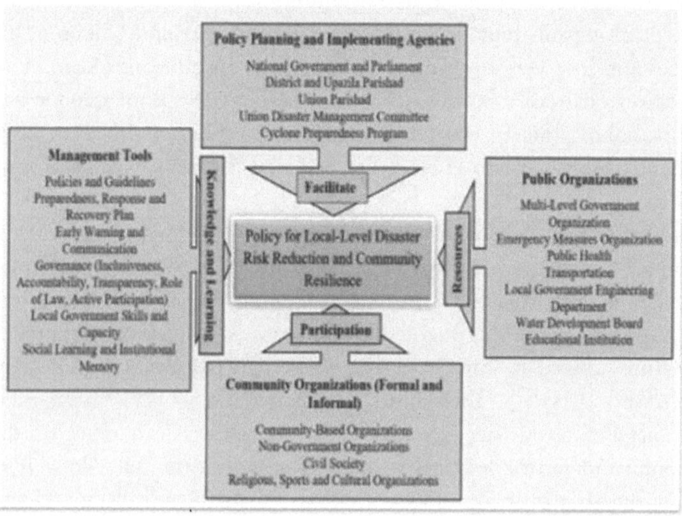

FIGURE 19.1 Critical tools, resources and mechanism for local-level policy implementation and community resilience

have been evolving over more than half a century. The conventional approaches to disaster management developed after World War II overwhelmingly emphasize top-down risk mitigation strategies, largely disregarding the role of stakeholders, good governance (Islam et al. 2019; Rajan 2002), and the participation of local actors (Azad et al. 2020; Fan2015). However, an emerging critical approach in recent decades has advocated for local community empowerment and the incorporation of local social and cultural knowledge and practices in order to improve local resilience and effective disaster governance (Fan 2015;Villagra and Quintana 2017). Further, critical scholars have argued that strengthening resilience and "community-based disaster risk reduction activities are more successful when they are institutionally embedded in local governance structures" (Jones et al. 2013, p. 442). It is thus apparent that delegated authority and good governance at the local level are necessary for effective disaster management, especially for fair and equitable risk mitigation, timely and efficient relief distribution, and strengthening resilience.

Notably, though there is a growing body of literature promoting the integration of disaster management and community resilience (Jones et al. 2013; Fan 2015; Villagra and Quintana 2017), the implementation of such an approach is still relatively limited. Despite some, albeit limited, efforts to engage local institutions in DRR policy implementation, there remain a variety of challenges to effective management of disaster risks locally, including insufficient resources, mismanagement (Islam et al. 2017), lack of coordination, and governance failure (van Niekerk 2015; Sovacool 2017; Ashu and van Niekerk 2019). In Bangladesh, additional barriers to disaster management at the local community level include a distorted perception of risk among inhabitants (Uddin et al. 2020; Sattar and Cheung 2019), lack of resources and intervention capacity among local institutions (Alam et al. 2011), lack of institutional collaboration (Sovacool et al. 2018), poor gender inclusivity in decision-making (Hasan et al. 2019), governance failures (Bhuiyan 2015), mismanagement of funds (Islam et al. 2017), and marginalization of local institutions (Choudhury et al. 2019).

As noted above, local institutions are characterized by formal (e.g., local governmental departments), quasi-formal (e.g., NGOs), and informal (social institutions, e.g., *samaj*, *panchayat*) structures, norms, and conventions, with a self-organizing capacity. As Berkes (2007)highlighted, in the context of nature-triggered disasters there are four notable specific aspects of self-organization: (i) the effectiveness of local first responders (e.g., firefighters, paramedics), who play a vital role in disaster response before outside assistance (e.g., the military) can arrive on the scene; (ii) horizontal and vertical linkages with other organizations, since local institutions and community-based organizations often have inadequate capacity to respond to disasters on their own; (iii) the ability of organizations and institutions to learn from past events and adjust their response strategies accordingly; and (iv) the sharing of information among institutions, which allows for more efficient management, mobilization, and use of shared resources.

Local institutions establish the rules within geographical and cultural spaces and interact with other institutional systems at various scales (Crane 2013), and over time become the repository of local and/or indigenous knowledge systems and the foundation upon which local society organizes itself. Networking, collaboration and partnerships, and linkages between various institutions reflect their strengths, degree of resource mobilization, capacity to execute, and commitment to societal welfare. As Agrawal et al. (2008) explain, *linkages to institutions*, referring to the degree to which different stakeholders are linked to various institutions in their locality and their access to resources and decision-making, play an instrumental role in disaster management. Under effective disaster governance, institutional collaboration, partnerships, and their linkages function as intertwined spheres affecting the conceptualization and operationalization of the various phases of disaster management (van Niekerk and Wisner 2014; van Niekerk 2015; Rahman 2018). Such collaboration also forges the links between development, disaster management, and humanitarian actions.

Study Area and Methods

Our focus in this study is on the roles of local institutions – particularly the Union *Parishad* and Union Disaster Management Committee (UDMC) – in disaster planning, interventions, governance, and NDMP implementation in the coastal communities of Bangladesh. We applied a qualitative research methodology following an Exploratory Case Study approach with multiple case designs, as advocated by Yin (2014). Data were collected from primary sources applying techniques of Participatory Rural Appraisal (PRA), as well as from secondary sources such as policy documents, guidelines, and reports.

Profile of the Study Area

The coastal regions of Bangladesh have been identified by IPCC-led scenario-building research as "hot spots" for environmental extremes associated with climate change (IPCC 2014). We conducted our field investigation in two cyclone-prone Unions, namely Nilganj and Dhulasar, in the Kalapara *Upazila* of Patuakhali District in the southern coastal region of Bangladesh (Figure 19.2; Ahmed et al. 2016). This area has experienced seven major cyclones during the past 40 years, two occurring within the last 15 years (Ahmed et al. 2016): Cyclone Sidr (Category V) in 2007 and Cyclone Aila (Category I) in 2009.

We selected Kalapara *Upazila* as it was among the areas most severely affected by Cyclones Sidr and Aila. Based on information gathered and consultation with residents and NGO officials from 12 Unions in Kalapara *Upazila*, the Unions of Nilganj and Dhulasar were selected for study. Nilganj Union is an inland community where the main occupation of the inhabitants is farming, while Dhulasar

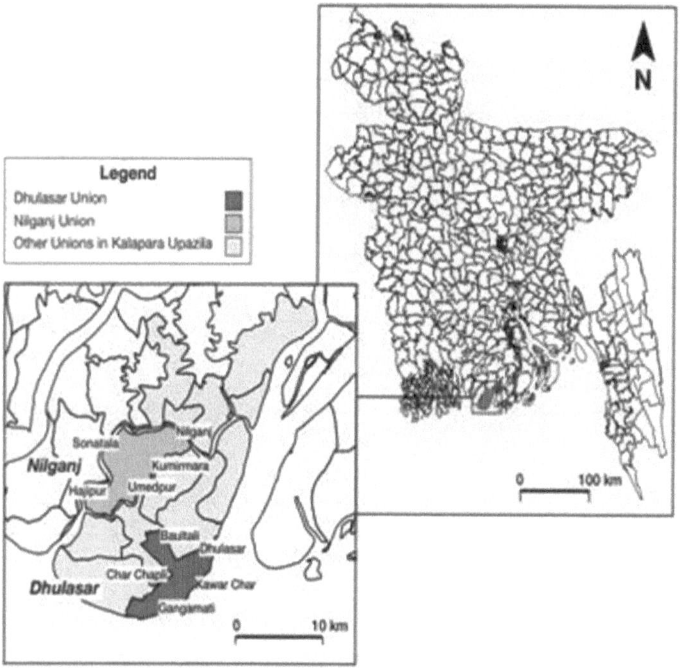

FIGURE 19.2 Study area map showing the location of the unions of Nilganj and Dhulasar (inset) within Bangladesh. The inset provides the names of the study villages (settlement units)

is located on the shore of the Bay of Bengal and has primarily afishing-based economy (Figure 19.2).

Data Collection Methods and Instruments

Using a questionnaire and various other PRA tools, field data were collected through a series of field visits between January 2014 and July 2015. To assess the applicability and validity of the collected data, a supplementary field visit was carried out in July–September 2019. Secondary information was collected through document review and assessment of reports related to loss and damage from cyclones and disaster policy implementation by local-level institutions.

A four-tier study design was followed to collect relevant data (Figure 19.3). Key methods used to collect data included: (i) focus group discussions (FGDs), (ii) key informant interviews (KIIs), and (iii) document review. To accumulate in-depth community information and explore the answers of specific questions, we employed semi-structured questionnaires.

We organized eight FGDs (four per Union) with participants of diverse gender, age, occupation, and social class, with each session lasting around three hours. Procedures followed for each session included: (i) group size was kept between 6–12 participants; (ii) professional and gender diversity was maximized; (iii) FGDs were conducted within the village and in places where participants felt most

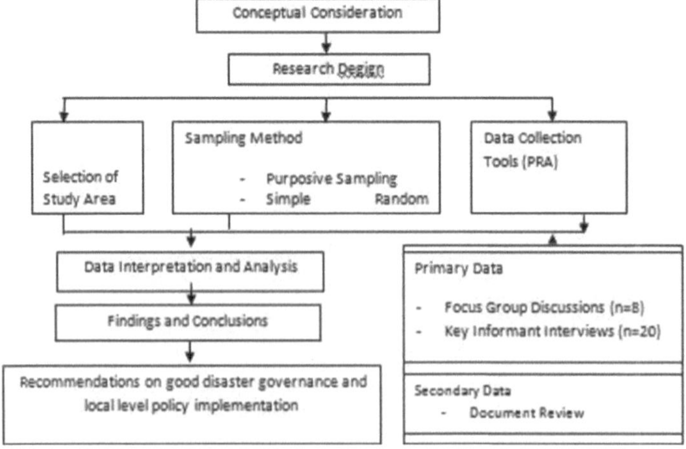

FIGURE 19.3 Multi-stage research design of the study

comfortable; (iv) one protocol writer, one moderator, and one organizer were present during each session; (v) sessions were recorded using audio devices with prior permission from the participants; and (vi) participation was voluntary, with participants permitted to withdraw at any time without adverse consequences. Collected data were coded deductively to trace out underlying themes running through the texts, and inductively based on *a priori* themes, such as good governance and policy impact.

Subsequently, a total of 20 KIIs were conducted using a semi-structured questionnaire with *Upazila Nirbahi* Officers, Union *Parishad* Chairmen and Members (both male and female), Union *Parishad* secretaries, Cyclone Preparedness Program (CPP) volunteers, school and *madrasa* teachers, retired government officers, knowledgeable residents, and social workers. Interviews were conducted at the interviewee's convenience and lasted between 1.5 and 2 hours. Participants were randomly selected from a comprehensive list including a broad range of sub-groups.

Finally, policy implementation and impact data were collected through FGDs, KIIs, case studies, and document reviews (hard and soft copies of reports, UDMC and Union *Parishad* registers, meeting minutes, previous disaster management documents, and records of disaster awareness, preparedness, response and recovery from Union *Parishads* and local NGOs).

As Patuakhali District, the administrative authority over Kalapara *Upazila*, did not have any formal institutional ethics review board, we procured approval from the local government authorities. We also strictly followed the University of Manitoba's Joint Faculty Research Ethics Board guidelines when obtaining primary data from participants. For each type of primary data collection, written consent was obtained from each participant.

Results

Findings from our empirical investigation in Dhulasar and Nilganj are presented below, where we examine the extent of implementation of national-level disaster management policies by local institutions (e.g., Union *Parishads* and local NGOs) and identify the major gaps in the implementation of national disaster management and DRR policy.

Local-Level Disaster Management Policy Implementation Institutions, Networks, and Partnerships

Disaster management by local institutions involves decision-making and implementation processes, with the active participation of local community people in, and ownership of relevant activities and programs. The aspects of disaster management here include the roles of the formal and informal actors in the spheres of the public and private sectors and civil society involved in decision-making. The relationships among the government, civil society, and the non-governmental sector are key determinants of a success in local institutional disaster management strategies.

The results of the document review, FGDs, and KIIs identified three major actors who shape local-level disaster management in Bangladesh: (i) Government, that is, Members of Parliament, *Upazila* administration, and Union *Parishads*; (ii) political parties and local elites; and (iii) NGOs, civil society, and community-based organizations (Figure 19.4).

For effective overall DRR, local-level implementation of national disaster management policies is critical. In order to determine the extent of policy implementation at the *Upazila* and Union *Parishad* levels, we examined the following documents: (i) NDMP (2015); (ii) SOD (2010); (iii) DMA (2012); and (iv) NPDM (2010–2015; 2016–2020). The SOD (2010) explicitly outlines the responsibilities for disaster management from the national to the Union level, as detailed in the country's core disaster policy documents, the NDMP (2015), NPDM (2010–2015; 2016–2020), and DMA (2012). The local government administration implemented many policy activities under the SOD, NDMP, NPDM, and DMA. Some examples of these functions included the formation of UDMCs, organizing various awareness and capacity building workshops, facilitating NGO activities to ensure coordination between different disaster management programs and preparation of local disastermanagement plans.

Our findings reveal that the respondents observed "medium" to "very high" impacts of the execution of national policies on local-level disaster management capacity. The formation of UDMCs following the SOD and aligned Union *Parishad* disaster management structure, for instance, generated "very high" impacts and significantly reduced the death toll and property losses due to disasters (Table 19.1). In line with the roles and responsibilities outlined in the SOD and NDMP, both

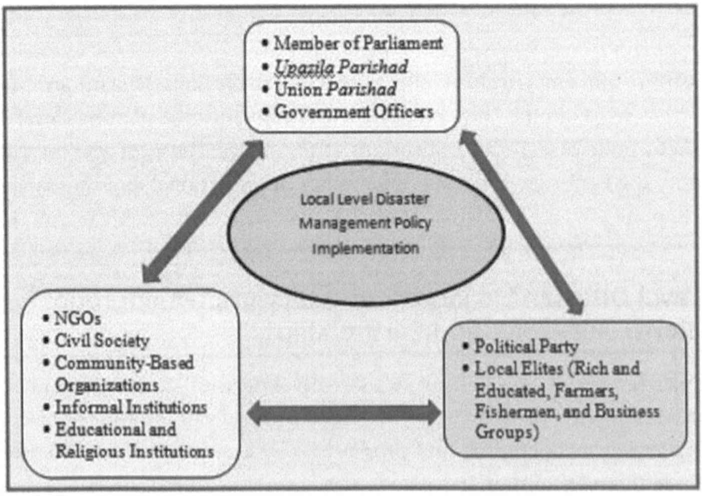

FIGURE 19.4 Actors of local-level policy implementation and disaster management

Source: Field investigation 2014–2015.

TABLE 19.1 Policy implementation, institutional adaptation, and impacts at the local level, Nilganj and Dhulasar Unions

Activity	*National policy*	*Action*	*Respondents' impact assessment*
Forming a UDMC	National Disaster Management Policy (2015) And Standing Order on Disaster (2010)	Adapted	+++++
Drafting a UDMC constitution		Not adapted	+
Organizing training and workshops on disaster issues		Adapted	+++
Preparing a comprehensive report incorporating a list of vulnerable communities and a location map based on the compilation of hazard, vulnerability, and risk assessment reports		Partially adapted	+++
Facilitating NGO activities to promote coordination and synergy between different disaster management programs		Adapted	+++++
Providing necessary support to the UDMC to generate funds for the implementation of risk reduction action plans		Not adapted	+
Building the capacity of local communities, volunteer organizations, and local institutions to help disaster-affected people in their home and livelihood reconstruction		Adapted	+++
Holding UDMC meetings at least once a month		Adapted	++
Storing medicine and emergency materials in Union communities		Not adapted	+
Counseling victims and organizing workshops on lessons learned from disaster experiences		Not adapted	+
Preparing a vulnerability map based on gender, age, and physical and social status		Not adapted	+
Preparing a Union Disaster Management Plan	National Plan for Disaster Management (2016)	Partially adapted	+++
Creating a Disaster Management Fund		Not adapted	+
Forming a UDMC	Disaster Management Act (2012)	Adapted	+++++
Organizing projects on vulnerability and DRR		Adapted	+++++
Establishing and strengthening community-based adaptation programs in disaster-prone areas	Bangladesh Climate Change Strategy and Action Plan (2009)	Partially adapted	+++

(continued)

TABLE 19.1 Cont.

Activity	*National policy*	*Action*	*Respondents' impact assessment*
Strengthening local government and community capacity to manage natural disasters.		Adapted	++++
Constructing, repairing, and rehabilitating disaster mitigation infrastructure, such as cyclone shelters, coastal and river embankments, and flood shelters		Partially adapted	+++
Formulating disaster vulnerability reduction strategies and comprehensive disaster management plans	Coastal Zone Policy (2005)	Partially adapted	+++
Measures to enhance the coping capacity of the poor during disasters, such as developing cyclone shelters, multi-purpose embankments, *killas*, road systems, and disaster warning systems		Partially adapted	+++
Adopting a gender-sensitive and participatory approach to reduce inequalities between men and women, such as closing the gender gap and giving priority to women's education, training, and employment		Partially adapted	+++
Ensuring the approach roads to cyclone shelters are user-friendly for all types of people and connected to the main road and villages	Cyclone Shelter Construction, Maintenance, and Management Policy (2011)	Not adapted	+
Ensuring cyclone shelters have sufficient light, safe water, food, toilet facilities, and sanitation during disasters		Partially adapted	++
Forming a Cyclone Shelter Management Committee responsible for managing the affairs of each of the cyclone shelters and earthen *killas* in line with the NPDM		Not adapted	+

Note: Impact: +++++=Very high, ++++=High, +++= Medium, ++=Low, +=Very low.

Source: Field investigations, 2014–2015, 2019.

Nilganj and Dhulasar Unions paid particular attention to DRR and mitigation, preparedness, emergency response, and recovery from disasters. They undertook measures to reduce risk at the household and community levels, enhanced local capacity by organizing training and workshops on disaster response, and prepared risk reduction maps and indices and disaster contingency plans with the help of NGOs working on disaster management in the locality.

TABLE 19.2 Participatory Strategic Vulnerability and Risk Reduction Action Plan developed by Nilganj and Dhulasar Unions

Goal/objective (outcome)	*Strategies*	*Executive agency*	*Target group/ stakeholders*
Enhancing disaster awareness Reducing loss of life and property	Awareness building	World Concern	Local community/ villagers
Safeguarding the community Reducing loss of life, property, and crops Restricting intrusion of saline water Saving seedbeds, trees, and domestic animals	Constructing and elevating coastal embankments	Speed Trust; Wave Foundation	
Reducing waterlogging duration Maintaining water level for agriculture Increasing fish and biodiversity	Excavation of local canals	Speed Trust; Wave Foundation	
Addressing crisis periods following disasters Reducing incidences of water-borne diseases	Water storage	AVAS	
Safeguarding communities from high tidal surges, soil erosion, and landslides Saving and maintaining ecosystem functions Generating additional sources of income and fuel wood	Roadside and embankment plantation	Muslim Aid; Wave Foundation; Speed Trust	
Saving homesteads from inundation Saving home gardens and domestic animals	Homestead elevation	World Concern	
Improving communication Ensuring easy access to shelters during disasters Safeguarding homesteads from tidal surges	Road construction and repair	Speed Trust; *Polli Gono Unnoin*	
Speeding post-disaster agriculture recovery Providing seed insurance	Seed and crop protection	Speed Trust; *Uddipan*	Farmers/ farmers' groups
Preserving wealth in the post-disaster period Reducing the burden of buying tools post-disaster	Keeping animals and agricultural tools safe	Speed Trust; *Uddipan*	
Reducing risk and loss of life and property Reducing accidents in waterways	Monitoring the weather and keeping emergency kits ready	WorldConcern; CODEC	Fishermen/ fisher groups
Enabling work to resume immediately after a disaster	Tool storage	Speed Trust	Day laborers/ seasonal laborers
Reducing loss of life and property Contributing to awareness building	Training on disaster management	AVAS; Wave Foundation	Women and Adolescents

Note: AVAS = Association of Voluntary Actions for Society; CODEC = Community Development Centre.

Source: Nilganj and Dhulasar Union Disaster Management Plan; field investigations, 2014–2015, 2019.

Examining the Union *Parishad* records and supporting NGO reports, we found that with the help of the NGO Web Foundation, the Nilganj UDMC organized training in disaster preparedness and livelihood diversification for 600 people between 2012 and 2014. Similarly, between 2014 and 2016, with active support of *Uddipan*– a local NGO in the Nilganj Union – the UDMC provided training on strategic adaptation to climate change–induced disasters to 387 people from 22 villages.

In Dhulasar Union, the UDMC formed 10 Friendship Coastal Development Committees for early warning dissemination and networks for disaster recovery and rehabilitation. In 2014, the UMDC also trained 370 people in disaster awareness and management. The impacts of these NGO–UDMC collaborations have been felt in many facets of local development, infrastructure, and DRR (Table 19.2), with the greatest impacts being in the areas of interactive governance through more participatory implementation of national policies and strategies.

Disaster Management Policy Implementation Gaps at the Local Level

Despite many notable successes in the implementation of national disaster management and DRR policies, serious gaps were identified in implementation at the local level (Table 19.3). For example, the drafting of a UDMC constitution, the organization of regular UDMC meetings, storage of medicine and emergency lifesaving materials in local communities, and organizing workshops on lessons learned from disaster experiences were not implemented by local governments in accordance with the SOD, NDMP, NPDM, and DMA.

We assessed the implementation process of UDMC responsibilities outlined in the SOD (2010) and NDMP (2015) [(Table 19.3)]. Based on Union *Parishad* records, local NGOs reports, and participants' responses, we observed some notable successes in implementing national policies and guidelines at the local level, for example, taking practical measures for DRR at the community level, arranging regular training and workshop for disaster preparedness (Plates 1 and 2) and livelihood adaptation, preparing risk maps, mobilizing CPP members and community volunteers for early warning and rescue operation during and after disasters, and coordinating with multi-level and multi-scale institutions for rapid recovery and rehabilitation after disasters (Table 19.3).

Although many UDMC responsibilities were fully or partially implemented, a noticeable number of policy activities were found unimplemented due to a lack of resources, autonomy, and technology, and rampant corruption. For example, many areas of disaster preparedness, response, and risk reduction were not addressed in accordance with national guidelines and policies. These included raising local funds for emergency use, storing lifesaving medicines and health care supplies, disseminating lessons learned and success stories, developing a comprehensive action plan for livelihood enhancement and rehabilitation, and preparing water purification technology and training local youths and community volunteers to use it (Table 19.3).

TABLE 19.3 Assessment of UDMC responsibilities implementation processes under the SOD and NDMP

Responsibilities/functions	*Observation: Success/failure*
A. Disaster Risk Reduction (DRR)	
1. Ensure that local people are kept informed and are capable of taking practical measures for the reduction of risk at household and community levels, and disseminate success stories of DRR at household and community levels widely among the local people	• Very instrumental in local-level DRR, most of the DRR initiatives were implemented by NGOs with the help of the UDMC • Dissemination of success stories were not found due to lack of social memory
2. Arrange training and workshops on disaster issues on a regular basis	• Very successful, with the help of NGOs
3. Conduct a hazard, vulnerability, and risk analysis at the Union level and prepare a risk reduction action plan and contingency plan for earthquakes and other hazards	• Disaster risk maps were found in most of the villages • Contingency plan was not found
4. Identify the most vulnerable or people at high risk by sex, age, physical ability, social status, occupation, and economic status	• Did not find identification of the most vulnerable people based on class
5. Prepare a short-, medium-, and long-term vulnerability reduction and capacity building action plan for the identified high-risk people with active participation of the at-risk people	• The UDMC failed to prepare any action plan for the people at high risk
6. Facilitate coordination among the development agencies and service providers through quarterly coordination meetings, and take decisions about implementation of the action plan for risk reduction as well as review the progress of the risk reduction action plan	• Found successful in coordination and mobilization of development agencies and service providers, with limited review of progress or revision of the action plan
7. Raise funds at the local level to implement the risk reduction action plan	• Not implemented
8. Prepare a Comprehensive Disaster Management Action Plan with a view to enabling local people, Union authority, and local organizations to increase the capacity of poor and vulnerable people to enhance their income and other assets for risk reduction, and also to take up security arrangements from the perspective of imminent danger-related warnings or occurrence of disasters	• The UDMC failed to prepare a comprehensive action plan focusing on livelihood enhancement for the affected people

(*continued*)

TABLE 19.3 Cont.

Responsibilities/functions	*Observation: Success/failure*
9. Take steps for capacity building of relevant persons and institutions, Union authority, volunteers, and local people such that they can forecast and publicize warnings relating to hazards (cyclones, storms, floods, droughts, tidal surges, tsunami, over- and under-rainfall, waterlogging, high tide, cold waves, etc.) in the quickest possible manner, and also inform people about their responsibilities for saving their lives and properties from disaster	• The UDMC was found very successful in early warning and volunteer mobilization and capacity building
10. Build the capacity of local institutions, volunteers, and local people such that they can help and motivate people to adopt disaster-resistant housing features	• Still in initial phase
11. Build the capacity of local institutions, volunteers, and local people such that they can help and motivate people to adapt via disaster-resistant agriculture and other livelihood options	• Found successful in capacity building and transferring technology for disaster-resilient agriculture and livelihoods
12. Determine specific safe centers/shelters where the population of certain areas will go in time of need, and assign responsibilities to different persons for various services and security at the shelters	• Many newly constructed cyclone shelters were found, with some issues regarding access roads and locations of the centers
13. Ensure a supply of safe water and if necessary other services from specific points near the shelters/centers with the help of the *Upazila* authority	• Few tube wells found within and near the cyclone shelters
14. Train students, youth, and local club members and volunteers on community-based water purification technology so that during disasters they can supply water-purifying technology in their community until external support reaches the high-risk people	• Partially implemented, but lacking in resources and technology
15. Plan for preparing some community-based highland, which can be serve as a playground in normal times and as a place of shelter during disasters, and where livestock, poultry, emergency food, kerosene, lamps, candles, matches, fuel wood, radios, and other important resources could be moved along with the evacuating people	• Three *killa* were found to use as shelter for livestock and people during disasters

TABLE 19.3 Cont.

Responsibilities/functions	*Observation: Success/failure*
16. Stock emergency lifesaving medicines at the Union level (Union Health and Family Welfare Centre) for use during disasters	• Failed to implement
17. Prepare relevant plans for rescue, primary relief operations, and restoration of communication with *Upazila* Headquarters, and local arrangements for rehabilitation of severely affected families	• Lack of a concrete plan, rather there were periodic rehabilitation plans and activities by donor agencies and NGOs
18. Arrange for rehearsals or drills on the dissemination of warning signals/ forecasts, evacuation, rescue, and primary relief operations	• Irregular rehearsals or drills were found in schools and colleges
B. Emergency Response: Warning Period	
1. Disseminate warning and security messages, evacuate vulnerable people as per evacuation plan, monitor the rescue team's last-moment check and preparation, and place high priority on taking effective measures to minimize gaps	• Very successful in early warning and evacuation
2. Engage trained institutions, volunteers, and people in the field for effective and rapid dissemination of early warning messages to vulnerable communities, and monitor all security and warning message dissemination activities	• Found successful in training and mobilization with the help of NGOs and CPP volunteers
3. Visit the pre-determined emergency shelter centers and ensure that relevant organizations and volunteers are alert and ready to provide essential services and security	• Partially failed regarding ensuring essential services and regularly monitoring cyclone shelters
4. Review the functionality of water supply sources near the shelters/centers, and, if necessary, fill gaps so that people can access a safe water supply from these sources during disasters	• Partially implemented
5. On a minimum scale, conduct a mock emergency or drill to ensure that the trained students, youth, clubs, and volunteers can prepare water purification technology in their locality to supply to victims during emergencies, and to monitor that adequate materials are ready to prepare such water purification technology	• Lack of mock emergencies or drills, and shortage of necessary materials and technology for water purification

(*continued*)

TABLE 19.3 Cont.

Responsibilities/functions	*Observation: Success/failure*
6. Review the stock of lifesaving medicines at the Union level and evaluate its adequacy for distribution among victims during disasters	• Failed
C. Emergency Response: During Disaster	
1. Organize emergency rescue work by using locally available facilities in times of need, and if directed, assist others in rescue work	• Found very successful in rescue operations during disasters with the help of local formal and informal institutions
2. Prepare water purification technology (tablet) at the local level with the help of trained students, youth, clubs, and volunteers, and distribute these products among at-risk people during emergencies before being affected by diarrhea or other water-borne diseases	• Failed to implement due to lack of technical knowledge and resources
3. Coordinate all relief activities (GO-NGO) in the Union in a manner that ensures social justice (on the basis of who is in need, what is needed, and how much is needed) in relief distribution	• Failed to ensure social justice, instead coordination was influenced by political ideology and nepotism
4. Provide accurate and timely information to protect people from becoming distressed by rumors during hazard periods	• Successful in early warning dissemination through CPP volunteers
5. Ensure the security of local and external relief workers during hazards	• Found successful
6. Ensure the security of women, children, and persons with disabilities during hazards	• Was not fully successful, instead security of these vulnerable people was left to individual and family evacuations
7. Take necessary actions to protect environmental degradation by arranging quick funerals and burying dead animal bodies	• Found successful in burial of deceased but failed to bury all dead animals
8. Help people to transfer their essential resources (livestock, poultry, essential food, kerosene, candles, matches, fuel, radio, etc.) to safe places	• Success in a limited scale, rather was done by self-help
D. Post-Disaster Rehabilitation	
1. Collect statistics of losses incurred following guidelines of the Disaster Management Bureau and *Upazila* DMC, and send to the *Upazila* DMC	• Found successful

TABLE 19.3 Cont.

Responsibilities/functions	*Observation: Success/failure*
2. Take steps for the distribution of rehabilitation materials from the locality, the Relief and Rehabilitation Directorate, and any other available sources	• Partially successful, but hampered by corruption and political biases
3. Ensure that people who were displaced due to hazards can return to their previous places; any disputes regarding displaced people's land should not be an obstacle to their return after a disaster	• Found successful
4. Arrange counseling for people suffering from psycho-trauma due to the disaster, with the collaborative support of experts and community elites	• Successful with the help of affected people's family and relatives
5. Ensure that injured people are getting fair and just treatment from health service providers	• Depends on the availability of health care facilities
6. Arrange a lesson-learned session with the participation of concerned institutions and individuals, focusing on learning during and after hazards	• Failed to arrange
E. Local Disaster Response Coordination	
Local-level coordination with *Upazila*, District and state department authorities, NGOs, and Civil Society Organizations for DRR, disaster preparedness-response-recovery, and disaster risk management	• Found very successful in coordination across vertical and horizontal organizational scales

Local-level disaster management and policy implementation largely depends on financial and technical capacity, decision-making power, and leadership by the Union *Parishad*. We asked particpants to identify and rank policy implementation problems. Research participants identified seven major problems, among which financial and technical capacity (93%), autonomy (87%), and corruption and political biases (83%) were ranked highest. Local leadership (67%) for networking and learning and dissemination of success stories (53%) were identified asmedium-level problems, and lack of gender inclusion in decision-making (47%) and poor project management (40%) were ranked lower by respondents (Figure 19.5).

Discussion

In the existing literature, numerous research results have revealed that disaster management policy implementation at the local level is vital to the success of disaster

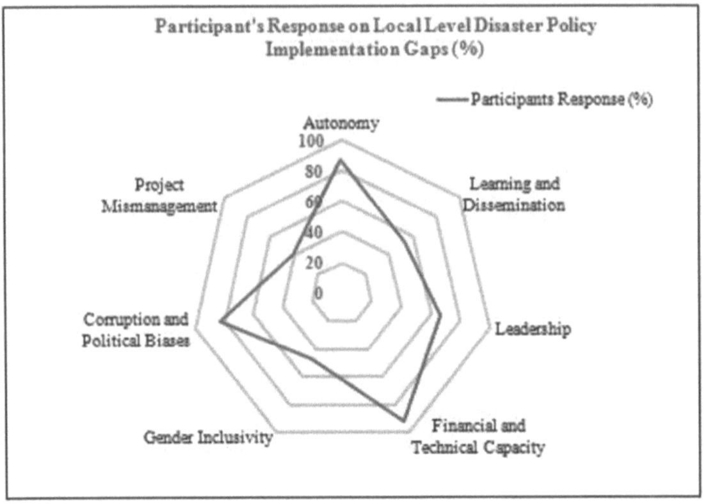

FIGURE 19.5 Participants' response on local-level disaster policy implementation

governance and effective disaster management (Haque and Uddin 2013;Thiupugazh 2014; Haque et al. 2017; Choudhury et al. 2019;Ahmed 2019). In many countries, decentralization of governing structures for disaster management has been pursued in recent years, including Bangladesh (Haque and Uddin 2013; Azad et al. 2020), South Korea (Bae et al. 2016), India (Rumbach 2016), and Indonesia (van Voorst, 2016). Successful policy implementation at the national level has also resulted in the mainstreaming of disaster risk and vulnerability reduction in Bangladesh (Ahmed 2019; Haque and Uddin 2013),Vietnam (Adger 2000), and Chile (Valdivieso et al. 2017). The success of these efforts has stemmed chiefly from the removal of major barriers to policy implementation and strengthening horizontal and vertical institutional linkages.

Operationalizing Disaster Governance and Comprehensive Disaster Management

Strong Leadership and Autonomy Is Critical for Local Institutions

Findings of our study revealed that emergency fund mobilization during the response and recovery periods, preparedness, planning, and coordination to deal with future cyclones were most effective where leadership of local institutions was strongest. Building and encouraging leadership skills through training and resource allocation should therefore be a national policy priority.

We observed that local institutions in both study Unions developed disaster management plans, risk maps, and vulnerability indices with no or limited participation from local communities. The prevalence of such top-down approaches may be attributed to the prevailing bureaucratic system and serious deficiencies in accountability and transparency. Maskrey (1990) and Ahrens and Rudolph (2006) observed that, in LMICs, a top-down decision-making system prevails due to the assumption of technological superiority and a lack of autonomy among local institutions (particularly in the selection of beneficiaries and the allocation of resources).

Gender Inclusivity and Women Empowerment Is the Key for Disaster Risk Reduction

The critical role of women in disaster management has often been neglected, with women traditionally being considered merely as victims of disasters rather than agents of change (Jayarathne 2014; Sohrabizadeh 2016; Hasan et al. 2019). In this regard, equity and inclusion of various genders and occupational and ethnic groups have been identified as important prerequisites for effective disaster governance (Aguilera and Cuervo-Cazurra 2009; UNESCAP 2009; Hasan et al. 2019). Our findings revealed that in Bangladesh's coastal communities the inclusion of women made the dissemination of early warnings and the subsequent response more effective. Similarly, based on his research on the 2012 earthquake and 2013 floods in Iran, Sohrabizadeh (2016)found that the destructive effects of disasters on women and girls can be ameliorated by mainstreaming their capacities in post-disaster phases. In the same vein, investigating the 2004 Tsunami in Sri Lanka, Jayarathne (2014) illustrated that women-centered public awareness and skills training programs can help increase women and children's disaster preparedness and equip them with the skills necessary to recover from disasters.

Compliance with Good Governance Criteria Can Enhance Effective Disaster Management

In the coastal communities of Bangladesh, we found that poor disaster governance, lack of transparency and accountability, corruption, the politicization of post-disaster relief operations, lack of attention to equity and justice issues, and lack of coordination all significantly hampered disaster response and recovery. Other studies conducted in coastal communities (Mahmud and Prowse 2012; Islam et al. 2017; Choudhury et al. 2019) and wetland areas (Choudhury and Haque 2016) of Bangladesh corroborate our findings. For example, Islam et al. (2017)documented that political affiliation, bribery, and corruption in relief and rehabilitation significantly diminish the equity and efficiency of recovery from disasters in the coastal areas of Bangladesh. Similarly, Valdivieso et al. (2017) identified that local municipal institutions with more transparent management practices and community-based participatory decision-making processes made tremendous advances in environmental disaster risk management in Chile.

Financial and Technical Capacity of Local Institutions Are Pivotal

We observed in rural Bangladesh that local governing bodies (i.e., Union *Parishads*) and their associated institutions (e.g., UDMC) lack sufficient financial and technical resources to organize training workshops for local residents on disaster awareness and management, develop projects for gender empowerment and social justice, or to implement required disaster management activities outlined in various disaster policies. We found that with such limited financial and administrative autonomy, it is difficult for the Union *Parishad*-led UDMCs to work as per the guidelines of Bangladesh's SOD. It is worth noting here that Marks and Lebel (2016) found in Central Thailand that incomplete decentralization impeded successful implementation of disaster governance. This clearly illustrates the differences in effectiveness between *de jure* and *de facto* decentralization (Hutchcroft 2001; Lessmann 2009).

Without local empowerment and resource provisions, the decentralization of disaster management often becomes a burden for local institutions (Allen 2006; Choudhury et al. 2019). In this context, Kooiman and Bavinck (2013) asserted that the governability of an institution depends on qualities of the institution itself, the governing system (e.g., the State), and interactions with other administrative levels. An analysis of interactive governance, such as consisting of such governing mixes and developing explanations of such emerging complexities, suggests it could find "answers" to ever growing societal diversity, dynamism, and complexity, and provide effective responses to major societal and environmental issues, such as poverty and climate change–induced environmental extremes (e.g., cyclones, storm surges) (Kooiman 2003).

Barriers against Effective Local-Level Policy Implementation and Challenges to Addressing Them

Disaster management policy implementation at the local level in accordance with national policy frameworks, along with good disaster governance, is essential for effective and successful DRR and disaster management (Blaikie et al. 1994; Liverman 2008; Berkes 2007; Chaudhury et al. 2019). We found that some local level (i.e., Union) disaster management programs under national policy frameworks were implemented with noticeable success. Such programs included the formation of the UDMC, organization of various awareness and capacitybuilding workshops, facilitation of NGO activities, and preparation of Union disaster management plans (Figure 19.6). The identified drivers of policy implementation were political

FIGURE 19.6 Empowerment and inclusivity

commitment, international DRR and mitigation policy ratification, donor and foreign aid-driven projects, awareness building, and resilience enhancement.

Plate 1: NGO rehabilitation project signboard **Plate 2:** Community consultation on gender

Institutional Adaptation Is a Key to Good Disaster Governance

Communities in other parts of the world have also succeeded in reducing their disaster risk and vulnerability through effective policy implementation; for example, Chau et al. (2014) found that local institutions changing their structure to align with policy was fruitful in integrating DRR and emergency management in Vietnam. Similarly, Valdivieso et al. (2017) observed in Chile that altering institutional settings in local municipal governments helped with proactive adaptation to climate change impacts and DRR. In Australia, Zurita et al. (2015) observed that empowering local governments and consensus-based inclusive decision-making strengthened disaster governance outcomes.

Multi-Level Networking and Capacity Building of Local Institution Is Prerequisite for DRR

However, we observed that many policy initiatives, such as drafting a UDMC constitution, organizing regular UDMC meetings, storing medicine and emergency lifesaving materials in local communities, organizing workshops on lessons learned from disaster experience, and preparing vulnerability maps based on gender, age, and physical and socioeconomic status were not executed effectively by the Union *Parishads* in accordance with the SOD, NDMP, NPDM, and DMA. Plausible explanations for these failures were lack of funding and training, insufficient technical and financial capability, poor leadership, top-down decision-making strategies, and corruption within local-level institutions. Yaro et al. (2015) made similar findings in Ghana, observing that the effectiveness of local institutions in implementing climate change policies was constrained by their limited spatial and temporal reach, limited financial and human resources, and faulty strategic design and implementation procedures. Ashu and van Niekerk (2019) identified that a lack of capacity both at the national and local levels has been hampering effective implementation of DRR initiatives in Cameroon.

Conclusion

Since the drafting of the UN Sendai Framework for Disaster Risk Reduction (2015–2030), the discourse on disaster management and DRR by local-level institutions has evolved significantly throughout the developing world, including Bangladesh. The success of multifaceted risk mitigation and disaster management, including preparedness and response, has been associated with the effectiveness of local governance structures and implementation of national disaster policy (Jones et al. 2013). The main purpose of our study was to determine the extent of compliance with and implementation of national DRR policies in Bangladesh, as well as to identify barriers to disaster governance and the successful implementation of disaster management policies.

The findings of our empirical investigation among Bangladesh's coastal communities reveal that interactive disaster governance, decentralization of disaster management, and implementation of national disaster policy in a participatory and transparent way are not only catalysts for rapid recovery, but they can also strengthen community resilience to disaster shocks. For efficient and sound management of disasters at the local level, the following interventions should be considered:

1. Policy priority should be given to empowering local communities and institutions with self-governance instruments, resource procurement and allocation, and decision-making capacity and authority.
2. Compliance with good governance criteria and ensuring that they follow the rule of law, are transparent and accountable, and engage in participatory and gender-inclusive risk identification and disaster management action should be required from the involved local institutions.
3. The capacity of local institutions in areas of technical and financial ability should be strengthened to enable them to appropriately execute policy actions, delegate disaster management activities, and build community awareness.
4. Coordination among non-governmental and private organizations, trust building within local communities, power sharing, and corruption-free relief, rehabilitation, and project management are necessary for effective disaster management.

The present study confirms that effective implementation of national policy by local-level institutions can improve disaster management and help to save lives and property. Implementation of national disaster policies by local institutions are also essential for effective disaster management and enhancing resilience to disasters.

References

Adger, W.N. (2000), "Institutional adaptation to environmental risk under the transition in Vietnam". *Annals of the American Association of Geographers*, Vol. 90, No. 4, pp. 738–758. doi: 10.1111/0004-5608.00220

Agrawal, A., Catherine, M., and Nicolas, P. (2008), *Local institutions and climate change adaptation*. Social Development Notes; No. 113. Washington, DC: World Bank. https://openknowledge.worldbank.org/handle/10986/11145 License: CC BY 3.0 IGO.

Aguilera, R.V. and Cuervo-Cazurra, A. (2009), "Codes of good governance". *Corporate Governance*, Vol. 17, No. 3, pp. 376–387. doi:10.1111/j.1467-8683.2009.00737.

Ahmed, B., Kelman, I., Fehr, H.K., and Saha, M. (2016), "Community resilience to cyclone disasters in coastal Bangladesh". *Sustainability*, Vol. 8, No. 805, pp. 1–29. doi:10.3390/su8080805

Ahmed, I. (2019), "The national plan for disaster management of Bangladesh: Gap between production and promulgation". *International Journal of Disaster Risk Reduction*, Vol. 37, No. 101179, pp. 1–8.doi: 10.1016/j.ijdrr.2019.101179

Ahrens, J. and Rudolph, P.M. (2006), "The importance of governance in risk reduction and disaster management". *Journal of Contingencies and Crisis Management*, Vol. 14, No. 4, pp. 207–220. doi: 10.1111/j.1468-5973.2006. 00497.x

Alam, K., Shamsuddoha, M., Tanner, T., Sultana, M., Huq, M.J., and Kabir, S.S. (2011), "The political economy of climate resilient development planning in Bangladesh". *IDS Bulletin*, Vol. 42, No. 3, pp. 52–61.

Allen, K.M. (2006), "Community-based disaster preparedness and climate adaptation: Local capacity-building in the Philippines". *Disasters*, Vol. 30, No. 1, pp. 81–101. doi: 10.1111/j. 1467-9523.2006. 00308.x

Ashu, R.E.A. and van Niekerk, D. (2019), "Building national and local capacity for disaster risk management in Cameroon". *Disaster Prevention and Management: An International Journal*, Vol. 28, No. 6, pp. 764–777. doi: 10.1108/DPM-06-2019-0176

Azad, M.A.K., Uddin, M.S., Zaman, S., and Ashraf, M.A. (2020), "Community-based disaster management and its salient features: A policy approach to people-centred risk reduction in Bangladesh". *Asia-Pacific Journal of Rural Development*, p. 101852911989803. doi: 10.1177/1018529119898036

Bae, Y., Joo, Y.M., and Won, S.Y. (2016), "Decentralization and collaborative disaster governance: Evidence from South Korea". *Habitat International*, Vol. 52, pp. 50–56. doi: 10.1016/ j.habitatint.2015.08.027.

Berkes, F. (2007), "Understanding uncertainty and reducing vulnerability: Lessons from resilience thinking". *Natural Hazards*, Vol. 41, No. 2, pp. 283–295. doi: 10.1007/ s11069-006-9036-7

Bhuiyan, S. (2015), "Adapting to climate change in Bangladesh: Good governance barriers". *South Asia Research*, Vol. 35, No. 3, pp. 349–367.

Blaikie, P., Cannon, T., Davis, I., and Wisner, B. (1994), *At risk: Natural hazards, people's vulnerability and disasters*. London: Routledge.

Bosomworth, K. (2018), "A discursive–institutional perspective on transformative governance: A case from a fire management policy sector". *Environmental Policy and Governance*, Vol. 28, No. 6, pp.415–425. doi: 10.1002/eet.1806

Chau, V.N., Holland, J., and Cassells, S. (2014), "Institutional structures underpinning flood management in Vietnam". *International Journal of Disaster Risk Reduction*, Vol. 10, pp. 341–348. doi: 10.1016/j.ijdrr.2014.10.008

Choudhury, M.U.I., and Haque, C.E. (2016), "We are more scared of the power elites than the floods: Adaptive capacity and resilience of wetland community to flash flood disasters in Bangladesh". *International Journal of Disaster Risk Reduction*, Vol. 19, pp. 145–158. doi: 10.1016/j.ijdrr.2016.08.004

Choudhury, M.U.I., Uddin, M.S., and Haque, C.E. (2019), "Nature brings us extreme events, some people cause us prolonged sufferings: The role of good governance in building community resilience to natural disasters in Bangladesh". *Journal of Environmental Planning and Management*, Vol. 62, No. 10, pp. 1761–1781. doi: 10.1080/ 09640568.2018.1513833

Crane, T.A. (2013), The role of local institutions in adaptive processes to climate variability. The cases of southern Ethiopia and Southern Mali. Oxfam Research Reports, Vol. 34.

Fan, M.F. (2015), "Disaster governance and community resilience: Reflections on Typhoon Morakot in Taiwan". *Journal Environmental Planning and Management*, Vol. 58, No. 1, pp. 24–38. Doi: 10.1080/09640568.2013.839444

Haque, A., Kamal, A.S.M.M., and Hassan, S.M.K. (2017), *Partnership, coordination, and accountability in urban disaster management: A review of policies in Bangladesh*. International Institute of Environment and Development, London.

Haque, C.E. and Uddin, M.S. (2013), "Disaster management discourse in Bangladesh: A shift from post-event response to the preparedness and mitigation approach through institutional partnerships". In: Tiefenbacher, J. (ed.), *Approaches to disaster management: Examining the implications of hazards, emergencies and disaster.* Croatia: InTech, pp. 33–54.

Hasan, M.R., Nasreen, M., and Chowdhury, M.A. (2019), "Gender-inclusive disaster management policy in Bangladesh: A content analysis of national and international regulatory frameworks". *International Journal of Disaster Risk Reduction*, Vol. 41 (September), p. 101324. doi: 10.1016/j.ijdrr.2019.101324

Hasan, M.T., Adhikary, G., Mahmood, S., Papri, N., Shihab, H.M., Kasujja, R., Ahmed, H.U., … Nasreen, M. (2020), "Exploring mental health needs & services among affected population in a cyclone affected area in coastal Bangladesh: A qualitative case study". *International Journal of Mental Health Systems*, Vol. 14, No. 12, pp. 1–9. doi: 10.1186/s13033-020-00351-0

Hutchcroft, P.D. (2001), "Centralization and decentralization in administration and politics: Assessing territorial dimensions of authority and power". *Governance*, Vol. 14, No. 1, pp. 23–53. doi:10.1111/0952-1895.00150

Ikeda, S. and Nagasaka, T. (2011), "An emergent framework of disaster risk governance towards innovating coping capability for reducing disaster risks in local communities". *International Journal of Disaster Risk Science*, Vol. 2, No. 2, pp. 1–9. doi: 10.1007/s13753-011-0006-7

IPCC. (2014), *Climate change 2014 synthesis report*. Cambridge: Cambridge University Press. doi:10.1073/pnas.1116437108.

IPCC. (2012), *Managing the risks of extreme events and disasters to advance climate change adaptation: Special report of the Intergovernmental Panel on Climate Change*. Cambridge University Press.

Islam, R. and Walkerden, G. (2017), "Social networks and challenges in government disaster policies: A case study from Bangladesh". *International Journal of Disaster Risk Reduction*, Vol. 22 (February), pp. 325–334. doi: 10.1016/j.ijdrr.2017.02.011

Islam, R., Walkerden, G., and Amati, M. (2017), "Households' experience of local government during recovery from cyclones in coastal Bangladesh: Resilience, equity, and corruption". *Natural Hazards*, Vol. 85, No. 1, pp. 361–378. doi: 10.1007/s11069-016-2568-6

Islam, S., Chu, C., Liew, L., and Smart, J.C.R. (2019) "Distributing flood shelters for disaster risk reduction: Exploring the practices in Bangladesh from a political economy perspective". *Disaster Prevention and Management: An International Journal*. doi: 10.1108/DPM-02-2019-0060

Jayarathne, S.S. (2014), "Women's potential in dealing with natural disasters: A case study from Sri Lanka". *Asian Journal of Women Studies*, Vol. 20, No. 1, pp. 125–136. doi: 10.1080/12259276.2014.11666175

Jones, S., Aryal, K., and Collins, A. (2013), "Local-level governance of risk and resilience in Nepal". *Disasters*, Vol. 37, No. 3, pp. 442–467. doi: 10.1111/disa.12006

Kooiman, J. (2003), *Governing the governance*. Thousand Oaks, CA: Sage Publications Ltd.

Kooiman, J. and Bavinck, M. (2013), "Theorizing governability—The interactive governance perspective". In: Bavinck, M. et al. (eds.), *Governability of fisheries and aquaculture: Theory and applications*. Dordrecht Heidelberg, New York, London: MARE Publication Series 7, Springer. doi: 10.1007/978-94-007-6107-0

Lessmann, C. (2009), "Fiscal decentralization and regional disparity: Evidence from cross-section and panel data". *Environment and Planning A*, Vol. 41, pp. 2455–2473. doi: 10.1068/a41296

Liverman, D. (2008), "Assessing impacts, adaptation and vulnerability: Reflections on the Working Group II Report of the Intergovernmental Panel on Climate Change". *Global Environmental Change*, Vol. 18, No.1, pp. 4–7.

Mahmud, T. and Prowse, M. (2012), "Corruption in cyclone preparedness and relief efforts in coastal Bangladesh: Lessons for climate adaptation"? *Global Environmental Change*, Vol. 22, No.4, pp. 933–943. doi: 10.1016/j.gloenvcha.2012.07.003

Marks, D. and Lebel, L. (2016), "Disaster governance and the scalar politics of incomplete decentralization: Fragmented and contested responses to the 2011 floods in Central Thailand". *Habitat International*, Vol. 52, pp. 57–66. doi: 10.1016/j.habitatint.2015.08.024

Maskery, A. (1990), "Disaster mitigation as a development tool". *The Futurist*, Vol. 24, No. 5, pp. 51.

Masud-All-Kamal, M. (2013), "Livelihood coping and recovery from disaster: The case of coastal Bangladesh". *Current Research Journal of Social Sciences*, Vol. 5, No. 1, pp. 35–44.

Nyamwanza, A.M. (2018), Local institutional adaptation for sustainable water management under increasing climatic variability and change: A case in the mid-Zambezi Valley, Zimbabwe. *International Journal of Climate Change Strategies and Management*, Vol. 10, pp. 453–471. doi: 10.1108/IJCCSM-03-2017-0078

Ostrom, E. (2008), "Institutions and the environment". *Economic Affairs*, Vol. 28, No. 3, pp. 24–31. doi: 10.1111/j.1468-0270. 2008.00840.x

Rahman, M. (2018), "Capacities and responses to disaster governance in Bangladesh: A reflection from the SDGs". *Civil and Environmental Research*, Vol. 10, No. 2, pp. 45–58.

Rahman, S., Islam, M.S., Khan, M.N.H., and Touhiduzzaman, M. (2019), "Climate change adaptation and disaster risk reduction (DRR) through coastal afforestation in South-Central Coast of Bangladesh". *Management of Environmental Quality: An International Journal*, Vol. 30, No. 3, pp. 498–517. doi: 10.1108/MEQ-01-2018-0021

Rajan, S. (2002), "Disaster, development and governance: Reflections on the 'lessons' of Bhopal". *Environmental Values*, Vol. 11, pp. 369–394.

Raju, E. and da Costa, K. (2018), "Governance in the Sendai: A way ahead?" *Disaster Prevention and Management: An International Journal*, Vol. 27, No. 3, pp. 278–291. doi: 10.1108/DPM-08-2017-0190

Rumbach, A. (2016), "Decentralization and small cities: Towards more effective urban disaster governance"? *Habitat International*, Vol. 52, pp. 35–42. doi: 10.1016/j.habitatint.2015.08.026

Sattar, M.A. and Cheung, K.K.W. (2019), "Tropical cyclone risk perception and risk reduction analysis for coastal Bangladesh: Household and expert perspectives". *International Journal of Disaster Risk Reduction*, Vol. 41(August), p. 101283. doi: 10.1016/j.ijdrr.2019.101283

Sohrabizadeh, S. (2016), "The neglect of women's capacities in disaster management systems in Iran: A qualitative study". *Indian Journal of Gender Studies*, Vol. 23, No. 3, pp. 467–480. doi: 10.1177/0971521516656080

Sovacool, B.K. (2017), "Don't let disaster recovery perpetuate injustice". *Nature*, Vol. 549, No. 7673, pp. 433–450.

Sovacool, B.K., Tan-Mullins, M., and Abrahamse, W. (2018), "Bloated bodies and broken bricks: Power, ecology, and inequality in the political economy of natural disaster recovery". *World Development*, Vol. 110, pp. 243–255.

Thiruppugazh, V. (2014), "Post-disaster reconstruction and institutional mechanisms for risk reduction: A comparative study of three disasters in India". In: Shaw, R. (ed.), *Disaster recovery: Used or misused development opportunity*. Springer.doi: 10.1007/978-4-431-54255-1_2

Tierney, K. (2012), "Disaster governance: Social, political, and economic dimensions". *Annual Review of Environment and Resources*, Vol. 37, pp. 341–363. doi: 10.1146/annurev-environ-020911-095618

Uddin, M. S., Haque, C.E., Walker, D., and Choudhury, M.U.I. (2020),"Community resilience to cyclone and storm surge disasters in coastal communities of Bangladesh". *Journal of Environmental Management*, Vol. 264, No. 110457.https://doi.org/10.1016/j.jenvman.2020.110457

UNDP. (2008), *Guidebook on promoting good governance in public-private partnerships*. United Nations, New York and Geneva: United Nations Economic Commission for Europe.

UNDP. (2010), *Disaster risk reduction, governance and mainstreaming*. New York: Bureau for Crisis Prevention and Recovery, pp. 1–4.

UNESCAP. (2009), "What is good governance"? Thailand: United Nations Economic and Social Commission for Asia and the Pacific, pp. 1–3. doi:10.1016/B978-012397720-5.50034-7

UNISDR. (2004), *Basic terms of disaster risk reduction*. Geneva: United Nations International Strategy for Disaster Reduction (UNISDR).

UNISDR (2005), Hyogo framework for action 2005–2015. Available at: www.unisdr.org/we/coordinate/hfa (accessed 9 January 2020).

UNISDR. (2015), *Sendai Framework for Disaster Risk Reduction 2015–2030*. Sendai, Japan: Third World Conference on Disaster Risk Reduction, 14–18 March 2015, pp. 1–25. doi: A/CONF.224/CRP.1

Valdivieso, P., Andersson, K.P., and Villena-Roldán, B. (2017), "Institutional drivers of adaptation in local government decision-making: Evidence from Chile". *Climate Change*,Vol. 143, No. 1, pp. 157–171. doi: 10.1007/s10584-017-1961-9

van Niekerk, D. (2015) "Disaster risk governance in Africa: A retrospective assessment of progress in against the hyogo framework for action (2000–2012)". *Disaster Prevention and Management*,Vol. 24, No. 3, pp. 397–416. doi: 10.1108/DPM-08-2014-0168

van Niekerk, D. and Wisner, B. (2014), "Integrating disaster risk management and development planning: experiences from Africa". In: Lopez-Carresi, A., Fordham, M., Wisner, B., Kelman, I., and Gaillard, J.C. (eds.), *Disaster management: International lessons in risk reduction, response and recovery*. New York: Routledge, p.330.

van Voorst, R. (2016), "Formal and informal flood governance in Jakarta, Indonesia". *Habitat International*,Vol. 52, pp. 5–10. doi: 10.1016/j.habitatint.2015.08.023

Vij, S., Biesbroek, R., Groot, A., and Termeer, K. (2018), "Changing climate policy paradigms in Bangladesh and Nepal". *Environmental Science and Policy*, Vol. 81 (November 2017), pp. 77–85. doi: 10.1016/j.envsci.2017.12.010

Villagra, P. and Quintana, C. (2017), "Disaster governance for community resilience in coastal towns: Chilean case studies". *International Journal of Environmental Research and Public Health*,Vol. 14, No. 1063, pp. 1–24. doi: 10.3390/ijerph14091063

Walker, B., Holling, C.S., Carpenter, S.R., and Kinzig, A. (2004), "Resilience, adaptability and transformability in social–ecological systems". *Ecology and Society*,Vol. 9, No. 2, p. 5.

Yaro, J.A., Teye, J., and Bawakyillenuo, S. (2015), "Local institutions and adaptive capacity to climate change/variability in the northern savannah of Ghana". *Climate and Development*, Vol. 7, No. 3, pp. 235–245. doi: 10.1080/17565529.2014.951018

Yin, R.K. (2014), *Case study research, design and methods*, 5th ed. London: Sage Publication.

Zurita, M.M. de L.,Cook, B., Harms, L., and March, A. (2015) "Towards new disaster governance: Subsidiarity as a critical tool". *Environmental Policy and Governance*,Vol. 25, No. 6, pp. 386–398. doi: 10.1002/eet.1681

20

PROTECTIVE EFFECTS OF EXISTING COASTAL BUFFERS

Syed Labib Ul Islam and G.M. Tarekul Islam

Introduction

The coastal zone of Bangladesh comprises 19 districts. This region is vulnerable due to the geographical positioning of the location, as well as the socio-economic condition of the coastal community. The coastal area is frequently ravaged by tropical cyclones. Such storms can be remarkably dangerous and cause enormous devastation (Grossman & Morgan, 2011). Almost every year in early summer during April–May or late rainy season during October–November, tropical cyclones, mostly originating from the Bay of Bengal, lash onto coastal Bangladesh and wreak havoc. Storm surges accompanying the cyclones kill many people living in the coastal areas and can cripple the local economy for a long time, thus exacerbating the existing socio-economic vulnerabilities. Climate change is expected to further increase the frequency of high-intensity tropical cyclones as ocean temperature and stronger cyclones have been correlated in studies (Mendelsohn et al., 2012).

Several techniques have been used in an attempt to mitigate such adverse natural phenomena. However, they can be classified into two main categories:

- Structural Measures
- Non-Structural Measures

The structural measures are often the concrete and steel-based large engineering structures whereas the non-structural measures are the strategically used interventions causing the least possible intrusion to the natural environment. Non-structural measures can be tangible, such as "building with nature", which was the original "traditional" engineering dating back to the first century BC, or can be intangible such as the development of warning systems, policies, laws, etc. The structural measures were used extensively to rebuild the cities ravaged by the horrors

DOI: 10.4324/9781003253495-26

and destructions of the two World Wars (Lehne& Preston, 2018). It was a strategy for quick recovery in the destroyed countries and initially, it was fruitful, paving the way for recovery in the short term. However, in doing so, several important aspects were overlooked. Environment, ecological balance, and climate took the backseat in the list of priorities, as economic development was given the top priority. This has not proven to be a sustainable strategy for development as evidenced by the numerous issues plaguing the modern world, including climate change. In the long term, many of the structural measures have also been proven to be the pitfalls to development (Hales et al., 2017).

To date, the structural measures undertaken at both national and local levels to mitigate cyclones have been inadequate and often economically inefficient for developing countries. Recently, some non-structural measures such as forecasting, warning, local action plans, etc. have been taken up on a short scale and found success in many countries including Bangladesh (Rahman et al., 2014; Smith et al., 2017; Baten et al., 2018). It is often strongly believed that structural mitigation measures need to be complemented by non-structural mitigation measures to modify or reduce the disaster effects (Rahman et al., 2002). Thus, the concept of using soft engineering measures is again coming into the limelight. The need for developing and utilising sustainable engineering techniques is gradually being understood all over the world to tackle various problems. This realisation has led to the development and improvement of various soft engineering measures. While it is not possible to prevent natural disasters like cyclones, coping with cyclones is possible (Ahamed et al., 2012). Coastal green buffer is one such soft engineering measure that is being used as an effective and sustainable form of coping mechanism.

Green buffers often offer effective protection against storm surges and intense wind speeds. Mangrove forests are efficient in reducing wind and storms. During cyclone Sidr, the thick growth of mangrove trees in Sundarbans successfully reduced the intensity of both the wind and the storm surge (Paul, 2009). Quantifiable monetary loss due to cyclones can be halved if there is mangrove protection and local people highly value the storm protection service of the Sundarbans mangrove forest (Akber et al., 2018). However, the protective impact of mangroves is supposed to be insignificant when the landfall location is outside the Sundarbans (Sakib et al., 2015). The effectiveness of such natural interventions in the context of Bangladesh can be clarified by analysing these indicative parameters such as the wind speeds at locations protected by coastal forests and locations that are exposed to the Bay of Bengal. The combined effect of mangrove afforestation and polder has the potential to significantly reduce thrust force from cyclones (Mita, 2019). Furthermore, a study in India showed that mangrove-protected villages had the lowest adverse factors, or the highest positive factors after the cyclone of 1999 (Badola & Hussain, 2005). Despite having extreme stress on land demand in Bangladesh, a combination of bioengineering and environmental measures could reduce the damage and act as a buffer against tides and cyclones (ADB, 2005).

In this chapter, the effectiveness of the existing mangrove buffer as a natural bioengineering intervention in Bangladesh has been presented based on a study conducted by the authors. Historical wind speed and water level data during previous cyclonic events were used for this purpose. Further, semi-structured interviews were utilised as a Participatory Rural Appraisal (PRA) tool to justify the findings from the data analysis. This provided an on-field perspective of the local residents and helped validate the secondary data as well as integrate their real-life experiences within this framework in a tangible way.

Methodology

The study was designed following the Mixed Research Method, which is appropriate for the mutual validation of data and findings as well as for the production of a more coherent and complete picture than mono method research (Kelle, 2008). The historical record of wind speed from the Bangladesh Meteorological Department (BMD) for different regions at different times was collected for the statistical analysis and investigation of the data with emphasis on cyclonic events. The statistical analysis, in this case, included sorting the data to identify the cyclone-induced spikes in values, and the variation of the amplitudes of maximum wind speed for different cyclonic and normal conditions. Attention was given to the locations behind coastal buffers and locations virtually exposed to the Bay of Bengal with a direct interface. Wind speed data were obtained as point data at specific locations where the stations are located. These data helped in identifying the variance of wind speed at different stations during cyclones and reflect the possible effects of coastal buffers in the form of trees impeding the direct lashing of winds. The selected stations for wind speed data are shown in Figure 20.1.

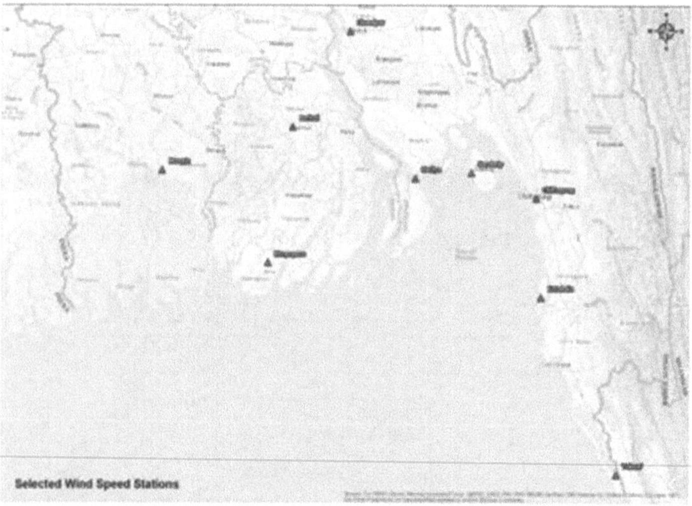

FIGURE 20.1 Map showing the selected wind speed stations

The quantitative analyses for the cyclonic storm surge were carried out by using the secondary recorded water level data focusing on selected cyclonic events from the Bangladesh Water Development Board (BWDB).

The historical record of consecutive five days water levels was collected focusing on selected individual past cyclonic events for statistical analysis of the data. Then the data were plotted chronologically for the five consecutive days on the vertical axis, and geographical location-wise (southwest to southeast) on the horizontal axis to provide a unique perspective to easily understand how the surge levels varied at different locations for individual cyclonic events. The variation of the amplitudes in water levels for both high tide and low tide was observed due to cyclones. This helped in investigating the variance of actual water levels in the coastal area during the same cyclonic event. Consequently, this variance reflected the mitigating effects of the coastal buffers. The selected stations for water level data are shown in Figure 20.2.

For primary data, semi-structured interviews (SSI) were utilised as a PRA tool. Some of the SSI was carried out through online mediums due to the restrictions

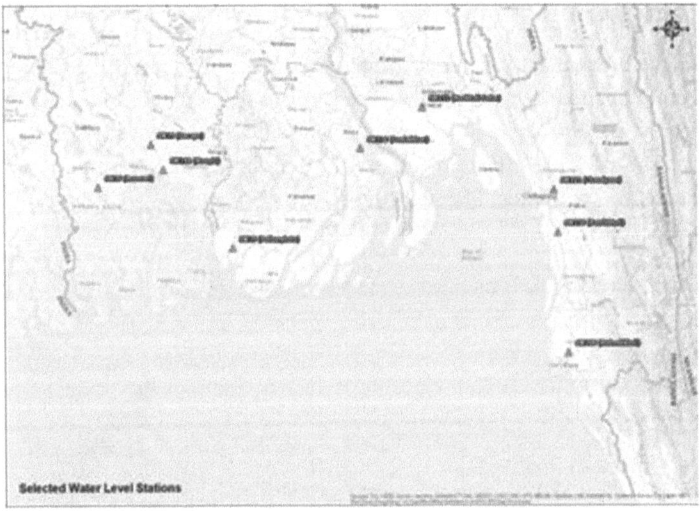

FIGURE 20.2 Map showing the selected water level stations

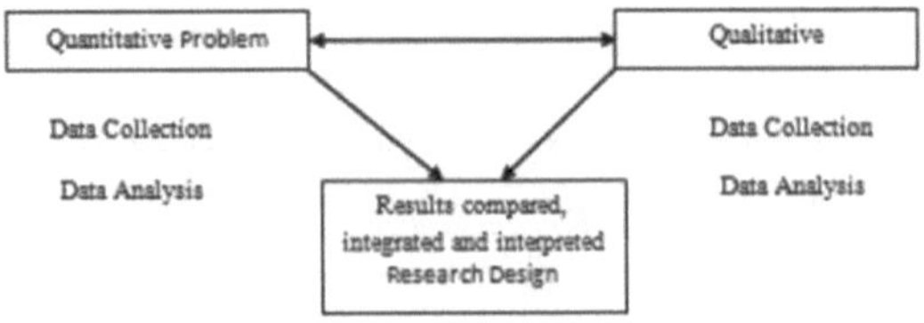

FIGURE 20.3 Mixed Research Method

Source: Kim (2019).

in place for the Covid-19 pandemic. The respondents were chosen randomly through peers and professionals and their acquaintances residing in the coastal zone of Bangladesh following the methods described by Cresswell (2007). The Mixed Research Method was utilised to integrate and interpret the qualitative and quantitative data as shown in Figure 20.3.

Mitigating Effects on the Cyclonic Winds

The historical daily maximum wind speed data of nine coastal stations (i.e. Mongla, Khepupara, Barisal, Chandpur, Hatiya, Sandwip, Chittagong, Kutubdia, and Teknaf) between the time periods of 1988 and 2019 were collected from BMD to carry out the wind speed secondary data analysis. The data for 2020 were unavailable at the time of analysis due to the Covid-19 pandemic. Among the stations selected, only the Mongla station is located behind a thick mangrove protective buffer in the form of Sundarbans. Barisal and Chandpur stations are situated more inland than the other coastal stations. The Chittagong, Kutubdia, and Teknaf stations have some green buffer strip areas around them, but they are not very dense. Whereas the remaining stations (i.e. Khepupara, Hatiya, and Sandwip) have very little protection in terms of green buffer zones as described in Table 20.1.

A list of major cyclones was collected from the official records of the BMD, and the number of cyclones with different ranges of maximum wind speeds were sorted by the authors as shown in Table 20.2 for the years between 1988 and 2019, to provide a basis for comparison. Following this, the daily recorded maximum wind speed data of the nine stations were compared for the date of the cyclones. The graphical representation of the maximum wind speeds for the nine stations is given in Figures 20.4, 20.5, and 20.6. It is to be noted that the BMD records daily maximum wind speed data in the unit "knots" where the conversion is taken as (1 knot = 1.858 km/hr).

It is evident from the visualisation that the Mongla station consistently records lower wind speeds during cyclonic events for cyclones with different maximum wind speed ranges. Mongla station falls within the western coastal zone where the

TABLE 20.1 Selected wind speed stations

Station name	*Location (decimal degrees)*	*Remarks*
Mongla	22.466670, 89.600000	Behind Thick Mangrove Cover
Khepupara	21.983300, 90.233300	Minimal Mangrove Cover
Barisal	22.716488, 90.366666	Comparatively Inland
Chandpur	23.233330, 90.700000	Comparatively Inland
Hatiya	22.450000, 91.100000	Minimal Mangrove Cover
Sandwip	22.483330, 91.433330	Minimal Mangrove Cover
Chittagong	22.350000, 91.816670	Moderate Mangrove Strip
Kutubdia	21.816670, 91.850000	Moderate Mangrove Strip
Teknaf	20.866670, 92.300000	Moderate Mangrove Strip

TABLE 20.2 Number of cyclones between 1988 and 2019 with different maximum wind speeds

Maximum wind speed (km/hr)	*Number of cyclones (1988–2019)*
≤100	8
101–150	5
≥151	6

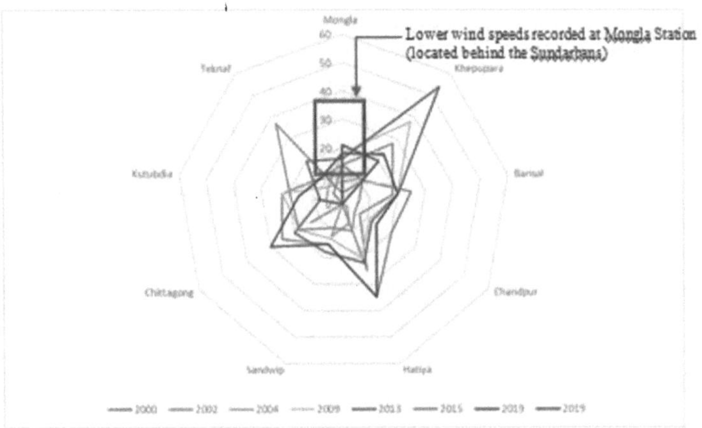

FIGURE 20.4 Radar chart showing the wind speeds at different stations for cyclones with maximum wind speed ≤100 km/hr between 1988 and 2019

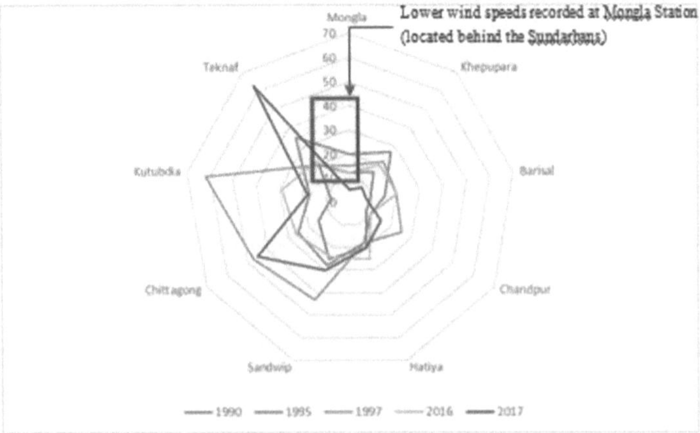

FIGURE 20.5 Radar chart showing the wind speeds at different stations for cyclones with maximum wind speed 101–150 km/hr between 1988 and 2019

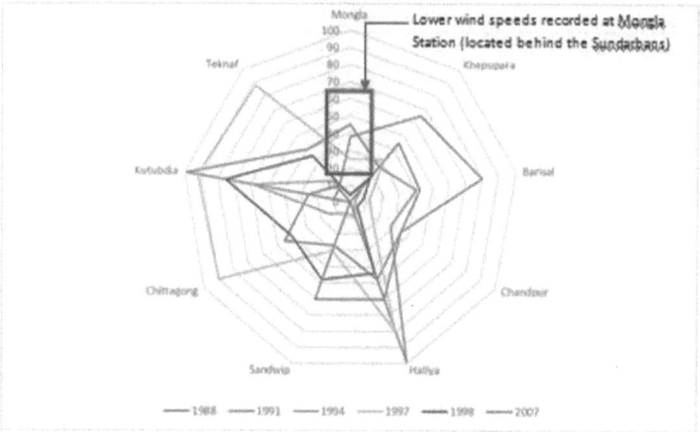

FIGURE 20.6 Radar chart showing the wind speeds at different stations for cyclones with maximum wind speed ≥151 km/hr between 1988 and 2019

station is located behind the Sundarbans, which provides it with vegetative buffer interventions impeding the cyclonic effects. Furthermore, during the cyclone Sidr in 2007, the Mongla station recorded a maximum wind speed of 38 knots while Khepupara station – a station in the central coastal zone with minimal vegetation cover – recorded 65 knots which are 41.5% higher (detailed tabulated data are available on request). On the other hand, the more recent cyclones in 2016 and 2017 recorded, respectively, 12 knots and 5 knots at the Mongla station, and 20 knots and 8 knots at the Khepupara station. These are increases of 40% and 37.5%, respectively. Hatiya and Sandwip recorded the highest wind speeds of 60 knots each, which are also lower than Khepupara, and may be attributed to the presence of comparatively more vegetation extent as this was observed for several other cyclones as well. Moreover, the highest recorded wind speeds for the 2016 and 2017 cyclones were 29 knots at Kutubdia, and 63 knots at Teknaf stations, which had moderate mangrove strip covers.

Mitigating Effects on Storm Surges

Storm surges during cyclonic storms are characterised by the historically recorded water level data at different water level gauge stations in the coastal region. Recorded water level data from nine coastal stations of BWDB were collected. The stations were selected in a way that a set of three stations represented each of the three distinct coastal regions and also kept in mind the position of the station – either behind a green buffer cover or somewhat exposed to the coast. The selected stations are shown in Table 20.3.

For each of these stations, the water level data of several specific cyclonic events, where data for all of the nine stations were available, were selected (Table 20.4) based on the BMD cyclone list. Water level data (both high tide and low tide) of

TABLE 20.3 Selected water level stations for secondary analysis

Region	*Station ID*	*Station*	*River*	*District*	*Upazila*
Western coastal zone	SW26	Protapnagar	Betna-Kholpetua	Satkhira	Assasuni
	SW29	SutarkhaliForest Office	Bhadra	Khulna	Dacope
	SW244	Mongla	Rupsa-Pasur	Bagerhat	Mongla
Central coastal zone	SW39	Patharghata	Bishkhali	Barguna	Patharghata
	SW278	Daulatkhan	Surma-Meghna	Bhola	Daulatkhan
	SW222	Noakhali	Noakhali Khal	Noakhali	NoakhaliSadar
Eastern coastal zone	SW121	Enayethat	Halda	Chattogram	Chandgaon
	SW250	Banigram	Sangu	Chattogram	Banshkhali
	SW200	Salafpur-Moheshkhali	Moheshkhali Channel	Cox's Bazar	Maheshkhali

TABLE 20.4 Selected cyclone events for secondary analysis of water level data

Landfall date	*Landfall area*	*Maximum wind speed (km/hr)*	*Tidal surge height (m)*
09.11.83	Cox's Bazar	136	0.91
24.05.85	Chattogram	154	4.57
29.11.88	Khulna	160	0.60–4.42
18.12.90	Cox's Bazar Coast	115	1.52–2.13
29.04.91	Chittagong	225	3.66–6.71
02.05.94	Cox's Bazar-Teknaf Coast	220	1.52–1.83
25.11.95	Cox's Bazar	140	3.05
19.05.97	Sitakundu	232	4.57
20.05.98	Sitakundu	173	0.91

five days (including the day of landfall, one day prior, and three days after) were taken for comparison.

The cyclonic events were chosen from a time period such that the effect and contrast of dense mangrove buffer and the exposed coastal area could be easily discernible. The stations were plotted on the horizontal axis, with the left to right showing the southern to eastern stations. On the vertical axis, the maximum water levels at both high and low tides for each of the five days were plotted. This shows the overall status of the water levels along the coast for the whole of the coastal zone of Bangladesh and brings a unique perspective of assessing the overall water levels including the storm surges and tidal effects. Figures 20.7 and 20.8 show the maximum water levels in this way for the cyclone in 1990 and 1998, where it is discernible that the southern gauge stations protected by the Sundarbans recorded lower water levels, while the central and eastern coastal zones, where the vegetation buffer is less dense, recorded comparatively higher water levels. This effect was evident for every cyclonic event analysed. Detailed tabulated data and analyses are available on request.

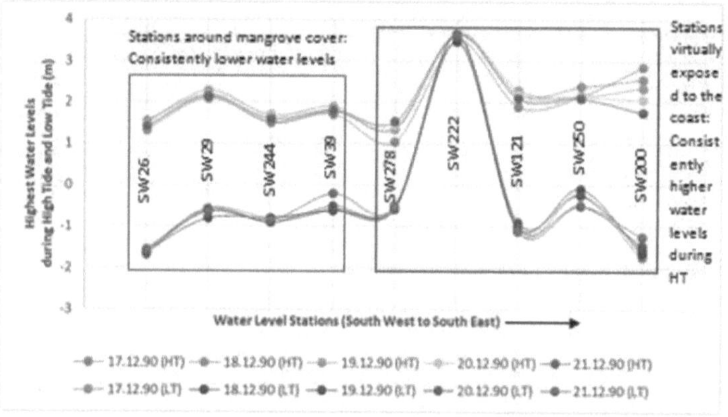

FIGURE 20.7 Maximum water levels during high tide and low tide at different stations for cyclone in 1990

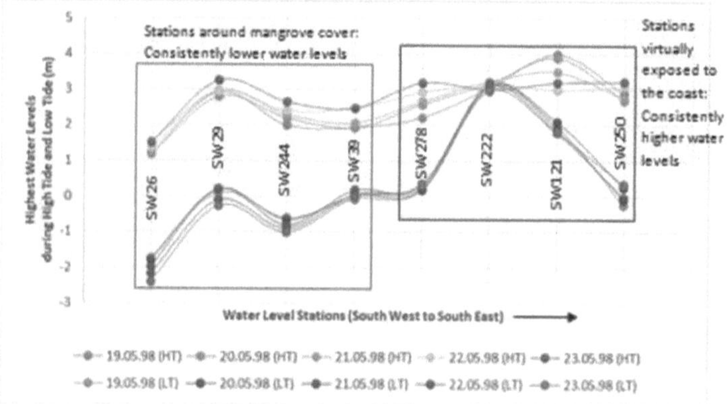

FIGURE 20.8 Maximum water levels during high tide and low tide at different stations for cyclone in 1998

Primary Data Analysis

The primary data for this analysis were collected using the PRA tools – specifically, the SSIs of the residents of the coastal areas. The SSI was chosen as the ideal tool because of its flexibility in extracting the information from the field level. Due to the Covid-19 pandemic, some of the responses were collected via online mediums. The SSI questions were designed to allow flexibility of the questions and answers so the participants can give their insights and observations that might otherwise have been missed in structured interviews with hard and fast questions.

The responses of the SSI were collected from people living in different representative districts of all three coastal areas of Bangladesh. Responses from Khulna

TABLE 20.5 Location and number of responses

Representative region	*District*	*Number*	*Subtotal*	*Total*
Western coastal area	Satkhira	3	9	40
	Khulna	6		
Central coastal area	Patuakhali	8	19	
	Bhola	5		
	Noakhali	6		
Eastern coastal area	Chattogram	8	12	
	Cox's Bazar	4		

TABLE 20.6 Background information of the respondents

Background Information		*Number*	*Percentage (%)*
Gender	Male	28	70
	Female	12	30
Age group	21–30	16	40
	31–40	7	18
	41–50	12	30
	51–60	5	12
Education level	School level	6	15
	Undergraduate level	27	68
	Postgraduate level	7	17

and Satkhira represented the western coastal area dominated by the mangrove forest of the Sundarbans. Responses from Patuakhali, Bhola, and Noakhali represented the central coastal area dominated by the broken coastline at the interface of lower Meghna and the Bay of Bengal. In addition, the responses from Chattogram and Cox's Bazar represented the eastern coastal area. The location and number of responses are shown in Table 20.5. Moreover, the background information of the respondents is given in Table 20.6.

The respondents of the southwestern coastal area were from Satkhira and Khulna, two large districts of Bangladesh. These districts are protected by the large mangrove forest – the Sundarbans. This region is frequently battered by tropical cyclones; the most notable of them in recent years was the severe cyclonic storm Sidr in 2007. Due to the disaster-prone nature of the region and the memories of recent disasters, most people are very much aware of the vulnerability and the threats. They were also found to be highly aware of the protective nature of the Sundarbans and the important role it plays in mitigating cyclonic winds and storm surges. People acknowledged the Sundarbans for bearing the brunt of the Sidr and protecting countless lives and properties. One of them mentioned,

> I was a school student when Sidr struck, and I can vividly remember the sound of the winds and the terror we felt. In the aftermath, everyone in

> the village came together for a recovery effort. I remember my grandfather recounting that without Sundarbans we would have perished.
>
> *Translated from Bangla by the Authors*

They also mentioned the cooperativeness of the local people in letting the Sundarbans recover from the damage on its ecosystem sustained due to the Sidr. The respondents acknowledged the importance of the mangrove and were concerned about the future of their protective barrier (Table 20.7).

The respondents from the central coastal area of Bangladesh also were familiar with natural disasters like cyclones. However, their main concern seemed to be coastal erosion because while the cyclones strike on average once a year, they are afflicted by coastal erosions almost throughout the year. Most were in support of large engineering works such as sea walls and large stretches of high embankments. One of the respondents mentioned, "Our embankments need to be improved, otherwise flood water will keep getting up and destroy our farm yields" (Translated from Bangla by the Authors). However, they also acknowledged the protective nature of green buffers and bioengineering solutions which have been effective in some places in protecting against erosion and slowing down the rate of loss of land – with the same person explaining, "River erosion situation is slightly better now as the banks are more or less stable. Still, when the water recedes quickly,

TABLE 20.7 Key responses from western coastal zone

Region	*Issues*	*Number (percentage) of responses*		*Remarks*
		Positive	*Negative*	
Western coastal zone	Existence of strategic green buffer at present	9 (100%)	0 (0%)	Acknowledged the strategic position of the Sundarbans
	Acknowledgement of the protective effect of green buffer and bioengineering	9 (100%)	0 (0%)	The respondents are aware of this from their experiences with severe cyclones
	Attitude towards an expansion of green buffer and bioengineering	5 (56%)	4 (44%)	Some feel that instead of expanding, the focus should be given to monitoring the existing buffers
	Responsibility of the public for protecting the green buffers	8 (89%)	1 (11%)	Most people feel they are responsible for protecting the Sundarbans
	Attitude towards adopting a combination of structural and non-structural measures	7 (78%)	2 (22%)	Most were in support

erosion is apparent". They shared experiences of plantation by the sides of the river embankments reducing the erosion and helping in stabilising the river reach at some stretches, especially upstream from the interface of the lower Meghna and the Bay of Bengal. Most of the undergraduate and post-graduate level educated respondents from this region recommended an integrated green barrier system, incorporating the structural and non-structural measures (Table 20.8).

The respondents from the eastern coastal area were familiar with the unbroken continuous shoreline of the region. People in this region are familiar with frequent cyclones, and the urban development of this area is mainly adjacent to the shoreline. This limits the available space to create a functional green barrier thick enough to reduce storm surges and mitigate cyclonic wind speeds. They also mentioned their acknowledgement of the effectiveness of green barriers on the southern coast, however, most of them were aware of the space requirements, and expressed their doubt of bioengineering being useful in this particular area. One of them mentioned, "Government is taking up Economic Zone initiatives, hilltop forests are being cut down, and I have no idea whether there will be any land left for forests for any buffer unless there are replantation programs". However, some people from Cox's Bazar mentioned the recent drive-in plantation programs in that region, especially near the Maheshkhali and Kutubdia region, and the environmental benefits of these

TABLE 20.8 Key responses from central coastal zone

Region	*Issues*	*Number (percentage) of responses*		*Remarks*
		Positive	*Negative*	
Central coastal zone	Existence of strategic green buffer at present	7 (37%)	12 (63%)	Respondents were unsure of the effective locations
	Acknowledgement of the protective effect of green buffer and bioengineering	11 (58%)	8 (42%)	Some felt the protective effects would not apply to their area
	Attitude towards an expansion of green buffer and bioengineering	10 (53%)	9 (47%)	More concerned about protective rigid, high embankments
	Responsibility of the public for protecting the green buffers	15 (79%)	4 (21%)	Most acknowledged the responsibility
	Attitude towards adopting a combination of structural and non-structural measures	13 (68%)	6 (32%)	Susceptive towards a combination, however, properly designed and regularly maintained embankments were their main concern

TABLE 20.9 Key responses from eastern coastal zone

Region	*Issues*	*Number (percentage) of responses*		*Remarks*
		Positive	*Negative*	
Eastern coastal zone	Existence of strategic green buffer at present	9 (75%)	3 (25%)	The respondents highlighted the artificial mangroves in Mirsarai and adjacent area
	Acknowledgement of the protective effect of green buffer and bioengineering	10 (83%)	2 (17%)	Most acknowledged positively
	Attitude towards an expansion of green buffer and bioengineering	8 (67%)	4 (33%)	Some were concerned about the space stress in the urban area
	Responsibility of the public for protecting the green buffers	9 (75%)	3 (25%)	Most acknowledged the responsibility
	Attitude towards adopting a combination of structural and non-structural measures	7 (58%)	5 (42%)	Space requirements remain an issue

programs. However, their opinion was split on the effectiveness of these programs in the mitigation of natural disasters, while focusing on the environmental – especially ecological – benefits, and the benefits against mudslides (Table 20.9).

Interpretation and Discussion

The secondary data analysis of wind speeds and water levels reiterated the effectiveness of a green barrier (natural bioengineering intervention) at the interface of the Bay of Bengal and the coastline in having significant positive impacts on mitigating the storm surge levels depending on the local conditions even if the coastal buffer is not very dense. Artificial elongated strip mangroves also have performed significantly in this regard and agree with the findings of Uddin and Hossain (2013) that the damage due to cyclones and floods was greatly reduced by them. This also agrees with the findings by Parvathy and Bhaskaran (2017), who stated that the energy reduction for waves due to mangroves and suitable bottom slopes can reach up to 98%. This entails that, through the application of mangroves, the vulnerabilities of the high-risk coastal areas can be significantly diffused (Banan-Dallallian et al., 2021).

In the case of secondary analysis of wind speeds, it has been revealed that the wind speed recording stations that are situated near or behind mangrove covers (i.e. Mongla Station) record consistently lower wind speeds for the cyclones with ≤100 km/hr, 101–151 km/hr, and ≥151km/hr maximum wind speeds. In the case of stations with moderate mangrove (strip) covers, the effectiveness in reducing the wind speeds is significant for the smaller cyclones with ≤100 km/hr maximum wind speeds. However, this reduction is less for the stronger cyclones with 101–151 km/hr and ≥151 km/hr maximum wind speeds. This signifies that dense mangrove cover is highly effective in reducing the destructive effects of the intense cyclonic winds. The moderate vegetation covers developed in the southeastern coastal zone, which are often elongated strip-shaped, are effective in reducing the wind speeds of weaker cyclones, but this effect diminishes with the increasing intensity of the cyclonic winds. These findings concur with the findings of McIvor et al. (2012), who concluded that the lower wind speeds experienced more than 85% reduction in wind speed while the reduction diminished to about 55% for higher wind speeds. Hence, there is a high potential of gaining significant benefits by increasing the forested area in the coastal zones by the means of bioengineering interventions such as creating green buffers with taller trees.

The water level data analysis has shown that the water level gauge stations located behind and around mangrove covers, or natural green buffers exhibit a consistently lower water level recorded for weak and moderately strong cyclones. This has been validated from both high tide and low tide water level data. Apart from the water level station SW222, located at Noakhali Khal, all other coastal water stations exhibit a highly tidal nature with large fluctuations in high tide and low tide water levels. Even when the landfall locations are directly around the water level station area, the water levels remain comparatively lower and stable. In these water level stations, the water level rises steadily and slowly during cyclonic events and during the ebb phase after the cyclones, the water drains out slowly as well, which has a moderating effect on the soil erosions in the area and the overall coastal erosion in the coastline – except for the severe cyclones. This reflects the findings of Mandal and Hosaka (2020), who found that only the largest cyclones affected a larger area of the mangrove forests through flooding or uprooting, and the protected region faced little effects below a breakpoint of 101.9 km/hr. On the other hand, in the exposed coasts, the water level sees a rapid spike in the rise and also a rapid drain out due to cyclonic events and the associated storm surges. Besides the direct threats of high-water levels, this also poses the risks of accelerated coastal and riverbank erosion, and rapid salinity intrusion. Dasgupta et al. (2019) also found similar results where they concluded that there was a significant reduction in flow velocity, and reduction in surge height protects the coastal embankments from erosion, breaching, and damages. Thus, the presence of an effective coastal green buffer demonstratively has positive effects in mitigating the cyclonic storm surge levels, especially for low to moderately strong cyclones.

The findings from the secondary data analysis have further been justified by the primary data analysis, which has brought forth the experiences and perceptions of the local community residing in the different coastal areas of Bangladesh. From this analysis, it has been found that the residents of areas with mangrove cover are aware of the beneficial effects of the green buffer in mitigating various natural disasters including cyclonic storm surges. This awareness can be attributed to the local knowledge and the experiences with prior cyclonic events. The importance of the local economic activities and creation of livelihood alternatives as highlighted by the interviews with the locals also underline the economic benefit of coastal forest buffers in addition to the disaster mitigation services. It was also highlighted and quantified by Akber et al. (2018), who found that villages not sheltered by mangroves faced twice as much economic losses – around $2050 per household.

Despite the land stress and scarcity in Bangladesh, the protective effects of existing natural and artificial green buffers – especially the mangroves are highly significant in reducing the damages induced by cyclonic storm surges. This is well recognised by the local community, and as such, the policymakers may focus on developing strategies incorporating soft and hard engineering measures to implement flexible, adaptive, and economically realistic sustainable solutions of coastal protection and management.

Moreover, the implementation of various complimentary bioengineering measures such as "Live Fencing" may yet enhance the performance of the exiting green buffers and provide a sustainable strategy for coastal zone management in the long term. This has the potential to further diversify the local economy if the green buffer is integrated effectively to create opportunities for alternate livelihoods. When used in conjunction with hard engineering measures such as embankments and sea walls, the soft engineering techniques can help in significantly reducing the required height and thickness of the protective hard engineering structures, thus creating a financially and environmentally optimal solution to the disaster management system in the coastal region of Bangladesh.

References

ADB. (2005). Coastal Greenbelt Project in the People's Republic of Bangladesh, Project Completion Report, Asian Development Bank.

Ahamed, S., RahmanM. M., &Faisal, M. A. (2012). Reducing Cyclone Impacts in the Coastal Areas of Bangladesh: A Case Study of Kalapara Upazila. *Journal of Bangladesh Institute of Planners*. 5:185–197.

Akber, M. A., Patwary, M. M., Islam, M. A., & Rahman, M. R. (2018). Storm Protection Service of the Sundarbans Mangrove Forest, Bangladesh. *Natural Hazards*. 94: 405–418. https://doi.org/10.1007/s11069-018-3395-8

Badola, R., & Hussain, S. A. (2005). Valuing Ecosystem Functions: An Empirical Study on the Storm Protection Function of Bhitarkanika Mangrove Ecosystem, India. *Environmental Conservation*. 32(1):85–92. https://doi.org/10.1017/S0376892905001967

Banan-Dallalian, M., Shokatian-Beiragh, M., Golshani, A., Mojtahedi, A., Lotfollahi-Yaghin, M. A., & Akib, S. (2021). Study of the Effect of an Environmentally Friendly Flood Risk Reduction Approach on the Oman Coastlines during the Gonu Tropical Cyclone (Case Study: The Coastline of Sur). *Eng*. 2(2):141–155. https://doi.org/10.3390/eng2020010

Baten, A., González, P. A., & Delgado, R. C. (2018). Natural Disasters and Management Systems of Bangladesh from 1972 to 2017: Special Focus on Flood. *OmniScience: A Multi-disciplinary Journal*. 8(3):35–47.

Cresswell, J. (2007). *Qualitative Inquiry and Research Design: Choosing among Five Approaches*. Thousand Oaks: Sage.

Dasgupta, S., Islam, M. S., Huq, M., Khan, Z. H., & Hasib, M. R. (2019). Quantifying the Protective Capacity of Mangroves from Storm Surges in Coastal Bangladesh. *PLoS ONE*. 14(3): e0214079. https://doi.org/10.1371/journal.pone.0214079

Grossmann, I., & Morgan, M.G. (2011). Tropical Cyclones, Climate Change, and Scientific Uncertainty: What Do We Know, What Does It Mean, and What Should Be Done? *Climatic Change*. 108: 543.https://doi.org/10.1007/s10584-011-0020-1

Hales, T. A., Pantelides, C. P., Sankholkar, P., & Reaveley, L. D. (2017). Analysis-Oriented Stress-Strain Model for Concrete Confined with Fiber-Reinforced Polymer Spirals. *ACI Structural Journal*. 114(5):1263–1272.

Kelle, U. (2008). Combining Qualitative and Quantitative Methods in Research Practice: Purposes and Advantages. *Qualitative Research in Psychology*. 3(4):293–311.

Kim, Y. (2019). New Opportunities for Sociological Research. *Journal of Asian Sociology*. 48(3):343–358. www.jstor.org/stable/26783836

Lehne, J., &Preston, F. (2018). *Making Concrete Change. Innovation in Low-carbon Cement and Concrete*. Cambridge: Chatham House, the Royal Institute of International Affairs.

Mandal, M. S. H., &Hosaka, T. (2020). Assessing Cyclone Disturbances (1988–2016) in the Sundarbans Mangrove Forests Using Landsat and Google Earth Engine. *Natural Hazards*. 102: 133–150. https://doi.org/10.1007/s11069-020-03914-z

McIvor, A. L., Spencer, T., Möller, I., & Spalding. M. (2012). Storm Surge Reduction by Mangroves. *Natural Coastal Protection Series: Report 2. Cambridge Coastal Research Unit Working Paper 41*. Published by The Nature Conservancy and Wetlands International. ISSN 2050-7941. www.naturalcoastalprotection.org/documents/storm-surge-reduction-by-mangroves

Mendelsohn, R., Emanuel, K., Chonabayashi, S., & Bakkensen, L. (2012). The Impact of Climate Change on Global Tropical Cyclone Damage. *Nature Climate Change*. 2: 205–209. https://doi.org/10.1038/nclimate1357

Mita, K. S. (2019). *Role of Mangrove Plantation in Improving Embankment Safety and Supporting Local Livelihoods*. M.Sc. Thesis, Institute of Water and Flood Management, Bangladesh University of Engineering and Technology.

Parvathy, K. G., &Bhaskaran, P. K. (2017). Wave Attenuation in Presence of Mangroves: A Sensitivity Study forVarying Bottom Slopes. *The International Journal of Ocean and Climate Systems*. 8(3):126–134. https://doi.org/10.1177%2F1759313117702919

Paul, B. K. (2009). Why Relatively Fewer People Died? The Case of Bangladesh's Cyclone Sidr. *Natural Hazards*. 50: 289.https://doi.org/10.1007/s11069-008-9340-5

Rahman, M. M., Haque, M. A., & Haque, M. M. (2002). Applicability of Bend Development Theory in Natural Alluvial Rivers. *International Journal of Sediment Research*. 17(3):211–219.

Rahman, M. M., Hossain, M. A., & Bhattacharya, D. A. K. (2014). An Analytical Study of Flood Management in Bangladesh. *IOSR Journal of Engineering (IOSR-JEN)*. 4(1): pp. 1–6. http://dx.doi.org/10.9790/3021-04170106

Sakib, M., Nahial, F., Haque, A., Rahman, M., & Jisan, M. A. (2015). Sundarban as a Buffer against Storm Surge Flooding. *World Journal of Engineering and Technology*. 3(3C):59–64. https://doi.org/10.4236/wjet.2015.33C009

Smith, P. J., Brown, S., & Dugar, S. (2017). Community-Based Early Warning Systems for Flood Risk Mitigation in Nepal. *Natural Hazards and Earth System Sciences*. 17(3):423–437. http://dx.doi.org/10.5194/nhess-2016-122

Uddin, M. M., & Hossain, M. K. (2013). Status and Protective Role of Mangrove Plantations: A Case Study of Mirsharai Coastal Forest, Bangladesh. *International Journal of Agricultural Science and Bioresource Engineering Research*. 2(2):47–59.

21

SUSTAINABLE LIVELIHOOD AND URBAN ENVIRONMENT

Learning from the Disaster-Affected Displaced Population in Khulna City, Bangladesh

Mosammat Rowshan Ara and Mahbuba Nasreen

Introduction

The livelihoods of the urban poor are defined in large part by the opportunities and constraints under which they are operating. As a result, it's crucial to comprehend the nature of long-term urban livelihoods. Livelihood refers to how people organize their means of subsistence, how they employ capabilities, assets, and activities to sniff around for opportunities, diversify by adding enterprises, and multiply activities and relationships for stability (Ellis, 1998, cited by Kazungu et al., 2014). According to Ellis (1998), enhanced livelihood outcomes include increased income, improved food security, improved shelter, education, and reduced vulnerability. A livelihood is sustainable if it can cope with and recover from stresses and shocks, as well as retain or improve its capabilities and assets in the present and future, without jeopardizing the natural resource–based livelihood activities (Chambers & Conway, 1991). About Scoones (1998), a livelihood comprises the skills, assets (such as material and social resources), and activities needed to make a living. Thus, sustainability is defined as the availability of natural and social resources to provide the necessary support for man's life, as well as the individual's ability to trade commodities (goods and services).

The urban poor are frequently congregated in great numbers in low-rent regions (Elliott, 2006). They often reside on polluted ground near industrial sites, where waste dumps are located and watercourses are contaminated, or on river plains prone to landslides and flooding. As a result, the urban poor are regularly exposed to a variety of environmental and health risks (Wratten, 1995). Khulna, located in the coastal zone and as an industrial city in Bangladesh, has been experiencing climate-induced vulnerability, rapid urbanization, and a tremendous rise in the population since the 1950s. This, in turn, means greater demand on the land, lack of information and knowledge, lack of access to renewable and affordable

DOI: 10.4324/9781003253495-27

energy, and use of unsustainable farming techniques. On the other hand, the south-western region of Bangladesh is severely affected by frequent climatic disasters like cyclones, tidal surges, coastal flooding, waterlogging, and salinity intrusion. People in this region, on the other hand, are subject to water-related natural hazards as well as the effects of climate change. Bangladesh is thought to be the third most susceptible country in terms of people at risk from rising sea levels (Pender, 2008). Sea level rise is becoming a greater threat to the coastal region while Khulna is one of the world's 15 most susceptible cities (Dasgupta et al., 2009). As a consequence, most of the respondents of the study area have migrated from surrounding areas namely Barishal, Satkhira, Faridpur, and Jessore. Climatic disasters adversely affect peoples' livelihood in Khulna and these disasters have increased the vulnerability of the local communities living in the area. Furthermore, the state-run Khulna Newsprint Mill, Daulatpur Jute Mill, People's Jute Mill, Khulna Textile Mill, privately owned Afil Jute Mill, and Dada Match Factory have remained closed, displacing nearly two lakh workers. The poor jobless workers are getting nothing but lip service from the government in the name of reopening these industries and are living in the marginal land of the city. Thus, the study was conducted in five areas of Khulna city, Bangladesh to explore the role of natural and social resources in livelihood including housing, income, food, health, and reducing the vulnerability of the disaster-affected urban population.

A healthy and sustainable living environment with basic services for all is required for sustained urban development. But it is evident that few studies have been conducted on the city environment that can positively contribute to improving the livelihood activities of the population in Khulna city. What is required therefore is to study the Khulna city environment and its role to ensure the livelihood of the urban population in the context of sustainable development.

The Sustainable Livelihood of Disaster-Affected People in Khulna City

The study recognized that the urban environment can positively contribute to ensuring the sustainable livelihood of Khulna city. Moreover, like many other developing countries, global warming is a challenging problem for Bangladesh. A sustainable urban environment can play a potential role in this regard. Efforts to reduce vulnerability in the South, especially for disaster-affected people in Khulna, have a far greater impact and are more sustainable if supported by its environment. The study has attempted to relate the theories with the urban environment and sustainable livelihood. To fulfill the objective of the study, socio-ecological approach, the basic needs approach to poverty measurement, and the capability approach have been followed.

Socio-ecological approach: To understand the role of natural resources in sustainable urban livelihood the socio-ecological approach has been followed in this study. The term "ecology" was coined by Ernest Haeckel in 1869 to describe the study of interactions between organisms as well as between the organisms and their surroundings (Bharatdwaj, 2006). Richard used the term human ecology as its first formal appearance in the field of sociology in the 1921 book *Introduction to the Science of Sociology* (Park & Burgess, 1921). The phrase "*survival of the fittest*" was coined by Herbert Spencer (Claeys, 2000), and it set the stage for the socio-ecological approach, which is more applicable to the link between urban environment and urban poor sustainability.

Basic needs approach: The World Bank and the Institute of Development Studies (IDS) collaborated on a paper entitled *Redistribution with Growth* in 1974 (Chenery et al., 1979). Building on this study, the International Labour Office (ILO) became one of the first proponents of the "basic needs" approach to development that established a formal link between development and previous poverty studies. Following the ILO (Riddell, 2004) the study defined basic needs. First, minimum requirements (for a family) for private consumption are adequate food, shelter, and clothing, as well as certain household equipment and furniture. Second, access to essential services provided by and for the community at large, such as safe drinking water, sanitation, public transport and health, educational and cultural facilities.

- **Capability approach:** The study highlighted entitlements and empowerment by adopting the capability approach, which was developed in part by Sen's writings (2003). The term "basic need" refers to a variety of approaches based on the idea that humans require a fundamental set of resources to exist (food, water, clothing, shelter, etc.). Sen distinguishes between basic wants and capacities (Sen, 2005), which are concerned with subjective well-being, such as excellent health, loving relationships, or the availability of means to a decent existence to which people have actual access.

Conceptual framework: Several agencies and donors are presently working on livelihood-based methods to use as the foundation for policy and practice development.

DFID, the UN Development Programme (UNDP), non-governmental organizations (NGOs) such as Oxfam and CARE, and research institutes such as the IDS are among them. The UNDP's sustainable livelihood approach, which has more in common with prior "basic needs" methods to poverty measurement and alleviation, was used in this study.

Sustainable livelihoods, one of UNDP's five corporate responsibilities, provides a conceptual and programming framework for long-term poverty reduction (Haan & Zoomers, 2005). Following these approaches, the conceptual framework has been constructed for this study (Figure 21.1).

The concept of livelihoods refers to the means, activities, entitlements, and assets that people use to make a living. Natural resources (i.e., land, water, common property resources, flora, and fauna) are defined as assets in this sense, but so are social (i.e., community, family, social connections, knowledge, and skills) and physical assets (i.e., roads, markets, health facilities, schools, and bridges). As a result, the

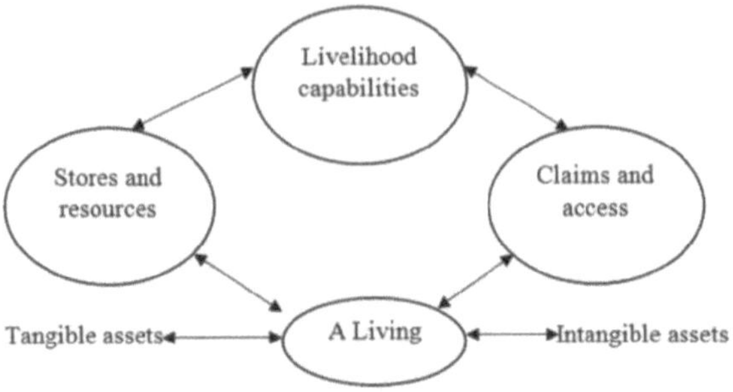

FIGURE 21.1 SL approach in promoting sustainable livelihoods
Source: Krantz (2001) (UNDP SL).

long-term viability of livelihoods is determined by how individuals of society use assets to suit their own needs without jeopardizing future generations.

Figure 21.2 indicates that a sustainable urban environment can be achieved. It also supports that the urban environment ensures different sources of household income and improved quality of life in urban livelihood. In this study, livelihood is described as the means of living involved with basic needs, available urban resources, and employment facilities with risk of vulnerability. The study includes the urban environment as natural, physical, and social resources and available employment opportunities. Thus, the sustainability of urban livelihood depends on the security of job, income, basic need fulfillment, and access to natural, financial, and social resources. The operational definitions of some concepts are stated below:

Livelihood: The dictionary definition of livelihood is "a means of subsistence; a source of income." The term "livelihood" refers to a person's ability to provide for his or her basic needs (English Oxford Living Dictionaries, 2014). It means that our livelihood is defined by what we eat, wear, and how we live in our homes as members of society. It indicates that we receive help from any organization, group, or other source that helps us meet our everyday needs.

Sustainability: The term "sustainability" refers to the long-term preservation or improvement of resource productivity. It includes environmental sustainability, economic sustainability, social sustainability, and institutional sustainability. When a baseline level of economic welfare can be reached and sustained, social exclusion is reduced, and social equality is maximized, sustainability is accomplished (Krantz, 2001).

Sustainable livelihoods: A livelihood is sustainable when it is able to cope with and recover from stresses and shocks, as well as manage to improve its capabilities and assets in the present and future without jeopardizing the natural resource base (Chambers & Conway, 1991). The study considers that a household is enabled to gain sustainable livelihood security through the city environment.

Sustainable urban livelihood: A sustainable livelihood approach must be sensitive to people and communities and appreciate the importance of social links

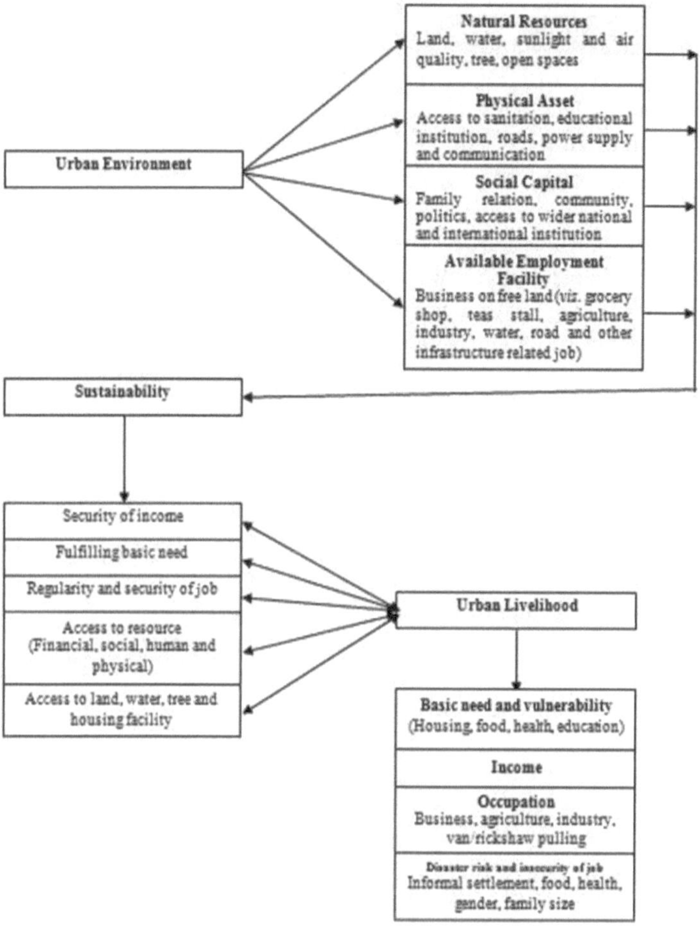

FIGURE 21.2 Relations between city environment and urban livelihood

Source: Compiled by the researcher based on literature review.

for livelihoods, focused on the need for equity and the participation of the people (Meikle et al., 2001). Here sustainable urban livelihoods model illustrates the relationship between key elements; natural, social, physical, and financial resources, means of livelihood, and disaster risk in the context of urban vulnerability. To measure sustainable livelihood, the study considers seven indicators: access to natural resources, access to financial resources, vulnerable situations, occupations, participation in disaster situations, access to social resources, and access to physical asset.

Specification of a typical family: Following Deptford et al. (2013) during the focus group discussion for this study, a typical home was identified to consist of four people: a man, a woman, and two children. According to key informants, a typical family in the city is the same size.

The Cost of the Diet technique identifies a family based on an energy requirement of 4 × 2,100 kcal, or 8,400 kcal overall. A nutritious diet for a typical family was defined as one that meets the estimated average requirement (EAR) for energy, the safe individual intake of protein, 30% of total energy intake from fat, the reference nutrient intake (RNI) of vitamins and minerals, and the safe intake of vitamin A, as defined by the World Health Organization (2008). Because the study's participants are vulnerable, a poor household that is unable to meet calorie requirements has been designated as a family.

City: A city is a permanent and generally significant settlement. Sanitation, utilities, land use, housing, and transportation are all complex systems in most cities (Goodall, 1987). In this context, the study includes five wards of Khulna city: (i) ward no. 15 (Khalishpur),(ii) ward no. 18 (Gollamari), (iii) ward no. 22 (Rupsha), (iv) ward no. 24 (Nirala), and (v) ward no 30 (Tutpara).

Vulnerability: The following is the direct link to assess the respondents' vulnerability, as determined by Moser's Vulnerability Matrix (1998): Members of a household have basic necessities such as food, water, housing, education, and so on. Household members employ resources or services such as land, water, food, shelter, healthcare, and power to meet these needs. Payment is required for majority of the access. Payment is guaranteed by engaging in productive activities using natural resources such as land, water, trees, and access to urban open areas.

The urban environment in disaster management: The study addresses the function of urban environment in disaster management, which is concerned with delivering services with community support; services are, in fact, the vehicle through which social assets are produced (Sanderson, 2000).

The hypothesis of this study was "Urban environment significantly contributes to the sustainable livelihood of the urban people." The independent variable of the study is the environment: natural, financial, and social resources and the dependent variable is sustainable livelihood whereas the intervening variables are land, water, tree, income, occupation, housing, health, technology, infrastructure, social capital, community participation, education, disaster, vulnerability, the security of job, income, and basic need fulfillment. Some studies also highlighted the issue of a sustainable urban environment.

Victor and Makalle (2003) conducted a study entitled "Poverty and Environment: Impact Analysis of Sustainable Dar es Salaam Project on Sustainable Livelihoods of Urban Poor." The purpose of the study was to identify the strengths and shortcomings in the institutional structure and related legislation that support initiatives to improve Dar es Salaam's environment. The study's focus is on the household, where socioeconomic changes occur and have an impact on the environment. According to the research, there are no longer any open places in the studied locations.

Nasreen et al. (2006) conducted a study on the "Interrelationship between Poverty, Environment and Sustainable Development in Bangladesh: An Overview." From the standpoint of environmental preservation and ecological balance, this study aims to focus on some of the concerns and problems associated to poverty and sustainable development in Bangladesh. Poverty reduction and environmental protection, it has been said, work in tandem to promote long-term development. The research intends to examine a development scenario aimed at eliminating poverty. Nonetheless, despite the best efforts of governments and non-governmental organizations, several development indicators in developing nations have already encountered challenges in meeting their goals and targets. Most development efforts, rather than contributing to sustainability, have a negative impact on the environment and society, according to this report. One of the primary measures to address both climate change adaptation and mitigation is the formation of the Climate Change Unit (CCU).

Ecosystems and resources that are healthy, well-managed, and diversified can help to mitigate future environmental concerns and improve lifestyles worldwide. As a result, it is critical to ensure that the future development plan takes into account the links between socioeconomic and environmental sustainability, as well as protects and strengthens the environmental pillar. Because the Sustainable Development Goals (SDGs) targets, and indicators were well-considered during the creation of Bangladesh's 7th FYP, the country can be considered an "early starter" in the SDGs' implementation process.

Hossain et al. (2010) studied "Urban Environmental Health in Bangladesh Slum: A Comparative Study of Two Metropolitan Cities." The study's goal is to give a comparison of the social, physical, and urban environmental health conditions in two Bangladeshi metropolises. Three slums in Khulna and five in Rajshahi were chosen for this study, with an emphasis on socioeconomic, physical, and environmental services, as well as urban health. The slums' environmental status has been established based on the state of sanitation, solid waste management, and drainage, as well as socioeconomic, political, and built environment issues. The study investigates how a community's environmental health is influenced by a variety of factors, including not only socioeconomic and physical condition, but also environmental services accessible in the area. The slums' environmental status has been established based on sanitation and solid waste management.

In "Urban agriculture for sustainable poverty reduction and food security," Hoornweg and Munro-Faure (2008) present an updated perspective on the importance of urban agriculture in achieving the Millennium Development Goals (MDGs), particularly MDGs 1 and 7 (poverty reduction, food security, and environmental sustainability). The study also demonstrates how ongoing programs are demonstrating how UPA can serve as a source of employment, money, and food in a variety of settings, all of which are indicators of poverty reduction and food security.

The world has the potential to agree on a new global sustainable development framework between September 25 and 27, 2015, during the 7th United Nations General Assembly (UNGA) in New York, according to General Economics Division (2015). The SDGs adopted during the 70th United Nations General Assembly (UNGA) replace the MDGs, which were established in 2000. Bangladesh has made commendable progress in achieving the majority of the MDGs when the study was being conducted.

Gender parity in elementary and secondary school enrollment has already been reached. The government is now executing a comprehensive National Education Policy (2010) in order to fulfill its broad goals. Meikle et al. (2001) studied "Sustainable Urban Livelihood: Concepts and Implications for Policy." According to the study, a sustainable livelihoods approach (SLA) takes a unique method of analyzing poverty and how to intervene to improve the poor's situation. Poverty is defined as a state of insecurity, not just a lack of resources, according to an SLA to poverty elimination.

In the book entitled *Women and Girls: Vulnerable or Resilient?*, it is examined whether women and girls are vulnerable or resilient. Nasreen (2012) focuses on rural women and girls' coping techniques and responses to a common calamity, flooding. It explained the main elements that shaped women's and girls' responses to the flood. A sociological perspective is used to address these difficulties, but not at the expense of other techniques. According to the report, floods in Bangladesh have been occurring for a long time, but the problem has become much worse over time. Because women have a broader range of duties than men, these issues affect them more severely. The main goal of the Environmental and Social Impact Assessment (ESIA) study was to identify, predict, and analyze the magnitude of environmental and social impacts, as well as to propose enhancement or mitigation measures for significant environmental and social effects that are likely to occur during the pre-construction, construction, and operation phases of the fruit and vegetables farm project.

"Selling to eat: Small trade and traders in peri-urban areas of sub-Saharan Africa" was a project Little (1999) worked on. According to the report, commerce is the most important non-farm activity and occupation in most African peri-urban economies. The extent and complexity of the unauthorized commerce sectors grew in tandem with the growth of metropolitan areas and the collapse of formal sector jobs. According to the findings, urban and peri-urban agriculture has transformed from a simple, traditional, and also informal activity into a commercial and professional endeavor as a result of urbanization.

The Food and Agricultural Organization (FAO) held an international meeting on urban and peri-urban forestry (UPF) in Bogotá, Colombia, in 2009, themed "Trees Connecting People: In Action Together" (hence referred to as the "FAO UPF Bogotá Meeting").

The main goal of this international conference on urban and peri-urban forestry was to gain a better understanding of the overall institutional, policy, and networking framework, develop strategic advice to raise the profile of forests and trees on national, regional, and global urban agendas, and identify strategic opportunities to implement an adaptable and efficient Urban and Peri-urban Forestry Program. Crabtree (2008) studied "the role of tenure, work and cooperativism in sustainable urban livelihoods." The purpose of this study is to address the difficulties of urban sustainability by looking at sustainable lives in the Western, industrialized environment of Australian cities.

The above studies have (Victor & Makalle, 2003; Meikle et al., 2001) highlighted the specific areas: the strengths and weaknesses in the institutional structure and poverty as the underlying causes of insecurity in urban life. On the other hand, some studies (Little 1999; Crabtree, 2008) have focused on the role of tenure, trading, and corporatism in sustainable urban livelihoods. The studies present the problems and prospects of different strategies of livelihood sources in the context of specific areas. The application of sustainable livelihood principles to urban settings is a job that has largely been overlooked in the literature. Efforts to reduce vulnerability in the South, especially in Khulna, Bangladesh have a far greater impact and be more sustainable if supported by its environment. In this regard, the study on the urban environment and sustainable livelihood would assist to understand the available natural resources in Khulna city and their contribution to the sustainability of the livelihood of its population.

Methodology of the Study

In this study, as identified by Shih (1998), triangulation has been used. Following Denzin (2006) and Patton (1999), the study used four types of triangulation: (a) method triangulation (the study has been conducted mainly through survey, IDIs, observation, and focus group discussion), (b) investigator triangulation (involves the participation of numerous researchers in a single investigation), (c) theory triangulation (three theoretical perspectives viz. basic needs approach, capability approach, and socio-ecological approach have been followed in the interpretation of the phenomenon of this study), and (d) data source triangulation (involves time, space, and persons).

The respondents were selected from the household heads residing in the study area for at least ten years. The respondents were selected using simple random sampling. To determine the exact sample after conducting a census, a household survey was conducted on respondents. Considering the required criterion for quantitative analysis, a total of 1414 household heads were selected as the study population. Moreover, five focus group discussions were conducted (each group contains eight members) and five IDIs have been studied. Data were collected from February 2015 to October 2016 from the selected study population.

The study has been carried out in urban areas of Khulna city in Bangladesh and has been conducted on five selected locations: (i) ward no. 15 (Khalishpur), (ii)ward no. 18 (Gollamari), (iii)ward no. 22 (Rupsha), (iv)ward no. 24 (Nirala), and (v)ward no. 30 (Tutpara) of Khulna city of Bangladesh.

Among the total, 752 heads of the households were selected as the sample of the study (male, 596; female, 156) who are residing at that locality for at least ten years and household income is not more than 10,000 BDT. To know the sustainability of the disaster-affected vulnerable people, the respondents were selected as having no permanent job and not skilled at work.

Data have been collected from two sources—primary source: field survey and secondary source: publications related to the study, official documents, research documents, etc. To analyze the access to and control over resources regarding sustainable livelihood, the study of Ara (2017), Maxwell (1995), Parveen and Leonhaeuser (2004) have been used.

To identify the disaster-affected poor vulnerable people, the estimates of household income for the typical family is based on family income, not more than $2 per day (World Bank, 2013). During the field survey, the conversion rate of US$1 was equivalent to almost 77.63 BDT (Bangladeshi currency). Collected data were processed by the SPSS 16.0 program and analyzed and interpreted by different statistical techniques like regression analysis, t-test, F- test, chi-square test, Pearson's "R" and Fisher's exact test.

Sustainable livelihood index (SLI): By constructing a cumulative food security index, Maxwell (1995) followed this process to measure the amount of food insecurity. Parveen and Leonhäuser (2004) used a similar approach to build a cumulative empowerment index to quantify women's empowerment.

SLI is calculated in this study using Maxwell's cumulative food security index (1995) and Parveen and Leonhäuser's cumulative empowerment index (2004). In this regard for constructing the SLI, seven indicators are used: (i) access to natural resources, (ii) security of financial resources, (iii) levels of vulnerability, (iv) occupation, (v) disaster situation, (vi) access to social resources, and (vii) access to the physical asset. Following the method of Wyatt and Meyers (1987), a Likert scale (1 for least important, 5 for most important) was used to weigh the indicators. Each indicator has a number of sub-indicators.

The quantitative section is divided into five categories, with 1 denoting extremely low and 5 denoting very high. According to the total score for access to and control over resources based on the field survey, each indicator was given a quantitative rating from 1 to 5. The qualitative dimension is used to rank the seven major indicators based on total responses and total responses from participants, with 7 representing "extremely significant" and 1 representing "less important."

The cumulative vulnerability index (CVI), cumulative empowerment index (CEI), and cumulative resource index (CRI) were used to create the SLI.

The CRI was developed to account for four different categories of resources: natural, financial, social, and physical (NRI, FRI, SRI, and PRI).

The method for constructing the SLI is shown in Table 21.1: Measurement of explained (dependent) variables. The SLI ranges from 28 to 140, with 28 indicating the lowest level of sustainability and 140 indicating the maximum level.

N.B* Qualitative ranking has been made on the basis of total scores attained from a ranking of the respondents such as 7=4942 score, 6=4247 score, 5=3537 score, 4=2820 score, 3= 2128 score, 2=1412 score, 1=710 score. For better understanding, the level of sustainability attained by the respondents the SLI range is further divided into five categories and labeled as not at all (28–50), very low sustainability (51–72), low sustainability (73–94), medium sustainability (95–117), and sustainable enough (118–140).

TABLE 21.1 Sustainable livelihood index

Indicators	*Quantitative rank*	*Qualitative rank**	*SLI range 28–140*
Natural resources	1–5	7	(1–5).7=7–35
Financial resources	1–5	6	(1–5).6=6–30
Situation of vulnerability	1–5	5	(1–5).5=5–25
Occupation	1–5	4	(1–5).4=4–20
Disaster situation	1–5	3	(1–5).3=3–15
Access to social resource	1–5	2	(1–5).2=2–10
Physical asset	1–5	1	(1–5).1=1–5

Source: Field survey, 2014–2015.

Indicators

Seven indicators have been used to measure the access to a resource which is discussed below:

> **Natural resource:** It is measured in terms of five resources viz. (i) land, (ii) water, (iii) air and sunlight, (iv) tree, and (v) open spaces (railway, riversides, garbage, and other free resources)
>
> **Financial asset:** It is measured in terms of money considering:(i) family income, (ii) no. of income earner, (iii) yearly family savings, (iv) regularity of wage-earning, and (v) access to an interest-free loan
>
> **Level of Vulnerability:** Seven indicators related to maintenance of basic needs are used to measure the level of vulnerability is (i) tenure security, (ii) income security, (iii) food security, (iv) workless situation, (v) literacy situation, (vi) access to health, and (vii) access to household goods
>
> **Occupation**: Five occupations have been categorized in terms of nature (i) business (fisheries, grocery shop, kitchen market), (ii) agriculture, (iii) industrial work (shrimp, wood, jute, ice), (iv) day labor (land and water-related), and (v) other services especially van or rickshaw pulling
>
> **The situation of disaster**: Reduction of disaster risk has been measured by access to and control over resources of seven basic needs during disaster viz. (i) access to housing, (ii) income security, (iii) food security, (iv) infrastructural facilities, (v) broader organizational support, (vi) knowledge about disaster, and (vii) emergency responses to disaster.
>
> **Social resource:** Social resource is assessed considering five types of social relations: (i) family relation, (i) relationship with outsider, (iii) involvement with politics, (iv) access to the wider institution, and (v) participation in a community-based organization
>
> **Physical asset:** It is measured by five types of infrastructural facilities in the study area: (i) sanitation, (ii) access to educational institution, (iii) communication facilities, (iv) access to the power supply, and (v) recreational facility

Regression model: The multiple regression model of this study is:

$$SLI = \beta_0 + \beta_1 Nat + \beta_2 Inc + \beta_3 Pov + \beta_4 Occ + \beta_5 Dis + \beta_6 SocRel + \beta_7 Inf + u$$

where SLI = Sustainable livelihood index
Nat = Access to natural resources of the respondents
Inc = Access to income of the respondents
Pov = Poverty situation of the respondents
Occ = Occupational status of the respondents
Dis = Access to and control over resources of the respondents during the disaster
SocRel = State of the social relation of the respondents
Inf = Access to infrastructural facilities of the respondents
β_0 = Intercept term
$\beta_1, \beta_2, \beta_3, \beta_4, \beta_5, \beta_6, \beta_7$ = Regression co-efficient
u = Stochastic disturbance term

Background of the Study Locations and Population

Globally, more people live in urban areas than in rural areas, with 54% of the world's population residing in urban areas in 2014 (United Nations, 2014). The urban population is increasing mainly due to rural–urban migration. Most of the migrants are rural poor who take shelter in slums, squatters, footpaths, railway stations, and other scattered places (Farhana et al., 2012). In Bangladesh, people of the coastal region are vulnerable to water-related natural hazards as well as climate change impacts. Khulna, the third-largest metropolitan city of Bangladesh, stands on the banks of the Rupsha and the Bhairab Rivers, located in the southwest of the country. Khulna is also an important river port city of Bangladesh. During the late 1950s and 1960s, Khulna became an important center for industrial development (Murtaza, 2001). All these developments have a far-reaching impact on the overall environment of Khulna city.

Major Challenges in the Study Area

Cities and climate change: The southern coastal cities of Bangladesh are frequently hit by monsoon rains and cyclones. Cyclones, tidal waves, floods, and other natural calamities frequently strike Bangladesh's coastal districts. In addition, man-made calamities such as arsenic, waterlogging, and salinity in water and agricultural land have occurred.

Disaster in the study area: Respondents in the study area are staying in hazardous areas such as riverbanks, railroad lines, and toxic waste dumps. Also, similar to other urban areas, the respondents noted fire accidents as the frequently occurring and serious disaster in the locality. Eviction is also responsible for the disruption of the livelihood of the respondents. Another disaster that causes vulnerability of the

TABLE 21.2 Frequency of disaster in the study area

Thoughtful disasters in the study area	*Percentage distribution of the respondents regarding the prevalence of disaster*				
Waterlogging	Waterlogged during the rainy season (59.0)	Waterlogged due to river flow (17.4)	Always flooded (17.1)	Never flooded (6.5)	-
Eviction	Threatened at present (32.2)	Threatened more than three times (19.2)	Threatened more than two times (28.7)	Threatened once (10.7)	No threat (9.1)
Fire accident	Never (35.1)	Once (14.9)	three times (20.7)	Four times (23.2)	Five times and more (6.2)
Coastal flooding	Never (46.2)	Once/twice (17.4)	Three times (6.5)	Four times (9.2)	Five times and more (20.7)

Source: Field survey, 2015–2016.

respondents is waterlogging. Among the respondents, females are affected more by disaster than males. The frequency of these disasters is presented in Table 21.2.

Among the respondents, 59 face waterlogging problems during the rainy season, whereas 20.7% informed that fire accidents occurred in their locality about three times and in 23.2% cases it was four times and 6.2% respondents talked about five or more times attacked by a fire accident. In the study area, especially in the slum areas, the lanes are so narrow that it is not possible for fire service authorities to go through their equipment to extricate the fire and this is why the fire disrupts the locality and create a disaster situation. About 90% of the respondents are under threat of eviction.

Disaster situation: Disasters are a complex mix of natural hazards and human action. All over the world, but especially in least developed countries (LDCs), vulnerable people often suffer repeated, multiple, mutually reinforcing, and sometimes simultaneous shocks to their families, their settlements, and their livelihoods. Disasters are the brake on economic and human development at the household level (when livestock, crops, homes, and tools are repeatedly destroyed) and at the national level when roads, bridges, hospitals, schools, and other facilities are damaged (Winser et al., 2003). The access to and control over resources during disaster, therefore, could reduce vulnerability and the impact of different hazards on them. Access to such resources is based on social and economic relations, including the social relations of production, gender, ethnicity, status, and age, meaning that rights and obligations are not distributed equally among all people. Especially women and girls in the disaster-prone areas face a number of problems due to their gender

TABLE 21.3 Nature and extent of disaster situation faced by the respondents

Situation of disaster	*Number*	*Percent*
Very highly disaster prone (27–36)	225	31.82
Highly disaster prone (37–46)	317	44.84
Moderately disaster prone (47–56)	96	13.58
Low disaster prone (57–66)	43	6.08
Very low disaster prone (67 and above)	26	3.68
Total (mean41.77)	707	100.00

Source: Field survey, 2015–2016.

identity. As a result, women and girls in poor and marginal households become more vulnerable and distressed (Nasreen, 2012).

In this study, the level of access to and control over resources during disaster has been attained by the respondents. The CEI range is further divided into five categories and labeled as very highly disaster prone (27–36), highly disaster prone (37–46), moderately disaster prone (47–56), low disaster prone (57–66), and very low disasterprone (67 and above). On the basis of this value, the situation of the disaster of the respondents has been assessed. Table 21.3 presents that about 76% of respondents were very highly and highly disasterprone, which means they had low access to the basic needs of food, shelter, education, and medical facilities. Whereas nearly 10% of the respondents were low and very low disasterprone. In this regard, natural resource has an influential role to reduce vulnerability in mitigating disaster situations.

Gender Difference in Access to and Control over Resources during Disaster

In practice, not all assets are owned by, or fully in the control of, men and women who are attempting to use them in their livelihood's strategies, but in practice institutional and practical barriers may limit the access to the poor to the benefits of such programs. It is therefore important to assess and distinguish between access to and control over assets.

Sen's concept of "entitlement" can clarify the extent of peoples' access to and control over resources (Sen, 1981). "Entitlement" is the ability to command access to different forms of capital assets usingas financial resources, formal and informal relationships with other groups and individuals, or legal rights. To identify the access and control over resources, collected data have been used to demonstrate this lack of secure access to resources, including land, income, food, infrastructural facility, different organizational support, information, and participation in emergency situations.

Men and women may have access to resources, either within or outside the household. They may not, however, always have control over these resources (United

TABLE 21.4 Gender difference in emergency response to a disaster situation (multi-response)

Emergency responses	*Male*	*Female*
	No. and percent	*No. and percent*
To save life	187 (28.90)	47(32.63)
To reduce health problems in family	155(23.96)	65(45.14)
Contribute to the household economy	224(34.62)	24(16.67)
To ensure public safety	34(5.26)	03(2.08)
To meet the subsistence needs of the people	47(7.26)	5(3.47)
Total (707:male 564 female 143)	647(100.00)	144(100.00)

Source: Field survey, 2015–2016.

Nations Human Settlements Programme, 2011). Women's and men's different roles, responsibilities, and access to resources influence how each will be affected by different hazards, and how they will cope with and recover from disaster (Ara & Nasreen, 2017). During the disaster, women face a number of problems due to their gender identity (Nasreen, 2008). It has been also observed that sometimes women face an array of physical, psychological, and social problems (Faisal & Kabir, 2005; Ahmed, 2010). Salinity, waterlogging, and frequently occurring natural disasters due to climate change have been making this scenario more difficult and complex. In a recent study Nasreen (2012) shows that climate change–induced disasters affect both women and men but the burden of coping with disasters falls heavily on women (Table 21.4).

In the southwest coast including the Khulna area, women are more vulnerable to the impacts of climate change since they are often not allowed to participate in public activities and discussions, and therefore are less likely to receive critical information for emergency preparedness (United Nations Development Programme, 2011). Sustainable cities support the equal participation of women and men, especially in urban planning and management, as well as governance. The study shows that in the study area women are participating in fulfilling basic needs, especially in their family economy (17%) and health (45%), but there is still persistent underrepresentation by saving their life and contributing to public safety at the time of emergency in disaster management.

Level of Vulnerability

Urban poor pursue their livelihood's objectives within political, social, economic, and environmental contexts that make them vulnerable to poverty. Often, vulnerability manifests inform of trends, shocks, and seasonality (Department for International Development, 1999). As revealed in this study the major trends that have an influence on urban poor livelihoods include tenure security, income, food security, and disaster. These threats, however, reinforce one another. Considering seven indicators related to maintenance of basic needs the CVI has been used to measure the level of vulnerability which are (i) tenure security, (ii) income security, (iii) food security, (iv) jobless situation, (v) literacy situation, (vi) access to health, and (vii) access to household goods. Therefore, based on CVI, the situation of vulnerability of the respondents has been assessed (Table 21.5).Table 21.5 shows that 32.5% of respondents were highly vulnerable, but 20.2% of respondents were in a low vulnerable situation. The mode (61) indicates that the majority of the respondents were highly vulnerable.

Household mitigation and coping strategies: Taking out a loan, followed by reducing household food expenditure at different times and in response to various types of shocks (such as a medical emergency) or, in the case of the financial strategies, to afford expenses such as weddings or ceremonies are the major coping strategies followed by the respondents (Figure 21.3).

Households that were fully affected by disaster (crops, land, house, or grain stores) were significantly more likely to use the above coping strategies. In most cases, taking out a loan was the most common choice, nearly half of the households fully affected, reporting taking out a loan. The only exception was if grain stores were affected, then 60 of households said that they also had to reduce household consumption, whereas the grains were a source of food. Often these strategies were taken in conjunction with each other, such as taking a child out of school and sending them to live or work elsewhere, reducing household food, and searching for an additional job.

TABLE 21.5 Percentage distribution of the respondents by the level of vulnerability

Vulnerability	*Percentage distribution of the respondents by the level of vulnerability*				
Mean: 72.96, mode: 61.00	Very high vulnerability (30–46) 5.9	High vulnerability (47–63) 32.5	Medium vulnerability (64–80) 37.1	Low vulnerability (81–97) 20.2	Very low vulnerability (98–114) 4.2

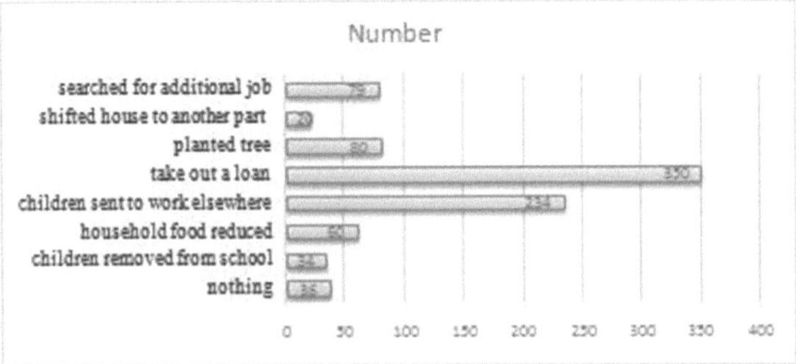

FIGURE 21.3 Household coping strategies (multi-response)

Source: Field survey, 2015–2016.

Households who planted vegetables (like papaya, brinjal, tomato or Solanum lycopersicum, gourd) were significantly more likely to have their land only partially affected compared to full effect. However, the study found evidence that community diversity (in wealth, ethnicity, and power) hinders participation in community-level mitigation strategies. Lack of interest from the non-affected households meant that no community-wide action has been taken.

Means of Livelihood

In the study area the respondents are mostly employed in self-managed low-paid jobs in the informal urban sectors like rickshaw pulling, day labor, factory work, agricultural activities, and domestic help. A small portion works in government and non-government organizations. The respondents have also faced underemployment frequently due to lack of employment opportunities, physical illness, or staying in their ancestral villages. Rickshaw pulling was the most popular because it requires no skills and is easily available (Table 21.6).

Among households whose members have been involved in agricultural activities, small-scale businessmen and day laborers have a significantly higher income than rickshaw pullers, industrial workers, and service holders. The value of chi-square tests (102.413) and Fisher's exact test (0.000) also supports that. Moreover, agricultural activities like homestead garden livestock and poultry rearing also have an important role in increasing income. The wood and shrimp-related industry also contribute to increasing income. For the study purpose, the occupation was further categorized into five (rickshaw pulling/not defined [100], industrial work [165], day labor/construction work [123], agriculture [61], and business [155/22%]). Most women in the Rupsha slum region work in the shrimp industry as a traditional source of income, and women have developed skills and competencies

TABLE 21.6 Occupation and income of the respondents

Occupation	*Income (in BDT) Mean: 4727, SD: 2271*		*No. of respondents*
	Mean	*Standard deviation*	
Van or rickshaw pulling	4964.00	1573.166	100
Labor in an industry	4930.61	1703.078	165
Small-scale business	5692.90	2137.000	155
Domestic help	2206.45	1271.989	31
Agricultural activities	5721.31	2532.332	61
Day labor	4975.20	1797.965	123
Service	4714.29	2884.689	7
Unemployed	1294.55	1731.347	55
Beggar	1600.00	1125.463	10
Total			707

Source: Field survey, 2015–2016, value of chi-square:102.413.

in the industry, particularly in shrimp/related occupations, through decades. The risk factor in the shrimp industry to the respondents is seasonal but nowadays very few are involved in almost all of the year/round in packing with cutting. Small-scale company activity in cities boosts the economy and, as a result, spreads socioeconomic benefits to a larger part of the population who were previously unemployed. Small-scale enterprises dealing with kitchen markets, grocery stores, cattle, and poultry are expected to earn income as a result of the findings of this study. Cultivators were able to meet fundamental human requirements such as food, shelter, and clothes through vegetable farming.

Urban agriculture has been characterized in a variety of ways that discriminate between different forms of activity and varying levels of urbanization (especially peri-urban and urban activity). In the case of Bangladesh, Khulna has described urban agriculture as a "micro-level or people" initiative to deal with the economic crisis as governments struggle to implement their structural adjustment programs.

The study looks at agricultural and gardening activities in five different locations of Khulna, including food production, livestock husbandry, and flower/ornamental gardens. This is supported by the FGD that respondents were very much convinced to provide land for livestock rearing and the development of shops. One woman named Aysha (63) in the Khalishpur slum area argued that urban agriculture contributes to establishing social networks, empowering women, and promoting community development, which symbolizes peoples' permanent roots in the city and their sense of stability. From IDI it is presented that a 40-year-old woman who lives in Khalishpur railway slum reported about her livelihood condition. In 2013, an NGO provided her with a goat. This goat is now grown and has given birth to six. By selling the milk of goats, the women earn money for their survival. She viewed that they were forced to move out from home many times, which was a great hindrance for their survival.

Livelihood Resources Portfolios of the Urban Disaster-Affected People

Livelihood resources of the selected areas and their residents are discussed under natural, financial, physical, and social assets as contained in SLI.

Natural Resources

In livelihood analysis, the importance and contributions of each component to natural capital differ between rural and urban environments. It is generally believed that natural capital is less important to the livelihoods of the urban poor. However, as noted by Farrington et al. (2002), natural capital and services are becoming important to urban poor's livelihood. Sustainable cities depend on a healthy ecosystem that influences both human well-being and numerous economic activities. Ellis (1999) defines natural capital as the land, water, and biological resources that are utilized by people to generate a means of survival. Both Ellis (1999) and Carney

TABLE 21.7 Index of natural resources

Indicators	*Quantitative rank*	*Qualitative rank*	*NRI range 15–53*
Land	1–5	5	(1–5).5=5–25
Water	1–5	4	(1–5).4=4–20
Sunlight and air quality	1–5	3	(1–5).3=3–15
Tree	1–5	2	(1–5).2=2–10
Open spaces	1–5	1	(1–5).1=1–5

(1998) state that natural capital is sometimes referred to as environmental resources, with the components, thought of as jointly comprising "the environment." Both emphasize that natural capital is not static and that it can comprise both renewable and non-renewable resources. In this context, the study reflects the relation between urban natural resources (land, water, tree, atmosphere, and other open spaces) and city sustainability is addressed through the investigation of the value of urban nature as a provider of social services essential to the quality of human life, which in turn, is a key component of sustainable development. In this study, the natural resources include five assets: land, water, sunlight, air quality, trees, and open spaces. Table 21.7represents the way of constructing the NRI.

The NRI varies from 15 to 53 in which15 denotes the lowest level of sustainability and 53 denotes the highest level of sustainability.

The level of natural resources accomplished by the respondents, the NRI range is further divided into five categories and labeled as very low (15–22), low (23–30), medium (31–38), high (39–46), and very high (47–53). Following this procedure, the index has been constructed for other resources (financial, social, and physical) and also for understanding vulnerability and access to resources during a disaster. In the study area, respondents are expected to ensure the security of food, housing, and income. Urban open spaces are defined in the study as land (private, public, or communal) that is used for conservation (e.g., reserves), recreation (e.g., parks and community gardens), or wasteland (e.g., dumps, verges, "brown field" sites, and so on). Table 21.8 shows that 8.2% of respondents have very low access to natural resources and 39.2% have medium access, the value range is 31–38. The mean and mode value are 32.62 and 28.00, respectively, which indicates majority of the respondents were in this group. On the other hand, 14.9% were in high access to land, water, trees, and other marginal resources.

Thus, the study suggests that the development of urban public park is an important mechanism for regenerating conserving, beautifying, and managing the environment sustainably from the effect of climate change. It also encourages healthy recreational pursuits and communication among the residents. It helps to retain green open spaces in urban areas adding value to properties thereby increasing their market value and enhancing the economic revival of cities and creating positive publicity for the business. Trees provide vital materials with which shelters can be constructed. Trees outside forests improve air quality and microclimate; provide

TABLE 21.8 Access to resources of the respondents

Resource types	*Percentage distribution of the respondents by level of access to resources*				
Natural resource (mean: 32.62, mode: 28.00)	15–22 Very low (8.2)	23–30 Low (32.2)	31–38 Medium (39.2)	39–46 High (14.9)	47–53 Very high (5.5)
Financial resources (mean: 33.07, mode: 31.00)	Very low (9–19) 7.2	Low (20–30) 33.5	Medium (31–41) 40.9	High (42–52) 15.1	Very high (53–63) 4.1
Physical asset (mean: 33.19, mode: 31.00, Max:54)	Very low (15–21) 4.0	Low (22–28) 20.2	Moderate (29–35) 43.3	Good (36–42) 27.9	Very good (43 and above) 4.7
Social capital (mean: 26.10, mode: 15.00, Max: 56)	Very low (15–21) 30.6	Low (22–28) 39.9	Moderate (29–35) 16.4	Good (36–42) 9.6	Very good (43 and above) 3.5

Source: Author's compilation on the basis of field survey, 2015–2016.

water protection, wildlife habitat, recreation, and increased carbon sequestration and biodiversity. They are also imbued with the symbolic cultural value being important parts of language, history, art, religion, medicine, politics, and so forth.

Rain trees retain their role as a community meeting place for discussion, consultation, and education. The coconut trees provide nuts, fruits, and a range of ingredients for medicines.

Tree as the Source of Livelihood

A kabiraj named Siddik Talukder (64) in the Rupsha slum area was asked for his medicine for the disease he provides for treatment. The name of the disease he treats is jaundice. The members of his family are eight, with six children. Kabiraji is his main occupation. Every day a good number of patients receives his treatment. The ingredient used for the treatment is the water of green coconut. The nut of coconut is a good source of food, especially for their children. It is also the source of fuel for his household. But nowadays he is worried about the sustainability of their livelihood. He argued that "now it is so tough to collect green coconut because of lack of its availability." He was asked about the cause of it and he answered that "the mobile network, lack of awareness, lack of space for tree plantation and waterlogging are the responsible factors. The livelihood of such a profession is being hampered due to this situation."

From IDI it is recognized that "urban forestry programs can help to strengthen urban community building. But community-based efforts are hampered for the lack of well-organized groups and lack of participation. There is also almost an absence of participation of women and children in the decision-making, planning, and management of urban trees. In this regard, from the choice of tree species to actual planting, tending, and (where appropriate) harvesting of tree products, urban dwellers should be able, and actively encouraged, to participate in decision-making and implementation as far as possible." From a woman, (Bilkis, 50-year-old) it is informed that, "in Tutpara if people do not have secure tenure, they are less likely to invest time and resources in planting trees that may only start to provide material benefits after a number of years." To them land ownership is unclear, but people actually plant trees to claim ownership of land themselves. Water is at the core of sustainable development and is critical for socioeconomic development, healthy ecosystems, and for human survival itself. It is vital for reducing the global burden of disease and improving the health, welfare, and productivity of populations (Veenhuizen, 2006). Water is a finite and irreplaceable resource that is fundamental to human well-being. It is only renewable if well managed. Today, more than 1.7 billion people live in river basins where depletion through use exceeds natural recharge, a trend that will see two-thirds of the world's population living in water-stressed countries by 2025 (United Nations Department of Economic and Social Affairs, 2017).

In this study, the term urban water resources is used to refer to freshwater used for productive and reproductive uses (including sanitation). As an industrial area, besides household level, the water is used in industrial and agricultural uses. The main sources of water in Khulna city are shallow and deep tubewell, open wells, rivers, lakes, ponds, and boreholes. Roof water harvesting is also of growing importance in household-level uses in Khulna city. Especially in Rupsha slum areas, the shrimp depots are developed for standing by the side of the river. Water is very essential for shrimp processing activities ready for its export. In that sense, water has an enormous contribution to earning foreign currency. Though, as noted by various respondents, fishing is not done at commercial quantity but is part of residents' livelihood strategies for meeting food needs. The water sources occupy a central position in the existence of sawmills, ice factories, and shrimp industries and the survival of its residents. Moreover, the soil collected from the riverbank is used for making *chula* made of clay. Many people choose this system as their profession. Waste water reuse is defined as using wastewater or reclaimed water from one application for another application (United States Environmental Protection Agency, 2004). In this context, the wastewater treatment of urban hot spots like industrial enterprises and hospitals combined with water reuse systems might be an interesting option to improve environmental protection and tap still unexplored water sources. In the Rupsha slum area, there is a government-regulated water purification system. Almost all the wastewater has been preserved in this place and after purification, the usable water is supplied for the purpose of different domestic and agricultural activities.

Financial Resources

In urban areas, possession of financial assets is important in accessing and accumulating other livelihood assets. In this study, financial capital is measured as a household's monthly income, savings and ability to meet basic needs, and also access to credit and sources of credit. Household income is essential for maintaining a sustainable livelihood. Living in an urban area is characterized by a high level of commoditization, which means urban dwellers have to pay for most of the goods and services consumed (Moser, 1998; Rakodi & Lloyd-Jones, 2002).

The FRI has been constructed to assess the access to financial resources of the respondents. The FRI varies from 9 to 63 in which 9 denotesthe very low access to financial resources and 63 denotes the higher to financial resources. For a better understandingof the level of financial resources attained by the respondents, the FRI range is further divided into five categories and labeled as very low (9–19), low (20–30), medium (31–41), high (42–52), and very high (53–63). It is observed that 33.5% of respondents had low access to financial resources and 40.9% had the medium access, the range is 31–41. The mean value (33.07) means that on average, the respondents have medium access to financial resources and 15.1% of respondents possessed high status with good access to credit facilities.

Social Resources

Social capital could come with inform of social networks, relationships, reciprocity, and kindred (Putnam, 1993). In this study, access to the social resource is assessed considering five types of social relations. Social capital (explained in Chapter 6) is computed by access to five types of resources: (i) family relation, (ii) relationship with neighbors, (iii) involvement with politics, (iv) access to the wider institution, and (v) participation in a community-based organization.

For a betterunderstandingof the level of social capital attained by the respondents, the SRI range is further divided into five categories and labeled as very low (15–21), low (22–28), medium (29–35), good (36–42), and very good (43 and above). In the study area, more than 16.4% were in moderate access to social capital. The mean value (26.10) indicates that the majority was in lower access to social capital. On the other hand, more than 13% had very good access to family relations and relations with outsiders.

Physical Asset

In metropolitan environment housing is one of the most essential physical assets as it can be used effectively to relieve financial stress (Schütte, 2016). Housing conditions in developing countries are often poor. Overcrowding is a problem both in home and the community. The results of this study have a negative impact on available utilities, housing, and general environmental circumstances. Sanitation facilities are frequently in bad condition.

The findings of this research put a strain on available utilities, housing, and overall conditions. Sanitation facilities are generally in poor status. The physical asset index

TABLE 21.9 Role of the social asset in reducing vulnerability

Level of social resource	*Number of respondents regarding the vulnerable situation*					
	30–46 Very high vulnerability	*47–63 High vulnerability*	*64–80 Medium vulnerability*	*81–97 Low vulnerability*	*98–114 Very low vulnerability*	*Total*
Very low (15–21)	28	102	69	15	2	216
Low (22–28)	11	100	115	52	4	282
Moderate (29–35)	3	21	48	37	7	116
Good (36–42)	0	5	21	29	13	68
Very good (43 and above)	0	2	9	10	4	25
Total (mean: 26.107, mode: 15.00, Max:56)	42	230	262	143	30	707

(PAI) varies from 15 to 54. Findings show that 4% of respondents had very low access to physical assets. The mean (33.19) and mode values (31.00) denote that the majority had moderate access to the physical asset. On the other hand, 27.9% had good access to the infrastructural facilities.

Though the respondents have little political participation, in some cases their community feelings are very high that is in many cases their strength to live. The social asset represents an important part of urban livelihood. In this context, the role of social resources in reducing vulnerability is presented in Table 21.9.

Table 21.10 indicates that the respondents having low natural resources are highly disaster prone than the respondents enriched with natural resources. Therefore, the SLI has been construed considering the above-mentioned explained variables: natural resource, financial asset, social resource, physical asset, occupation, vulnerability, and disaster situation.

Level of Sustainability

See Table 21.11.

Sustainable Livelihood Index

Table 21.12 shows that the mean value (74.61) indicates that on average the respondents were at a low level of sustainability, but 15.8% were in a moderate level of sustainability, and only 3.5% were sustainable enough. In this regard, the regression model has also been drawn (Table 21.11).

TABLE 21.10 Role of natural resources in reducing vulnerability during disaster

Level of natural resource	*Number of respondents*					*Total*
	Disaster situation					
	27–36 Very highly disaster prone	*37–46 Highly disaster prone*	*47–56 Moderately disaster prone*	*57–66 Low disaster prone*	*67 and above Very low disaster prone*	
Very low (15–22)	35	12	9	2	0	58
Low (23–30)	155	45	22	5	1	228
Medium (31–38)	113	71	59	27	7	277
High (39–46)	10	29	33	20	13	105
Very high (47–53)	5	7	9	11	7	39
Total (mean: 32.62, mode: 28.00)	318	164	132	65	28	707

TABLE 21.11 Percentage distribution of the respondents regarding sustainable livelihood

Indicators of sustainable livelihood	*Percentage distribution of the respondents (with level)*					*Total*
	1	*2*	*3*	*4*	*5*	
Natural resource (mean: 19.4059, SD: 6.9)	Very low (8.2)	Low (32.2)	Medium (39.2)	High (14.9)	Very high (5.5)	100.00
Financial resource (mean: 16.5233, SD: 5.6)	Very low (7.2)	Low (33.5)	Medium (40.0)	High (15.1)	Very high (4.1)	100.00
Vulnerability occupation (mean: 14.2150, SD: 4.77)	Very high (5.9)	High (32.5)	Medium (37.1)	Low (20.2)	Very low (4.2)	100.00
	Rickshaw pulling/ not defined/ (28.7)	Industry (18.7)	Construction (22.1)	Agriculture (8.8)	Business (21.8)	100.00
Disaster situation (mean: 41.7680, SD: 11.75)	Very low (45.0)	Low (23.2)	Medium (18.7)	High (9.2)	Very high (4.0)	100.00
Social asset (mean: 26.1075, SD: 9.39)	Very low (30.6)	Low (39.9)	Medium (16.4)	High (9.6)	Very high (3.5)	100.00
Physical asset (mean: 33.1881, SD:7.01)	Very low (4.0)	Low (20.2)	Medium (43.3)	High (27.9)	Very high (4.7)	100.00

TABLE 21.12 Level of sustainability of the respondents

Level of sustainability	*Number and percentage distribution of the respondents*	
	Number	*Percent*
Not at all (28–50)	87	12.3
Very Low (51–72)	274	38.8
Low (73–94)	209	29.6
Moderate (95–117)	112	15.8
Sustainable enough (118–140)	25	3.5
Total (mean:74.61, SD:20.91)	707	100

TABLE 21.13 Effects of determinants of livelihood on sustainability

Independent variables	*Dependent variable: Sustainable livelihood index (SLI)*		
	Coefficients	*t*	*Significance level*
(Constant)	1.122	3.335	0.001
Natural resource	0.350	91.477	0.000
Income	0.266	54.962	0.000
Poverty situation	0.249	50.565	0.000
Occupation	0.297	85.555	0.000
Disaster	0.190	43.465	0.000
Infrastructural facility	0.036	11.031	0.000

N= 707; adjusted R^2 = 0.993; F= 15911.975 (significance level: 0.000) significant at 1% level of significance.

The value of R^2 (0.993) means that the level of sustainability increased by 0.993 units for changing 1 unit of the above-mentioned variables. Among the variables natural asset (land, water, sunlight and air quality, and tree) has contributed (0.35) more than a financial asset (income, working situation, number of income earner, and savings), which is 26.6% following low vulnerable situation (25%), occupational status (0.30), fulfilling the basic need in a disaster situation (0.19), and physical asset (0.036). The role of the natural environment on the basis of SLI and NRI is presented in Table 21.13.

Resource, Vulnerability, and Sustainable Livelihood

Table 21.14 shows that the respondents enriched with natural resources are in a medium level of sustainability (60) and sustainable enough but the respondents with very poor natural resources are in very low and low levels of sustainability. The value of the chi-square test (593.567) with Fisher's exact test of 16 degrees of freedom is also significant (0.000). Not only the natural resources but the social resource also contributes to this regard.

Table 21.15 presents that the respondents who have a good stock of social capital are sustainable enough, but the respondents with a low level of social capital are not sustainable at all or in a very low level of sustainability. The level of vulnerability is a significant indicator of sustainable livelihood.

Both male (67.20%) and female members (71.32%) like long-term livelihood patterns than the short-term patterns, 32.80% and 27.68%, respectively (Table 21.16). The preferable long-term livelihood patterns are child education and permanent job facility, tree plantation, agriculture, development of tea stalls and grocery shops whereas, micro-credit facilities and access to interest-free loan are short-term livelihood facilities, among others.

In this study, the natural resources include five assets: land, water, sunlight and air quality, trees, and open spaces. Among all, the respondents having free access to land can earn more from a land-related business-like grocery shop kitchen market, fish business, and agricultural activities. The water bodies are used for different

TABLE 21.14 Access to natural resource and sustainability (cross-tabulation)

Level of natural resource	*Number of respondents*					*Total*
	Level of sustainability					
	28–50 Not at all	*51–72 Very low sustainability*	*73–94 Low sustainability*	*95–117 Medium sustainability*	*118–140 Sustainable enough*	
Very low (15–22)	25	25	8	0	0	58
Low (23–30)	54	140	32	2	0	228
Moderate (31–38)	8	107	128	33	1	277
Good (39–46)	0	1	37	60	7	105
Very good (47–53)	0	1	4	19	15	39
Total (value of chi-square-593.567)	87	274	209	114	23	707

TABLE 21.15 Contribution of social resources in sustainable livelihood (cross-tabulation)

Level of social asset	*Number of respondents*					*Total*
	Level of sustainability on the basis of SLI					
	Not at all	*Very low sustainability*	*Low sustainability*	*Medium sustainability*	*Sustainable enough*	
Very low	54	113	42	7	0	216
Low	22	133	99	28	0	282
Medium	10	21	47	34	4	116
High	1	3	17	36	11	105
Very high	0	4	4	9	8	39
Total	87	274	209	114	23	707

TABLE 21.16 Expected livelihood pattern

Pattern of livelihood	*Male choice*		*Female choice*	
	Number	*Percent*	*Number*	*Percent*
Long-term	379	67.20	102	71.32
Short-term	185	32.80	41	28.68
Total	564	100.00	143	100.00

purposes such as bathing, sanitary, washing (clothes and plates), and fishing. Fishing is an important part of residents' livelihood strategies for meeting food needs. The river water is used for ice factories and for transporting shrimps and logs from the hinterland. Vegetable cultivation and tree plantation in Khulna city appear to play a significant role in job security as well as improving social capital, health, and nutrition. Therefore, it contributes to reducing vulnerability and resilience with disasters that have a positive effect on the sustainable livelihood of vulnerable people in Khulna city. In the study area, respondents are expected to ensure the security of food, housing, and income. The natural resource–based livelihood activities to them have long gestation periods and are tied to activities that do not yield immediate returns for their families. The study indicated that the poor in the study area rely on short-term livelihood activities such as daily casual labor, involvement in the shrimp industry, most of which are only dictated by the season and therefore may not be sustainable on an all-year-round basis, if not overexploited. The longest "gestation" period for the natural resource–based livelihood activities identified is land-based. The shortest is the wood industry. The problem the poor face with these long gestation periods is that they often lack the capital to sustain the livelihood investment at the same time enough food to sustain the family while waiting, which often becomes impossible when it comes to the issue of survival.

Conclusion

The study exposes that the urban poor have unstable and inadequate access to livelihood assets and infrastructural facilities such as food, housing, health, education, sanitation, and job security. Major livelihood shocks are the threat of eviction, fire accident, loss of property, waterlogging, loss of social network, unemployment, flooding, violence and crime, pollution, and health problems. These shocks, however, reinforce one another thereby worsening the condition of the urban poor. What's more, their living is founded on shaky and insufficient assets. The study contains comprehensive indicators of livelihoods on disaster-affected vulnerable individuals in Khulna city, as seen through the lens of the SLI. As a result, this study suggests that a policy framework that ensures the equitable distribution of land, trees, and other natural resources while also minimizing livelihood vulnerability is an important component in alleviating poverty.

Therefore, on the basis of empirical findings, it can be concluded that natural resources and their related occupation and infrastructural development contribute to increasing income, job security, and social capital; reducing vulnerability and resilience with the disaster has a positive effect on the sustainable livelihood of vulnerable people in Khulna city.The poor people enriched with natural resources are less vulnerable than those who have low access to natural resources. The poor communities cope with urban life through "household strategies" such as putting more family members into the workforce, through small-scale business, increasing their household size by inducting more relatives, withdrawing their children from education, and establishing patron–client relationships with local leaders.Vegetable cultivation and tree plantation in Khulna city appear to play a significant role in improving social capital, health, and nutrition. Households with more than one earning member, access to urban land, and access to social capital for higher possibilities of well-being show intra-household heterogeneity in coping with poverty and deprivation. Home-based income-generating activities, on the other hand, may help to improve livelihoods.

References

Ahmed, A. U. (2010). *Desakota Phenomenon Observed in Satkhira-Khulna-Jessore-Dhaka Corridor in the Southwestern Bangladesh*. Dhaka: Centre for Global Change, pp. 4–13.

Ara, M.R. (2017). *Urban Environment and Sustainable Livelihood: The Case of Khulna City*. Bangladesh Doctoral (Dissertation). Dhaka: University of Dhaka, pp. 1–229.

Ara, M.R., & Nasreen, M. (2017). Gender Difference in Access to and Control over Resources during Disaster: The Case of Khulna City Corporation, Southwest Bangladesh. *Khulna University Studies*, *14*(1&2), 229–241.

Bharatdwaj, K. (2006). *Physical Geography*. New Delhi: Discovery Publishing House, p. 33.

Carney, D. (1998). *Sustainable Rural Livelihoods: What Contribution can We Make?* London, United Kingdom:Department for International Development (DFID), pp. 7–213.

Chambers, R., &Conway, G. R. (1991). *Sustainable Rural Livelihood: Practical Concepts for the 21st Century*. Brighton: Institute of Development Studies (IDS), pp. 1–33.

Chenery, H., Ahluwalia, M. S., Bell, C. L. G., Duloy, J. H., &Jolly, R. (1979). *Redistribution with Growth*. Oxford: Oxford University Press, pp. 3–22.

Claeys, G. (2000). The Survival of the Fittest and the Origins of the Social Darwinism. *Journal of the History of Ideas*, *61*(2), 223–240.

Crabtree, L. (2008). The Role of Tenure, Work and Cooperativism in Sustainable Urban Livelihoods. *An International E-Journal for Critical Geographies*, 7(2), 260–282.

Dasgupta, S., Laplante, B., Murray, S., & Wheeler, D. (2009). *Climate Change and the Future Impacts of Storm-Surge Disasters in Developing Countries*. Washington, DC: Center for Global Devlopment, pp. 17–19.

Denzin, N. (2006). *Sociological Methods: A Sourcebook*. Chicago: Aldine Transaction, pp. 1–600.

Department for International Development. (1999). *Sustainable Livelihoods Guidance Sheets*. London: Department for International Development, pp.1–150.

Deptford, A., Louis, C., &Hall, A. (2013). *A Cost of the Diet Analysis in Khulna District of Bangladesh*. Dhaka: Save the Children, p. 19.

Elliott, J. A. (2006). *An Introduction to Sustainable Devlopment* (3rd ed.). London: Routledge, pp. 1–302.

Ellis, F. (1999). *Rural Livelihood Diversity in Developing Countries: Evidence and Policy Implications.* London, United Kingdom: Overseas Development Institute (ODI), pp. 2–10.

Ellis, F. (1998). Household Strategies and Rural Livelihood Diversification. *Journal of Development Studies, 35*(1), 1–38. doi: 10.1080/002

English Oxford Living Dictionaries. (2014, August 28). *Livelihood.* Retrieved from Oxford Dictionaries: https://en.oxforddictionaries.com/definition/livelihood

Faisal, I. M., & Kabir, M. R. (2005). An Analysis of Gender–Water Nexus in Rural Bangladesh. *Journal of Developing Societies, 21*(12), 175–194.

Farhana, K. M., Rahman, S. A., & Rahman, M. (2012). Factors of Migration in Urban Bangladesh: An Empirical Study of Poor Migrants in Rajshahi City. *Bangladesh E-Journal of Sociology, 9*(1), 105–117.

Farrington, J., Ramasut, T., &Walker, J. (2002). *Sustainable Livelihoods Approaches in Urban Areas: General Lessons, with Illustrations from Indaian Cases.* Stockholm: Overseas Development Institute, pp. 1–60.

Food and Agricultural Organization (2009). *Trees Connecting People: In Action Together.* Bogota: Food and Agriculture Organization of the United Nations (FAO), pp.1–58.

General Economics Division. (2015). *Millennium Development Goals: Bangladesh Progress Report 2015.* Dhaka, Bangladesh: Bangladesh Planning Commission, pp. 1–138.

Golden, S. (2013). *Urban Agriculture Impacts: Social, Health, and Economic: A Literature Review.* New York: University of California Agriculture and Natural Resources, pp. 1–22.

Goodall, B., & King, K. (1987). *The Penguin Dictionary of Human Geography.* London: Puffin, pp. 1–481.

Haan, L. J. D., & Zoomers, A. (2005). *How to Research the Changing Outlines of African Livelihoods.* Paper presented at the 11th General Assembly of CODESRIA, Maputo, pp.1–8.

Hossain, M. A., Moniruzzaman, M., & Islam, M. A. (2010). Urban Environmental Health in Bangladesh Slum: A Comparative Study of two Metropolitan Cities. *Journal of Scientific Foundation, 8*(1&2), 67–76.

Hoornweg, D., & Munro-Faure, P. (2008). *Urban Agriculture: For Sustainable Poverty Alleviation and Food Security.* Rome: Food and Agricultural Organization, pp. 1–84.

Kazungu, I., Ndiege, B. O., Mchopa, A., & Moshi, J. (2014). Improving Livelihoods through Micro and Small Agribusiness Enterprises: Analysis of Contributions, Prospects and Challenges of Nursery Gardens in Arusha Tanzania. *European Journal of Business and Management, 6*(9), 142–148.

Krantz, L. (2001). *The Sustainable Livelihood Approach to Poverty Reduction: An Introduction.* Stockholm: Swedish International Development Cooperation Agency, pp. 1–44.

Little, P. D. (1999). Selling to Eat: Petty Trade and Traders in Peri-urban Areas of Sub-Saharan Africa. Land Tenure Center (LTC). University of Wisconsin-Madison. www.ies.wisc.edu/ltc/live/bashorn9910a.pdf

Maxwell, D. G. (1995). *Measuring Food Insecurity: The Frequency and Severity of "Coping Strategies".* Washington, DC: International Food Policy Research Institute, pp. 1–26.

Meikle, S., Ramasut, T., & Walker, J. (2001). *Sustainable Urban Livelihoods: Concepts and Implications for Policy.* London: Working Paper 112, pp. 1–35.

Moser, C. O. N. (1998). The Asset Vulnerability Framework: Reassessing Urban Poverty Reduction Strategies. *World Development, 26*(1), 1–19.

Murtaza, G. (2001). Environmental Problems in Khulna City, Bangladesh: A Spatio-Household Level Study. *General Block Exemption Regulations, 1*(2), 32–37.

Nasreen, M. (2008). *Violence aganist Women during Flood and Post-Flood Situations in Bangladesh.* Dhaka: Women's Rights & Gender Equality Sector, pp. 1.

Nasreen, M. (2012). *Vulnerable or Resilient? Women and Girls.* Dhaka: Institute of Disaster Management and Vulnerability Studies, pp. 1–121.

Nasreen, M., Hossain, K. M., &Kundu, D. K. (2006). The Interrelationship between Poverty, Environment and Sustainable Development in Bangladesh: An Overview. *E-Journal of Sociology*, 4–21.

National Institute of Population Research and Training. (2013). *Bangladesh Demographic and Health Survey 2011*. Dhaka: National Institute of Population Research and Training (NIPORT) & Mitra and Associates, pp.1–458.

Park, R. E., & Burgess, E. W. (1921). *Introduction to the Science of Sociology*. Chicago: The University of Chicago Press, pp. 161–216.

Parveen, S., & Leonhauser, I.U. (2004). *Empowerment of Rural Women in Bangladesh: A Household Level Analysis*. Paper presented at the Rural Poverty Reduction through Research for Development and Transformation, Deutscher Tropentag-Berlin, pp. 1–10.

Patton, M. Q. (1999). Enhancing the Quality and Credibility of Qualitative Analysis. *Health Services Research, 34*(5), 1189–1208.

Pearson. (2011). *The Urban Environment: Creating Livable and Sustainable Cities*. London: Pearson Education Incorporation, pp. 195–208.

Pender, J. S. (2008). *What Is Climate Change? And How It Will Affect Bangladesh*. Dhaka: Church of Bangladesh Social Development Programme, pp. 1–75.

Putnam, R. D. (1993). The Prosperous Community: Social Capital and Public Life. *The American Propect, 13*, 35–42.

Rakodi, C., & Lloyd-Jones, T. (2002). *Urban Livelihodds: A People-Centred Approach to Reducing Poverty*. London: Routledge, pp 23–34.

Riddell, R. (2004). Approaches to Poverty: A Note from the Development Perspective. Geneva: International Council on Human Rights Policy (ICHRP), pp. 1–4.

Sanderson, D. (2000). Cities, Disasters and Livelihoods. *Environment and Urbanization, 12*(2), 93–102.

Schutte, S. (2016, April 24). *Livelihoods of the Urban Poor in Afghanistan – Conceptional Issues and Review of Literature*. Retrieved from Afghanistan Research and Evaluation Unit (AREU). https://areu.org.af/wp-content/uploads/2015/12/439E-Livelihoods-of-the-Urban-Poor-WP-2004.pdf

Scoones, I. (1998). *Sustainable Rural Livelihoods: A Framework for Analysis*. Brighton: Institute of Development Studies (IDS), pp. 1–22.

Sen, A. (1981). *Poverty and Famines: An Eassy on Entitlement and Deprivation*. Oxford: Clarendon Press, pp.1–8.

Sen, A. (2003). Capability and Well-being. In: M. Nassbaum& A. Sen (Eds.), *The Quality of Life*. Oxford: Oxford University Press, pp. 30–53.

Sen, A. (2005). Human Rights and Capabilities. *Journal of Human Development, 6*(2), 151–166.

Shih, F. J. (1998). Triangulation in Nursing Research: Issues of Conceptual Clarity and Purpose. *Journal of Advanced Nursing, 28*(3), 631–641.

United Nations Department of Economic and Social Affairs. (2017, January 11). *In Focus: We're Finally at the End of the UN Decade for Water 2005–2015 – It Is Time to Say Good-bye*. Retrieved from United Nations Department of Economic and Social Affairs (UNDESA). www.un.org/waterforlifedecade/index.shtml

United Nations Development Programme. (2011). *Africa Adaptation Programme Experiences Gender and Climate Change: Advancing Development through an Integrated Gender Perspective*. New York: United Nations Development Programme, pp. 1–20.

United Nations. (2014). *World Urbanization Prospects*. New York: United Nations, pp. 1–32.

United States Environmental Protection Agency. (2004). *Guidelines for Water Reuse (EPA/625/R-04/108 September 2004)*. Washington, DC: Environmental Protection Agency, pp. 1–478.

Veenhuizen, R.V. (2006). *Cities Farming for the Future Urban Agriculture for Green and Productive Cities*. Ottawa: International Institute of Rural Reconstruction, pp. 1–500.

Victor, M. A. M., & Makalle, A. M. P. (2003). *Poverty and Environment: Impact Analysis of Sustainable Dar es Salaam Project on Sustainable Livelihoods of Urban Poor*. Dar Es Salaam, Tanzania: Mkuki Na Nyota Publishers, pp. 1–40.

Wisner, B., Blaikie, P., Cannon, T., &Davis, I. (2003). *At Risk: Natural Hazards, People's Vulnerablility and Disasters* (2nd ed.). London: Routledge, pp. 3–123.

World Bank. (2013). *World Development Indicators*. Washington, DC: The World Bank, p. 28.

Wratten, E. (1995). Conceptualizing Urban Poverty. *Environment and Urbanization*, 7(1), 11–38.

Wyatt, R. C., & Meyers, L. S. (1987). Psychometric Properties of Four 5-point Likert Type Response Scales. *Educational and Psychological Measurement*, 47(1), 27–35.

22

A CONCEPTUAL APPROACH FOR UNDERSTANDING THE DISASTER-PANDEMIC NEXUS

Rupkatha Priodarshini, Bishawjit Mallick and Saleemul Huq

Introduction

2020, a year defined by the corona virus pandemic, also saw a devastating array of natural disasters from historic California and Australia 'mega-fires', to 'super-cyclones' and flooding in record-breaking storm seasons across the globe, to disastrous earthquake events. The global crisis and continuing threat to public safety posed by SARS-Cov-2 and the resultant Covid-19 pandemic have exacerbated crises of the economy and environment, including natural disasters that have struck during the pandemic's course. By the end of 2020, there had been a total of 84.5 million Covid-19 cases globally and 1.8 million deaths (WHO, 2021), along with an estimated cost of US$28 trillion in lost economic output (IMF, 2020). TheCovid-19 response had both positive and negative environmental impacts; restricted mobility and economic activity reduced air pollution, noise pollution and greenhouse gas (GHG) emissions, whereas precautions taken to prevent contamination reduced recycling and increased waste (Zambrano-Monserrate et al., 2020). Socio-cultural impacts are also extensive. The psychological toll of the Covid-19 crisis on individuals and family dynamics has been explored (Serafini et al., 2020; Prime et al., 2020). The pandemic's social impacts further include a gendered dimension, as the risk of unemployment and financial injury to women increased disproportionately as their caregiving duties increased during lockdown (Collins et al., 2020).

Natural disasters have had comparable impacts to the Covid-19 pandemic. While directly responsible for 8,200 deaths, they cost the global economy a further US$210 billion in 2020 (MunichRe, 2021). These disasters had widespread environmental impacts; notably, unprecedented environmental degradation resulted from the 2020 wildfires (Bowman et al., 2020). The mental health effects observed in the wake of natural disasters have drawn parallels with the pandemic (Agarwal

DOI: 10.4324/9781003253495-28

et al., 2020;North et al., 2021). Likewise, female victims of natural disaster are understood to experience ongoing gendered impacts, including the aforementioned economic dispensability and subsequent poverty also seen under Covid-19 lockdowns (Bradshaw & Fordham, 2015). Thus, alongside the demographic and economic costs, both crises yielded concurrent and widespread environmental and socio-cultural effects. Hence, they may be conceptualized as a 'double strike' or 'double disaster', a multi-hazard event whose interface compounds the adverse effects experienced by those affected.

The Sendai Framework for Disaster Risk Reduction (SFDRR) (UNDRR, 2015, p. 10) identifies 'pandemics and epidemics' as a 'disaster risk driver'; in other words, the global spread of SARS-Cov-2 constitutes a hazard to the populations it encounters. Selby and Kagawa (2020) frame the Covid-19 pandemic and natural disasters, specifically those linked to climate change, a shaving 'not only confluence but collision' (p.18)—sharing similarities in their provenance and impacts, but some contradicting response efforts as well. In light of their commonalities, the logic of approaching the corona virus crisis as a natural disaster when it comes to response strategies has been noted (Seddighi, 2020; Sakurai &Chughtai, 2020). For example, Sakurai and Chughtai (2020) use insights from the response to the 2011 Tohoku earthquake in Japan to reflect on the potentials of information systems and technological transformation during the Covid-19 crisis. Moreover, countries' resource strain while grappling with the 'double strike' provides a basis for placing Covid-19 and other epidemiological crises in a similar conceptual category to conventional natural disaster. In Queensland, Australia, for example, local authorities urged for federal relief funds aside for natural disasters to be provided to those crippled by the pandemic, arguing that the Covid-19 crisis was no different to the litany of natural disasters the region experiences, and this was where the funds were urgently needed (Remeikis, 2020).

Emerging from our observations of the pandemic-cyclone double strike in Bangladesh, this article provides a snapshot of the mutual influence and various commonalities observed between the impacts of Covid-19 and recent natural disasters, reviewing how the pandemic has impacted the effects, responses and outlook for natural disaster events. A state-and-transition model (Westoby et al., 1989; Mallick, 2014) is used to visualize the impact of Covid-19 and disaster on livelihoods at the household level, with Cyclone Amphan in Bangladesh as a case study. Furthermore, we explore how the response to the 'double strike' of the corona virus pandemic and natural disaster events may be reconfigured to better account for their intertwined, often novel effects, now and for similar events in the future. To this end, we contribute an approach for conceptualizing the path to improved pandemic-disaster response through greater collaboration, drawing upon the '4Cs' (communication, cooperation, coordination and collaboration) of disaster partnering (Martin et al., 2016). We posit that a holistic perspective of public crises is essential going forward. Finally, having reviewed the present discourses, this paper highlights some of the possible strategies for an integrated approach at various scales, including the inclusion of pandemic preparedness in key multilateral climate agendas.

Compounding Complexities: Covid-19 and Disasters

The disruptive effects of natural disasters and the corona virus pandemic on social and organizational systems and practices are comparable (Sakurai &Chughtai, 2020). Their co-occurrence presents a multiplicity of implications for physical and psychological health, community resilience, disaster response and management and more. As indicated, these overlaid consequences may be classified as demographic, economic, environmental, socio-cultural and political in nature. This pentagon lens for viewing these topical consequences is roughly adapted from the five capitals for sustainable livelihoods outlined in the sustainable livelihoods framework (DFID, 2005; Mallick et al., 2020), precipitating the conceptual model of how the joint pandemic-disaster influences livelihoods that follow in this paper. Significantly, the most marginalized communities are affected first and worst by both disasters (Seddighi, 2020; Manzanedo& Manning, 2020), illuminating the lines of social and economic vulnerability that define susceptibility to disaster phenomena, which have frequently been characterized as undiscriminating in their destruction (Mulholland, 2020). Thus, many of the 'frontline' communities of the ongoing climate change crisis, shepherded to this position by a host of socio-economic and political realities as well as ecological ones, have also borne the brunt of the corona virus crisis. The case of coastal Bangladesh serves to illustrate this.

In Bangladesh, Covid-19 lockdown restrictions endangered the livelihoods of rickshaw pullers and other informal urban service sector laborers, prompting an exodus to rural home villages (Savage & Farhad, 2020). This cut off stipends to economically precarious rural areas, much of whose income derives from urban sources, in part due to the debilitating effects of climate change on local farmland (ibid). Indeed, across South Asia, so-called climate-induced migrants found their vulnerable livelihoods in jeopardy due to the restricted mobility, industry shutdowns and other effects of Covid-19 (Shabbir & Naeem, 2020). When supercyclone Amphan made landfall in May—followed by an unprecedented monsoon flooding season—it battered an already struggling rural populace whose ability to bounce back economically was impeded (Ober, 2020). Analysis of coastal regions has found that the resilience of local livelihoods had been significantly weakened by the pandemic, setting the stage for losses from Amphan that would have been less injurious otherwise (Priodarshini et al., 2021). Furthermore, some Bangladeshi return migrants encountered additional social challenges such as women's inability to work and study as they had in the city, due to rural cultural norms (Savage &Farhad, 2020). Overall, livelihood resilience was deeply impacted by the double strike, with average incomes reduced by over half and around seven out of ten livelihoods lost in one coastal region (Priodarshini et al., 2021). Islam et al. (2020) further note the need for research into whether Amphan aggravated psychological stress experienced during Bangladesh's lockdown.

Comparable scenarios are found around the world. Around the same time as Amphan, the US indigenous communities of the Navajo Nation and surrounding regions, who were already grappling with disproportionately high

Covid-19 morbidity and mortality rates, faced a succession of damaging wildfires that devastated the local environment (Nelson, 2020). Many of the Navajo characterized their vulnerability to the corona virus as an undeniably political issue, opining that improved relations with the state would ameliorate risk factors such as lack of public health and water services (Mulvaney et al., 2020). Quality of life was further impaired in the wildfire-stricken areas, with some residents unable to step outside without coughing from the smoke, on alert for evacuation orders (Nelson, 2020). Thus, environmental changes caused by natural disaster exacerbated the existing physical and psychological stresses of Covid-19. At the level of biology, the wildfire and bushfire disasters that ravaged many communities in 2020may even have worsened the outlook for those who contracted the virus, as exposure to such air pollution is linked with increased chances of severe complications and death from Covid-19 (Wu et al., 2020).As in Bangladesh, Cyclone Yasa in Fiji, Typhoons Goni and Vamco in the Philippines and Hurricanes Eta and Iota in Central America, all evoked similar concerns that these disasters would curtail the recovery of—if not destroy—thousands of livelihoods, leaving developing economies already stretched thin by the pandemic reeling (FAO, 2020; McDonald, 2021; Tucker &Pindado, 2020).

Considering the potential for multi-hazards to deepen existing inequalities, it is essential to heed the differential vulnerabilities and unique challenges faced by particular groups in place (Philips et al., 2020). Therefore, we present a state-and-transition model (Mallick, 2014) for conceptualizing these double strikes at the household level, considering the particular context and capabilities of the household as it proceeds through the event. This concept integrates the pandemic and natural disaster events to visualize how they have interacted as a joint force on the individual household (Figure 22.1).

Figure 22.1 shows the livelihood capitals of the individual household at the center. These include demographic or human capital, that is, the health and ability of the household members; economic capital, that is, employment, income, assets and other financial resources; socio-cultural capital, that is, social networks, culturallyspecific roles and abilities; environmental capital, that is, access to natural resources and local environment; and political or institutional capital, that is, access to political resources and representation (Mallick et al., 2020). Broadly, these five categories encompass the influences that a disaster event may have on the household. The five capitals are subject to disruption and change from the effects of both the Covid-19 pandemic and natural disaster as the household experiences the double strike and its aftermath. The temporal stages of this experience are pre-/post-event, preparedness, coping, rehabilitation and adaptation. These reflect how the disaster is anticipated and managed. As the household transitions through the stages, its capital characteristics and the progressive stages of the experience mutually influence one another. For example, the household's ability to cope with the disaster influences its capitals, which in turn shape how rehabilitation can proceed. In the double strike scenario, the multi-hazard exerts new, compounded pressures on these capacities. To draw on the capabilities approach, it limits capabilities and

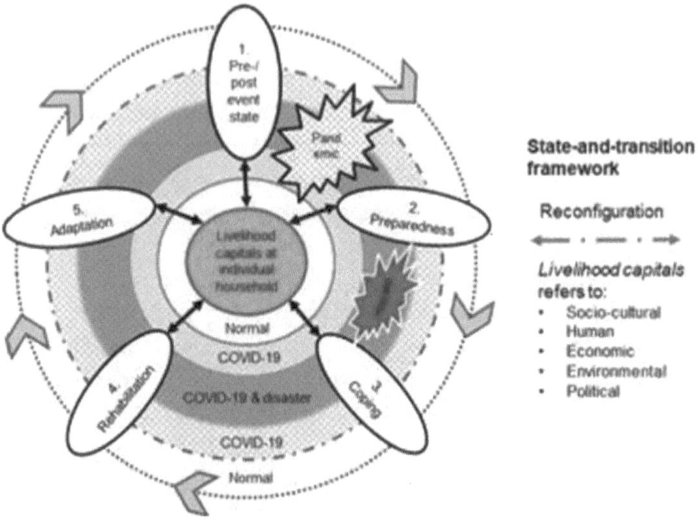

FIGURE 22.1 State-and-transition model of the pandemic-disaster double strike at the household level (authors' illustration)

poses a greater risk of capability failures when compared with single disasters (Sen, 1985). In the Bangladeshi context, we observed that the double strike has significantly affected the economic capital (income, working days) of a broader range of occupations as compared to single coastal disasters, where resource-dependent livelihoods are primarily affected (Priodarshini et al., 2021).

Following the livelihood capitals represented in the center of Figure 22.2, the concentric areas represent the temporal progression of the disaster itself. The Covid-19 pandemic disrupts a 'normal' state for the livelihood, continuing through a period of co-occurrence with the natural disaster. This indicates how the pandemic has preceded and exceeded the active period of recent natural disasters, which are typically more contracted in duration. It defines the co-occurrence, 'Covid-19 & disaster', as a phenomenon of its own with common consequences. Furthermore, the pandemic event initiates following the pre-event stage of the overall double strike event, prior to preparedness, as preparedness for the pandemic existed minimally, if at all, in comparison to that for natural disasters. Thus, the pandemic occurs without warning, lacking the critical mechanisms generally in place to mitigate disaster risks, whereas the natural hazard is shown to initiate following the preparedness stage. In the coastal region we observed, the far greater proportion of losses followed the pandemic, not the cyclone. Unlike the weather events to which they are somewhat accustomed, the novelty and scale of the pandemic left coastal people without recourse. With the pandemic persisting, a full return to the pre-disaster state is as yet undetermined, and the prospect of how this will look remains hazy. This is indicated by the perforated line bounding the post-disaster 'normal' state in the concentric visualization. We noted that lack of preparedness and low levels of diversification in the coastal regions of Bangladesh primarily resulted in widespread livelihood loss, rather than recovery and adaptation soon after the double strike (Priodarshini et al., 2021).

Adapting to compound disaster pressures is thus an iterative process. The 'new normal' these pressures produce may entail additional vulnerabilities and livelihood loss if the pressures are not surmounted, or strengthened resilience if the recovery process is successful. If the double strike is overcome through effective and innovative coping mechanisms, future resilience may be improved, however, this is not anticipated to be a common outcome due to the novel challenges double strikes pose. When operationalized according to the context of a specific disaster or community, this conceptual model may yield either outcome. As mentioned, the outcome is shaped by the existing scale of vulnerability and resilience, which differs across socio-economic and demographic status (Cutter et al., 2008). Diversified livelihood strategies is a socio-economic attribute that has positively influenced resilience in the face of the pandemic-cyclone Amphan double strike, as is proximity to services such as markets (Priodarshini et al., 2021). In the process of coping with the double strike, these factors support recovery by providing additional means of income generation if the primary strategy is disrupted and streamlining access to disaster relief. 'Reconfiguration', represented by the dot-dash line, alludes to the reappraisal of the compound disaster and its management, which occurs in the disaster's aftermath. As the following section suggests, it may entail reflection on potential changes to the management approach in line with the reconceptualization of the joint events as effectively one disaster on the ground with overlaying dimensions. This model presents 'reconfiguration' as a *post*-disaster strategy in light of the reality that prior to the Covid-19 pandemic, the virus had not been in the remit of existing disaster preparedness and management strategies. Moreover, co-occurrences have thus far been understood and treated in ways that fail to fully resolve the conflicts and address the commonalities between the pandemic and the natural disaster. These management challenges, and the potential for a reconfigured holistic approach, are elaborated in the following sections.

Challenges and Potentials

Multi-hazard conditions can pose unique management challenges (Rahman et al., 2021). With regard to disaster preparedness and response, the co-occurrence of natural disasters and corona virus outbreaks have often presented not merely a double strike, but a double bind: 'We have two parallel crises that contradict each other', as Croatian prime minister Plenkovic put it following the historic March 2020 earthquake (Perry, 2020). Natural disasters have often left people with no choice but to abandon Covid-19 precautions flouting stay-at-home mandates to flee hazards and cramming into shelters where social distancing is impossible and the risk of transmission high. Like the Zagreb quake, the 2020 California wildfires compelled authorities to issue evacuation orders that contradicted existing corona virus guidelines (Sottile & Arkin, 2020). Likewise, victims of Cyclone Amphan were faced with the 'impossible choice' of whether to remain in place and withstand the cyclone or risk infection while seeking refuge in shelters (Ellis-Petersen & Ratcliffe, 2020). Disinfecting aid donations was yet another concern. Australian supplies sent to Vanuatu in the wake of category 5 Cyclone Harold were required to await clearance for days, with officials confronting the quandary of whether to bend established wait time protocols in the face of a crisis where time was of the essence (McGarry, 2020). The diversion of various emergency resources, money and human resources alike, into Covid-19 relief stoked concerns. In Australia, to take just one example, worries that the $76m fund set aside for bushfire recovery might be siphoned into corona virus relief efforts incited tension among communities affected by fires (ABC Radio National, 2020).

Across the board, therefore, responding agencies must perform a balancing act. An ethnographic study of stakeholders by Palinkas et al. (2021) found that hurricane disaster management in Louisiana was affected in 'function and form' by the pandemic. California officials mitigated the dire double strike situation by introducing virus screenings and PPE to emergency fire shelters which had been drastically reduced in capacity to limit the chances of infection (Sottile& Arkin, 2020). Similarly, Bangladeshi authorities more than doubled the number of emergency cyclone shelters in an effort to maintain social distancing among evacuees (Ellis-Petersen & Ratcliffe, 2020). Resource strain and the unprecedented nature of the pandemic has resulted in an inevitable trade-off between natural disaster preparedness and recovery, and corona virus response (Cardil & de-Miguel, 2020). The rapid generation and adjustment of new safety guidelines, risk assessments, resource allocation and mobilization protocols and more have been crucial—but highly demanding and often ad hoc. As new variants of the corona virus evolve, such improvization and ingenuity remains critical to the effective containment and treatment of Covid-19 over two years since the pandemic was declared (Penta et al., 2021).

As Philips et al. (2020) highlight, existing governance and institutional structures tend to be vulnerable to compartmentalization, impeding efficient multi-hazard response. Yet, disasters of this compound nature are in fact increasing, with the pandemic-disasters of the present representing a 'sobering signal' of what we may expect in the future. As climate change itself propels the spread of infectious disease, through its effects on the viability and geographic range of vector and intermediary species, mainstreaming climate epidemiology and public health concerns into disaster management holds significant potential for anticipating and managing future crises (Anderson et al., 2019; Carlson et al., 2020). Overall, it is clear that thorough assessment, strategic planning and an emphasis on preparedness is crucial for the effective management of such multi-hazard scenarios.

A Holistic Approach: The Reconfiguration of the Commonalities

What are the broader implications of this for conceptualizing natural disasters and their impacts? The complicating effects of the pandemic imply that models of natural disaster impacts and decision-making that can accommodate a range of influential variables, actors and scenarios have greater utility under the circumstances of a double or multi-strike event. As the present situation has proven, biological vulnerability underscores all interactions and plans in the time of a pandemic. This is a novel aspect of the dual crisis that has caught out the existing disaster response strategies of many countries and agencies—'that is a scenario we hadn't even contemplated', said Mexico's former disaster response director of his five-year tenure (Perry, 2020). Taking the long view, such an issue may be ameliorated by a conceptual approach to natural disasters that is fully equipped for the possibility of such novelties. This means *reconfiguring* increasingly co-occurring public crises as multi-pronged disasters, not merely overlapping events which only require the extension of contingency planning and response capacity in quantitative terms.

Folke et al. (2010) stress the importance of such transformability for resilient social-ecological systems. Crucially, an integrated view of public crises lays the conceptual groundwork for a more holistic and efficient disaster response from day one. In this mode, Rahman et al. (2021), for example, present scenario assessments for multi-hazard events in Bangladesh that encompass infectious diseases like Covid-19 and dengue fever, and climate hazards. They outline various possible management strategies for such compound disasters. Importantly, they begin to identify policy and information gaps at the country level, making subsequent recommendations. These include advocating cross-sector data accessibility, as well as strengthening the study of local environmental diseases so that this expertise might be readily incorporated into climate-related disaster management strategies. More broadly, the World Health Organization's recent Health EDRM Framework presents an important springboard for foregrounding and funding pandemic-related disaster preparedness, as it explicitly advocates for improved health infrastructure, knowledge management, emergency response, capacity building and more at the intersection of health and disaster risk management (WHO, 2019).

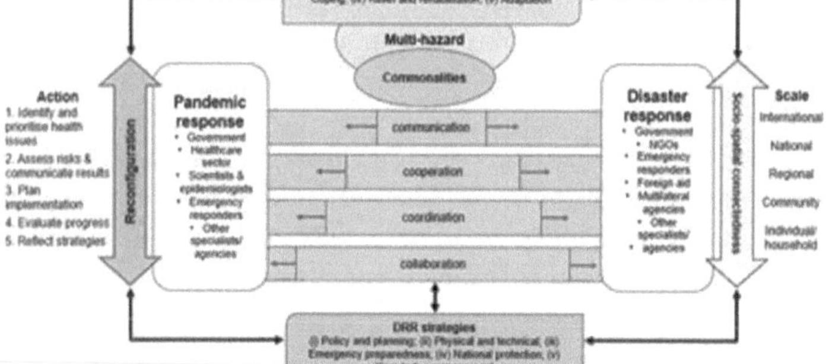

FIGURE 22.2 A conceptual approach to reconfiguration of pandemic and natural disaster risk management

In Figure 22.2, we draw upon the work of Martin et al. (2016) to visualize how improved management of double strike events can be facilitated when they are reconfigured as multi-hazards. Martin et al. (2016) identify the 4Cs of disaster partnering between organizations: communication, cooperation, coordination and collaboration. These are sequential steps in inter-organizational embeddedness, with each category in this typology defined as having qualitative distinctions from the previous one. In brief, 'communication' involves minimal inter-organizational embeddedness, with potential partners touching base on an infrequent and ad hoc basis. In the proposed framework, double disasters are conceived as distinct events at this stage, with separate management protocols that have minimal reference to multi-hazard scenarios. Hence, their co-occurrence necessitates ad hoc inter-organization, reliant on communication technologies, which may be compromised by the disasters. 'Cooperation' involves greater recognition of shared aims and a subsequent desire to operate alongside one another, especially in the field. In the double strike context, this necessitates a familiarity between operators working on disaster and pandemic response, respectively, even if it is informal. It requires a shared 'picture' of the multi-hazard scenario that enables responders to avoid duplication or interference between their strategies that might impede efficient relief. 'Coordination' moves the partners beyond recognition of shared aims to a measure of cooperation that actually yields savings; disaster and pandemic responses may achieve such alignment through joint pre-planning and resource sharing. Finally, 'collaboration' implies a deeper, longer-term partnering that may necessitate structural change within the organizations (Martin et al., 2016). This is posited as the end goal of 'reconfiguration', the process whereby response to disaster scenarios is reconceptualized to account for the complex interface of multiple disasters.

Figure 22.2 borrows this conception of increasing embeddedness to imagine what multi-hazard 'reconfiguration' can entail in practice—progressive embeddedness between pandemic response and disaster response planning where apposite, improving efficiency, response time and results in the event of double strikes. While the framework of Martin et al. seeks to define levels of embeddedness for the purpose of improving clarity of expectations and weaknesses in inter-organizational partnering, we borrow their sliding scale to explicitly suggest that greater embeddedness between pandemic and disaster response agencies is desirable and necessary. At the level of management, 'reconfiguration' is envisioned as a process whereby the relevant response agencies assess current strategies for their compatibility with double strike scenarios, reformulate accordingly, implement them, evaluate their progress and reflect upon them to continue maximizing benefits in the case of a multi-hazard event. This reconfiguration necessarily entails a deeper interface with parallel sectors (whether of knowledge or practice) and their expertise. Thus, as reconfiguration proceeds, pandemic and natural disaster response are increasingly collaborative and integrated.

Figure 22.2 also incorporates socio-spatial scale as a key aspect of reconfiguration, as these actions should occur at multiple scales from the international to the individual, promoting inter-scale as well as cross-sector cooperation. The state-and-transition of disaster at the household level (Figure 22.1) will be inflected by the emerging reconfigured strategies; furthermore, strategies should be developed with a view to fortifying vulnerable livelihoods against the double strike. Ultimately, at the 'collaboration' level, we should see detailed and specific advance planning at a range of scales for high-risk, low-frequency biological multi-hazards like the pandemic-natural disaster events. The common concerns and impacts of the compound disasters should be accounted for without compartmentalization that risk compromising the response. The risks that demand contradictory strategies, such as social distancing versus emergency sheltering, have been thoroughly assessed to determine the best protocols. Figure 22.2 also indicates the interrelationship with disaster risk reduction (DRR), as we believe these initiatives should permeate all dimensions of DRR, aiming to address compound pandemic disasters as a part of sustainable development, given the dramatic effects of the pandemic seen so far. DRR strategies may also guide the anchoring of reconfiguration initiatives to concerns at various socio-spatial scales, especially for improving the resilience of the most vulnerable at different phases of disaster-related livelihood challenges.

To summarize, Figure 22.1 conceives the pandemic and natural disaster crises we have seen as compound disasters with interrelating effects on livelihoods. As Figure 22.2 summarizes, for these compound disasters to be efficiently managed, the extension of this reconfigured conception into long-term strategic planning for joint crises, setting in motion deeper and earlier cooperation across sectors, is key. Besides, the parallels seen between natural disasters and the coronavirus crisis can provide an instructive starting point for such cooperation across various domains. A few such potentials for confluence are noted in the following section, along with other broad strategies to bolster and streamline management of double strike events. Furthermore, the models can also provide a conceptual outlook to undergird further knowledge generation, especially empirical research, regarding the vulnerabilities and responses to double strikes among disaster-vulnerable communities.

Outlook

Throughout the pandemic, the application and utility of concepts used to analyze natural disaster for understanding Covid-19 pandemic impacts have been noted (Penta et al., 2021). The conceptual map put forth here must be operationalized according to the nature and scale of the specific context to which it is applied. Some of the potentials it reflects for exchange and integration between different communities and domains of crisis management as outlined in this section. Pandemic-specific perspectives on mental health, for example, are thin on the ground (Galea et al., 2020). Existing concepts of psychological preparedness during a natural disaster could be applied to the pandemic context to study resilience and bolster it at collective and individual levels, helping to minimize psychological fallout (Agarwal et al., 2020). Such insights can have particular resonance for contexts where grave psychological injury has emerged due to simultaneous disasters, natural and biological. Similarly, in the critical arena of public administration, Dzigbede et al. (2020) demonstrate how national survey data on local governments' preparedness for weather-related disasters in the United States may also be employed to inform and improve policy related to the management of biological disasters like the Covid-19 pandemic.

In future, aggregated crisis relief resources—with the necessary measures for assessing appropriate allocation—could provide a more efficient way of dealing with ever-changing double strike events and the dilemmas of distribution they pose. Likewise, pooled funds for charity donations to communities affected by multiple disasters could help solve the issue seen in Vanuatu, where sorely needed cyclone relief donations were very low compared to previous years because public attention had already been diverted toward pandemic relief (McGarry, 2020). On the ground once the cyclone hit, however, these crises were hardly extricable. Aid agencies should reflect this interconnectedness from the outset to provide relief as swiftly as possible in the case of co-occurring and complex disasters. In addition to contingency planning and long-term research needs to see where investment may be the most fruitful, such an integrated approach could improve the response time and scope for agencies handling double strike scenarios. As mentioned, cross-sector mechanisms are crucial. Lessons on the ground have been swift, providing insight into the direction in which such mechanisms as mapped in Figure 22.2 should be developed. Considering the case of the Covid-19 pandemic and cyclones, for example, it would be valuable for emergency response agencies to cooperate with the health sector early. This may involve medically training responders to reduce burdens on hospitals and health professionals, stocking PPE with other first response equipment and using the insights of epidemiologists to produce informed, emergency-ready procedures for biological risk assessment and safer evacuation during the natural disaster (Kishore,2020). The unanticipated effects of Covid-19 meant existing defenses against Cyclone Amphan such as evacuation plans and shelters could not be fully operationalized (Priodarshini et al., 2021); better understanding and integration of biohazards can improve such procedures.

As with exchange across divergent sectors of expertise, improved international cooperation can provide a path to better management of the double strike burden. Studying Indian participants, Agarwal et al. (2020) suggest that the chronic challenges faced by many developing world populations may foster a personal and cultural resilience that has aided people's capacity to mentally cope in the face of the coronavirus pandemic, despite their greater physical vulnerability. Other studies suggest that the tight, embedded social support networks seen in such places (KC et al., 2019), and previous experiences of managing traumatic adversities (Palinkas et al., 2021), better equip communities to bear new disasters. As KC et al. (2019) suggest, this may be a 'take-home message for other disaster-prone countries' (p. 14), with the informal traditions of mutual aid and community organizing providing possible models for countries that are less accustomed to public disasters at the scale of the coronavirus pandemic. For a yet more specific case, the multi-hazard scenario assessments and policy recommendations presented by Rahman et al. (2021) for the Bangladeshi context concern the same multi-hazard—dengue fever, coronavirus, and tropical storms in combination—that has already presented grave management challenges for Honduras in 2020 (Tucker & Pindado, 2020). Thus, there is potential for shared insight here.

Multilateral agencies represent an important pathway for facilitating international cooperation on compound disaster management, and for elevating public health as a climate- and disaster-related concern. Within the UN Sustainable Development Goals, for example, health and well-being are promoted under Goal 3, and climate action under Goal 13, but there is scant mention of novel public health risks and how they might interact with increasing climate risks, either in their provenance or their impacts (UN, 2015). Likewise, the SFDRR, despite mentioning disease outbreaks and the need to 'build back better', provides little further guidance on risk management in pandemic scenarios (UNDRR, 2015). A need for 'sustainable health decisions' is palpable in the (post-)pandemic age, where the impacts of the Covid-19 crisis are prompting a reappraisal of sustainable development parameters (Heggen et al., 2020). We suggest that future agendas prioritize public health as a climate concern, foregrounding the outlook for epidemics and biological disasters, and underscoring healthcare's relevance to disaster management and resilience. Thus far, UNEP has called for countries to consider 'future environmental health disasters like this pandemic' in their Nationally Determined Contributions (NDCs) to COP26 (UNEP, 2020); such concerns should now be cemented in longer-term agenda setting.

In the sober light of the coronavirus pandemic, Mulholland (2020) reminds us of the interconnected nature of the world, of disasters and their impacts on the most marginalized—as well as the fact that such crises present windows of opportunity to compel creativity and critical shifts in policy. As evidenced by intertwined impacts of the pandemic and Cyclone Amphan in Bangladesh alongside other double strikes across the world, coastal DRR is a highly relevant area that can expand in scope and flexibility. Climate change, the force that underscores much of the devastating natural disaster unfolding, is not an isolated issue in its cause or effects. Neither, as its disproportionate impact on the poor, sick, climate-vulnerable and others has shown, is the coronavirus. Increasingly, with the interrelating factors of global warming, disease epidemics, social strife, economic crisis and the decay of other systems, we are coming to see complex joint crises and double strikes on communities as the norm. With its arresting geographical and temporal scale, the coronavirus has proven to be the ultimate example. Embracing this interconnectedness in theoretical and practical realms will yield the most efficient and innovative responses to these challenges.

References

ABC Radio National. (2020). 'The effect of COVID-19 on Australia's bushfire recovery efforts' [online]. *ABC News* (17 March). Available at: www.abc.net.au/radionational/programs/breakfast/the-effect-of-covid-19-on-australias-bushfire-recovery-efforts/12062078. Accessed 6 Jan 2021.

Agarwal, V., Sharma, S., Gupta, L., Misra, D., Davalbhakta, S., Agarwal, V., Goel, A., & Aggarwal, S. (2020). COVID-19 and psychological disaster preparedness –an unmet need. *Disaster Medicine and Public Health Preparedness, 14*(3), 387–390. doi:10.1017/dmp.2020.219

Anderson, G. B., Barnes, E. A., Bell, M. L., & Dominici, F. (2019). The future of climate epidemiology: Opportunities for advancing health research in the context of climate change. *American Journal of Epidemiology, 188*(5), 866–872. https://doi.org/10.1093/aje/kwz034

Bowman, D. M., Kolden, C. A., Abatzoglou, J. T., Johnston, F. H., van der Werf, G. R., & Flannigan, M. (2020). Vegetation fires in the Anthropocene. *Nature Reviews Earth & Environment, 1*(10), 500–515. https://doi.org/10.1038/s43017-020-0085-3

Bradshaw, S., & Fordham, M. (2015). Double disaster: Disaster through a gender lens. In: Collins, E., Jones, S., Manyena, B., & Jayawickrama, J. (Eds.), *Hazards, risks and disasters in society*. Amsterdam: Elsevier, 233–251.

Cardil, A., & de-Miguel, S. (2020). COVID-19 jeopardises the response to coming natural disasters. *Safety Science*, 130, 104861.https://doi.org/10.1016/j.ssci.2020.104861

Carlson, C. J., Albery, G. F., Merow, C., Trisos, C. H., Zipfel, C. M., Eskew, E. A., ..., & Bansal, S. (2020). Climate change will drive novel cross-species viral transmission. *BioRxiv*. https://doi.org/10.1101/2020.01.24.918755

Collins, C., Landivar, L. C., Ruppanner, L., &Scarborough, W. J. (2020). COVID-19 and the gender gap in work hours. *Gender, Work & Organization* [online]. https://doi.org/10.1111/gwao.12506

Cutter, S. L., Barnes, L., Berry, M., Burton, C., Evans, E., Tate, E., &Webb, J. (2008). A place-based model for understanding community resilience to natural disasters. *Global Environmental Change, 18*(4), 598–606.

DFID. (2005). *Sustainable livelihoods framework*. London: DFID.

Dzigbede, K. D., Gehl, S. B., &Willoughby, K. (2020). Disaster resiliency of US local governments: Insights to strengthen local response and recovery from the COVID-19 pandemic. *Public Administration Review, 80*(4), 634–643. https://doi.org/10.1111/puar.13249

Ellis-Petersen, H., & Ratcliffe, R. (2020). 'Super-cyclone Amphan hits coast of India and Bangladesh'. [online]. *The Guardian* (20 May). Available at: www.theguardian.com/world/2020/may/20/super-cyclone-amphan-evacuations-in-india-and-bangladesh-slowed-by-virus. Accessed 6 Jan 2021.

Folke, C., Carpenter, S. R., Walker, B., Scheffer, M., Chapin, T., & Rockström, J. (2010). Resilience thinking: Integrating resilience, adaptability, and transformability. *Ecology and Society, 15*(4), 20. [online] Available at: www.ecologyandsociety.org/vol15/iss4/art20

Food and Agricultural Organization of the United Nations. (2020). The Philippines—Super Typhoon Goni and Typhoon Vamco [online]Available at: www.fao.org/3/cb2476en/cb2476en.pdf

Galea, S., Merchant, R. M., & Lurie, N. (2020). The mental health consequences of COVID-19 and physical distancing: the need for prevention and early intervention. *JAMA Internal Medicine, 180*(6), 817–818. doi:10.1001/jamainternmed.2020.1562

Heggen, K., Sandset, T. J., & Engebretsen, E. (2020). COVID-19 and sustainable development goals. *Bulletin of the World Health Organization, 98*(10), 646–646. doi: http://dx.doi.org/10.2471/BLT.20.263533

International Monetary Fund. (2020). *World economic outlook: A long and difficult ascent*. Washington, DC: IMF.

Islam, M. S., Potenza, M. N., & van Os, J. (2020). Posttraumatic stress disorder during the COVID-19 pandemic: Upcoming challenges in Bangladesh and preventive strategies. *International Journal of Social Psychiatry*, 0020764020954469. https://doi.org/10.1177/0020764020954469

KC, A., Gan, C.C.R., & Dwirahmadi, F. (2019). Breaking through barriers and building disaster mental resilience: A case study in the Aftermath of the 2015 Nepal Earthquakes. *International Journal of Environmental Research and Public Health, 16*(16), 2964.https://doi.org/10.3390/ijerph16162964

Kishore, K. (2020). 'Managing tropical storms during COVID-19: Early lessons learned and reflections from India' [online]. World Bank Blogs (27 July). Available at: blogs.worldbank.org/climatechange/managing-tropical-storms-during-covid-19-early-lessons-learned-and-reflections-india. Accessed 8 Jan 2021.

Mallick, B. (2014). *Der gesellschaftliche Umgang mit zunehmender Verwundbarkeit: Eine Analyse der sozialen Bedingungen für vulnerabilitätsorientierte räumliche Planung in den Küstenzonen von Bangladesch* (Vol. 36). KIT Scientific Publishing.

Mallick, B., Sultana, Z., & Bennett, C. M. (2020). How do sustainable livelihoods influence environmental (non-) migration aspirations?*Applied Geography*, 124, 102328.https://doi.org/10.1016/j.apgeog.2020.102328

Manzanedo, R. D., & Manning, P. (2020). COVID-19: Lessons for the climate change emergency. *Science of the Total Environment, 742,* 140563.https://doi.org/10.1016/j.scitotenv.2020.140563

Martin, E., Nolte, I., & Vitolo, E. (2016). The four Cs of disaster partnering: Communication, cooperation, coordination and collaboration. *Disasters, 40*(4), 621–643. https://doi.org/10.1111/disa.12173

McDonald, J. (2021). 'Fiji cleaning up after Cyclone Yasa' [online]. The Diplomat (4 January). Available at: https://thediplomat.com/2021/01/fiji-cleaning-up-after-cyclone-yasa/. Accessed 8 Jan 2021.

McGarry, D. (2020). 'Cyclone Harold: Relief for Vanuatu delayed by coronavirus contamination fears' [online]. *The Guardian* (14 April). Available at: www.theguardian.com/world/2020/apr/14/cyclone-harold-relief-for-vanuatu-delayed-by-coronavirus-contamination-fears. Accessed 6 Jan 2021.

Mulholland, M. (2020). A moment of intersecting crises: Climate justice in the era of coronavirus.*Development*, 1–5. doi.org/10.1057/s41301-020-00259-9

Mulvaney, D., Busby, J., &Bazilian, M. D. (2020). Pandemic disruptions in energy and the environment. *Elementa: Science of the Anthropocene, 8*(1). https://doi.org/10.1525/elementa.052

MunichRe. (2021). 'Record hurricane season and major wildfires – The natural disaster figures for 2020' [online] (7 Jan). Available at: www.munichre.com/en/company/media-relations/media-information-and-corporate-news/media-information/2021/2020-natural-disasters-balance.html Accessed 6 Jan 2021.

Nelson, C. (2020). 'Navajo Nation faces twin threats as wildfires spread during pandemic' [online]. *The Guardian* (6 July). Available at: www.theguardian.com/us-news/2020/jul/06/arizona-wildfires-coronavirus-navajo-nation. Accessed 6 Jan 2021.

North, C. S., Surís, A. M., & Pollio, D. E. (2021). A nosologicalexploration of PTSD and trauma in disaster mental health and implications for the COVID-19 pandemic. *Behavioral Sciences, 11*(1), 7. https://doi.org/10.3390/bs11010007

Ober, K. (2020). 'Complex road to recovery: COVID-19, Cyclone Amphan, monsoon flooding collide in Bangladesh and India' [online], Refugees International. Available at: www.refugeesinternational.org/reports/2020/10/5/complex-road-to-recoverynbspcovid-19-cyclone-amphan-monsoon-flooding-collide-in-bangladesh-and-india. Accessed 7 Jan 2021.

Palinkas, L. A., Springgate, B. F., Sugarman, O. K., Hancock, J., Wennerstrom, A., Haywood, C., ... & Wells, K. B. (2021). A rapid assessment of disaster preparedness needs and resources during the COVID-19 pandemic. *International Journal of Environmental Research and Public Health, 18*(2), 425.https://doi.org/10.3390/ijerph18020425

Penta, S., Kendra, J., Marlowe, V., & Gill, K. (2021). A disaster by any other name?: COVID-19 and support for an All-Hazards approach. *Risk, Hazards & Crisis in Public Policy, 12*(3), 240–265.

Perry, N. (2020). 'Pandemic hurts ability of nations to face natural disasters' [online]. *ABC News* (4 April). Available at: https://abcnews.go.com/Health/wireStory/pandemic-hurts-ability-nations-face-natural-disasters-69973439. Accessed 6 Jan 2021.

Phillips, C.A., Caldas, A., Cleetus, R. et al. Compound climate risks in the COVID-19 pandemic. (2020).*Nature Climate Change*, 10, 586–588. https://doi.org/10.1038/s41558-020-0804-2.

Prime, H., Wade, M., & Browne, D.T. (2020). Risk and resilience in family well-being during the COVID-19 pandemic. *American Psychologist*, 75(5), 631–643. http://dx.doi.org/10.1037/amp0000660

Priodarshini, R., Biswas, B., Higuera, A. M. S., & Mallick, B. (2021). Livelihood challenges of 'double strike' disasters: Evidence from rural communities of southwest coastal Bangladesh during the Covid-19 pandemic and cyclone Amphan. *Current Research in Environmental Sustainability, 3*, 100100.

Rahman, M. M., Bodrud-Doza, M., Shammi, M., Islam, A. R. M. T., & Khan, A. S. M. (2021). COVID-19 pandemic, dengue epidemic, and climate change vulnerability in Bangladesh: Scenario assessment for strategic management and policy implications. *Environmental Research, 192*, 110303.https://doi.org/10.1016/j.envres.2020.110303

Remeikis, A. (2020). 'Businesses affected by coronavirus should have access to disaster relief funds, says Queensland' [online]. *The Guardian* (11 March). Available at: www.theguardian.com/australia-news/2020/mar/12/businesses-affected-by-coronavirus-should-have-access-to-disaster-relief-funds-says-queensland. Accessed 7 Jan 2021.

Sakurai, M., & Chughtai, H. (2020). Resilience against crises: COVID-19 and lessons from natural disasters. *European Journal of Information Systems*, 29(5), 585–594. doi: 10.1080/0960085X.2020.1814171

Savage, S., & Farhad, T. (2020). 'Mass exodus from Dhaka as the economic impact of Covid-19 forces Bangladeshis to flee cities' [online]. *The Telegraph* (22 July) Available at: www.telegraph.co.uk/global-health/science-and-disease/mass-exodus-dhaka-economic-impact-covid-19-forces-bangladeshis/. Accessed 8 Jan 2021.

Seddighi, H. (2020). COVID-19 as a natural disaster: Focusing on exposure and vulnerability for response. *Disaster Medicine and Public Health Preparedness, 14*(4), E42–E43. doi:10.1017/dmp.2020.279

Selby, D., & Kagawa, F. (2020). Climate change and coronavirus. In: Carmody, P., McCann, G., Colleran, C., & O'Halloran, C. (Eds),*COVID-19 in the global South: Impacts and responses.* Bristol: Bristol University Press, 17–27.

Sen, A. (1985). *Commodities and capabilities.* Amsterdam & New York: North-Holland. ISBN 9780444877307.

Serafini, G., Parmigiani, B., Amerio, A., Aguglia, A., Sher, L., & Amore, M. (2020). The psychological impact of COVID-19 on the mental health in the general population. *QJM: An International Journal of Medicine, 113*(8), 531–537. https://doi.org/10.1093/qjmed/hcaa201

Shabbir, M., & Naeem, K. (2020). Apathy towards climate-induced migrants amid COVID-19: Insights from Bangladesh, India & Pakistan. *Sustainable Development Policy Institute.* Available at: https://sdpi.org/wp-content/uploads/2020/09/climate-migration-policy-review.pdf

Sottile, C., & Arkin, D. (2020). 'Disaster inside a disaster': California wildfires and COVID-19 form twin crises' [online]. *NBC News* (26 August). Available at: www.nbcnews.com/news/us-news/disaster-inside-disaster-california-wildfires-covid-19-form-twin-crises-n1238031. Accessed 8 Jan 2021.

Tucker, D., & Pindado, E. (2020). 'When it rains it pours: The devastating impact of hurricanes Eta and Iota in Honduras' [online]. *Amnesty International* (13 December). Available at: www.amnesty.org/en/latest/news/2020/12/devastating-impact-hurricanes-eta-iota-honduras/. Accessed 12 Jan 2021.

UN. (2015). *Transforming our world: The 2030 agenda for Sustainable Development*, A/RES/70/1. Available at:www.un.org/en/development/desa/population/migration/generalassembly/docs/globalcompact/A_RES_70_1_E.pdf

UNDRR. (2015). *Sendai Framework for Disaster Risk Reduction 2015–2030.* Available at: www.preventionweb.net/files/43291_sendaiframeworkfordrren.pdf

UNEP. (2020). *Nationally determined contributions.* Available at: https://wedocs.unep.org/bitstream/handle/20.500.11822/32941/FAQsNDCs. pdf?sequence=1&isAllowed=y

Westoby, M., Walker, B., & Noy-meir, I. (1989). Opportunistic Management for Rangelands Not at Equilibrium. *Journal of Range Management*, 42, 266–274. http://dx.doi.org/10.2307/3899492

WHO (2019). *Health Emergency and Disaster Risk Management Framework*. Geneva: World Health Organization. Available at: www.who.int/hac/techguidance/preparedness/health-emergency-and-disaster-risk-management-framework-eng.pdf

WHO. (2021). WHO Coronavirus Disease (COVID-19) Dashboard (public dashboard). Available at: https://covid19.who.int/. Accessed 6 Jan 2021.

Wu, X., Nethery, R. C., Sabath, M. B., Braun, D., & Dominici, F. (2020). Air pollution and COVID-19 mortality in the United States: Strengths and limitations of an ecological regression analysis. *Science Advances*, *6*(45), eabd4049. doi: 10.1126/sciadv. abd4049

Zambrano-Monserrate, M. A., Ruano, M. A., & Sanchez-Alcalde, L. (2020). Indirect effects of COVID-19 on the environment. *Science of the Total Environment*, *728*, 138813.https://doi.org/10.1016/j.scitotenv.2020.138813

23

CONCLUSIONS

Khondoker Mokaddem Hossain, Mahbuba Nasreen and Mohammed Moniruzzaman Khan

This book has drawn a nexus between climate change and disaster risk reduction (DRR) issues in coastal region of Bangladesh from risk management perspectives. Different chapters of this book have discussed about most common disasters including floods and cyclone that affect the country and increase vulnerability of the people living in different hazardous locations. Major focus of the book is the nature of vulnerability and risk in coastal region of Bangladesh with particular focus on factors shaping vulnerability, and experiences of different kinds of disasters that community people confront in their everyday life. This book also explores some crucial indicators of empowerment, agency of the community and local people in disaster management in the coastal Bangladesh, where vulnerable communities, that is, women, poor, marginal and intersect members of society lack access to resources, income and employment opportunities and others. It has further addressed the nature of sustainable capacity and agency building for adaptation along with investigating the dynamics of resilience, challenges and opportunities among the coastal people of Bangladesh.

The book demonstrates that extreme climate variability causes disasters like cyclone and floods in coastal Bangladesh and creates epidemiological burden of coastal communities spreading both communicable and non-communicable diseases. Moreover, there are shortages of medicine supplies and skilled human resources in addressing those diseases. Most healthcare providers recommended training on 'coastal area-specific disease controlling mechanisms and supply of medicines' as per requirements of community people. However, the community clinics (CCs) in the coastal area have the potentiality to reduce the level of healthcare-related vulnerability and risk for hard-to-reach coastal people as CCs ensure easy access to primary healthcare services. Moreover, the health education program for adolescents is changing the behavior of the community people. However, the research findings of this book indicate that an inadequate supply of medicine is making it hard for

DOI: 10.4324/9781003253495-29

care providers to deliver adequate services to the disaster-affected people in the community. Furthermore, healthcare providers recommended providing medicines based on the actual needs and requirement of disaster-affected coastal community.

The book also analyzed the nature of salinity intrusion due to cyclonic storm and tidal surge during cyclones and vulnerability and health risk of southern coastal belt people. It is evident that the saltwater or salinity intrudes in the coastal areas from the flow of sea water through the environmental exposures and disasters like cyclone, tidal surge, flood, river flow, estuarine circulation and so on. Besides, climate change, sea-level rises, precipitation and water and land management practices exacerbate the salinity. Surface water, groundwater and soil contaminated by salinity in coastal areas. About half of the coastal areas are affected by salinity in Bangladesh. Salinity causes unfavorable environment and hydrological situation that restrict the normal crop production throughout the year. Climate change can be impacted negatively in the future as the land becomes increasingly saline, health risk becomes greater, loss of livelihood and hardship may be arisen, the productivity of the land decreases, future food insecurity becomes a possibility and the people become more vulnerable. In Bangladesh salinity intrusion is a time-varying event, maximum in dry season and minimum during the monsoon (June–October) by pushback of the rivers discharge at the salinity front in estuarine and floodplains.

This book also reflects to the riverbank erosion, which is a slow and silent disaster with a devastating nature that damages lands, homes, other properties and changes people's life permanently. Eroded people become displaced from their origin. It changes people's identity along with changing socio-economic conditions. Solvent families are degraded to the poor group through losing everything into the river while the return remains uncertain and unsecured. Dependency on agricultural lands, lower livelihood opportunities in the region, lack of education and skills make recovery difficult for the erosion-hit families. They are collectively called *Nadi Bhanga Manush*, which has severe social and psychological impacts. Erosion survivors require economic, health and mental support for recovery. It is a matter of concern that the response plan provides relief to retaliate this disaster. Survivors of riverbank erosion fight against this severe disaster of their own and most of the time they change their occupation. Poor housings, lower education and healthcare accessibility with increasing insecurity and frustration among the erosion-hit residents show the alarming situations. But the eroded people gradually develop strategies to cope with and adapt, finding alternative means of livelihoods.

The book further unearths the nature of community participation and indigenous knowledge of community people in the coastal area, which is quite effective in managing disasters with particular focus on community-based DRR (CBDRR), including faith and religious value-based organizations' roles and activities, etc. It is evident from the findings of this book that indigenous knowledge of the coastal people and their practices over time made them enabled to survive with existing natural hazards and disasters. Their very nurtured and practiced indigenous knowledge including beliefs and traditions, etc. help them to secure and protect their community and lives in disastrous time. This book also highlights how the contribution of indigenous knowledge to DRR in coastal Bangladesh is functional and effective for the DRR that the community people have gained over the generations and how they are gradually combing the technical, scientific and institutional knowledge with their indigenous wisdom and practices.

This book also discusses the nature of institutional arrangement that includes integrated disaster management approach of the government including policies and regulations to reduce the disaster vulnerability of people, especially the poor, from the effects of natural, environmental and human-induced disasters to a manageable and acceptable humanitarian level. It also overviewed the policies, regulations and action plans taken by GOs and NGOs to explore how they contribute to mainstreaming disaster, risk and vulnerability management. It further explores what changes have been made in local practices, lifestyle and livelihood and what are the gaps at community level or level of changes at community level contributed to formulate comprehensive policies, regulations and action plans.

This book has identified how infrastructure, innovation and technology have been placed to support DRR and to reduce disasters in coastal region of Bangladesh. Specifically, this book provides an overview of the nature of disaster risks and vulnerabilities faced by the coastal people in Bangladesh, and has identified the links between infrastructure, innovation, technology and disaster risk and risk reduction. It has further discussed how infrastructure and technology are intervened to move from crisis response to crisis management as well as explored how does vulnerability differ for infrastructure and technology.

This book has figured out the nature of exclusion and inclusion of some disadvantaged socio-economic groups in DRR process in coastal Bangladesh. The book has also discussed the socio-economic and demographic aspects, nature of vulnerability and coping, adaptation and resilience mechanisms of vulnerable women including widow, solo mother, female-headed households, abandoned women, divorced women, elderly people, persons with disability and minority and how they play crucial role during disastrous times in the coastal region of Bangladesh. It has also explored the nexus between disaster and gender relations, that is, do women leadership and contributory roles of women in disaster management to enhancing disaster resilience. It is evident from the findings of different writers of this book that many stakeholders including government, NGOs, development partners, local and international voluntary organizations, private sectors and disaster-affected community people pay their attention to the disaster management practices in coastal Bangladesh.

This book has focused on how the state policies have provided a significant impetus to the rise in production and export of shrimping global markets from coastal Bangladesh. Moreover, it has also discussed how social and environmental effects of shrimp aquaculture has impacted the vulnerable coastal communities. It is reflected in the findings of this book that the shrimp farming in the coastal area has resulted in exploitation, injustice and degradation of the coastal environment that have intensified the existing vulnerability of many disaster-stressed coastal population.

Finally, the book has sketched out the adaptation and resilient mechanisms of the community people that they adopt to earn their livelihood. Salinity intrusion, waterlogging, lack of safe drinking water and other phenomenon emerged due to extreme climatic events, which forced people to develop alternative arrangements. It is evident that the coastal people have introduced some coping and adaptable strategies in addressing the salinity. It includes rainwater harvesting for pure drinking water, cultivation of salinity-tolerant crop species, floating gardening, crab fattening by women, adaptive agricultural production system, alternative crop diversity, etc. Some of the adaptation practices have been continuing for long with or without supports from government or other organizations; more attention of policy makers is required in order to make them sustainable.

INDEX

Note: Where an indexed item is included in the Glossary of Terms the glossary page number is included. Figures are indicated by *italics* and tables are in **bold type**. Endnotes are indicated by the page number followed by "n" and the note number e.g., 86n9 refers to endnote 9 on page 86. The term "disaster risk reduction" is listed as "DRR" in some entries.